T0334533

# Heat Pipes

Theory, Design and Applications

# Heat Pipes
## Theory, Design and Applications

**Seventh Edition**

**Hussam Jouhara**
Brunel University London, United Kingdom; Vytautas Magnus University

**David Reay**
David Reay and Associates, United Kingdom; Newcastle University; Brunel University, London, United Kingdom

**Ryan M[c]Glen**
Boyd Technologies Ltd, Ashington, United Kingdom

**Peter Kew**
Heriot-Watt University, United Kingdom; David Reay and Associates

**Jonathan McDonough**
Newcastle University, United Kingdom

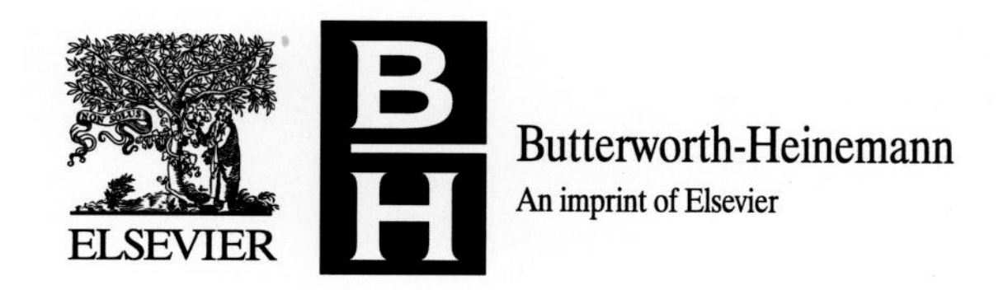

Butterworth-Heinemann is an imprint of Elsevier
The Boulevard, Langford Lane, Kidlington, Oxford OX5 1GB, United Kingdom
50 Hampshire Street, 5th Floor, Cambridge, MA 02139, United States



ISBN: 978-0-12-823464-8

For Information on all Butterworth-Heinemann publications
visit our website at https://www.elsevier.com/books-and-journals

*Publisher:* Joseph P. Hayton
*Acquisitions Editor:* Fran Kennedy-Ellis
*Editorial Project Manager:* Teddy A. Lewis
*Production Project Manager:* Paul Prasad Chandramohan
*Cover Designer:* Matthew Limbert

Typeset by MPS Limited, Chennai, India

# Contents

# About the authors

**Prof. Hussam Jouhara** is a professor of thermal engineering and a director of research in the College of Engineering at Brunel University London. He is also the editor-in-chief of the International Journal of Thermofluids, executive editor of Thermal Science and Engineering Progress and associate editor of the International Journal of Heat and Mass Transfer, all published by Elsevier. In addition, Prof. Jouhara is the technical director of Econotherm Limited (United Kingdom), a world-leading British heat pipe heat exchanger manufacturing company. Throughout his career, he has extensive expertise in designing and manufacturing various types of heat pipes and heat pipe—based heat exchangers for low, medium and high temperature applications. His work in the field of heat pipe—based heat exchangers resulted in novel designs for recouperators, steam generators and condensers and flat heat pipes. These have been implemented across various industries including, but not limited to, food, electronics thermal management, solar energy harvesting, industrial waste heat recovery and Energy from Waste. Over the last few years, he has successfully managed to achieve new designs for industrial waste heat recovery and many thermal systems that have enhanced the performance of various industrials processes in the United Kingdom, Europe and worldwide. Prof. Jouhara is a fellow and a chartered engineer in both the United Kingdom (FIMechE) and Ireland (Engineers Ireland - FIEI), a fellow of the Chartered Institution of Building Services Engineering (FCIBSE) and a senior fellow of the Higher Education Academy (SFHEA), United Kingdom. He is also a chief researcher in Vytautas Magnus University and a visiting professor at City, the University of London.

**Prof. David Reay** manages David Reay & Associates, United Kingdom, and he is a visiting professor at Northumbria University, Emeritus professor at Newcastle University and honorary professor at Brunel University London, United Kingdom. His main research interests are compact heat exchangers, process intensification and heat pumps. He is also the editor-in-chief of Thermal Science and Engineering Progress and associate editor of the International Journal of Thermofluids, both are published by Elsevier. Prof. Reay is the author/co-author of eight other books.

**Dr. Ryan McGlen** is the advanced technology manager at Boyd's UK facility, where he leads research and development of future heat pipe technologies and hi-tech commercial applications. His research interests include patented additive manufactured heat pipe technology, heat pipe fluids and material combinations, novel heat pipe geometry, wick construction and heat pipe functionality. He has more than 20 years of experience in commercial electronic thermal management application, with main focus areas in the space, aerospace and defence and automotive application. Dr. McGlen is a chartered engineer (MIMechE) and is a Royal Academy of Engineering Visiting Professor in Practice at Newcastle University.

**Peter Kew** first became involved in heat pipes in the late 1970s as a research officer with International Research and Development (IRD) working on a range of heat transfer and energy conservation projects, including heat pipe development which was then led by David Reay. He has maintained this interest in this area for 40 years at IRD and then as a lecturer and senior lecturer at Heriot-Watt University and as associate head of the University's School of Engineering and Physical Sciences, responsible for the School's activities on the Dubai Campus of the University. On retirement from Heriot-Watt, he has been active as a consultant.

**Dr. Jonathan McDonough** is a lecturer in chemical engineering in the School of Engineering at Newcastle University and is an associate member of IChemE. His expertise resides in additive manufacturing and its application to areas such as reaction engineering, flow chemistry, fluid mechanics, heat transfer, fluidisation, and carbon capture. One of Dr. McDonough's key research themes is to exploit additive manufacturing for the fabrication of new and novel reactor geometries that can unlock previously unobtainable operating windows, paving the way for potentially new chemistries and processes. He is actively involved in several complementary projects that explore different aspects of this goal.

# Preface

It seems strange to realise that nearly 10 years have passed since the publication of the previous edition, a time span where so much has happened and yet, in concept, not so. We still have a world of even greater reliance on electronic equipment – and the problems associated with cooling devices of ever decreasing size but increasing complexity and component density.

With such emphasis in mind, this seventh edition has been updated to reflect current critical thinking relevant to realising optimal solutions to thermal problems. For example, advances in space applications are demanding more and more from heat pipes that have already proved themselves as the 'go to' technology for providing thermal management of the critical systems that are used in all space vehicles, settings and satellites. But in addition, the drive for decarbonising our industries and society has led to heat pipe–based solutions for waste heat recovery systems, HVAC and renewable energy systems, all of which now regularly utilise them to cater for applications that were, in the past, deemed impossible due to the limitations of conventional heat exchangers.

Looking to the future, another topic of interest also covered in this edition is the exciting promise (and challenges) of additive manufacturing which has seen a steady increasing use of technologies such as 3D printing and laser melting to produce more complex constructions than were ever possible with conventional manufacturing techniques. However, those new to the whole subject area have not been forgotten. The edition still retains its discussions on the simply physical principles on which these devices operate and examine the mathematics and science theories relevant to the topic.

In short, this seventh edition explores heat pipe developments and across its chapters offers practical advice on the selection of suitable materials together with information to guide the user/designer's choice between the diverse forms of heat pipe, their prototyping, construction, testing and perhaps most important of all – their application. And definitively, despite the age of some of the data sets, this edition still acts as a collection and repository of relevant practical information and advice to provide the necessary background to both users and designers of heat pipes and researchers within this field.

Finally, to repeat what was noted within the previous edition, it perhaps goes without saying that the authors hope readers find this latest edition as useful as the earlier editions.

# Preface

# Acknowledgements

The authors would like to thank all those who assisted in the compiling and revision of this edition and would notably like to acknowledge:

- Brad Whitney, Kevin Lynn, Nelson Gernert, Sukhvinder Kang and Amie Jeffries of Boyd, for substantial data on a range of heat pipes and case studies, in particular those related to thermal management of electronics systems;
- Drummond Hislop for initiating discussions on using additive manufacturing for the fabrication of heat pipes, and for David Reay and Associates for funding the initial review of the literature;
- Les Norman, Bertrand Delpech, Amisha Chauhan (Brunel University London) and Sulaiman Almahmoud (Spirax Sarco Limited) for their support and input;
- Richard Meskimmon and Allan Westbury (S & P Coil Products Limited), Mark Boocock (Econotherm (UK) Limited), Stephen Lester and Mark Robinson (Flint Engineering Limited), and Roy Presswell (Kool Technology Limited), for their input and guidance.

# Nomenclature

|  |  |
|---|---|
| $A_C$ | circumferential flow area |
| $A_w$ | wick cross-sectional area |
| $C_P$ | specific heat of vapour, constant pressure |
| $C_V$ | specific heat of vapour, constant volume |
| $D$ | sphere density in Blake–Kozeny equation |
| $H$ | constant in the Ramsey–Shields–Eotvös equation |
| $J$ | 4.18-J/g mechanical equivalent of heat |
| $K$ | wick permeability |
| $L$ | enthalpy of vaporisation or latent heat of vaporisation |
| $M$ | molecular weight |
| $M$ | Mach number |
| $M$ | figure of merit |
| $N$ | number of grooves or channels |
| $Nu$ | Nusselt number |
| $Pr$ | Prandtl number |
| $P$ | pressure |
| $\Delta P$ | pressure difference |
| $\Delta P_{C\ max}$ | maximum capillary head |
| $\Delta P_l$ | pressure drop in the liquid |
| $\Delta P_v$ | pressure drop in the vapour |
| $\Delta P_g$ | pressure drop due to gravity |
| $Q$ | quantity of heat |
| $R$ | radius of curvature of liquid surface |
| $R_0$ | universal gas constant $8.3 \times 10^3$ J/K kg mol |
| $Re$ | Reynolds number |
| $Re_r$ | radial Reynolds number |
| $Re_b$ | a bubble Reynolds number |
| $S$ | volume flow per second |
| $T$ | absolute temperature |
| $T_C$ | critical temperature |
| $T_V$ | vapour temperature |
| $\Delta T_S$ | superheat temperature |
| $T_W$ | heated surface temperature |
| $V$ | volume |
| $V_c$ | volume of condenser |
| $V_g$ | volume of gas reservoir |
| $We$ | Weber number |
| a | groove width |
| $a$ | radius of tube |
| $b$ | constant in the Hagen–Poiseuille equation |
| $c$ | velocity of sound |
| $d_a$ | artery diameter |
| $d_w$ | wire diameter |
| $f$ | force |
| $g$ | acceleration due to gravity |
| $g$ | heat flux |
| $g_c$ | Rohsenow correlation |
| $h$ | capillary height, artery height, coefficient of heat transfer |
| $k$ | Boltzmann constant = $1.38 \times 10^{-23}$ J/K |
| $k_w$ | wick thermal conductivity–$k_s$ solid phase, $k_l$ liquid phase |
| $l$ | length of heat pipe section defined by subscripts |
| $l_{eff}$ | effective length of heat pipe |
| $m$ | mass |
| $m$ | mass of molecule |
| $m$ | mass flow |
| $n$ | number of molecules per unit volume |
| $r$ | radius |
| $r$ | radial co-ordinate |
| $r_e$ | radius in the evaporator section $r_c$ radius in the condensing section |
| $r_H$ | hydraulic radius |
| $r_v$ | radius of vapour space |
| $r_w$ | wick radius |
| $u$ | radial velocity |
| $v$ | axial velocity |
| y | co-ordinate |
| $z$ | co-ordinate |
| $\alpha$ | heat transfer coefficient |
| $\beta$ | defined as $(1 + k_s/k_l)/(1 - ks/k_l)$ |
| $\delta$ | constant in Hsu formula – thermal layer thickness |
| $\varepsilon$ | fractional voidage |
| $\theta$ | contact angle |
| $\varphi$ | inclination of heat pipe |
| $\varphi_c$ | function of channel aspect ratio |
| $\lambda$ | characteristic dimension of liquid–vapour interface |
| $\mu$ | viscosity |
| $\mu_l$ | dynamic viscosity of liquid |
| $\mu_v$ | dynamic viscosity of vapour |
| $\gamma$ | ratio of specific heats |
| $\rho$ | density |
| $\rho_l$ | density of liquid |
| $\rho_v$ | density of vapour |
| $\sigma$ | $\sigma_{LV}$ used for surface energy where there is no ambiguity |
| $\sigma_{SL}$ | surface energy between solid and liquid |
| $\sigma_{LV}$ | surface energy between liquid and vapour |
| $\sigma_{SV}$ | surface energy between solid and vapour |

Other notations are as defined in the text.

# Introduction

The heat pipe is a device of very high thermal conductance. The idea of the heat pipe was first suggested by Gaugler [1] in 1942. It was not, however, until its independent invention by Grover [2,3] in the early 1960s that the remarkable properties of the heat pipe became appreciated and serious development work took place.

The heat pipe is similar in some respects to the thermosyphon and it is helpful to describe the operation of the latter before discussing the heat pipe. The thermosyphon is shown in Fig. 1a. A small quantity of water is placed in a tube from which the air is then evacuated and the tube sealed. The lower end of the tube is heated causing the liquid to vapourise and the vapour to move to the cold end of the tube where it is condensed. The condensate is returned to the hot end by gravity. Since the latent heat of evaporation is large, considerable quantities of heat can be transported with a very small temperature difference from end to end. Thus the structure will also have a high effective thermal conductance. The thermosyphon has been used for many years and various working fluids have been employed. (The history of the thermosyphon, in particular the version known as the Perkins Tube, is reviewed in Chapter 1.) One limitation of the basic thermosyphon is that in order for the condensate to be returned to the evaporator region by gravitational force, the latter must be situated at the lowest point.

The basic heat pipe differs from the thermosyphon in that a wick, constructed for example from a few layers of fine gauze, is fixed to the inside surface and capillary forces return the condensate to the evaporator (see Fig. 1b). In the heat pipe the evaporator position is not restricted and it may be used in any orientation. If, of course, the heat pipe evaporator happens to be in the lowest position, gravitational forces will assist the capillary forces. The term 'heat pipe' is also used to describe high thermal conductance devices in which the condensate return is achieved by other means, for example centripetal force, osmosis or electrohydrodynamics.

Several methods of condensate return are listed in Table 1. A review of techniques is given by Roberts [4], and others are discussed by Reay [5], Leyadeven et al. [6] and Maydanik [7].

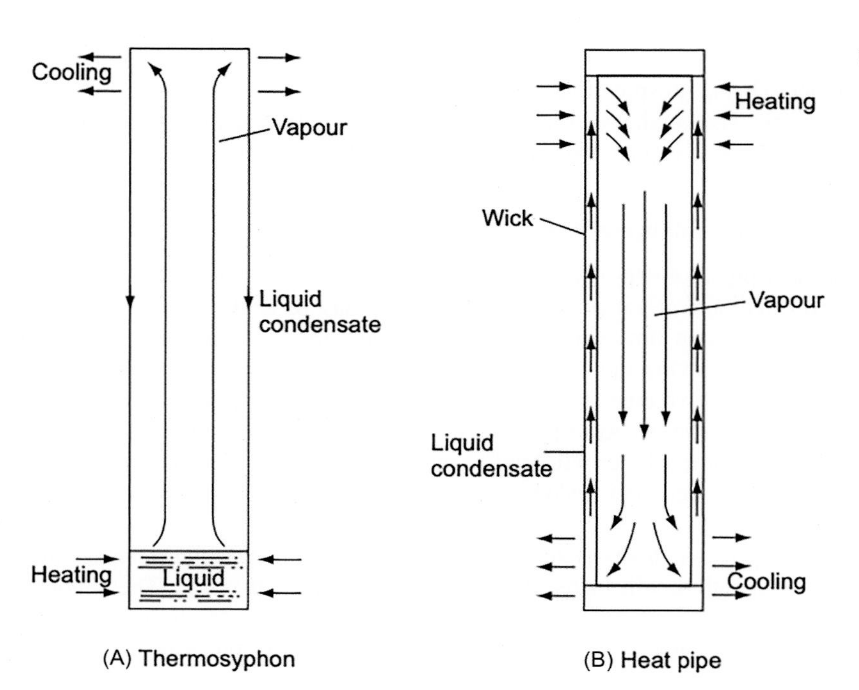


**FIGURE 1** The heat pipe and thermosyphon.

**TABLE 1 Methods of condensate return.**

| Method | |
|---|---|
| Gravity | Thermal syphon |
| Capillary force | Standard heat pipe loop heat pipe |
| Centripetal force | Rotating heat pipe |
| Electrokinetic forces | Electrohydrodynamic heat pipe, electroosmotic heat pipe |
| Magnetic forces | Magnetohydrodynamic heat pipe, magnetic fluid heat pipe |
| Osmotic forces | Osmotic heat pipe |
| Bubble pump | Inverse thermal syphon |

# 1 The heat pipe construction, performance and properties

The main regions of the standard heat pipe are shown in Fig. 2. In the longitudinal direction (see Fig. 2a), the heat pipe is made up of an evaporator section and a condenser section. Should external geometrical requirements make this necessary, a further, adiabatic, section can be included to separate the evaporator and the condenser. The cross section of the heat pipe, Fig. 2b, consists of the container wall, the wick structure and the vapour space.

The performance of a heat pipe is often expressed in terms of 'equivalent thermal conductivity'. A tubular heat pipe of the type illustrated in Fig. 2, using water as the working fluid and operated at 150°C, would have a thermal conductivity several hundred times that of copper. The power handling capability of a heat pipe can be very high. Pipes using lithium as the working fluid at a temperature of 1500°C will carry an axial flux of 1020 $kW/cm^2$.

By suitable choice of working fluid and container materials, it is possible to construct heat pipes for use at temperatures ranging from 4K to in excess of 2300K.

For many applications, the cylindrical geometry heat pipe is suitable, but other geometries can be adopted to meet special requirements.

The high thermal conductance of the heat pipe has already been mentioned; this is not the sole characteristic of the heat pipe.

The heat pipe is characterised by the following:

1. Very high effective thermal conductance.
2. The ability to act as a thermal flux transformer. This is illustrated in Fig. 3.
3. An isothermal surface of low thermal impedance. The condenser surface of a heat pipe will tend to operate at uniform temperature. If a local heat load is applied, more vapour will condense at this point, tending to maintain the temperature at the original level.

Special forms of heat pipe can be designed having the following characteristics:

1. Variable thermal impedance A form of the heat pipe, known as the gas-buffered heat pipe, will maintain the heat source temperature at an almost constant level over a wide range of heat input. This may be achieved by maintaining a constant pressure in the heat pipe but at the same time varying the condensing area in accordance with the change in thermal input. A convenient method of achieving this variation of condensing area is that of 'gas buffering'. The heat pipe is connected to a reservoir having a volume much larger than that of the heat pipe. The reservoir is filled with an inert gas that is arranged to have a pressure corresponding to the saturation vapour pressure of the fluid in the heat pipe. In normal operation, the heat pipe vapour will tend to pump the inert gas back into the reservoir and the gas—vapour interface will be situated at some point along the condenser surface. The operation of the gas buffer is as follows:
   a. Assume that the heat pipe is initially operating under steady-state conditions. Now let the heat input increase by a small increment. The saturation vapour temperature will increase and with it the vapour pressure. The vapour pressure increases very rapidly for very small increases in temperature, for example the vapour pressure of sodium at 800°C varies as the 10th power of the temperature. The small increase in vapour pressure will cause

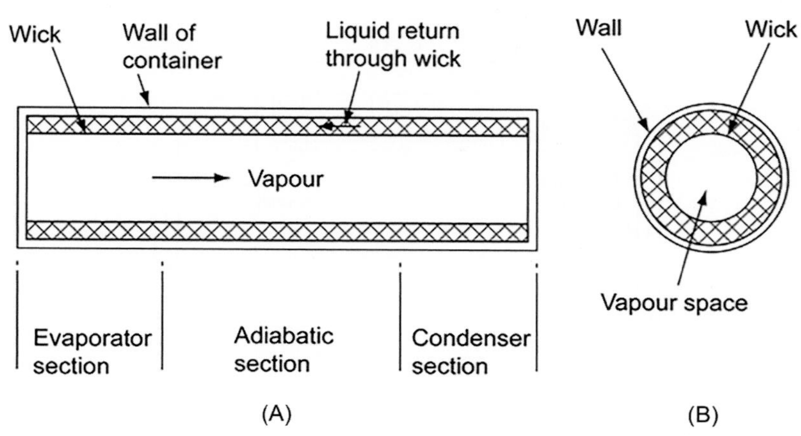


**FIGURE 2** Main regions of the heat pipe.

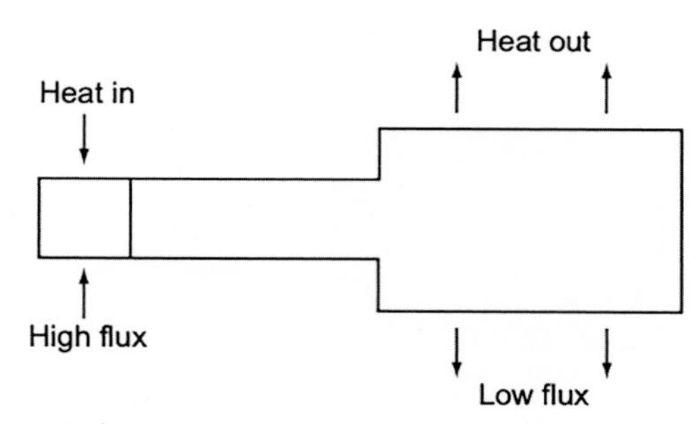


**FIGURE 3** The heat pipe as a thermal flux transformer.

the inert gas interface to recede, thus exposing more condensing surface. Since the reservoir volume has been arranged to be large compared to the heat pipe volume, a small change in pressure will give a significant movement of the gas interface. Gas buffering is not limited to small changes in heat flux but can accommodate considerable heat flux changes.

**b.** It should be appreciated that the temperature, which is controlled in the simpler gas-buffered heat pipes, as in other heat pipes, is that of the vapour in the pipe. Normal thermal drops will occur when heat passes through the wall of the evaporating surface and also through the wall of the condensing surface.

A further improvement is the use of an active feedback loop. The gas pressure in the reservoir is varied by a temperature-sensing element placed in the heat source:

**1.** Loop heat pipes (LHPs): The LHP, illustrated in Fig. 4, comprises an evaporator and a condenser, as in conventional heat pipes, but differs in having separate vapour and liquid lines, rather like the layout of the single-phase heat exchanger system used in buildings for heat recovery, the run-around coil. Those who recall the technical efforts made to overcome liquid-vapour entrainment in heat pipes and, more importantly, in thermosyphons will know that isolation of the liquid path from the vapour flow (normally countercurrent) is beneficial. In the LHP, these flows are cocurrent in different parts of the tubing.

A unique feature of the LHP is the use of a compensation chamber. This two-phase reservoir helps to establish the LHP pressure and temperature as well as maintain the inventory of the working fluid within the operating system. The LHP, described fully in Chapter 2, can achieve very high pumping powers, allowing heat to be transported over distances of several metres. This overcomes some of the limitations of other 'active' pumped systems that require external power sources.

**1.** Thermal diodes and switches: The former permit heat to flow in one direction only, while thermal switches enable the pipe to be switched off and on.

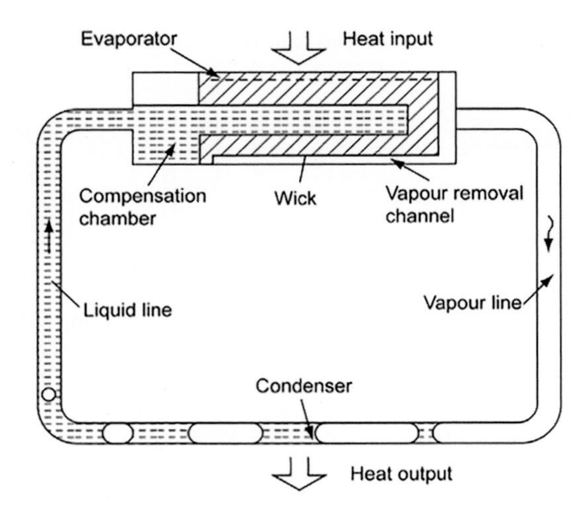


FIGURE 4 Loop heat pipe.

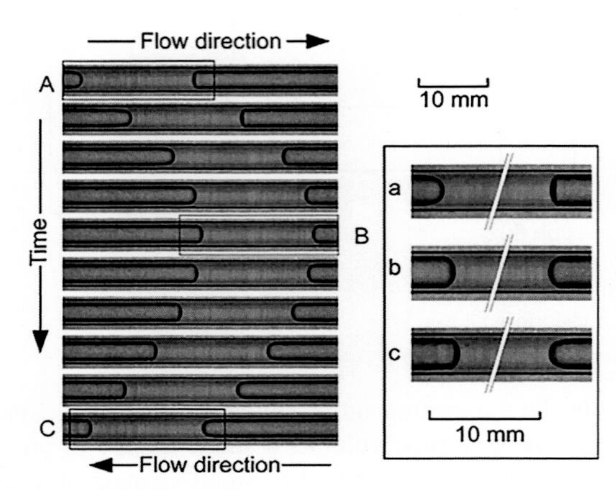


FIGURE 5 Position of the oscillating slug in a PLP tube at different times [8].

2. Pulsating or oscillating heat pipes (PHPs and OHPs): The pulsating heat pipe (PHP, sometimes called the OHP), discussed in Chapter 2, is like the LHP, a relative newcomer to the heat pipe field, but one that is receiving substantial attention because of its lack of reliance on capillary action. The PHP consists of a long, small diameter, tube that is 'concertined' into a number of U-turns. There is no capillary structure within the tube, and the liquid distributes itself in the form of slugs between vapour sections, as shown in Fig. 5. Heat transfer is via the movement (oscillation) of the liquid slugs and vapour plugs between the evaporator and the condenser.

The oscillation of the slug, in this case in a single tube at different times, is shown. The time step between two frames is equal to 20 ms. The authors [8] state: 'Advancing and receding menisci refer to the liquid plug motion: advancing corresponds to the leading edge and receding to the tail of the liquid plug. Enlargements (a), (b) and (c) show a strong dissymmetry of left and right interfaces: the curvature radius of advancing menisci (right interface in (a) and left interface in (c)) is smaller than that of receding menisci. This dissymmetry is a view of the pressure difference between both sides of the liquid slug. In case (b), the interface velocity is equal to zero and the liquid slug is then symmetric'.

Increasingly in the literature, one notes the addition of nanoparticles, to create nanofluids, in an attempt to improve the performance of most of the above types of heat pipe or thermosyphon. Some examples are given elsewhere in the text, but the word 'nano' does not in our opinion warrant its use in a new category of heat pipe. A nano-sized heat pipe would be a different kettle of fish, however (as would a carbon nanotube heat pipe).

## 2 The development of the heat pipe

Initially, Grover was interested in the development of high-temperature heat pipes, employing liquid metal working fluids, suitable for supplying heat to the emitters of thermionic electrical generators and removing heat from the collectors of these devices. Shortly after Grover's publication [3], work was started on liquid metal heat pipes by Dunn at Harwell and Neu and Busse at Ispra where both establishments were developing nuclear-powered thermionic generators. Interest in the heat pipe concept developed rapidly both for space and terrestrial applications. Work was carried out on many working fluids, including metals, water, ammonia, acetone, alcohol, nitrogen, and helium.

At the same time, the theory of the heat pipe became better understood; the most important contribution to this theoretical understanding was by Cotter [9] in 1965. The manner in which heat pipe work expanded is seen from the growth in the number of publications, following Grover's first paper in 1964. In 1968 Cheung [10] lists 80 references; in 1971 Chisholm in his book [11] cites 149 references and by 1972 the NEL Heat Pipe Bibliography [12] contained 544 references. By the end of 1976, in excess of 1000 references to the topic were available.

By the early 21st century, the heat pipe became a mass-produced item necessary in 'consumer electronics' products such as laptop computers that are made by the tens of millions per annum. It also remained, judging by the literature appearing in scientific publications, a topic of substantial research activity and which, as a result, the authors have introduced of a new chapter (Chapter 7) within this edition on the cooling of electronic comments.

The most obvious pointer to the success of the heat pipe is the wide range of applications where its unique properties have proved beneficial. Some of these applications are discussed in detail in Chapters 6 and 7, but they include the following: electronics cooling, nuclear and chemical reactor thermal control, heat recovery and other energy conserving uses, de-icing duties, cooking, control of manufacturing process temperatures, thermal management of spacecraft and in automotive systems and, increasingly, in renewable energy devices.

There is one aspect of heat pipe manufacture that must not pass unnoticed in this introduction – the use of 3D printing (selective laser remelting, rapid prototyping or whatever you may wish to call it). Already used for heat exchangers [13], the method is now being evaluated for making heat pipes see Chapter 3.

## 3 The contents of this book

Chapter 1 describes the development of the heat pipe in more detail. Chapter 2 covers heat pipe types and range of special types, including LHPs. Chapter 3 discusses the main components of the heat pipe and the materials used and includes compatibility data. Chapter 3 also sets out design procedures and worked examples, details how to make and test heat pipes and covers a range of special types, including LHP and PHP. Chapter 4 gives heat transfer and fluid flow theory relevant to the operation of the classical wicked heat pipe and details analytical techniques that are then applied to both heat pipes and thermosyphons. Chapter 5 describes additive manufacturing used in heat pipes. Chapter 6 presents applications of heat pipe heat exchangers and their contribution in the waste heat recovery domain. Heat pipes applications are fully discussed in Chapter 7 and covers the expertise of the leading heat pipe manufacturer of electronics thermal control systems for terrestrial and aerospace systems.

As in previous editions, a considerable amount of data are collected together in appendices for reference purposes.

## References


[1] R.S. Gaugler, US Patent 2350348. Applied 21 December 1942, 6 June 1944.

[2] G.M. Grover, US Patent 3229759, 1963.

[3] G.M. Grover, T.P. Cotter, G.F. Erickson, Structures of very high thermal conductance, J. App. Phys. 35 (1964) 1990.

[4] C.C. Roberts, A review of heat pipe liquid delivery concepts, Advances in Heat Pipe Technology. Proceedings of Fourth International Heat Pipe Conference, Pergamon Press, Oxford, 1981.

[5] D.A. Reay, Microfluidics Overview. Paper presented at Microfluidics Seminar, East Midlands Airport, UK, April 2005. TUV-NEL, East Kilbride, 2005.

[6] B. Jeyadevan, H. Koganezawa, K. Nakatsuka, Performance evaluation of citric ion-stabilised magnetic fluid heat pipe, J. Magnet. Magn. Mater. 289 (2005) 253256.

[7] Y.F. Maydanik, Loop heat pipes (Review article), Appl. Therm. Eng. 25 (2005) 635657.

[8] S. Lips, A. Bensalem, Y. Bertin, V. Ayel, C. Romestant, J. Bonjour, Experimental evidences of distinct heat transfer regimes in pulsating heat pipe (PHP), Appl. Therm. Eng. 30 (2010) 900907.

[9] T.P. Cotter, Theory of heat pipes. Los Alamos Scientific Laboratory Report No. LA-3246-MS, 1965.
[10] H. Cheung, A critical review of heat pipe theory and application. UCRL 50453, 15 July 1968.
[11] D. Chisholm, (M & B Technical Library, TL/ME/2) The Heat Pipe, Mills and Boon Ltd., London, 1971.
[12] J. McKechnie, The heat pipe: a list of pertinent references, National Engineering Laboratory, East Kilbride, Applied Heat ST. BIB, 1972, pp. 272.
[13] http://www.within-lab.com/case-studies/index11.php. (accessed 02.01.13).

Chapter 1

# Historical development

The heat pipe differs from the thermosyphon (a method of passive heat exchange based on natural convection) by virtue of its ability to transport heat *against gravity* by an evaporation–condensation cycle. It is, however, important to realise that many heat pipe applications do not need to rely on this feature and the Perkins tube, which predates the heat pipe by several decades, is basically a form of thermosyphon and is still used in heat transfer equipment. The Perkins tube should therefore be regarded as an essential precursor in the historic development of the heat pipe.

## 1.1 The Perkins tube

Angier March Perkins (the son of the engineer Jacob Perkins [1]) was born in Massachusetts, United States at the end of the 18th century, but in 1827 he came to England where he subsequently carried out much of his development work on boilers and other heat distribution systems. His work on the Perkins tube – a two-phase flow device – is attributed in the form of a patent to Ludlow Patton Perkins (his son) in the mid-19th century, but A.M. Perkins also worked on single-phase heat distribution systems (with some considerable success), and although the chronological development is somewhat difficult to follow from the available papers, his single-phase systems preceded the Perkins tube and so some historical notes on both systems seem appropriate.

A catalogue (published in 1898) describing the products of A.M. Perkins & Sons Ltd states that in 1831 A.M. Perkins took out his first patent for what is known as the 'Perkins' system of heating by small bore wrought iron pipes. This system is basically a hermetic tube boiler in which water is circulated through tubes (in single phase and at high pressure) between the furnace and a steam drum providing an indirect heating system. The boiler, described in UK Patent No. 6146, used hermetic tubes and was produced for over 100 years on a commercial scale. The specification describes this closed cycle hot water heater as being adaptable to suit: sugar making and refining evaporators, steam boilers, and also for various processes requiring molten metals for alloying or working of other metals at high temperatures – suggesting that the tubes in the Perkins system operated with high-pressure hot water at temperatures well in excess of 150°C.

The above patent also describes an 'ever full' water boiler, the principle of which was devised in the United States by Jacob Perkins to prevent the formation of a film of bubbles on the inner wall of the heat input section of the tubes.

> As water expands about one-twentieth of its bulk being converted into steam, I provide about double that extra space in the 'expansion tube' which is fitted with a removable air plug to allow the escape of air when the boiler is being filled. With this space for the expansion of the heated water the boiler is completely filled and will at all times be kept in constant contact with the metal however high the degree of heat such apparatus may be submitted to; and at the same time there will be no danger of bursting the apparatus with the provision of the sufficient space as named for the expansion of the water.

In 1839 most of the well-known forms of A.M. Perkins' hot water hermetic heating tubes were patented within UK Patent No. 8311, and in that same year a new invention, a concentric tube boiler, was revealed. The hot water closed circuit heating tubes in the concentric tube system were fork-ended and dipped into two or more steam generation tubes. These resembled superheater elements as applied on steam locomotives and a large boiler operating on this principle would consist of many large firetubes, all sealed off at one end and traversed by the inner hot water tubes, externally connected by U bends. Of all the designs produced by the Perkins Company, this proved to be the most rapid producer of superheated steam manufactured and was even used as the basis for a steam-actuated rapid-firing machine gun offered to the US Federal Government at the time of the Civil War. Although not used, these were '*guaranteed to equal the efficiency of the best Minié rifles of the day, but at a much lower cost for coal than for gun powder*'. Notwithstanding, the system was however used in marine engines, '... *it gives a surprising economy of fuel and a rapid*

Heat Pipes. DOI: https://doi.org/10.1016/B978-0-12-823464-8.00010-3

*generation, with lightness and compactness of form; and a uniform pressure of from 200 lbs to 800 lbs per sq. in., may be obtained by its use*'.

Returning to the Perkins hermetic tube single-phase water circulating boiler, as illustrated in Fig. 1.1, some catalogues describe these units as operating at pressures up to 4000 psi and of being pressure-tested in excess of 11 000 psi. In addition, operators were quick to praise the cleanliness, both inside and outside, of the hermetic tubes even after prolonged use.

The first use of the Perkins tube, that is one containing only a small quantity of water and operating on a two-phase cycle, is described in a patent by Jacob Perkins (UK Patent No. 7059, April 1936). The general description is as follows [2]:

> One end of each tube projects downwards into the fire or flue and the other part extends up into the water of the boiler; each tube is hermetically closed to prevent escape of steam. There will be no incrustation of the interior of the tubes and the heat from the furnace will be quickly transmitted upwards. The interior surfaces of the tubes will not be liable to scaleage or oxidation, which will, of course, tend much to preserve the boiler so constructed.' The specification also says 'These tubes are: each one, to have a small quantity of water depending upon the degree of pressure required by the engine; and I recommend that the density of the steam in the tubes should be somewhat more than that intended to be produced in the boiler and, for steam and other boilers under the atmospheric pressure, that the quantity of water to be applied in each tube is to be about 1:1800 part of the capacity of the tube; i.e., for a pressure of 2 atm to be two 1:1800 parts; for 3 atm, three 1:1800 parts, and so on, for greater or less degrees of pressure, and by which means the tubes of the boiler when at work will be pervaded with steam, and any additional heat applied thereto will quickly rise to the upper parts of the tubes and be given off to the surrounding water contained in the boiler – for steam already saturated with heat requires no more (longer) to keep the atoms of water in their expanded state, consequently becomes a most useful means of transmitting heat from the furnace to the water of the boiler.

The earliest applications for this type of tube were in locomotive boilers and in locomotive firebox superheaters (in France, 1863). Again, as with the single-phase sealed system, the cleanliness of the tubes was given prominent publicity in many papers on the subject and at the Institution of Civil Engineers in February 1837, Perkins stated that following a 7-month life test with such a boiler tube operating under representative conditions, there was no leakage nor incrustation, in fact, no deposit of any kind occurring within the tube.

## 1.2 Patents

Reference has already been made to several patents taken by A.M. Perkins and J. Perkins on hermetic single-phase and two-phase heating tubes normally for boiler applications; however, the most interesting patent relating to improvements

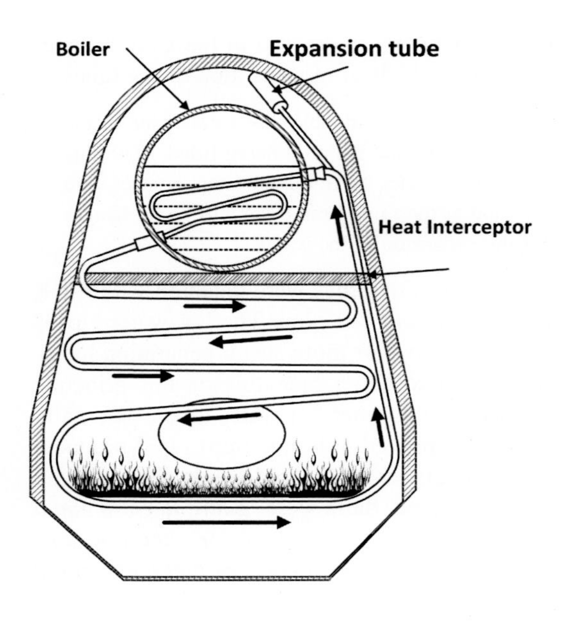


**FIGURE 1.1** Perkins boiler.

in the basic Perkins tube is UK Patent No. 22272, dated 1892, and granted to L.P. Perkins and W.E. Buck: 'Improvements in Devices for the Diffusion or Transference of Heat' [3].

The basic claim, containing a considerable number of modifications and details referring to fluid inventory and application, is for a closed tube or tubes of suitable form or material partially filled with a liquid, and whilst water is given as one specific working fluid, the patent also covers the use of antifreeze type fluids as well as those having a boiling point higher than water.

It is obvious that previous work on the Perkins tube had revealed that purging of the tubes of air, possibly by boiling off a quantity of the working fluid before sealing, was desirable, as Perkins and Buck indicate that this should be done for optimum operation at low temperatures (hence low internal pressures) and when it is necessary to transmit heat as rapidly as possible at high temperatures.

Safety and optimum performance were also considered in the patent, where reference is made to the use of 'suitable stops and guides' to ensure that tubes refitted after external cleaning were inserted at the correct angle and with the specified amount of evaporator section exposed to the heat source (normally an oil, coal or gas burner). Some form of entrainment had also probably occurred in the original straight Perkins tube, particularly when transferring heat over considerable distances. As a means of overcoming this limitation, the patent provides for U bends so that the condensate return occurs in the lower portion of the tube, vapour flow to the heat sink taking place in the upper part.

Applications cited included heating of greenhouses, rooms, vehicles, dryers, and as a means of preventing condensation on shop windows, the tubes providing a warm convection current up the inner face of the window. Indirect heating of bulk tanks of liquid is also suggested. The use of the device as a heat removal system for cooling dairy products, chemicals and heat exchanging with the cooling water of gas engines is also proposed, as its use in waste heat recovery, the heat being recovered from the exhaust gases from blast furnaces and other similar apparatus, and used to preheat incoming air.

On this and other air/gas heating applications of the Perkins tube, the inventors have neglected to include the use of external finning on the tubes to improve the tube-to-gas heat transfer. Although not referring to the device as a Perkins tube, such modifications were proposed by F.W. Gay in US Patent No. 1725906, dated 27 August 1929, in which a number of finned Perkins tubes or thermosyphons are arranged as in the conventional gas/gas heat pipe heat exchanger, with the evaporator sections located vertically below the condensers; a plate sealing the passage between the exhaust and inlet air ducts, as shown in Fig. 1.2. Working fluids proposed included methanol, water and mercury depending upon the likely exhaust gas temperatures.

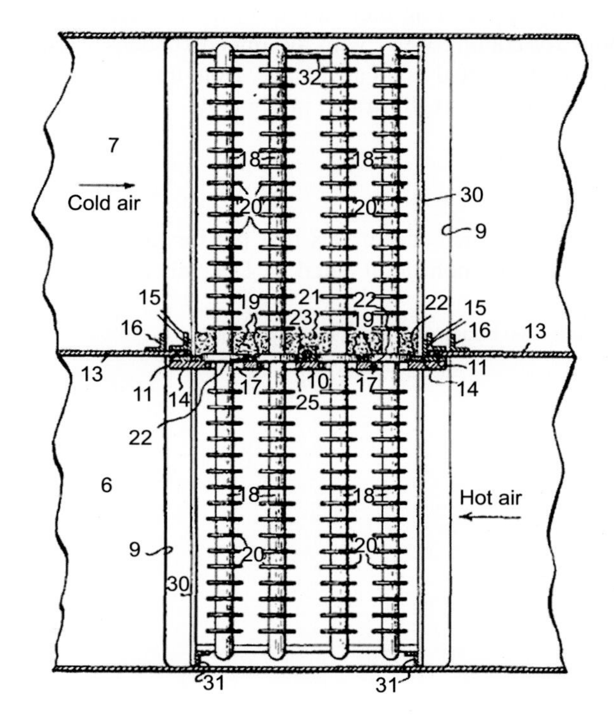


FIGURE 1.2 Thermosyphon heat exchanger proposed by F.W. Gay.

## 1.3 The baker's oven

The primary application for the Perkins tube was in baking ovens, and one of the earliest forms of baking oven to which the Perkins tube principle was applied was a portable bread oven supplied to the British army in the 19th century. In common with static ovens that employed the Perkins tube, the firing was carried out remotely from the baking chamber, the heat then being transferred from the flames to the chamber by the vapour contained within the tubes. The oven operated at about 210°C, and it was claimed that the fuel savings using this type of heating were such that only 25% of the fuel typically consumed by conventional baking ovens was required [4].

A more detailed account of the baking oven is given in Ref. [5]. This paper, published in 1960 by the Institution of Mechanical Engineers, is particularly concerned with failures of the tubes used in these ovens, and the lack of safety controls which led to a considerable number of explosions within the tube bundles.

By this time, gas or oil firing had replaced coal and coke in many of the installations, resulting in the form of oven shown in Fig. 1.3, in which 80 tubes are heated at the evaporator end by individual gas flames, the illustration showing the simplest form of Perkins tube oven in which straight tubes are used. Other systems employing U-tubes or a completely closed loop, as put forward by Perkins and Buck, as a method of overcoming entrainment. In the particular oven shown the maximum oven temperature was of the order of 230°C.

One feature of interest is the very small evaporator length, typically less than 5 cm for many of these ovens, and this is in contrast to later designs with overall tube lengths around 3 m and a condenser section of about 2.5 m depending upon the thickness of the insulating wall between the furnace and the oven. The diameter of the tubes is typically 3 cm and the wall thickness can be considerable, in the order of 5–6 mm. Solid drawn tubes were used for the last Perkins ovens constructed, with one end closed by swaging or forging before charging with the working fluid although originally, seam-welded tubes or wrought iron tubes were used. The fluid inventory in the tubes is typically about 32% by volume, a very large proportion when compared with normal practice for heat pipes and thermosyphons, and which, in addition, generally has much longer evaporator sections. This larger fluid mass might be explained because Perkins and Buck, in proposing a larger fluid inventory than that originally used in the Perkins tube, indicated that dryout had been a problem in earlier tubes, leading to overheating, caused by complete evaporation of the relatively small fluid inventory.

The inventory of 32% by volume is calculated on the assumption that just before the critical temperature of saturated steam is reached (374°C), the water content would be exactly 50% of the total tube volume, and it can be shown by calculation that up to the critical temperature, if the water inventory at room temperature is less than 32%, it will all evaporate before the critical temperature is reached resulting in overheating. If, however, the quantity of water is greater than 32% by volume, the tube will be completely filled with water and unable to function before the critical temperature is reached.

Another example application for the thermosyphon cited within early engineering literature from the last century is the Critchley–Norris car radiator. This employed 110 thermosyphons for cooling water with the finned condenser sections projecting into the airstream – ease of replacement of burst tubes being cited as an advantage. Note that the maximum speed of the vehicle on which they were used was 12 miles/h! The radiator is illustrated in Fig. 1.4.

## 1.4 The heat pipe

As mentioned in the introduction, heat pipes differ from thermosyphons and the concept was first put forward by R.S. Gaugler of the General Motors Corporation, Ohio, United States, in a patent application dated 21 December 1942 and

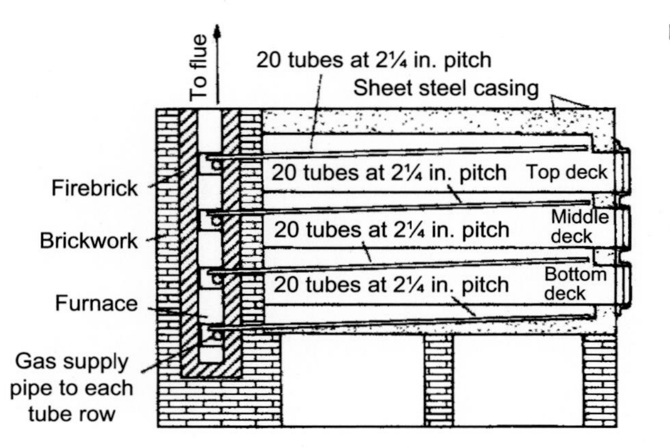


FIGURE 1.3 Gas-fired baking oven using 80 Perkins tubes.

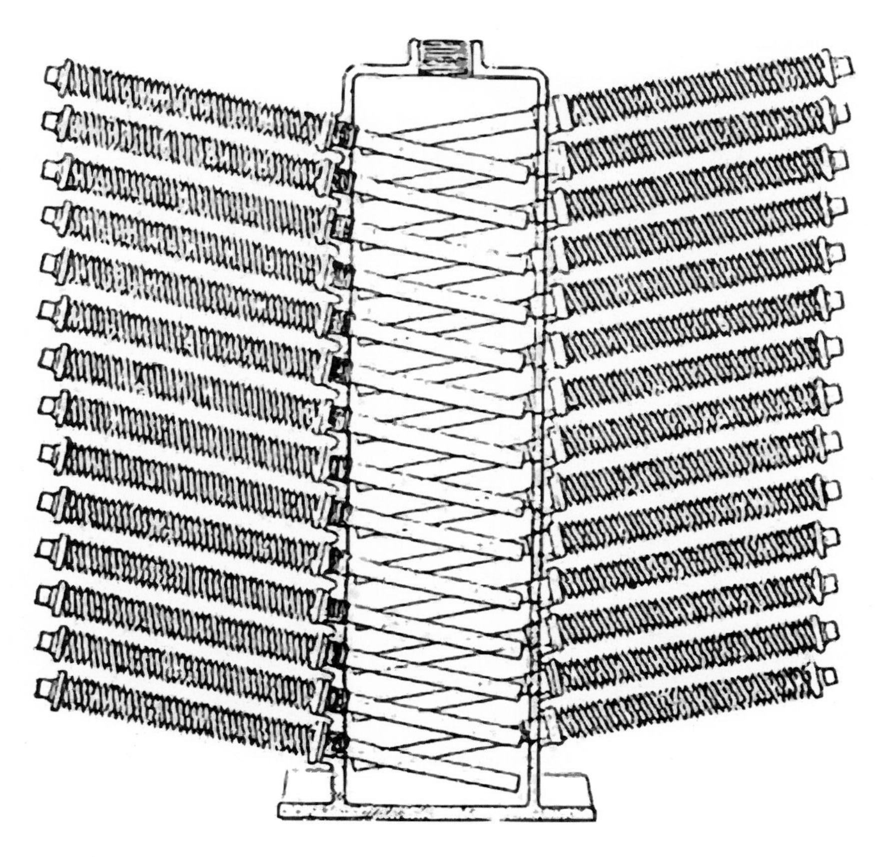

**FIGURE 1.4** The Critchley–Norris Radiator.

published [5] as US Patent No. 2350348 on 6 June 1944; the heat pipe described being applied to a refrigeration system.

According to Gaugler, the object of the invention was to '... *cause absorption of heat, or in other words, the evaporation of the liquid to a point above the place where the condensation or the giving off of heat takes place without expending upon the liquid any additional work to lift the liquid to an elevation above the point at which condensation takes place*'. A feature of the patent was the capillary structure which was proposed as the means for returning the liquid from the condenser to the evaporator, and Gaugler suggested that one form of this structure might be a sintered iron wick. The wick geometries proposed by Gaugler are shown in Fig. 1.5, and it is interesting to note the comparatively small proportion of the tube cross-section allocated to vapour flow in all three of his designs.

One form of refrigeration unit suggested by Gaugler is shown in Fig. 1.6, in which the heat pipe is employed to transfer heat from the interior compartment of the refrigerator to a pan below the compartment containing crushed ice. To improve heat transfer from the heat pipe to the ice, a tubular vapour chamber with external fins is provided into which the heat pipe is fitted. This also acts as a reservoir for the heat pipe working fluid.

Despite the granted patent, the heat pipe proposed by Gaugler was not developed beyond the patent stage because another technology (available at that time) was applied to solve General Motor's particular thermal problem.

Grover's patent [6], filed on behalf of the US Atomic Energy Commission in 1963, coins the name '*heat pipe*' to describe devices essentially identical to that in the Gaugler patent; however, Grover's work includes a limited theoretical analysis and presents results of experiments carried out on stainless steel heat pipes incorporating a wire mesh wick with sodium as the working fluid although lithium and silver are also mentioned as working fluids.

Under Grover, an extensive programme was conducted on heat pipes at the Los Alamos Laboratory in New Mexico, and preliminary results were reported in the first recorded publication on heat pipes [7]. Following this, the Harwell-based UK Atomic Energy Laboratory started similar work on sodium and other heat pipes [8]; the Harwell interest being primarily their application to nuclear thermionic diode converters. Around this time, a similar programme commenced at the Joint Nuclear Research Centre, Ispra, Italy under Neu and Busse – the work at Ispra building up rapidly until the laboratory became the most active centre for heat pipe research outside the United States [9,10].

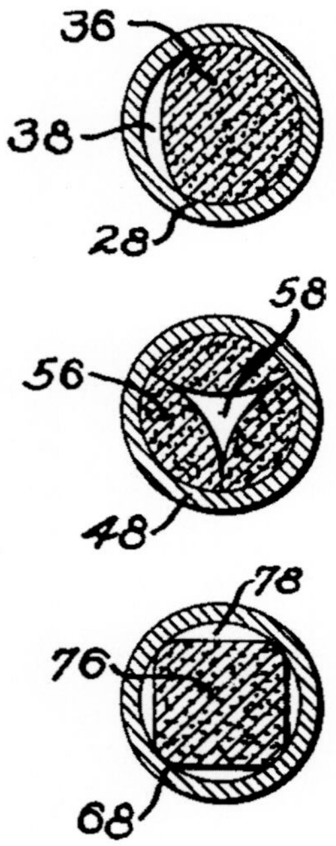

**FIGURE 1.5** Gaugler's proposed heat pipe wick geometries.

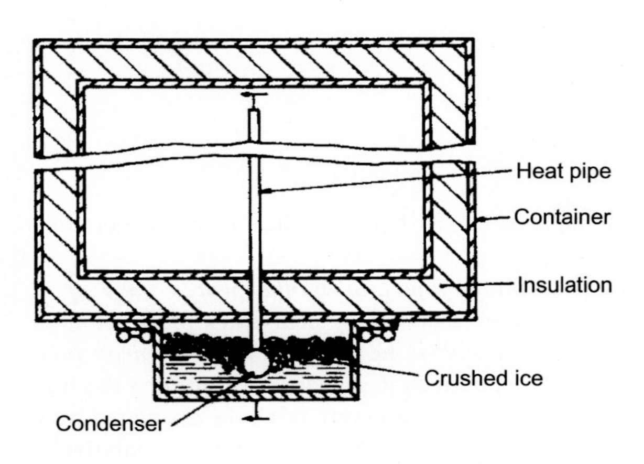


**FIGURE 1.6** The refrigeration unit suggested by Gaugler in his patent published in 1944.

The work at Ispra was concerned with heat pipes for carrying heat to emitters (operating between 1600°C and 1800°C) and for dissipating waste heat from collectors (operating at 1000°C), with the emphasis being on emitter heat pipes which posed more difficult problems concerning reliability over extended periods of operation.

Radio Corporation of America (RCA) [11,12] was the first commercial organisation to work on heat pipes and most of their early support came from US Government contracts. During the 2-year period, mid-1964 to mid-1966, they made heat pipes using glass, copper, nickel, stainless steel, molybdenum and titanium zirconium molybdenum (TZM) molybdenum for wall materials with various working fluids which included water, caesium, sodium, lithium and bismuth; maximum operating temperatures of 1650°C being achieved.

Not all of the early heat pipe studies involved high operating temperatures. Deverall and Kemme [13] developed a heat pipe for satellite use incorporating water as the working fluid, and their first proposals for a variable conductance heat pipe were again made for a satellite [14].

1967 and 1968 saw the publication of several articles in the scientific press (most originating in the United States), that indicated a broadening of the areas of heat pipe application to electronics cooling, air-conditioning, engine cooling and other areas [15–17]. These revealed developments such as flexible and flat plate heat pipes, but one point the articles also stressed was the vastly increased thermal conductivity of a heat pipe compared to a solid conductor such as copper. A water heat pipe with a simple wick has an effective conductivity several hundred times greater than that of a copper rod of similar dimensions.

Work at Los Alamos Laboratory continued apace with a continuing emphasis on satellite applications, the first flights using heat pipes technology taking place in 1967 [13]. To demonstrate that heat pipes would function normally in a space environment, a water/stainless steel heat pipe (electrically heated at one end) was launched into an earth orbit from Cape Kennedy on an Atlas–Agena vehicle, and once orbit was achieved, the heat pipe was automatically activated. Telemetry data (received at five tracking stations) over 14 orbits indicated that the heat pipe operated successfully.

At this stage, heat pipe theory was well developed, based largely on the analysis work of Cotter [18] who was also working at Los Alamos. So active were laboratories both within the United States and at Ispra, that in his critical review of heat pipe theory and applications [19], Cheung was able to list over 80 technical papers on all aspects of heat pipe development. His review endorsed the reliability of liquid metal heat pipes under long-term operation (9000 h) at elevated temperatures (1500°C) and that heat pipes were capable of transferring axial fluxes of 7 $kW/cm^2$.

Cheung also referred to various forms of wick, including an arterial type illustrated in Fig. 1.7, which was developed by Katzoff [20] and tested in a glass heat pipe using an alcohol as the working fluid. The function of the artery – by then a common feature of heat pipes developed for satellite use – being to provide a path with a low-pressure drop to aid the transportation of liquid from the condenser to the evaporator where it is redistributed around the heat pipe circumference using a fine-pore wick structure, sited around the heat pipe wall.

Following the successful heat pipe test in space in 1967 [13], the first use of heat pipes for satellite thermal control was on GEOS-B (Geodetic Earth Orbiting Satellite 2 – part of NASA's Explorer Programme), which was launched from Vandenburg Air Force Base in 1968 [21]. Two heat pipes were used, located as shown in Fig. 1.8, and these were constructed using 6061 T-6 aluminium alloy with 120-mesh aluminium as the wick material and Freon 11 as the working fluid. Their purpose was to minimise the temperature differences between the various transponders located within the satellite, and based on an operating period of 145 days, the maximum to minimum transponder temperature ranges were considerably smaller than those for similar transponder arrangements in GEOS-A – an earlier satellite not employing heat pipe technology. The heat pipes performed well over the complete period of observation, maintaining the transponders at near iso-thermal conditions.

In 1968 Busse wrote a paper [22] summarising the heat pipe activities in Europe at that time, and it is notable that the Ispra Laboratory of Euratom was still the focal point for European activities. Other laboratories, however, were also making useful contributions to knowledge, and these included Brown Boveri, the Karlsruhe Nuclear Research Centre, the Institüt für Kernenergetic, and the Stuttgart and Grenoble Nuclear Research Centre. The experimental programmes of these laboratories were mainly performed on heat pipes using liquid metals as their working fluids and focussed on life tests and measurements of the maximum axial and radial heat fluxes although theoretical aspects of heat transport limitations were also studied. By this time, the results of basic studies on separate features of heat pipes were also beginning to appear. For example: wick development, factors affecting the limiting of evaporator heat flux, and the influence of noncondensable gases on performance.

In Japan, the Kisha Seizo Kaisha Company [23] conducted a limited experimental programme but one with considerable impact of energy conservation. Presenting a paper on this work in April 1968 to an audience of air-conditioning and refrigeration engineers, Nozu et al. described an air heater utilising a bundle of finned heat pipes. The significance of this heat pipe heat exchanger application to environmental protection being that it could also be used to recover heat from hot exhaust gases and could be applied to both industrial and domestic air-conditioning systems. These heat exchangers are now available commercially and are referred to in Chapter 6.

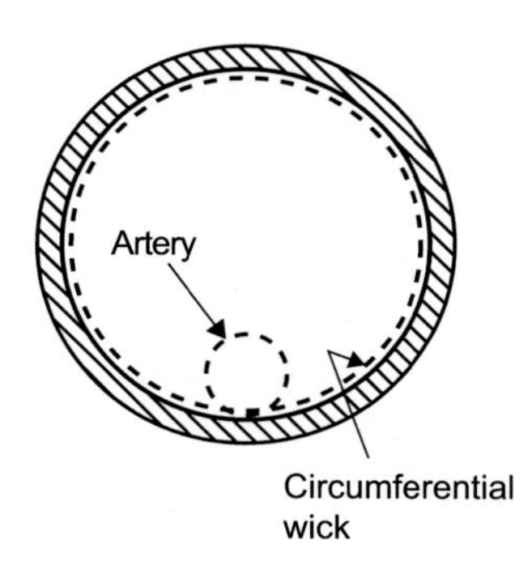


**FIGURE 1.7** An arterial wick form developed by Katzoff [20].

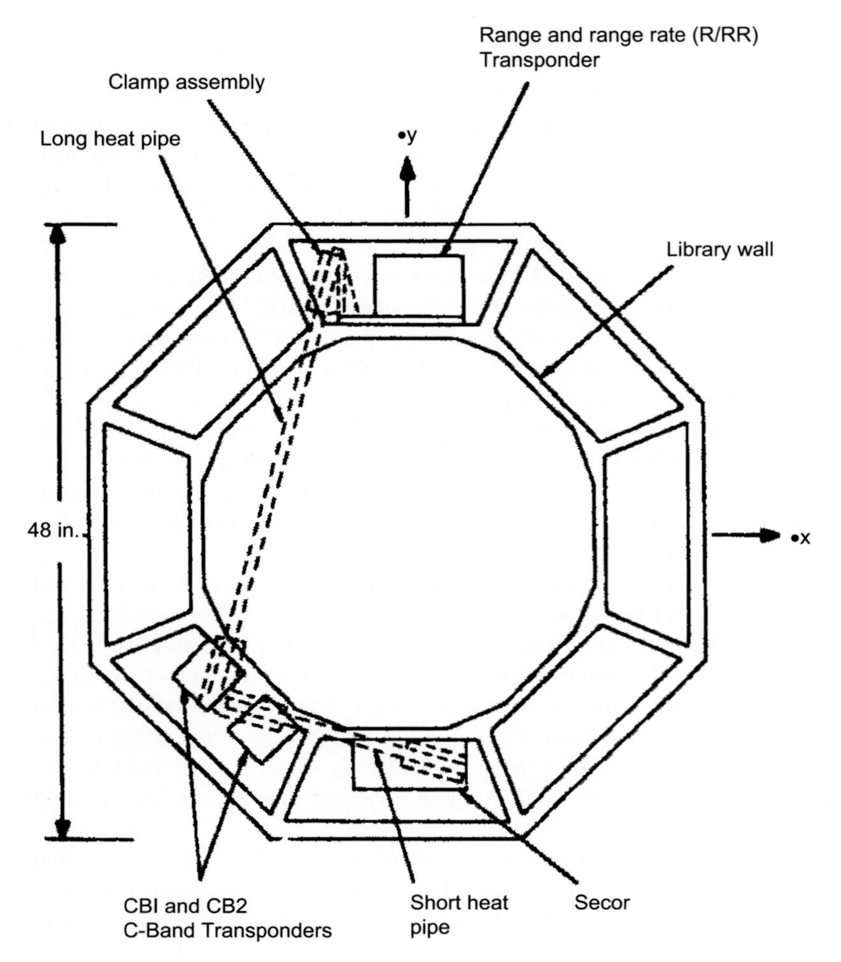


**FIGURE 1.8** Heat pipes used in space – the GEOS-B satellite.

As evidenced by published literature on heat pipes, 1969 saw establishments in the United Kingdom taking increasing interest in their potential, with such companies as the British Aircraft Corporation (BAC) and the Royal Aircraft Establishment, Farnborough (RAE) [24] evaluating heat pipes and vapour chambers for the thermal control of satellites.

It was also during 1969, that Reay (then at IRD), commenced his work on heat pipes – initially in the form of a survey of potential applications – followed by an experimental programme concerned with the manufacture of flat plate and tubular heat pipes. At this time, Dunn, co-author of the first four editions of *Heat Pipes*, was also carrying out work at Reading University where several members of the staff had experience of the heat pipe activities at Harwell described earlier.

A bumper year for heat pipe research 1969 saw a developing research contribution from many agencies, for example:

- The National Engineering Laboratory (now the TUV-NEL) at East Kilbride entered the field.
- The publication of further work on variable conductance heat pipes – the principle contributions being made by Turner [25] at RCA and Bienert [26] at Dynatherm Corporation. All of which added to the theoretical analyses of variable conductance heat pipes (vCHPs) to determine parameters such as reservoir size and consider other practical aspects such as reservoir construction and their susceptibility to external thermal effects.
- The former USSR published an article in the Russian journal '*High Temperature*' [27], although much of the information described a summary of work published elsewhere.
- A new type of heat pipe – in which the wick is omitted – was developed by NASA [28], in which a rotating heat pipe utilised centrifugal acceleration to transfer liquid from the condenser to the evaporator which could then be used for cooling motor rotors and turbine blade rotors. Gray [28] also proposed an air-conditioning unit based on the rotating heat pipe and this is illustrated in Fig. 1.9. The rotating heat pipe is described more fully in Chapter 2, but its

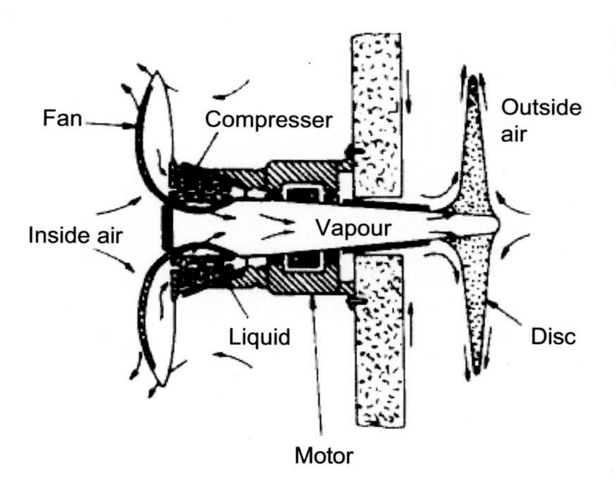


**FIGURE 1.9** A compact air-conditioning unit based on the wickless rotating heat pipe. *Courtesy: NASA.*

advantage is that it does not suffer from the limitations of capillary pumping that occur in conventional heat pipes and so its transport capability is superior to that of a wicked pipe.

- 1969 also saw the application of heat pipes to areas other than satellites beginning to receive attention. For example, their use for electronics cooling, with heat pipes of the rectangular section being proposed by Sheppard [29] for the cooling of integrated circuit packages, and the design, development and fabrication of a heat pipe to cool a high-power airborne travelling-wave tube as described by Calimbas and Hulett of the Philco-Ford Corporation [30].

Most of the work on heat pipes described so far has been associated with liquid metal working fluids and, for lower temperatures, water, acetone, alcohols etc. Nonetheless, around this time, cryogenic heat pipes also began to receive particular attention [31,32] with the need to cool the detectors within satellite infrared scanning systems being but one application.

At the time, although the most common working fluid for these heat pipes was nitrogen (which was acceptable for temperatures between 77 and 100K with liquid oxygen also being used for this temperature range), the Rutherford High Energy Laboratory (RHEL) was the first organisation in the United Kingdom to operate cryogenic heat pipes [33] employing liquid hydrogen units for cooling targets at the RHEL, and later a helium heat pipe operating at 4.2K [34].

By 1970 a wide variety of heat pipes were commercially available from a number of companies in the United States: RCA, Thermo-Electron and Noren Products being notable amongst several firms marketing a range of 'standard' heat pipes and offering the facility to construct 'specials' for specific customer applications. Over the next few years, several manufacturers were established in the United Kingdom (see Appendix 3), with a number of companies specialising in heat pipe heat recovery systems (based primarily on technology from the United States) entering what was becoming an increasingly competitive market [35].

Moreover, the early 1970s saw an increasing momentum in their development for spacecraft thermal control, when the European Space Agency (ESA) commenced the funding of research and development work with companies and universities in Britain and Continental Europe for the purpose of producing space-qualified heat pipes for ESA satellites. Other companies, notably Dornier, sABCA, Aerospatiale and Marconi, also putting considerable effort into development programmes, initially independent of ESA, to keep abreast of the technology. As a result of this, a considerable number of European companies were able to compete effectively in the field of spacecraft thermal control using heat pipes. In the 1990s, the role of European companies supplying heat pipes to national and pan-European space programmes was continuing apace, with Alcatel Space consolidating their early research on axially grooved heat pipes with new high-performance variants [36]. At the same time, Astrium was pioneering its capillary pumped fluid loop systems for the French technological demonstrator spacecraft called STENTOR [37]. Capillary pumped loops are used by NASA on the Hubble Space Telescope, and in December 2001 the Space Shuttle carried out an experiment to test a loop system with more than one evaporator in order to overcome a perceived limitation of single evaporators [38]. The experiment, known as CAPL-3, was photographed from the International Space Station and is visible in Fig. 1.10.

The early 1970s also saw considerable growth in the application of heat pipes to solve terrestrial problems, and although the application of heat pipes to spacecraft may seem rather esoteric to the majority of potential users of these devices, the 'technological fallout' has been considerable. Both in terms of applications to space flight and in nuclear engineering, the work has contributed significantly to the design procedures, reliability and performance improvements of all such commercial products over the past 20 years. For example, the effort expended in developing vCHPs, initially

**FIGURE 1.10** CAPL-3 visible in the Space Shuttle Bay on STS-108 [38].

to meet the requirements of the US space programme, led to one of the most significant types of heat pipe, which is able to effect precise temperature control [39]. In fact, the number of techniques available for the control of heat pipes is now very considerable, and some are discussed in Chapter 6.

One of the major engineering projects of the 1970s – the construction and operation of the trans-Alaska oil pipeline – makes crucial use of heat pipe technology to prevent thawing and melting of the permafrost around the pipe supports for the elevated sections of the pipeline. The magnitude of the project necessitated McDonnell Douglas Astronautics Company producing 12,000 heat pipes per month, over a total contract number of nearly 100,000 heat pipes, the pipes themselves ranging in length between 9 and 23 m. At the end of the thaw season, the permafrost remained almost 0.5°C below its normal temperature and this cooling permitted a 10–12 m reduction in pile length required to support the pipe structure.

During this era [40], the Sony Corporation integrated heat pipes into 'heat sinks' of its tuner amplifier products, and these proved to be 50% lighter and 30% more effective than conventional extruded aluminium units. The number of units delivered approached 1 million, and this is probably the first large-scale use of heat pipes in consumer electronic equipment.

The 1970s also saw the first use of heat pipes in combination with phase change materials (PCMs) for heat storage applications. The heat pipes help to overcome difficulties such as poor thermal conductivity and freezing and melting profiles of PCMs. Lee and Wu [41] implemented a water thermosyphon in a paraffin wax PCM and concluded that the total heat transfer coefficient of the system increased with an increase in the superheat of the PCM, which allowed the PCM to exceed the saturation temperature.

Whilst much development work has concentrated on 'conventional' heat pipes, the last few years have seen increasing interest in other systems for liquid transport. The proposed use of 'inverse' thermosyphons and an emphasis on the advantages (and possible limitations) of gravity-assisted heat pipes have stood out as areas of considerable importance, and the reason for the growing interest in these topics is not too difficult to find. Heat pipes in terrestrial applications have proved (in the majority of cases) particularly viable when gravity, in addition to capillary action, has aided condensate return to the evaporator. This is best seen in heat pipe heat recovery units, where slight changes in the inclination of the long heat pipes used for such heat exchangers can cut off heat transport completely if the wick is solely relied on for the condensate return.

To date, the chemical heat pipe has not yet been commercially applied but 'conventional' very long heat pipes have been developed. '*Very long*' in this case means 70–110 m, the units being tested at the Kyushu Institute of Technology in Japan [42]. These were expected to be used for underfloor heating/cooling and for snow melting or de-icing of roads, and long flexible heat pipes have also recently been used in Japan for cooling or isothermalising buried high-voltage cables.

At the time of writing this edition of Heat Pipes, the dominant application, in terms of numbers of units produced, remains the same as that for the 5th Edition (2006), and that is for the cooling of computer chips – desktops and laptops. The demands of computer hardware manufacturers dictate that thermal control systems are compact and of low cost. The heat pipes used, typically 2–3 mm diameter, cost about $1(US) each. Mass production methods have been developed to achieve these low costs, with millions of units being produced per month in conjunction with their manufacture in countries where labour costs currently remain relatively low – predominantly China.

Nonetheless despite this manufacturing dominance, with the ever-increasing processing power of portable computing, the increasing demand on cooling systems will eventually lead to loads that may not be manageable with conventional heat pipe technology. So we are forced to ask at this stage of their development, '*Can heat pipes address our future thermal challenges?*', because although liquid cooling (pioneered by IBM) is becoming more common in data centres, it is not suitable for portable computers.

## 1.5 Can heat pipes address our future thermal?

Some computer and chip manufacturers have indicated that by the time of this book publication, heat dissipation (thermal design power) will have increased to around 400 W for both desktop and server computers and, more specifically, the critical area is dissipating heat from the CPUs. The CPU size is little more than 72 $mm^2$, and thus the equivalent heat flux is approaching 555 $W/cm^2$. Moreover, the advances in very large-scale integrated manufacturing and in nanotechnology have led to a significant reduction in the size of electronic chips so requiring high thermal transfer solutions with minimum size and the smallest thickness possible. And, in this respect, one such application of powerful CPU chips is within a modern smartphone or tablet pc where active cooling is a challenge. Consequently, addressing the heat dissipation challenge by passive cooling would also open the door for much more powerful portable devices within the same size and weight restrictions.

Workers in Australia and Japan are currently examining methods for overcoming the perceived limitations of conventional heat pipe systems when heat fluxes are too high, but until heat pipe technology comes up with a passive solution, a pumped system such as that shown in Fig. 1.11 is being proposed. This uses a combination of impinging jet cooling, linked to a heat sink formed of a highly compact heat exchanger [43].

NASA has also highlighted the challenges of thermal management in a different environment – in robotic spacecraft [38], with Swanson and Birur (based at the Goddard Space Flight Center and the Jet Propulsion Laboratory respectively), having both indicated that loop heat pipes and capillary pumped loops will not satisfy future thermal control requirements. These include both very low-temperature operation (40K) and/or operation in very high temperature environments; tight temperature control; high heat fluxes ($>100$ $W/cm^2$) and the need to minimise mass, for example for nanosatellites.

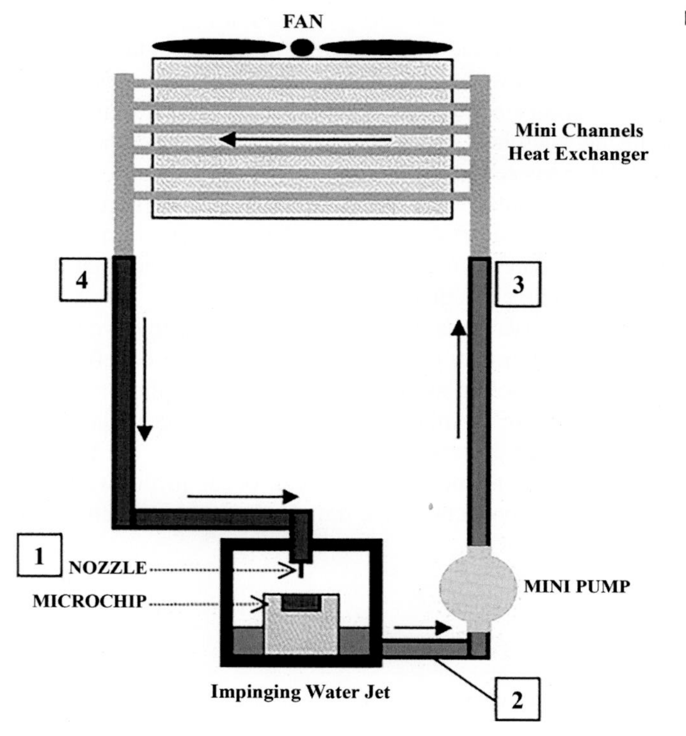


**FIGURE 1.11** Impinging water jet cooling system for CPUs [43].

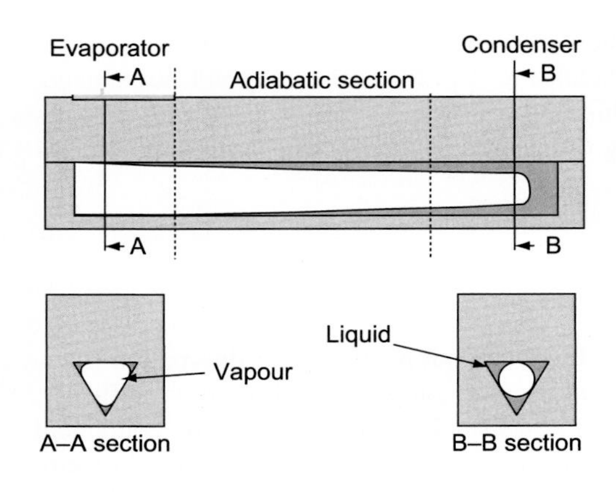


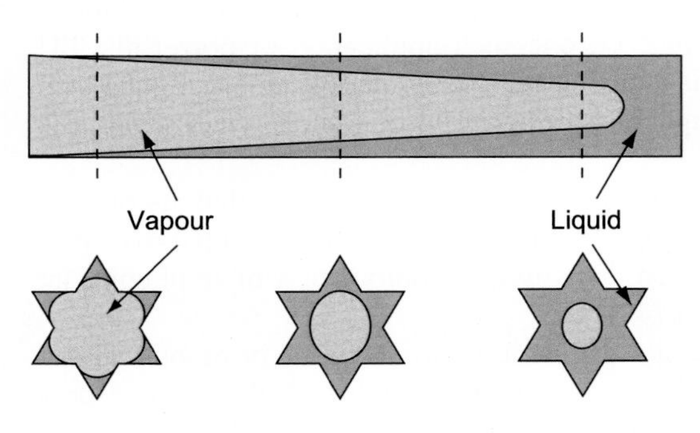


**FIGURE 1.12** Microheat pipe cross sections showing capillary paths. *Courtesy: UCL.*

NASA sees possible answers in the use of heat-pipe-based two-phase thermal control systems combined with spray cooling (as in the example cited above), that is phase change thermal storage (that can be integrated with heat pipes – see Chapter 6) and mechanically pumped units.

If we are to address this challenge – which has implications for the use of heat-pipe-based technologies in a large number of emerging systems – it is worth examining the opportunity to use unconventional liquid transport systems such as electrokinetics. *(There is nothing new here, such systems were discussed in the first edition of Heat Pipes but were deemed not to be necessary at the time).* Electrokinetics is often discussed in the context of microfluidics – fluid flow at the microscale, for example in heat pipe wick pores or in channels with dimensions around 100 μm or less.

Within two-phase systems (liquid–vapour) the influence of capillary action can be high, from both negative and positive viewpoints. Where force is used to drive liquid, as in the microheat pipes illustrated in Fig. 1.12, there is a wealth of literature examining optimum configurations of channel cross-section, pore sizes etc. However, the impression is that for many microfluidic devices, capillary action is a disadvantage that needs to be overcome by external 'active' liquid transport mechanisms. Perhaps there is room for a more positive examination of passive liquid transport designs where the use of electric fields and other more advanced procedures can provide the control necessary that may not always be available using passive methods alone. Some of these 'active' methods are described below.[1]

Two-phase flow and heat transfer and capillary action do not necessarily imply that a heat pipe is involved, and porous structures are extensively used to promote boiling heat transfer in many other areas. However, the benefits that such structures bring to heat pipe evaporators may also enhance other cooling systems such as those applied to light-emitting diode (LED) devices that are taxing thermal control engineers in a number of static and mobile applications (see for example Ye et al. [44]). Encouragingly, one of the premier heat pipe laboratories in the world – the Luikov Heat and Mass Transfer Institute in Minsk – has recently examined the phenomenon that can occur in the wicks of microheat pipes that can be used to enhance evaporation and two-phase convection heat transfer in porous structures in general [45].

1. A rotating heat pipe, see Chapter 2, is an active heat transfer device, in that external energy is used to rotate the system, so the borderline between 'active' and 'passive' can be blurred.

## 1.6 Electrokinetics

Electrokinetics (the study of the dynamics of electrically charged fluids) is the name given to the electrical phenomena that accompany the relative movement of a liquid and a solid. These phenomena are ascribed to the presence of a potential difference at the interface between any two phases at which movements occur, and thus, if the potential arises from the existence of electrically charged layers of opposite sign at the interface, then the application of an electric field should result in the displacement of one layer with respect to the other. Hence, if the solid phase is fixed and the liquid is free to move – as in a porous material – then the liquid will tend to flow through the pores as a consequence of this applied field. This movement is known as electro-osmosis and is discussed in detail in Chapter 2.

Electro-osmosis using an alternating current (ac) has also been used as the operating basis for microfluidic pumps, allowing normal batteries to be used at modest voltages.

For those wishing to follow up the numerous techniques for influencing capillary forces, the paper by Le Berre et al. [46] on electrocapillary force actuation in microfluidic elements is worthy of study.

## 1.7 Fluids and materials

Over the past few years, heat pipes have seen significant changes regarding a number of aspects and many of these may be classed as changes to 'fluids and materials'. In the case of the former, the necessity to replace ozone-depleting fluids and the need to phase out fluids that contribute to greenhouse gases are likely to influence the design and operation of future heat pipe systems operating at moderate/low temperatures. Chlorofluorocarbons (CFCs) (hopefully) disappeared several years ago from the heat pipe manufacturer's inventory and hydrofluorocarbons (HFCs) may also be on their way out. One only has to look at the legislation proposed for heat pumps and refrigeration/air-conditioning systems – heavy users of such fluids – to see what might happen to their availability for heat pipes. For many refrigeration systems, hydrocarbons, ammonia and carbon dioxide are now the preferable choices, and, of course, all except carbon dioxide have already found applications in heat pipes or thermosyphons. Although over a decade has now passed since its publication, the thought-provoking paper by Harvey and Laesecke at the National Institute of Standards and Technology (NIST) in the United States introduced a number of ideas relating to the challenges of using fluids in systems, such as those at the microscale [47]. They pointed out that '*... microscale technology will drive fluid property research into relatively unexplored areas ...*' and that some thermophysical properties, for example viscosity have more important roles to play at such scales – for example where wall effects become dominant in narrow passages.

Within the context of 'fluids and materials', nanofluids – fluids containing nanosized particles – cannot be ignored. Nanoparticles have been introduced into most fluid applications, in some contexts with more success than others. In the case of heat pipes and thermosyphons – see Chapter 2 – the general principal role of a nanoparticle is to improve the thermal conductivity of the working fluid and/or promote nucleation sites within it, and quite marked improvements in performance with thermal resistances decreasing by as much as 60% have been reported. Putra et al. [48] used aluminium oxide (5%) and water in a mesh-wicked unit and suggested that the nanoparticles coated the surface of the mesh and thus promoted good capillary action. Nanoparticles do not follow the flow of the vapour, and if a heat pipe is operating against gravity with the condenser below the evaporator, one might hypothesise as to how the nanoparticles migrate to the evaporator on start-up, and their impact on wick pressure drop.

## 1.8 The future?

This chapter has looked at the heat pipe from its historical perspective and seen how the technology has adapted to changing demands – examples in Chapters 6 and 7 will illustrate these, and the concepts described in Chapter 2 will show how new ideas are addressing the challenges. Nevertheless, there are limitations to heat pipes (and thermosyphons), and these can be calculated using the theory discussed in Chapter 4. Of these, the counter-current flow limit (entrainment) can be mitigated using loop heat pipes and the boiling limit can also be addressed using wick/porous surface designs. However, increasingly these are likely to be important for applications involving very high heat flux/low-temperature applications and, as one example, the very long thermosyphons proposed for use to minimise oil viscosity during extraction from wells. For the future, the heat pipe has a continuing role to play in many applications including both domestic equipment and industrial applications. Its role in spacecraft is also guaranteed by its unique features, and its growing use for renewable energy systems, which can benefit from mass-produced passive heat transport systems, must auger well for the technology in the future. In addition, heat pipe technology has been investigated for various applications such as lithium battery cooling, fuel cell thermal management, waste heat recovery applications, and other challenging thermal management applications. The technology has many desirable characteristics such as scalability, near iso-thermal surface operation, quick response, and passive heat transfer. It is therefore a promising technology to help enable the achievement of energy decarbonisation and sustainability strategies targets, but it is probable (as indicated with one or two examples earlier in this chapter), that much of the R&D associated with heat pipe technology will also find uses in other heat transfer areas.

In conclusion, without doubt, heat pipes will remain a viable and cost-effective solution for many existing and evolving technologies and the research community will continue to support these applications by improving (perhaps incrementally) performance, whilst reducing unit sizes and costs.

## References


[1] Anon, Memoirs of Angier March Perkins, Proc. Inst. Civ. Eng. 67 (1882) 417–419.

[2] C.R. King, Perkins' hermetic tube boilers, Engineering 152 (1931) 405–406.

[3] L.P. Perkins, W.E. Buck, Improvements in devices for the diffusion or transference of heat, UK Patent 22, 1892.

[4] Anonymous, The Paris Exhibition-Perkins' portable oven, Engineering (1867) 159.

[5] R. Gaugler, Heat transfer device, Patent no. US2350348, 1944.

[6] M.G. George, Evaporation-condensation heat transfer device, Patent no. US3229759, 1966.

[7] G.M. Grover, T.P. Cotter, G.F. Erickson, Structures of very high thermal conductance, J. Appl. Phys. 35 (1964) 1990–1991.

[8] K.F. Bainton, Experimental heat pipes. AERE-M1610, Harwell, Berks, Atomic Energy Research Establishment, Applied Physics Divison, June 1965.

[9] G.M. Grover, J. Bohdansky, C.A. Busse, The use of a new heat removal system in space thermionic power supplies, European Atomic Energy Community, 1965.

[10] C.A. Busse, R. Caron, C. Capelletti, Prototypes of heat pipe thermionic converters for space reactors, in: IEE1st Conf. Thermion. Electr. Power Gener. Conf. Thermion. Electr. Power Gener., London, 1965.

[11] B.I. Leefer, Nuclear thermionic energy converter, in: Proc. 20th Annu. Power Sources Conf, Atlantic City, NJ, 1966, pp. 172–175.

[12] J.F. Judge, RCA test thermal energy pipe, Missiles Rocket. 18 (1966) 36–38.

[13] J.E. Deverall, J.E. Kemme, Satellite heat pipe, Los Alamos Scientific Laboratory, University of California, N. Mex., 1965.

[14] T. Wyatt, A controllable heat pipe experiment, Applied Physics. Laboratory. Johns Hopkins University, Baltimore, Maryland, USA. SDO-1134., 1965.

[15] K.T. Feldman Jr, G.H. Whiting, The heat pipe, University of New Mexico, Albuquerque, 1967.

[16] G.Y. Eastman, The heat pipe, Sci. Am. 218 (1968) 38–47.

[17] K.T. Feldman, Applications of the heat pipe (heat pipe for thermal control of terrestrial and aerospace energy conversion systems), Mech. Eng. 90 (1968) 48–53.

[18] T.P. Cotter, Theory of heat pipes, Los Alamos Scientific Laboratory of the University of California, 1965.

[19] H. Cheung, A critical review of heat pipe theory and applications, California University, Livermore, Lawrence Radiation Labortory, United States, 1968.

[20] S. Katzoff, Notes on heat pipes and vapor chambers and their application to thermal control of spacecraft, United States, 1966.

[21] D.K. Anand, Heat pipe application to a gravity gradient satellite, in: Proc. ASME Annu. Aviat. Sp. Conf., 1968, pp. 634–658.

[22] C.A. Busse, Heat pipe research in Europe, in: Second Int. Conf. Thermion. Electr. Power Gener., 1968, pp. 461–475.

[23] S. Nozu, Studies related to the heat pipe, Trans. Soc. Mech. Eng. Japan 35 (1969) 392–401.

[24] C.J. Savage, Heatpipes and vapour chambers for satellite thermal balance, Royal Aircraft Establishment, Farnborough (England), 1969.

[25] R. Turner, The "constant temperature" heat pipe-A unique device for the thermal control of spacecraft components, in: 4th Thermophys. Conf., 1969, p. 632.

[26] W. Bienert, Heat pipes for temperature control, in: Proc. Fourth Intersoc. Energy Convers. Conf. Wash., DC, 1969, pp. 1033–1041.

[27] Y.V. Moskvin, Y.N. Filinnov, Heat tubes, High Temp. 7 (1969) 704.

[28] V.H. Gray, The rotating heat pipe-A wickless, hollow shaft for transferring high heat fluxes, in: Heat Transf. Conf., American Society of Mechanical Engineers, 1969.

[29] T.D. Sheppard Jr, Heat pipes and their application to thermal control in electronic equipment, in: Proc. Nat. Electron. Packag. Prod. Conf, Anaheim, CA, 1969, pp. 11–13.

[30] A.T. Calimbas, R.H. Hulett, An avionic heat pipe. ASME Paper 69-HT-16 New York, Am. Soc. Mech. Eng. ASME 69-HT, 1969.

[31] P.E. Eggers, A.W. Serkiz, Development of cryogenic heat pipes. ASME 70-WA/Ener-1, Am. Soc. Mech. Eng. ASME 70-WA, 1970.

[32] P. Joy, Optimum cryogenic heat pipe design, Advances in Cryogenic Engineering, Springer, 1972, pp. 438–448.

[33] A.R. Mortimer, The heat pipe, Engineering Note-Nimrod, NDG/70-34, Harwell, Rutherford Laboratory, Nimrod Design Group, 1970.

[34] J.A. Lidbury, A helium heat pipe, in: Nimrod Des. Gr. Rep. NDG 72-11, Rutherford Laboratory, 1972.

[35] D.A. Reay, Industrial Energy Conservation: A Handbook for Engineers and Managers, Pergamon Press Inc., Oxford, 1977.

[36] C. Hoa, B. Demolder, A. Alexandre, Roadmap for developing heat pipes for ALCATEL SPACE's satellites, Appl. Therm. Eng. 23 (2003) 1099–1108. Available from: https://doi.org/10.1016/S1359-4311(03)00039-5.

[37] C. Figus, L. Ounougha, P. Bonzom, W. Supper, C. Puillet, Capillary fluid loop developments in Astrium, Appl. Therm. Eng. 23 (2003) 1085–1098. Available from: https://doi.org/10.1016/S1359-4311(03)00038-3.

[38] T.D. Swanson, G.C. Birur, NASA thermal control technologies for robotic spacecraft, Appl. Therm. Eng. 23 (2003) 1055–1065. Available from: https://doi.org/10.1016/S1359-4311(03)00036-X.

[39] M. Groll, J.P. Kirkpatrick, Heat pipes for spacecraft temperature control: an assessment of the state-of-the-art, Heat. Pipes (1976) 167–181.

[40] T. Osakabe, T. Murase, T. Koizumi, S. Ishida, Application of heat pipe to audio amplifier, Advanced Heat Pipes Research & Technology, Elsevier, 1982, pp. 25–36.

[41] Y. Lee, C.Z. Wu, Solidification heat transfer characteristics in presence of two-phase closed thermosyphons in latent heat energy storage systems, in: Proc. 6th Int. Heat Pipe Conf. Grenoble, 25Á/29 May, 1987.

[42] O. Tanaka, H. Koshino, H. Sakai, R. Furukawa, Y. Inada, Heat transfer characteristics of super heat pipes, Heat Pipe Technology. Vol. 1. Fundamentals and Experimental Studies, Begel House Inc., 1993.

[43] J.S. Bintoro, A. Akbarzadeh, M. Mochizuki, A closed-loop electronics cooling by implementing single phase impinging jet and mini channels heat exchanger, Appl. Therm. Eng. 25 (2005) 2740–2753. Available from: https://doi.org/10.1016/j.applthermaleng.2005.01.018.

[44] H. Ye, M. Mihailovic, C.K.Y. Wong, H.W. Van Zeijl, A.W.J. Gielen, G.Q. Zhang, et al., Two-phase cooling of light emitting diode for higher light output and increased efficiency, Appl. Therm. Eng. 52 (2013) 353–359. Available from: https://doi.org/10.1016/J.APPLTHERMALENG.2012.12.015.

[45] L.L. Vasiliev Jr, A. Zhuravlyov, A. Shapovalov, L.L. Vasiliev Jr, Microscale two phase heat transfer enhancement in porous structures, in: Int. Heat Transf. Conf. 13, Begel House Inc., Sydney, Australia, 2006.

[46] M. Le Berre, Y. Chen, C. Crozatier, Z.L. Zhang, Electrocapillary force actuation of microfluidic elements, Microelectron. Eng. 78–79 (2005) 93–99. Available from: https://doi.org/10.1016/j.mee.2004.12.014.

[47] A.H. Harvey, A. Laesecke, Fluid properties and new technologies: connecting design with reality, Chem. Eng. Prog. 98 (2002) 34–41.

[48] N. Putra, W.N. Septiadi, H. Rahman, R. Irwansyah, Thermal performance of screen mesh wick heat pipes with nanofluids, Exp. Therm. Fluid Sci. 40 (2012) 10–17. Available from: https://doi.org/10.1016/J.EXPTHERMFLUSCI.2012.01.007.

Chapter 2

# Heat pipe types and developments

In terms of their geometry, function and/or the methods used to transport the liquid from the condenser to the evaporator, there are many types of heat pipe; some of which have become less popular over the decades or have ceased to be regarded as 'special' (e.g. flat plate heat pipes and vapour chambers – see Chapters 5, and 7 for examples of the latter types). Other types have seen a reawakening of interest brought about by the demands for increased miniaturisation and the need to enhance performance – in particular, liquid return flow rates. Interestingly, the use of electrokinetic forces is one technique in which the 'wheel has gone full circle'!

This chapter includes and describes the following:

- variable-conductance heat pipes
- thermal diodes
- pulsating (oscillating) heat pipes
- loop heat pipes (LHPs) and capillary-pumped loops (CPLs)
- microheat pipes
- the use of electrokinetic forces
- rotating heat pipes
- miscellaneous types – sorption heat pipe (SHP); magnetic fluid heat pipes

(*Note: The types of heat pipe that were present in earlier editions have been omitted from this chapter and some suggestions for further reading have been retained in Appendix 3*).

## 2.1 Variable-conductance heat pipes

The variable-conductance heat pipe (VCHP), sometimes called in a specific variant the gas-controlled or gas-loaded heat pipe, has a unique feature that sets it apart from other types of heat pipe, and that is its ability to maintain a device (mounted at the evaporator) at a near constant temperature, independent of the amount of power being generated by the device.

VCHPs are now routinely used in many applications (see also Chapter 6). These applications range from thermal control of components and systems on satellites to precise temperature calibration duties and conventional electronics temperature control.

The temperature control functions of a gas-buffered heat pipe were first examined as a result of noncondensable gas generation within a sodium/stainless steel basic heat pipe. It was observed [1] that as heat fed into the evaporator section of the heat pipe the hydrogen generated was swept to the condenser section and an equilibrium situation (shown in Fig. 2.1) was reached.

Subsequent visual observation of high-temperature heat pipes and temperature measurements indicated that the working fluid vapour and the noncondensable gas were segregated so that a sharp interface existed between the working fluid and the noncondensable gas. Hence the noncondensable gas effectively blocked off the condenser section it occupied, stopping any local heat transfer.

Significantly, it was also observed that the noncondensable gas interface moved along the pipe as a function of the thermal energy being transported by the working fluid vapour, and it was concluded that suitable positioning of the gas interface could be used to control the temperature of the heat input section within close limits.

Much of the subsequent work on heat pipes containing noncondensable or inert gases has been in developing the means for controlling the positioning of this gas front, and in ensuring that the degree of temperature control achievable is sufficient to enable components adjacent to the evaporator section to be operated at essentially constant temperatures, independent of their heat dissipation rates, over a wide range of powers.

Heat Pipes. DOI: https://doi.org/10.1016/B978-0-12-823464-8.00014-0

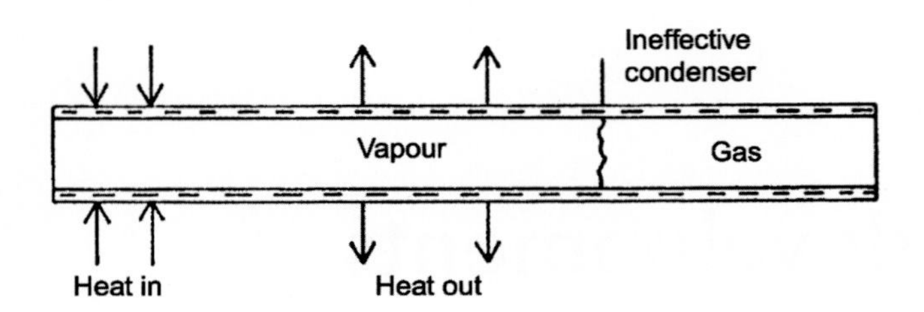


FIGURE 2.1 Equilibrium state of a gas-loaded heat pipe.

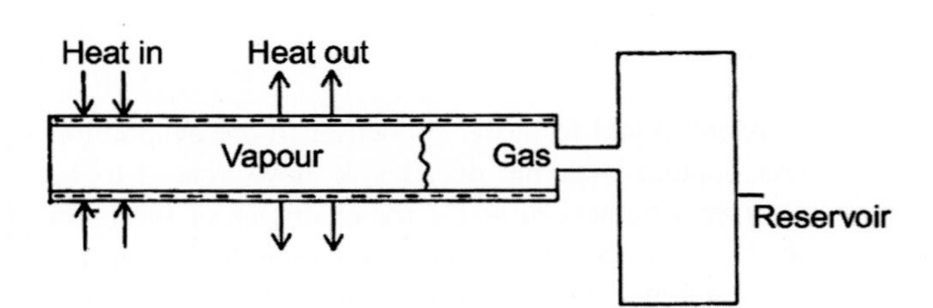


FIGURE 2.2 Cold-reservoir variable-conductance heat pipe.

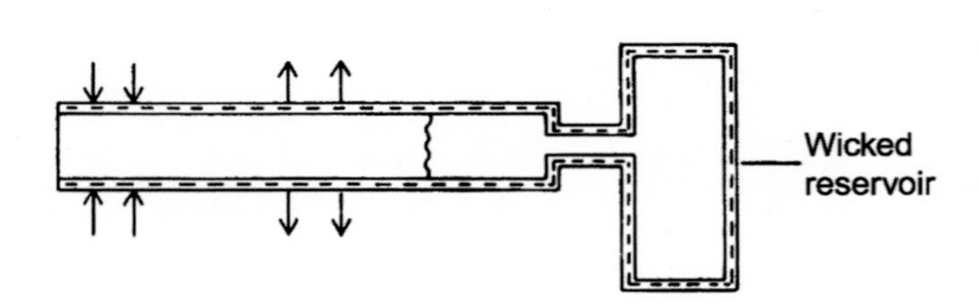


FIGURE 2.3 Cold wicked reservoir variable-conductance heat pipe.

The first extension of the simple form of gas-buffered pipe shown in Fig. 2.1 was the addition of a reservoir downstream of the condenser section (Fig. 2.2). This was added to allow the whole of the heat pipe length to be effective when the pipe was operating at maximum capability and to provide more precise control of the vapour temperature, but in addition, the reservoir could also be conveniently sealed using a valve.

The early workers in the field of cold-reservoir VCHPs were troubled by vapour diffusion into the reservoir followed by condensation even if liquid flow into the gas area had been arrested. In practice, it is necessary to wick the reservoir of a cold-reservoir unit to enable the condensate to be removed, and the partial pressure of the vapour in the reservoir will then be at the vapour pressure corresponding to its temperature (Fig. 2.3).

The type of VCHP described above is of the passively controlled type where the active condenser length varies in accordance with temperature changes in various parts of the system. An increase in evaporator temperature causes an increase in vapour pressure of the working fluid, which causes the gas to compress into a smaller volume, releasing a larger amount of active condenser length for heat rejection. Conversely, a drop in evaporator temperature results in a lower vapour surface area. The net effect being to produce a passively controlled variable condenser area that increases or decreases heat transfer in response to the heat pipe vapour temperature.

A successful cold-reservoir VCHP was constructed by Kosson et al. [2]. An arterial wick system of high liquid transport capability was used in conjunction with ammonia to carry up to 1200 W. The nominal length of the heat pipe, including reservoir, approached 2 m and the diameter was 25 mm. Nitrogen was the control gas.

An important feature of this heat pipe was a provision for subcooling the liquid in the artery. This was to reduce inert gas bubble sizes and help the liquid to absorb any gas in the bubbles. This phenomenon is discussed in Section 3.8.11.2 in Chapter 3.

### 2.1.1 Passive control using bellows

Wyatt [1] proposed as early as 1965 that a bellows might be used to control the inert gas volume, but he was not specific in suggesting ways in which the bellows' volume might be adjusted.

It is impractical to insert a wick into a bellows unit, and consequently it is necessary to have a semipermeable plug between the condenser section and the bellows to prevent the working fluid from accumulating within the storage volume. The plug must be impervious to both the working fluid vapour and the liquid but permeable to the noncondensable gas. Marcus and Fleischman [3] proposed and tested a perforated Teflon plug with success, preventing liquid from entering the reservoir during vibration tests.

However, Wyatt did put forward proposals that would have the effect of overcoming one of the major problems associated with the 'cold-reservoir' VCHP, although he did not appreciate the significance at the time. His proposal was to electrically heat the bellows that would be thermally insulated from the environment. His argument for doing this was to ensure that stray molecules of working fluid would not condense in the bellows if the bellows temperature was maintained at about 1°C above the heat pipe vapour temperature. In fact, controlling the temperature of the noncondensable gas reservoir helps eliminate the most undesirable feature of the basic cold-reservoir VCHP, namely the susceptibility of the gas to environmental temperature changes that can upset the constant temperature performance.

Turner [4] investigated the use of bellows to change the reservoir volume and/or pressure. He proposed a mechanical positioning device to control the bellows between two precisely determined points but listed several disadvantages of this type of control including the fact that mechanical devices require electrical energy for their activation and are also subject to failure due to jamming and friction. In proposing ammonia as the working fluid, he also felt that the associated pressure might add considerably to the weight of the bellows for containment reasons and also that fatigue failure was a possibility.

### 2.1.2 Hot-reservoir variable-conductance heat pipes

The cold-reservoir VCHP is particularly sensitive to variations in sink temperature which could affect the reservoir pressure and temperature, and in an attempt to overcome this drawback the hot-reservoir unit was developed.

One attractive layout for a hot-reservoir system is to locate the gas reservoir adjacent to, or even within, the evaporator section of the heat pipe because thermally coupling the reservoir to the evaporator minimises gas temperature fluctuations that can limit controllability. Fig. 2.4 illustrates one form that this concept might take.

As has been previously stated, it is undesirable to have working fluid inside the hot reservoir, and a semipermeable plug has been suggested as one way of preventing diffusion of large amounts of vapour or liquid into the gas volume. If the reservoir has a wick and contains working fluid, there will be within it a vapour pressure corresponding to its temperature, which — in the case of the hot-reservoir pipe having a gas reservoir within the evaporator — would be essentially the same as the temperature throughout the whole pipe interior, and so there would not be gas in the reservoir.

However, an alternative technique [5] (applied to the hot-reservoir VCHP) is to provide a cold trap between the condenser and the reservoir. This effectively reduces the partial pressure of the working fluid vapour in the gas, and the system provides temperature control which is relatively independent of ambient radiation environments, but this system is not applicable to a hot reservoir located inside the evaporator.

### 2.1.3 Feedback control applied to the variable-conductance heat pipe

The cold-reservoir VCHP is particularly sensitive to variations in sink temperature which could affect the reservoir pressure and temperature, and in an attempt to overcome this drawback, the hot-reservoir unit was developed, as described above.

Ideally, each of these forms of heat pipe is, at best, capable of maintaining its own temperature constant, but this is true only if an infinite storage volume is used. As a result, if the thermal impedance of the heat source is large, or if the power required to be dissipated by the component is liable to fluctuations over a range, the temperature of the sources would not be kept constant and severe fluctuations could occur, making the system unacceptable.

Opportunely, the development of the feedback-controlled VCHPs has enabled absolute temperature control to be obtained, and this has been demonstrated experimentally [6]. These heat pipes are representative of the third generation of thermal control devices incorporating the heat pipe principle.

Two forms of feedback control are available: active (electrical) and passive (mechanical).

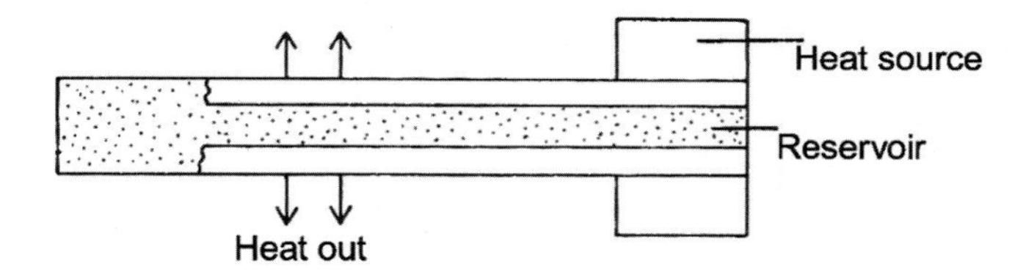


FIGURE 2.4 Hot-reservoir variable-conductance heat pipe.

### 2.1.3.1 *Electrical feedback control (active)*

An active feedback-controlled VCHP is shown diagrammatically in Fig. 2.5 and includes a temperature sensor, an electronic controller and a heated reservoir (internal or external heaters) used to adjust the position of the gas—vapour interface such that the source temperature remains constant. As in the cold-reservoir system, the wick is continuous and extends into the storage volume so saturated working fluid is always present in the reservoir. The partial pressure of the vapour in equilibrium with the liquid in the reservoir is determined by the reservoir temperature and this can be varied by the auxiliary heater.

The two extremes of control required in the system are represented by the high-power/high-sink and low-power/low-sink conditions. The former case necessitates operation at maximum conductance, whereas the low-power/low-sink condition is appropriate for operation with minimum power dissipation. But by using a temperature-sensing device at the heat source and connecting this via the controller to the heater at the reservoir, the auxiliary power and thus the reservoir temperature can be regulated so that precise control of the gas—buffer interface occurs, maintaining the desired source temperature at a fixed level.

### 2.1.3.2 *Mechanical feedback control (passive)*

Much of the work on mechanical feedback control has involved the use of a bellows reservoir, as advocated in several earlier proposals for nonfeedback-controlled passive VCHPs. A proposed passive feedback system utilising bellows was designed by Bienert et al. [7,8] and is illustrated in Fig. 2.6. The control system consists of two bellows and a sensing bulb located adjacent to the heat source. The inner bellows contains an auxiliary fluid (generally an incompressible liquid) and is connected to the sensing bulb by a capillary tube.

Variations in source temperature will cause a change in the pressure of the auxiliary fluid resulting in a displacement of the inner bellows. This displacement causes a movement of the main reservoir bellows and by relating the displacement of the bellows system (and therefore that of the vapour/gas interface in the heat pipe) to the heat source, a feedback-controlled system that regulates the source temperature is obtained.

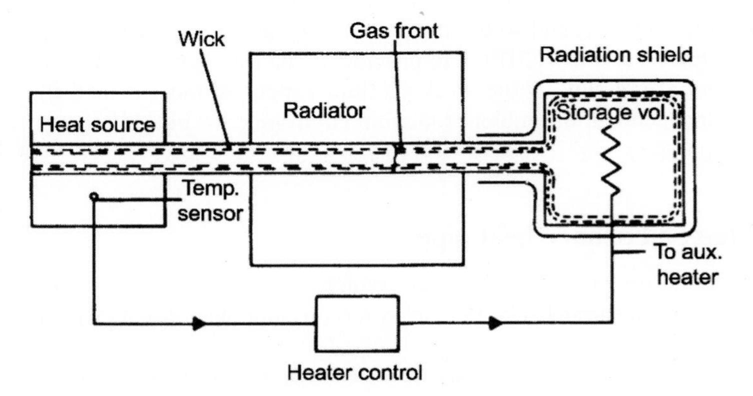


**FIGURE 2.5** Active feedback-controlled variable-conductance heat pipe.

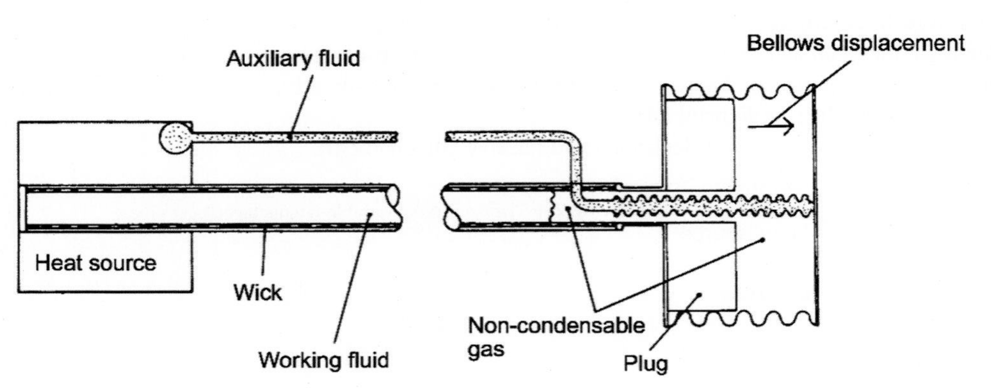


**FIGURE 2.6** Passive feedback-controlled variable-conductance heat pipes.

The construction and testing of a passive control VCHP using methanol as the working fluid and nitrogen as the gas has been reported by Depew et al. [9]. The system was operated over a power input range of 2–30 W, with the heat source at ambient temperature and the heat sink at a nominal value of 0°C. Control was obtained with a metal bellows gas reservoir that was actuated by an internal liquid-filled bellows. The liquid bellows was pressurised by expanding the liquid methanol in an auxiliary reservoir in the evaporator heater block. Temperature variation of the heat source was restricted to ±4°C with this design.

The use of VCHPs for primary temperature measurement has proved attractive because of the ability to set temperatures within ±0.5 mK, an accuracy difficult to achieve with other temperature control systems. The CNR, *Istituto di Metrologia 'G. Colonnetti'*, (IMGC) in Turin, Italy, has had considerable success in applying liquid metal VCHPs for accurate temperature measurement [10]. in the range from 100°C to 962°C (sodium being used as the working fluid in the latter case, and employing a mercury unit to study the vapour pressure relationship for this fluid [11]).

Illustrated in Fig. 2.7, the heat pipe (constructed using stainless steel) was 400 mm long and 30 mm in diameter. The capillary structure (with helical knurling) covered all of the inner surfaces and the inert gas reservoir was connected to the heat pipe via a water-cooled limb. Mercury was filled to about 280 g.

Initial tests found that temperature response times to changes in pressure were unsatisfactory, so modifications were made to the capillary structure that resulted in excellent temperature stability (within less than 1 mK) and a rapid response to pressure changes. Further units were constructed using sodium as the working fluid (for high-temperature tests) and the 3 M fluid 'Fluorinert FC-43' for operation around 177°C.

With the reawakening interest in nuclear power generation, and spurred by concerns about global warming and $CO_2$ emissions, researchers have been studying methods for controlling heat removal from reactors, in particular the modular high-temperature gas-cooled reactor. Should a loss of forced circulation through the core occur, or the primary system suffers depressurisation, passive decay heat removal becomes necessary, and one way of implementing this is to use VCHPs [12].

Experiments on a laboratory-scale simulation of the decay showed that the VCHP could effectively compensate for a threefold increase in decay heat rate by allowing only a modest rise in system temperature. In the tests, nitrogen was used as the control gas and water as the working fluid. It was concluded, should an accident occur, that the VCHP system could potentially control the reactor vessel temperature within safety limits – the passive nature of the system having an added safety attraction.

An analogy may be made with the work of the Tokyo Electric Power Company and others [13] where a VCHP has been studied for load levelling on sodium–sulphur (NAS) batteries. During operation such batteries need to be kept

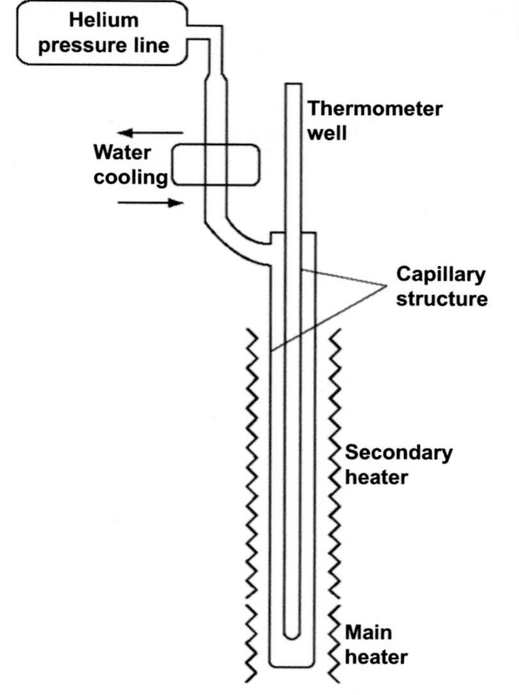


**FIGURE 2.7** The stainless steel mercury gas-controlled heat pipe developed at CNR-IMGC, Italy [10].

**TABLE 2.1 Comparison of temperature-controlled heat pipe systems.**

| System | Advantages | Disadvantages |
|---|---|---|
| Wicked cold | Reliable without moving parts.<br>No auxiliary power needed.<br>Sensitivity is governed by vapour within the reservoir. | Very sensitive to sink conditions. Large storage volume. Only heat pipe temperature controlled. |
| Nonwicked hot reservoir<br>Passive control | Reliable without moving parts.<br>Less sensitive to sink conditions than cold reservoir.<br>No auxiliary power needed. | Sensitive to heat carrier diffusion into the reservoir. Only heat pipe temperature controlled. |
| Passive feedback-controlled bellows system | Heat source control. Slight sensitivity to sink conditions. No auxiliary power needed. | Complex and expensive system. Sensitive to heat carrier diffusion into the reservoir. Moving parts used. |
| Active electrical feedback-controlled system | Heat source control. Best adjustability to various set points. Sensitivity is governed by vapour pressure within the reservoir. Minimum storage volume of all concepts. Relative insensitivity to gas generation. No moving parts. | Auxiliary power needed. |

above 300°C, but if they overheat by more than 10°C, their lifetimes are reduced. The VCHP was shown to provide a satisfactory level of heat control, thus contributing to improved charge/discharge efficiency (see information on aerospace LHPs in Chapter 7).

#### *2.1.3.3 Comparison of systems*

The most common VCHP systems are compared in Table 2.1, and when comparing the two types of feedback control (active and passive), better temperature control is obtained using an active system. In an active system, all the noncondensable gas will be in the condenser when the low-power sink condition is attained. Whilst in the passive arrangement, noncondensable gas will be present in the reservoir, regardless of the use of a plug, and the excess gas must be accommodated where the bellows are already at their maximum size (high-power/sink conditions). Hence, generally, storage requirements will be greater for a passive system having the same degree of temperature control as an active system. In addition, better temperature control can be achieved with an active system than with the equivalent volume passive system. Plus, the necessity to incorporate a semipermeable plug and the use of moving parts in the form of a bellows also adds to the complexity of passive systems.

In some applications, the use of additional electrical power to heat the reservoir, or the increased complication of a bellows system, may be unacceptable, and in such cases the choice lies between simple hot-reservoir and cold-reservoir VCHPs.

## 2.2 Heat pipe thermal diodes and switches

### 2.2.1 The thermal diode

Here, we are concerned with on/off rather than proportional functioning.

The simplest thermal diode is the thermosyphon in which gravity provides the asymmetry but, of course, with the restriction on positioning. Gravity will also give a diode effect in the wicked heat pipe since

$$\Delta P_c = \Delta P_1 + \Delta P_v \pm \Delta P_g$$

Reversal of direction of flow will reverse the sign of $\Delta P_g$ and provided that $|\Delta P_g| > \Delta P_c$, the pipe will behave as a diode.

Kirkpatrick [14] describes two types of thermal diode, one employing liquid trapping and the second liquid blockage. Referring to Fig. 2.8 with the heat flow shown in Fig. 2.8a, the heat pipe will behave normally. If the relative positions of the evaporator and condenser are reversed then the condensing liquid is trapped in the reservoir whose wick is not connected to the pipe wick on the left-hand side of the diagram and hence cannot return. The pipe will then not operate, and no heat transfer will occur.

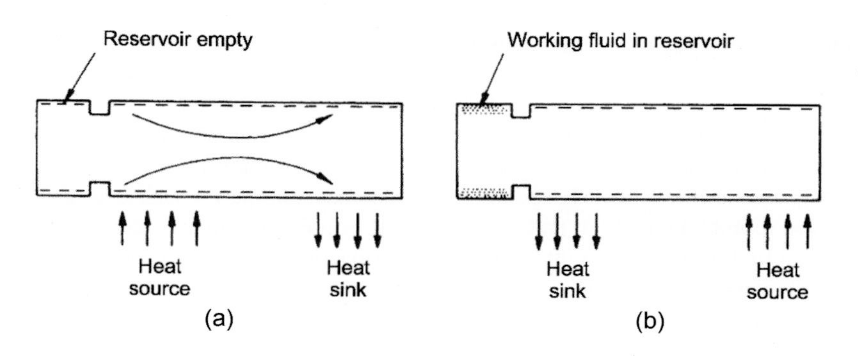


**FIGURE 2.8** Liquid trap diode: (a) normal mode and (b) reverse mode.

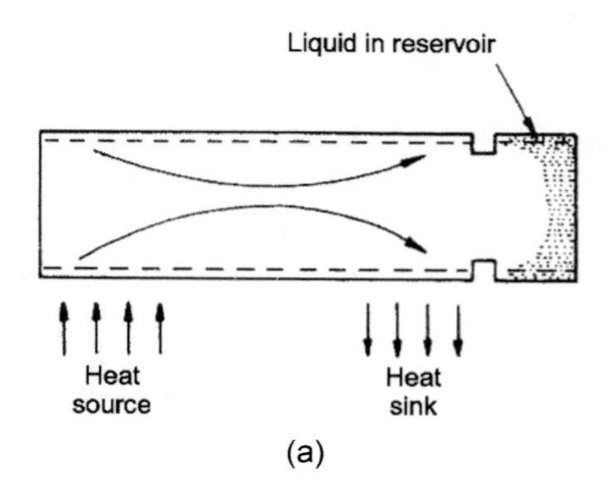


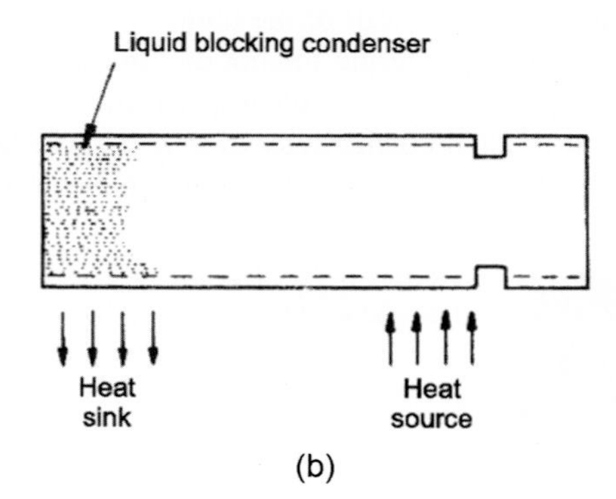


**FIGURE 2.9** Liquid blockage diode: (a) normal mode and (b) reverse mode.

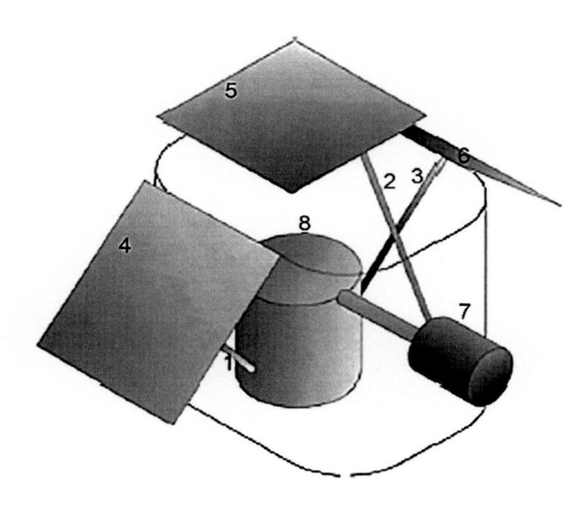


**FIGURE 2.10** The three-way cooler for an HPGe spectrometer proposed for a lunar lander. The HPGe detector (7) is connected to a Stirling microcooler (8) and radiator (5) using a diode heat pipe (2). The microcooler compressor (8) is connected to radiators (4 and 6) by two other thermal diodes (1 and 3) [15].

Fig. 2.9 shows a similar arrangement but in this case excess liquid is placed in the pipe. In Fig. 2.9a, this liquid will accumulate in the reservoir at the condensing end and the pipe will operate normally. In Fig. 2.9b, the positions of the evaporator and condenser are reversed and the excess liquid blocks off the evaporator and the pipe ceases to operate.

Spacecrafts have employed thermal diodes for several decades. In Russia, a comparison of cryogenic heat pipe thermal diodes with active systems, such as Stirling coolers, suggested that high-purity Germanium (HPGe) detectors, used in gamma-ray spectroscopy, could reap several benefits using the thermal diode. Of course the passive nature is attractive, but in addition weight reduction and reliability are featured. Where Stirling coolers are necessary to provide active cooling, the heat pipe thermal diode is proposed, as shown in Fig. 2.10 to link the HPGe to a space radiator [15]. This particular model is a liquid trap diode, as additionally described for the Russian project MARS-96 in Ref. [16].

Thermal diodes in terrestrial applications have been important components of renewable energy systems where heat/coolth transport in one direction is essential under some circumstances such as when collecting solar gain in the winter for space heating. In hot climates, the use of the clear night sky as a radiative heat sink can provide useful cooling duties via thermal diodes [17]. The unit developed in Nigeria employed four methanol working fluid thermal diodes, with their condensers linked to an external radiator and their evaporators inserted into the walls of a cooler box. The radiator had a cooling capacity of over 600 kJ/m$^2$ per night and was able to cool the chamber to about 13°C when the ambient (at night) was 20°C.

Although commercialisation of many of the systems proposed for building heating/cooling that rely on heat pipe thermal diodes is difficult because of the perceived high cost and difficulties in retrofitting such systems, interest does continue. Work involving a number of universities and companies in the United Kingdom and continental Europe on such a system for room heating in winter or cooling in summer was recently carried out under a European Commission-funded study [18].

The thermal diode panels, illustrated in Fig. 2.11a and b were manufactured by Thermacore Europe and each panel comprised nine copper—water heat pipes, with diameters of 12.7 mm, welded to two aluminium sheets spaced in the configuration shown in Fig. 2.11b. Wicks were put in the lower sections of the heat pipes to assist liquid distribution over what were the evaporator sections.

Using the left-hand side of the panel (where the heat pipe evaporators are fixed), as the inside wall of the building, and the outer condenser sections adjacent to the outside wall of the building, in summer the unit provides a night-time cooling role. The diode effect prevents the ingress of heat during the hot day in Southern Europe, where the system was tested. It was found that the panels with a thickness of 10 cm were able to perform as well as those with a thickness of 4 cm of thermal insulation material, with an apparent thermal conductivity of 0.07 W/mK. When heat transfer was required the apparent thermal conductivity was three to five times this value, depending upon the temperature difference.

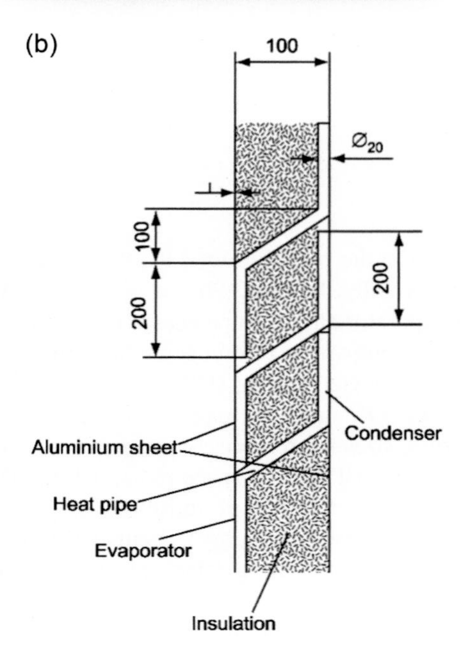


**FIGURE 2.11** Thermal diode wall panels developed as a passive thermal control feature for buildings [18]. (a) Wall panel, (b) Wall structure.

For readers interested in other similar uses of heat pipes as thermal diodes in buildings, for heating or cooling, see references [19,20]. In addition, another development of interest might be the low-volume solar-powered water heating system, because of the low volume of the working fluid reduces the risk of rupture due to freezing [21].

### 2.2.2 The heat pipe switch

A number of methods for switching off the heat pipe have been referred to. Some examples are discussed by Brost and Schubert [22] and Eddleston and Hecks [23]. Fig. 2.12a shows a simple displacement method in which the liquid working fluid can be displaced from an unwicked reservoir by a solid displacer body [22]. Fig. 2.12b shows an interruption of the vapour flow by means of a magnetically operated vane. The working fluid may be frozen by means of a thermoelectric cooler.

A thermal switch employing bellows was developed in the United States in the early 1980s [24]. Illustrated in Fig. 2.13, this device was capable of transporting about 100 W horizontally and the distance between the source and the sink was approximately 10 cm. Later, Peterson [25] modelled such devices for use with electronic components or multichip modules. One advantage of such a design was the use of the internal pressure in the bellows to ensure good thermal contact between the heat pipe and the heat sink when the switch was 'on'. Similar bellows systems were developed in the United Kingdom in the 1970s, the bellows being used to undertake a mechanical function in addition to the basic switching mode created by the presence or absence of a thermal contact.

The heat pipe switch can be particularly useful in cryogenic applications, where heat leakage can affect sensitive instruments in avionics and in spacecraft. This was the reasoning behind the development by the Los Alamos National Laboratory [26].

## 2.3 Pulsating (oscillating) heat pipes

Pulsating, or oscillating, heat pipes comprise a tube of capillary diameter, evacuated and partially filled with the working fluid. These have since been developed having been originally described in a series of Patents [27–29]. Pulsating heat pipe configurations are shown schematically in Fig. 2.14 and their implementation is shown in Fig. 2.15. Typically, a pulsating heat pipe comprises a serpentine channel of capillary dimension which has been evacuated and partially filled with the working fluid. Surface tension effects result in the formation of slugs of liquid interspersed with bubbles of vapour. The operation of pulsating heat pipes was outlined in [30]. When one end of the capillary tube is heated (the evaporator), the working fluid evaporates and increases the vapour pressure, thus causing the bubbles in the evaporator zone to grow. This pushes the liquid towards the low-temperature end (the condenser). Cooling of the condenser results in a reduction of vapour pressure and condensation of bubbles in that section of the heat pipe. The growth and collapse of bubbles in the evaporator and condenser sections, respectively, results in an oscillating motion within the tube. Heat is transferred through latent heat in the vapour and through sensible heat transported by the liquid slugs.

Closed-loop pulsating heat pipes (CLPHPs) perform better than open-loop devices because of the fluid circulation that is superposed upon the oscillations within the loop. It has been suggested that further performance improvements may result from the use of check valves within the loop; however, due to the inherently small nature of the device, it is difficult and costly to install such valves [28,33]. Therefore a closed-loop device without a check valve is the most practicable implementation of the pulsating heat pipe.

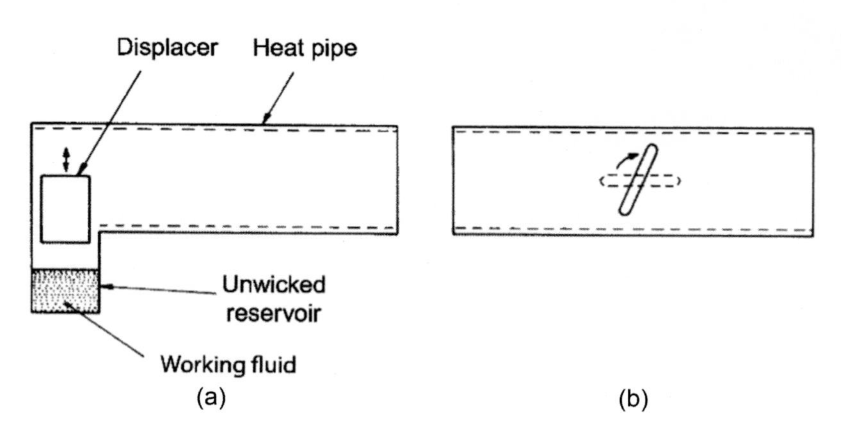


**FIGURE 2.12** Thermal switches: (a) displacement type and (b) vane type.

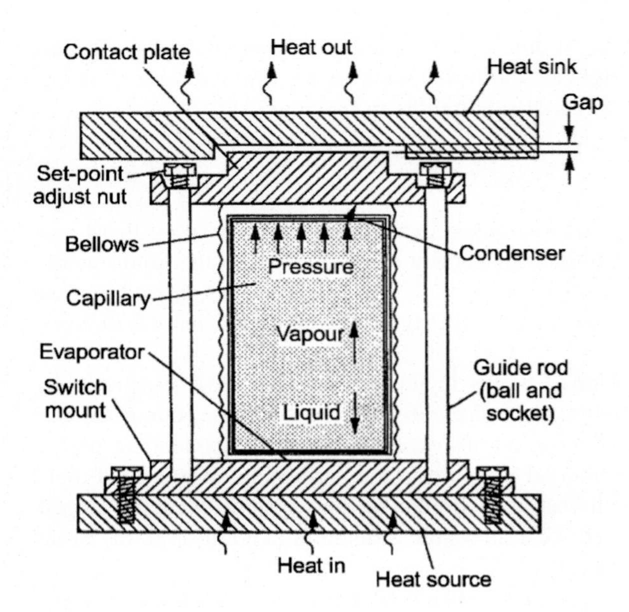


FIGURE 2.13 Heat pipe thermal switch.

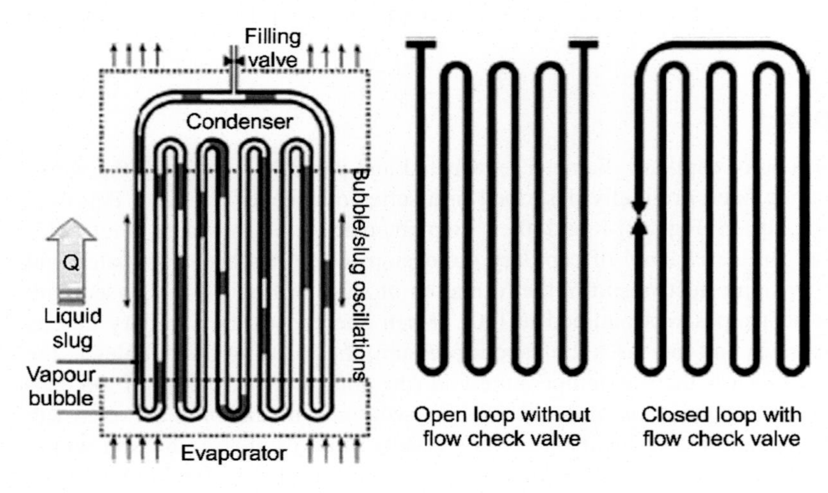


FIGURE 2.14 Schematic representation of pulsating heat pipe [31].

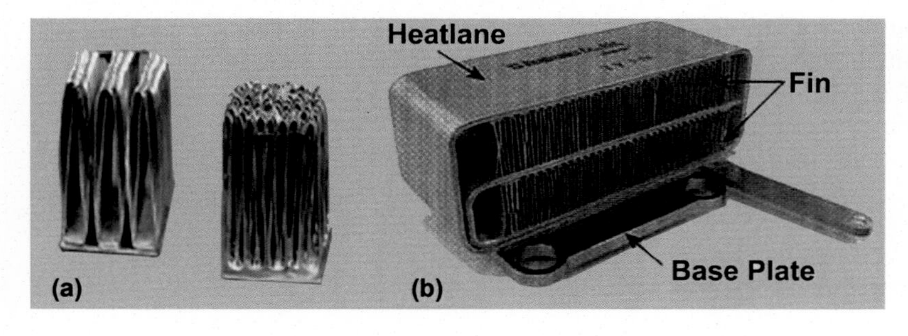


FIGURE 2.15 Practical implementation of pulsating heat pipe: (a) fin, Ref. [31] and (b) complete module, Ref. [32].

The parameters affecting the performance of CLPHP have been summarised by Charoensawan and coworkers [31] as

- working fluid
- internal diameter
- total tube length
- length of condenser, evaporator and adiabatic sections
- number of turns or loops
- inclination angle

**TABLE 2.2** Matrix of parameters tested [30].

| Working fluids | $D_i$ (mm) | $L_{total}$ (m) | $L_e = L_a = L_c$ (m) | $N$ (number of turns) |
|---|---|---|---|---|
| Water–ethanol–R-123 | 2.0 | ≈5 | 0.15 | 5 |
| | 2.0, 1.0 | ≈5 | 0.10 | 7 |
| | | ≈10 | 0.10 | 16 |
| | | | 0.15 | 11 |
| | | ≈15 | 0.10 | 23 |
| | | | 0.15 | 16 |

*Note:* Fill ratio is always maintained at 50% in all configurations. All configurations were tested at inclinations of 0 degree (horizontal) to +90 degrees (vertical, evaporator down).

Kandhekar et al. [34] have examined the fundamental flow phenomena within pulsating heat pipes and conclude that it is not yet possible to mathematically model the pulsating heat pipe even though the steady-state behaviour of Taylor bubbles is well understood.

A comprehensive test programme based upon the values of these parameters listed in Table 2.2 was undertaken and this has been used to support the development of a semiempirical correlation to assist in the design of CLPHPs [35].

The correlation was based on three dimensionless groups, the Karman number, the liquid Prandtl number and the Jacob number, as defined in Eqs (2.1a), (2.1b), and (2.1c), respectively, combined with the inclination angle ($\beta$, measured in radian) and number of turns, $N$.

$$Ka_{liq} = f \cdot Re_{liq}^2 = \frac{\rho_{liq} \cdot (\Delta P)_{liq} \cdot D_i^2}{\mu_{liq}^2 \cdot L_{eff}} \quad \text{where } L_{eff} = 0.5(L_e + L_c) + L_a \tag{2.1a}$$

$$Pr_{liq} = \left( \frac{C_{p.liq} \cdot \mu_{liq}}{k_{liq}} \right) \tag{2.1b}$$

$$Ja = \left( \frac{h_{fg}}{C_{p.liq} \cdot (\Delta T)_{sat}^{e-c}} \right) \tag{2.1c}$$

where $\Delta T_{sat}^{e-c}$ is the temperature difference between the working fluid in the evaporator and condenser and $\Delta P_{liq}$ the corresponding pressure difference.

$$\dot{q} = \left( \frac{\dot{Q}}{\pi D_i \cdot N \cdot 2L_e} \right) = 0.54(\exp(\beta))^{0.48} Ka^{0.47} Pr_{liq}^{0.27} Ja^{1.43} N^{-0.27} \tag{2.2}$$

This correlation holds provided the surface tension forces are large relative to gravitational forces. For this to be true, the Bond number, defined by Eq. (2.3) must be less than ~2 [36].

$$Bo = \frac{D_i}{\sqrt{\frac{\sigma}{g}(\rho_l - \rho_g)}} \tag{2.3}$$

This has been shown to be the condition for discrete bubbles and liquid slugs sustained in a tube when the fluid is stationary and thus determines whether confined bubbles exist during the two-phase flow in a channel [37]. At higher heat fluxes with inclination angles greater than 30 degrees from the horizontal, annular flow predominates. The transitions are shown in Fig. 2.16.

Eq. (2.2) fits the data of [31], as shown in Fig. 2.17 [36].

As indicated in Table 2.2, these data were from tests with a 50% filling ratio. The studies carried out by Groll et al. indicated that the CLPHP operated satisfactorily with volumetric filling ratios in the range of 25%–65% as indicated in Fig. 2.18. With greater filling ratios, the pumping action of bubbles was insufficient for good performance. At very low filling ratios, partial dryout of the evaporator was detected. It was also noted that at zero fill, the thermal resistance was effectively that of the copper tube, whilst at 100% fill the loop operated as a single-phase thermosyphon loop.

Another study, on a similar system [39], suggested that the optimum filling ratio is approximately 70%.

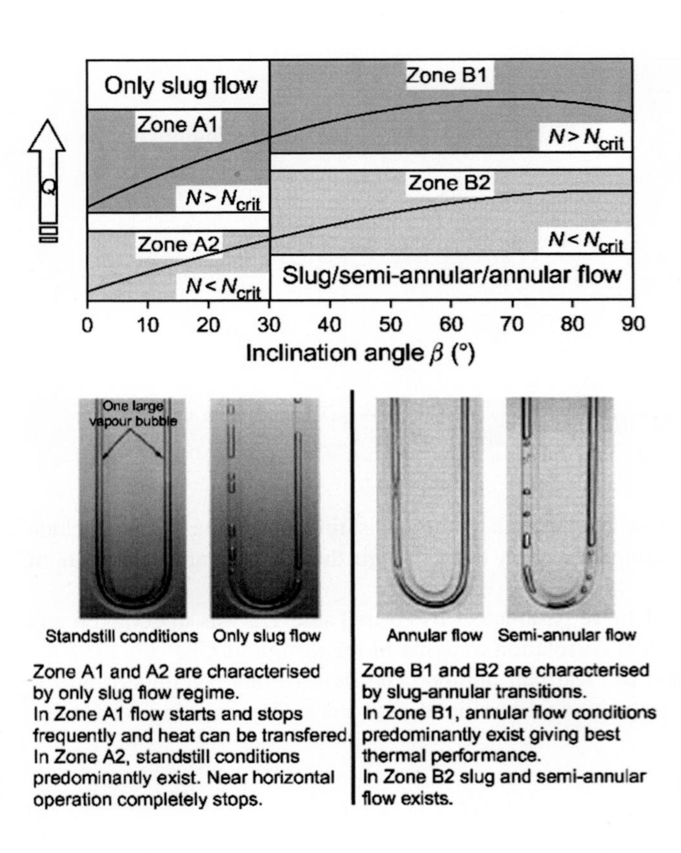


**FIGURE 2.16** Operating zones in pulsating heat pipes [35].

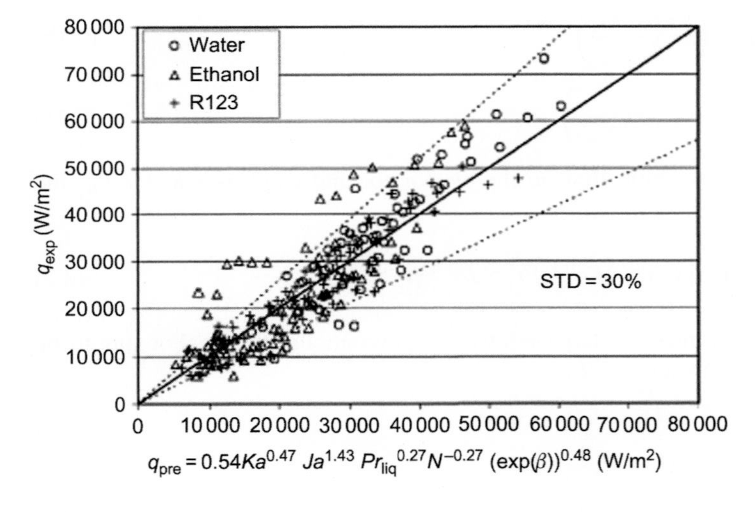


**FIGURE 2.17** Comparison of Eq. (2.2) with experimental results [35].

Sakulchangsatjatai et al. [40] have devised a numerical model applicable to closed- and open-loop pulsating heat pipes. This underpredicted the heat transferred, but followed the correct trends, as shown in Fig. 2.19, indicating that the following parameters have the greatest influence on the heat pipe performance.

- Evaporator length – as the evaporator length is increased, the amplitude of pulsation increases and the frequency decreases. The net result is a reduction in heat transfer capacity with increasing evaporator length.
- Internal diameter – the heat transfer capacity increases with pipe diameter, provided the diameter is less than the value specified by Eq. (2.3).
- Working fluid – the analysis of Sakulchangsatjatai et al. indicates that latent heat is the most important fluid property and that low latent heat is desirable since this results in greater vapour generation and hence fluid movement for a given heat flux. This conclusion conflicts with the trends predicted by Eq. (2.2) and the results of other workers [39–41].

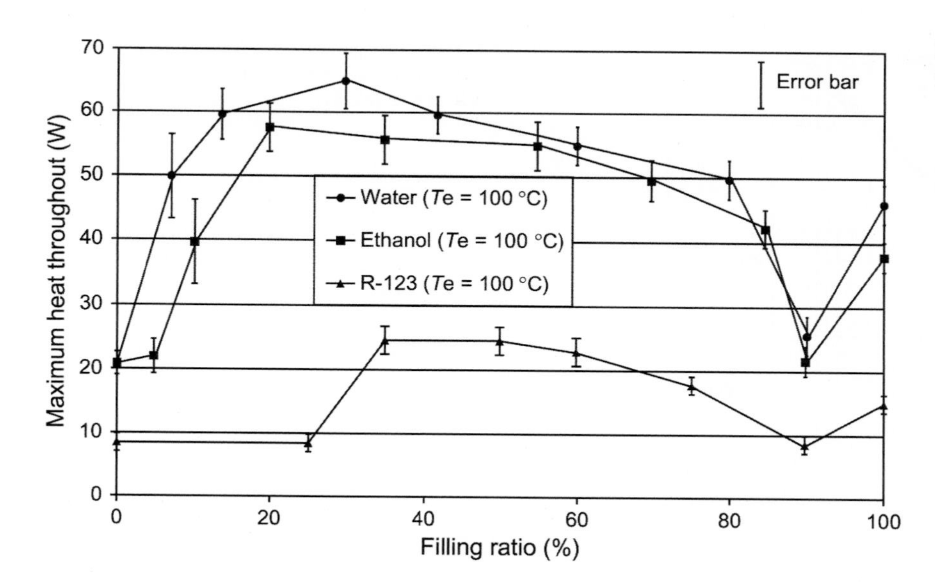


**FIGURE 2.18** Influence of filling ratio on performance of pulsating heat pipe [38].

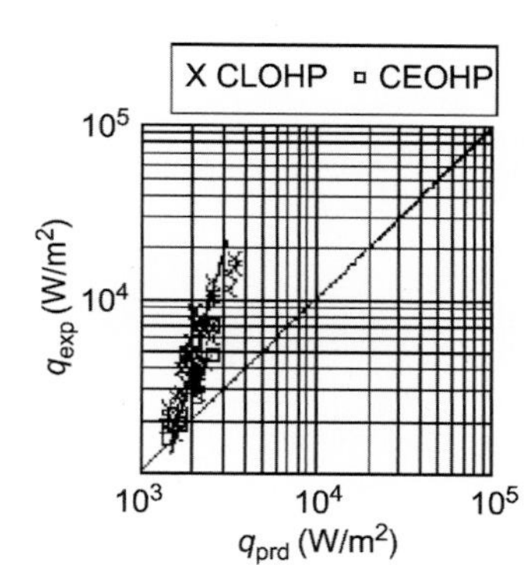


**FIGURE 2.19** Comparison of experimental and predicted heat transfer [40].

A correlation has been formulated for the maximum heat flux in closed-end oscillating heat pipes (CEOHP) by Akbarzadeh and coworkers [41]. This may be stated as follows:

$$Ku_0 = 53680 \times \left[\frac{D_i}{L_e}\right]^{1.127} \times \left[\frac{C_p \Delta T}{h_{fg}}\right]^{1.417} \times \left[D_i\left[\frac{g(\rho_1-\rho_v)}{\sigma}\right]^{0.5}\right]^{-1.32} \tag{2.4a}$$

$$Ku_{90} = 0.0002 \times \left[\frac{D_i}{L_e}\right]^{0.92} \times \left[\frac{C_p \Delta T}{h_{fg}}\right]^{-0.212} \times \left[D_i\left[\frac{g(\rho_1-\rho_v)}{\sigma}\right]^{0.5}\right]^{-0.59} \times \left[1+\left(\frac{\rho_v}{\rho_1}\right)^{0.25}\right]^{13.06} \tag{2.4b}$$

where $Ku$ is the Kutateladze Number $\{= q/h_{fg}\rho_v^{0.5}[\sigma g(\rho_1-\rho_v)]^{0.25}\}$, $h_{fg}$ the latent heat, and the subscripts 0 and 90 refer to horizontal and vertical orientation, respectively (Fig. 2.20).

Groll and coworkers [42] have further investigated the effect of orientation and fill ratio on the performance of CLPHPs with channel diameters of 1 and 2 mm using R123 as a working fluid. They demonstrated that both CLPHPs performed satisfactorily in all orientations. The 2 mm diameter unit had the lower thermal resistance (by some 10%) but the 1 mm diameter unit was able to carry a higher heat flux by a factor of approximately 50% (based on radial heat flux). The best performance was obtained when operating in a vertical orientation with heating at the bottom, as shown in Fig. 2.21 and Fig. 2.22; however the effect of orientation was negligible for the 1 mm diameter CLPHP. The factor which most significantly affected both thermal resistance and dryout was the fill ratio, as illustrated in Fig. 2.23.

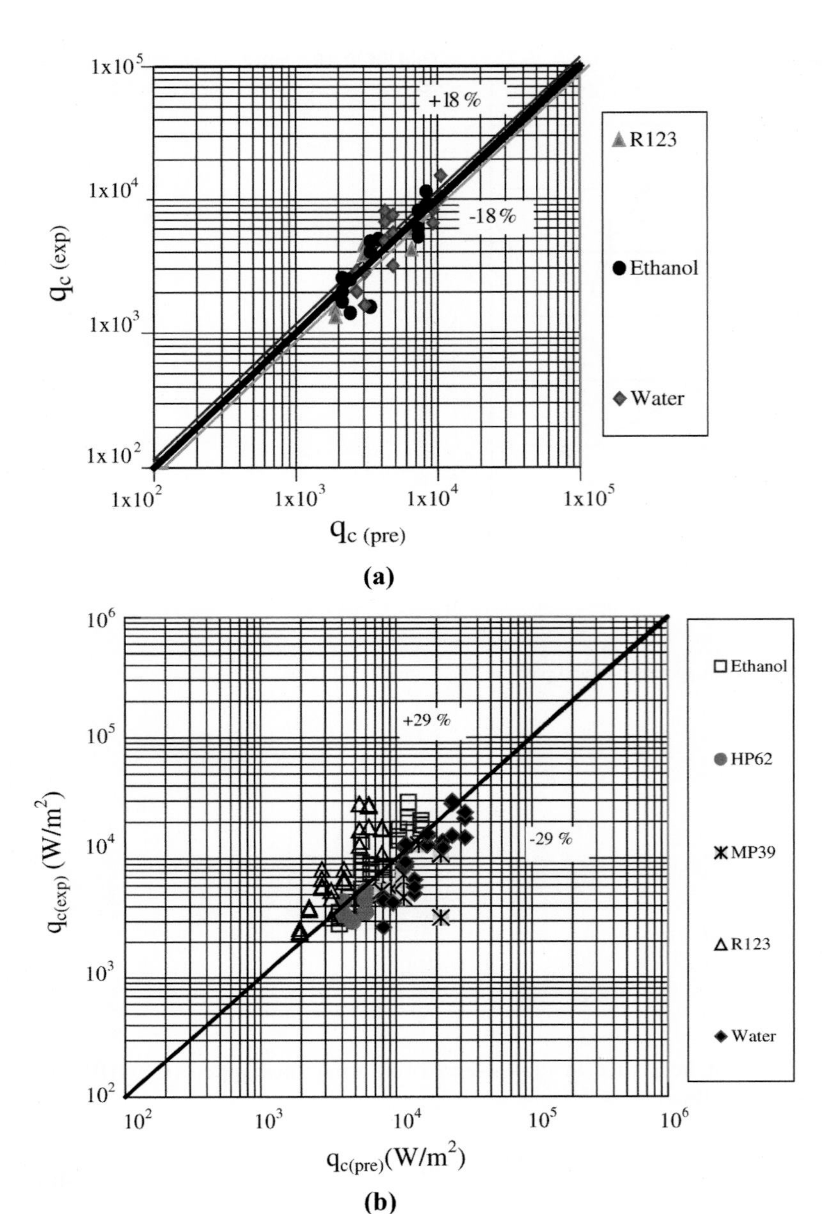


**FIGURE 2.20** Comparison of maximum heat flux with prediction: (a) horizontal CEOHP (Eq. 2.4a) and (b) vertical CEOHP (Eq. 2.4b) [41].

Chien et al. have suggested that using a combination of different size channels in a pulsating heat pipe (in their study they used square channels of 2 × 2 mm alternating with channels of cross-section 1 × 2 mm) can improve the performance of a pulsating heat pipe (PHP), particularly in a horizontal orientation.

The use of nanofluids and treated surfaces is of increasing interest in all heat transfer applications. As discussed in Chapter 2, nanofluids have been shown to give significant performance improvements in pulsating or oscillating heat pipes. The mechanism of the performance enhancement is not fully explained. Liu and Li [43] in reviewing studies involving the use of nanofluids suggest that the layer of nanoparticles formed on the wall modifies the boiling characteristics of the surface, but the mechanism is not fully understood. Ji et al. [44] have shown that an oscillating heat pipe will function satisfactorily, albeit with a higher thermal resistance, if the wall is hydrophobic (i.e. nonwetting)

The pulsating heat pipe clearly has potential applications, particularly in electronic cooling systems, where it is suggested [45] that the spreading resistance may be 50% of that of a conventional heat sink and the effective conductivity 24 kW/mK, compared to 0.4 kW/mK for metallic copper.

At Oakland University in the United States, PHPs have been investigated in the context of fuel cell cooling [46]. Test on the PHP, illustrated schematically in Fig. 2.24, proved that it was functional and was able to self-regulate and

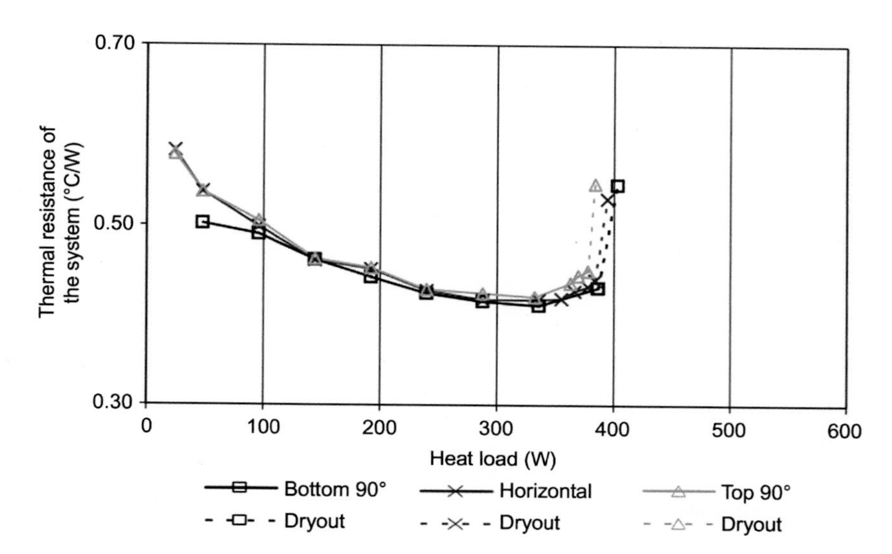


**FIGURE 2.21** Thermal performance of CLPHP for three heat modes (ID = 1 mm, fill ratio = 50%) [42].

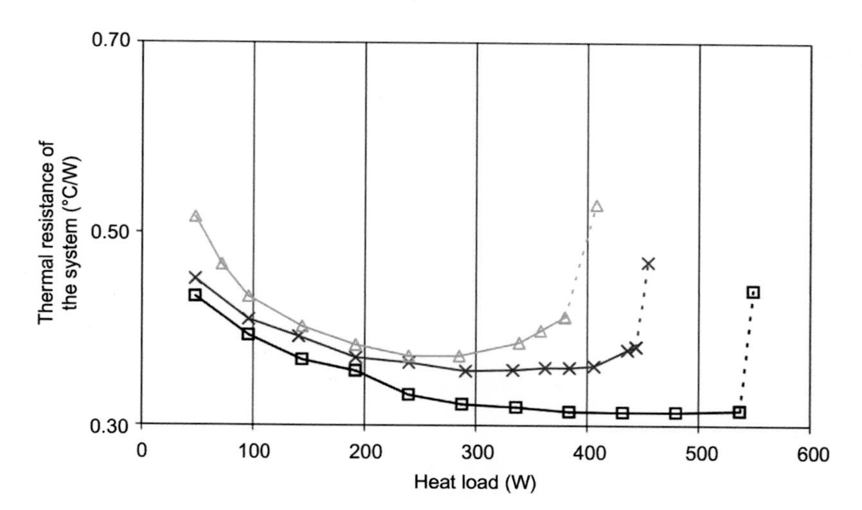


**FIGURE 2.22** Thermal performance of CLPHP for three heat modes (ID = 2 mm, fill ratio = 50%) [42].

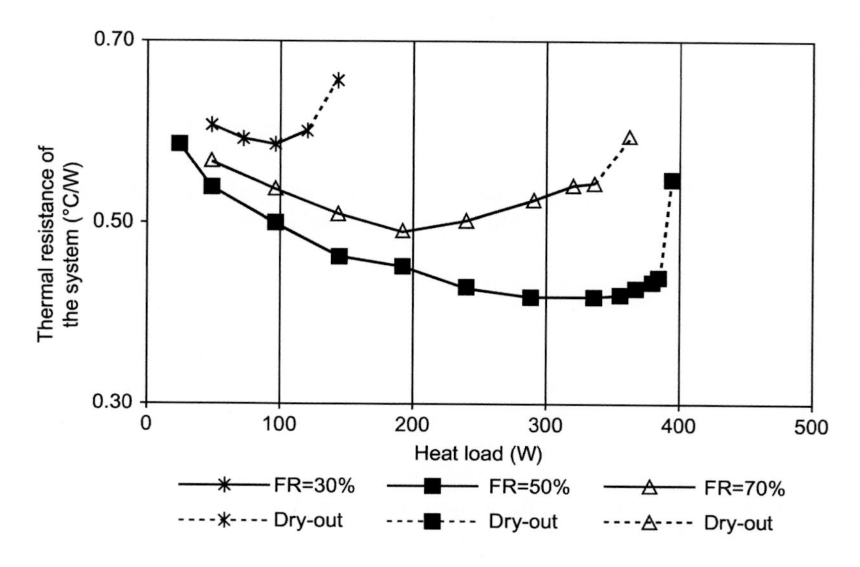


**FIGURE 2.23** Thermal performance of CLPHP for three filling ratios (ID = 1 mm/horizontal heat mode) [42].

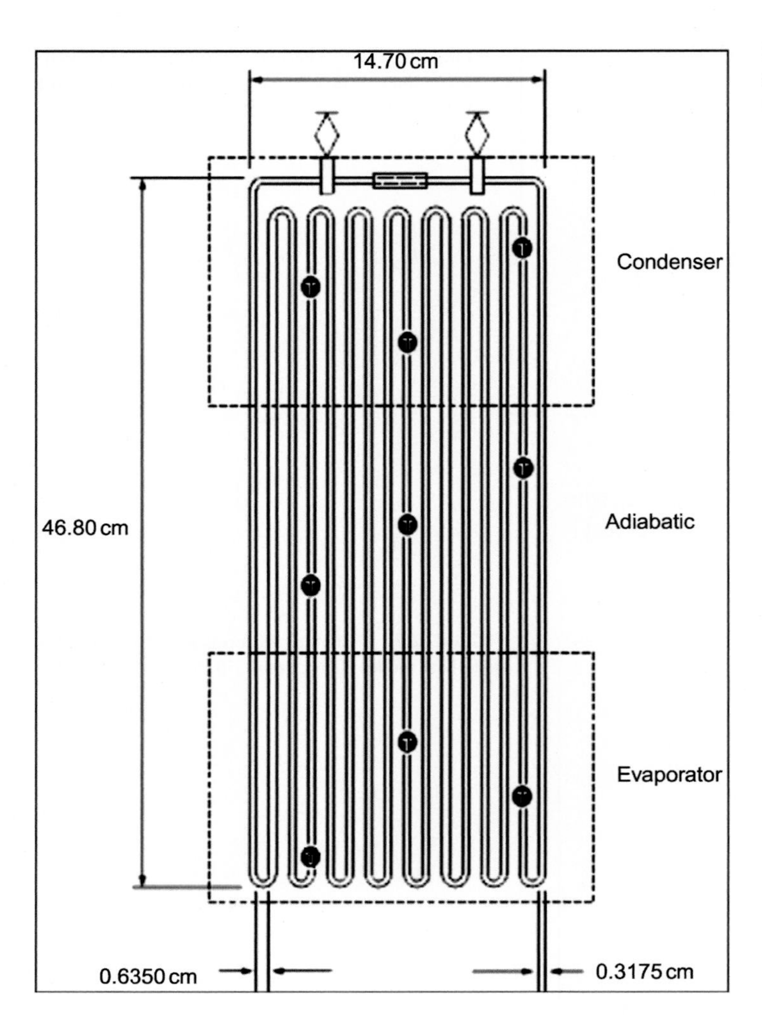


**FIGURE 2.24** PHP design developed for fuel cell thermal management ('T' denotes thermocouple positions in the test unit) [46].

maintain a constant evaporator temperature within a small range. Methanol as a working fluid outperformed both acetone and water, and the best performance achieved was using methanol with a fill ratio of 45% and power input around 110 W. Thermal resistance calculations indicated that a smaller resistance was associated with a higher power input to the system – the reduction ranging from 0.22 C/W at 100 W to less than 0.16 C/W at 150 W. The authors stated that 'the pulsating heat pipe concept has potential to be utilised in a bottom-heated vertical orientation within a fuel cell. A PHP evaporator inserted within bipolar plates of a horizontally configured stack would allow for the condenser to protrude from the stack and transfer heat to its surroundings. Several pulsating heat pipes aligned between cells would remove heat directly from the source and act as an active fin array'. It was believed that this particular PHP could be suitable for use as a passive cooling technique in a 200 $cm^2$ polymer electrolyte membrane fuel cell.

Current design methods are incomplete due to a lack of understanding of the mechanisms involved in pulsating capillary flow; however, empirical design equations have been presented which yield results in agreement with the experiment.

## 2.4 Loop heat pipes and capillary-pumped loops

As discussed in Chapter 2, the operation of a heat pipe relies upon the capillary head within the wick which is sufficient to overcome the pressure drops associated with the liquid and vapour flow and the gravitational head. To operate with the evaporator above the condenser in a gravitational field, it is necessary for the wick to extend for the entire length of a conventional heat pipe. The capillary head $\Delta P_c$ is inversely proportional to the effective pore radius of the wick but independent of length, whilst the hydraulic resistance is proportional to the length of the wick and inversely proportional to the square of the pore radius. Thus if the length of a heat pipe operating against gravity is to be increased, a reduction in pore radius is required to provide the necessary capillary head, but this results in an increase in the liquid pressure drop. The conflicting effects of decreasing the pore size of the wick therefore inherently limit the length at which heat pipes operating against gravity can be successfully designed. In the same way, the necessity for the liquid to flow through the wick limits the total length of the classic wicked heat pipe.

LHPs and CPLs were developed to overcome the inherent problem of incorporating a long wick with a small pore radius in a conventional heat pipe. The CPL was first described by Stenger [47], working at the NASA Lewis Research Centre in 1966 and the first LHP was developed independently in 1972 by Gerasimov and Maydanik [48] of the Ural Polytechnic Institute. Development of these heat pipe types has been largely driven by their potential for use in space where initially they were considered for applications requiring high (0.5–24 kW) transport capabilities, but the advantages of the two-phase loops with small-diameter piping systems and no distributed wicks have led to their exploitation at lower powers [49].

A simple LHP is shown schematically in Fig. 2.25.

The operation of an LHP may be summarised as follows. At start-up, the liquid load is sufficient to fill the condenser and the liquid and vapour lines and there is sufficient liquid in the evaporator and compensation chamber to saturate the wick (Level A–A, Fig. 2.26). When a heat load is applied to the evaporator, fluid evaporates from the surface of the wick, and to a lesser extent in the compensation chamber, but as the wick has an appreciable thermal resistance the temperature and pressure within the compensation chamber are less than that in the evaporator. The capillary forces in the wick prevent the flow of the vapour from evaporator to compensation chamber. As the pressure difference between evaporator and compensation chamber increases, the liquid is displaced from the vapour line and the condenser and returned to the compensation chamber.

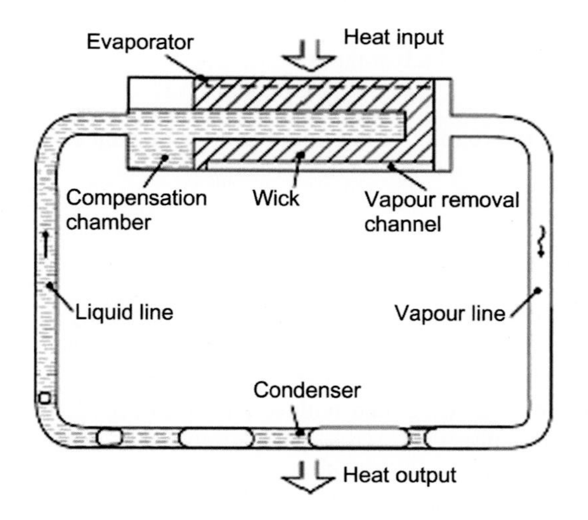


**FIGURE 2.25** Operating principle of loop heat pipe. [50].

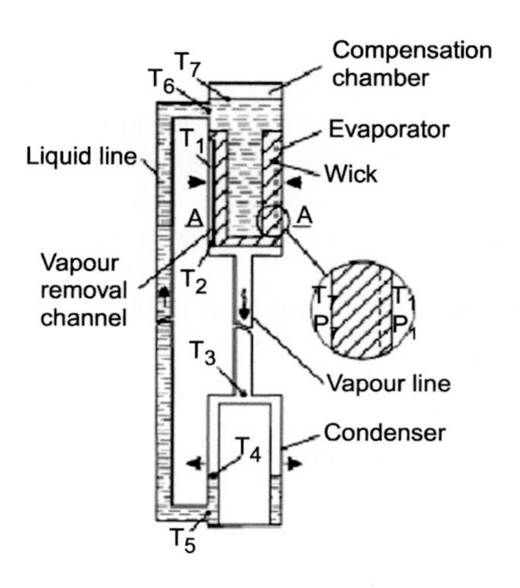


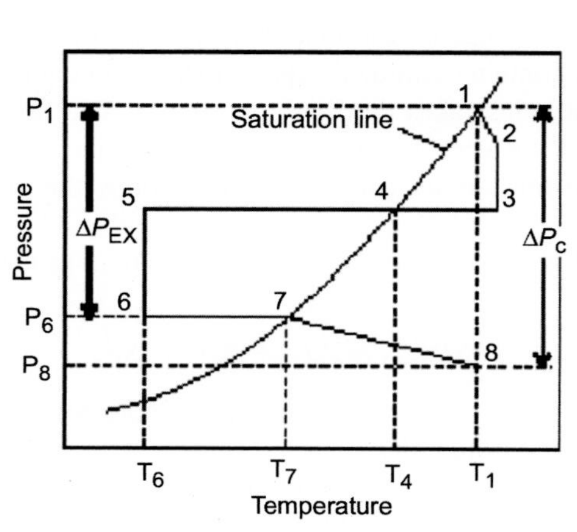


**FIGURE 2.26** The loop heat pipe and its thermodynamic cycle.

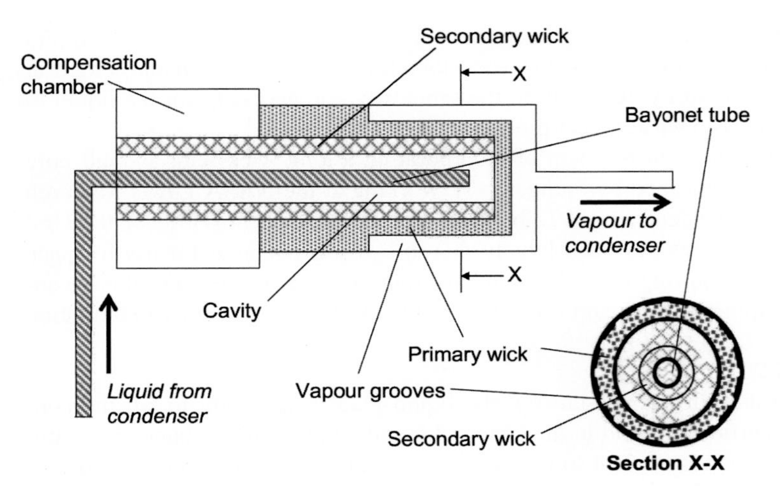


**FIGURE 2.27** Typical evaporator of loop heat pipe [50].

A typical LHP evaporator is shown schematically in Fig. 2.27. The evaporation in LHPs takes place on the surface of the wick adjacent to the evaporator wall. Vapour removal channels must be incorporated in the wick or evaporator wall to ensure that the vapour can flow from the wick to the vapour line with an acceptable pressure drop. A secondary wick is used to ensure uniform liquid supply to the primary wick and to provide liquid to the wick in the event of transient dryout.

The LHP has the features of the classical heat pipe with the following additional advantages:

- high heat flux capability
- capability to transport energy over long distances without restriction on the routing of the liquid and vapour lines
- ability to operate over a range of 'g' environments
- no wick within the transport lines
- vapour and liquid flows separated, therefore no entrainment
- may be adapted to allow temperature control, typically in the form of a CPL (as discussed below)

Mechanically pumped loops, or run-around coils, share the features listed above but have the added complexity of the pump and require an auxiliary power supply. That said, Anderson et al. [51] have developed a hybrid looped heat pipe (passive primary loop, pumped secondary loop) that operates with high heat fluxes over long distances.

When operating at steady state, the thermodynamic cycle of the LHP may be described with reference to Fig. 2.26. Point 1 represents the vapour immediately above the meniscus on the evaporator wick, this vapour is saturated. As the vapour passes through the vapour channels in the evaporator, it is superheated by contact with the hot wall of the evaporator and enters the vapour line at state 2. Flow in the vapour line results in an approximately isothermal pressure drop to state 3 at the entry to the condenser. The pressure drop in the condenser is generally negligible and heat transfer results in condensation to saturated liquid (state 4) and subcooling to state 5. This subcooling is necessary to ensure that the vapour is not generated in the liquid return line either as the pressure drops in the liquid or due to heat transfer from the surroundings. The liquid enters the compensation chamber at state 6 and is heated to the saturation temperature within the compensation chamber that contains saturated liquid at state 7 in equilibrium with the saturated vapour filling the remaining space. Flow through the wick takes the liquid to state 8. The liquid within the wick is superheated, but evaporation does not occur because of the very small pore size and absence of nucleation sites. At the surface of the wick the capillary effect results in the formation of menisci at each pore, the pressure difference across these menisci being $\Delta P_c$.

There are three conditions for an LHP to function. The first is identical to that of a classical heat pipe, as discussed in Chapter 4,

$$\Delta P_{c,max} \geq \Delta P_1 + \Delta P_v + \Delta P_g \tag{2.2}$$

and the pressure drops are evaluated using similar techniques, but note that the liquid pressure drop is due to flow through the liquid line in addition to that through the wick.

The second condition applicable to LHPs is that the pressure drop between the evaporating surface of the wick and the vapour space in the reservoir, $\Delta P_{EX}$, corresponds to the change in saturation temperature between states 1 and 7.

This is the condition for liquid to be displaced from the evaporator to the compensation chamber at start-up. For the relatively small temperature changes in a heat pipe, the slope of the $(dP/dT)_{sat}$ may be regarded as constant and the characteristics of the LHP must be such that

$$\left(\frac{dP}{dT}\right)_{sat} \times \Delta T_{1-7} = \Delta P_{EX} \tag{2.5}$$

The third requirement for correct operation is that the liquid subcooling at the exit from the condenser (state 5) must, as stated above, be sufficient to prevent excessive flashing of liquid to vapour in the vapour line. This implies

$$\left(\frac{dP}{dT}\right)_{sat} \times \Delta T_{4-6} = \Delta T_{5-6} \tag{2.6}$$

The claimed advantages of LHPs emphasise their suitability for transmitting heat over significant distances, but their ability to operate with high heat fluxes and tortuous flow paths also makes the use of miniature LHPs attractive in electronic cooling applications. However, miniaturisation presents the problem of meeting the condition given in Eq. (2.6), in that the necessary temperature difference between the evaporating meniscus and the compensation chamber is difficult to maintain with the thin wick inherent in a miniature system. One solution to this problem is to use wick structures having low thermal conductivity, but this option makes the design of the evaporator to minimise the temperature drop between the evaporator case and the wick surface more problematic.

The CPL differs from the LHP in that the compensation chamber is separate from the evaporator as illustrated in Fig. 2.28.

Before the start-up of a CPL, the two-phase reservoir must be heated to the required operating temperature. The entire loop will then operate at this nominal temperature with slight variations owing to some superheating or subcooling. Upon setting an operating temperature in the reservoir, the internal pressure will rise and the entire loop will be filled with liquid. This process is referred to as pressure priming. After priming, the CPL is ready to start operating. Initially, when heat is applied to the capillary evaporator, its temperature rises as sensible heat is transferred to the working fluid. When the evaporator temperature reaches the reservoir temperature the latent heat is transferred to the working fluid and evaporation commences. As evaporation proceeds a meniscus is formed at the liquid/vapour interface which then develops the capillary pressure that results in fluid circulation. Steady-state operation is then similar to that of an LHP with an integral compensation chamber.

The LHP benefits at start-up because the evaporator and compensation chamber are integral, whilst the separate compensation chamber of the CPL must be maintained at the appropriate temperature. This can be advantageous if control of the operating conditions of the device is required. Packaging of the evaporator and compensation chamber in a single envelope can be problematic in some applications, in which case the CPL would be preferable. A comprehensive discussion of the relative merits of the two technologies is given in Ref. [52].

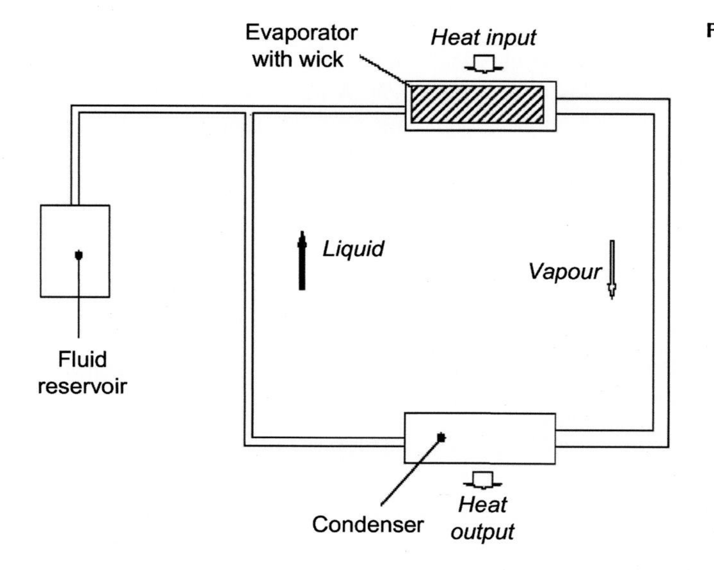


**FIGURE 2.28** Capillary-pumped loop.

Correct sizing of the compensation chamber is an essential part of the design of an LHP or CPL. This is relatively simple in principle. Since mass is conserved, the mass, $M_{charge}$, must be the same at start-up and during operation. Therefore

$$M_{charge} = \rho_{1.cold}\left(\sum V_{1.cold} + aV_{cc}\right) + \rho_{v.cold}\left(\sum V_{v.cold} + (1-a)V_{cc}\right) \quad (2.7a)$$

$$M_{charge} = \rho_{1.hot}\left(\sum V_{1.hot} + bV_{cc}\right) + \rho_{v.hot}\left(\sum V_{v.hot} + (1-b)V_{cc}\right) \quad (2.7b)$$

where $\sum V$ is the total volume occupied by the liquid or vapour excluding the compensation chamber, and $V_{cc}$ is the volume of the compensation chamber.

The compensation chamber has a filling ratio $a$ when cold and $b$ when hot. The liquid and vapour distribution can be determined before and during operation, and appropriate values of $a$ and $b$ are chosen, thus permitting the solution of Eqs (2.7a) and (2.7b) to yield $V_{cc}$.

Maydanik [50] has classified LHPs in terms of their design configuration, the design of the individual components, their temperature range and control strategy. This results in a range of potential LHP systems as summarised below:

| LHP design | LHP dimensions | Evaporator shape | Evaporator design |
|---|---|---|---|
| • Conventional (diode) | • Miniature | • Cylindrical | • One butt-end compensation chamber |
| • Reversible | • All the rest | • Flat disc-shaped | • Two butt-end compensation chambers |
| • Flexible<br>• Ramified | | • Flat rectangular | • Coaxial |
| 1–4 Condenser design | Number of evaporators and condensers | Temperature range | Operating-temperature control |
| • Pipe-in-pipe<br>• Flat coil<br>• Collector | • One<br>• Two and more | • Cryogenic<br>• Low temperature<br>• High temperature | • Without active control<br>• With active control |

Many of the papers published in the area of LHPs describe the design, operation and performance of a particular unit, some, for example Ref. [53], include additional features such as a wick structure within the condenser. A parametric study is presented by Launay et al. [54], however, they conclude that whilst the principal limitations are the boiling and capillary limits in most cases, a full parametric analysis is not possible because of the interrelationship between the many factors in the design. A later review by Ambirajan et al. [55] summarises design approaches, suggesting the use of proprietary thermohydraulic modelling software suitable for designing LHP.

Typically, the structural elements of LHPs are stainless steel but aluminium and copper are also used. Sintered nickel and titanium are commonly used in the manufacture of the wick; these materials are attractive because of their high strength and wide compatibility with working fluids as well as their wicking performance.

The properties of sintered metal wicks are summarised in Table 2.3.

Working fluids are selected using broadly the same criteria as for the classical wicked heat pipe. However, LHPs are more tolerant to the presence of noncondensable gases than the classic heat pipe [49]. A high value of $(dP/dT)_{sat}$ is also desirable, to minimise the temperature difference required at start-up. Maydanik suggests that ammonia is the optimum fluid for LHPs operating in the temperature range − 20°C–80°C, whilst water is best suited to temperatures in the range 100°C–150°C.

Typical applications of LHPs in aerospace and electronics cooling are discussed in Chapter 7.

**TABLE 2.3 Properties of typical sintered wicks [49].**

| Material | Porosity (%) | Effective pore radius (μm) | Permeability, $\times 10^{13}$ ($m^2$) | Thermal conductivity (W/mK) |
|---|---|---|---|---|
| Nickel | 60–75 | 0.7–10 | 0.2–20 | 5–10 |
| Titanium | 55–70 | 3–10 | 4–18 | 0.6–1.5 |
| Copper | 55–75 | 3–15 | – | – |

A representative small CPL is described in Fig. 2.29. This was designed on the basis of the pressure drops in each component using equations of the form developed in Section 2.3. The calculated maximum pressure drop obtained for either working fluid was less than 150 Pa, which was well below the capacity of the capillary evaporator, which was demonstrated to produce 2250 or 1950 Pa of capillary pressure with acetone and ammonia, respectively, at 30°C.

Examples of the performance of this unit are given in Fig. 2.30a and b. It can be seen that the temperature difference when operating at an equal load is higher for acetone than for ammonia. The loop was demonstrated under start-up conditions and with intermittent load. In the latter case, some temperature spikes were observed when acetone was used for the working fluid (as indicated in Fig. 2.31). However, the unit rapidly returned to correct operation.

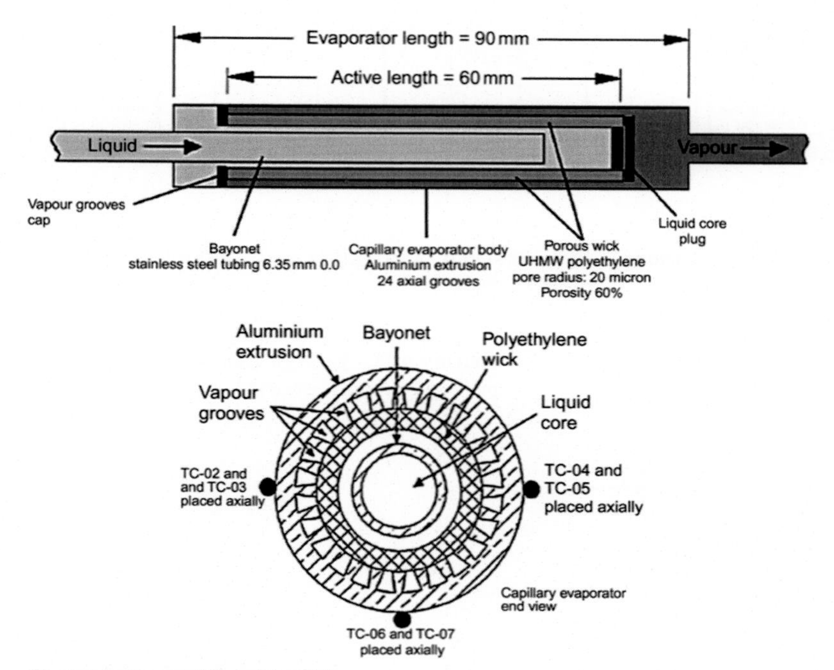


Geometric characteristics of the CPL

| | | | |
|---|---|---|---|
| *Evaporator* | | *Two-phase reservoir* | |
| Total length (mm) | 90 | Outer diameter (mm) | 25.4 |
| Active length (mm) | 60 | Length | 200 |
| Diameter (outer/inner)(mm) | 19.0/12.7 | Volume (cm$^3$) | 85 |
| Material | Aluminum alloy 6063 (ASTM) | Screen mesh | #200 Stainless steel grade 304L (ASTM) |
| Number of grooves | 24 | Material | Stainless steel grade 316L (ASTM) |
| Grooves height/width (top and base)/angle (mm) | 1.5/1.0/0.66/29° | | |
| *Polyethylene Wick* | | *Condenser* | |
| Mean pore radius (μm) | 20 | Outer diameter (mm) | 6.35 |
| Permeability (m$^2$) | $10^{-12}$ | Inner diameter (mm) | 4.35 |
| Porosity (%) | 60 | Length (mm) | 800 |
| Diameter (outer/inner)(mm) | 12.7/7.0 | Material | Stainless steel grade 316L(ASTM) |
| *Vapour line* | | *Liquid line* | |
| Outer diameter (mm) | 6.35 | Outer diameter (mm) | 6.35 |
| Inner diameter (mm) | 4.35 | Inner diameter (mm) | 4.35 |
| Length(mm) | 200 | Length (mm) | 200 |
| Material | Stainless steel grade 316L (ASTM) | Material | Stainless steel grade 316L (ASTM) |

**FIGURE 2.29** Dimensions of a capillary-pumped loop [56].

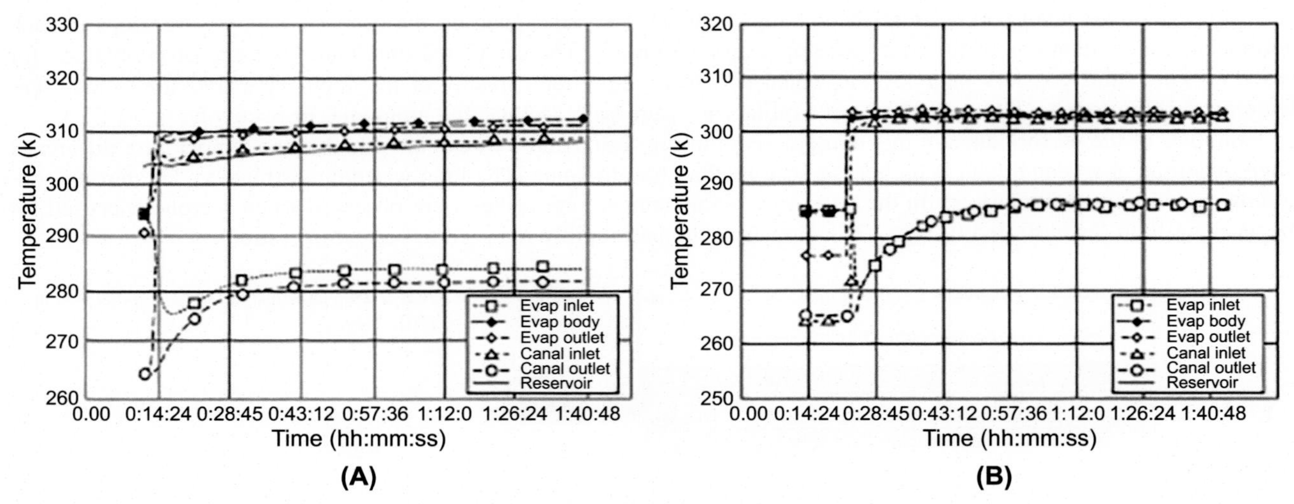


**FIGURE 2.30** Performance of capillary-pumped loop with acetone and ammonia: (a) power = 50 W, working fluid acetone and (b) power = 50 W, working fluid ammonia [56].

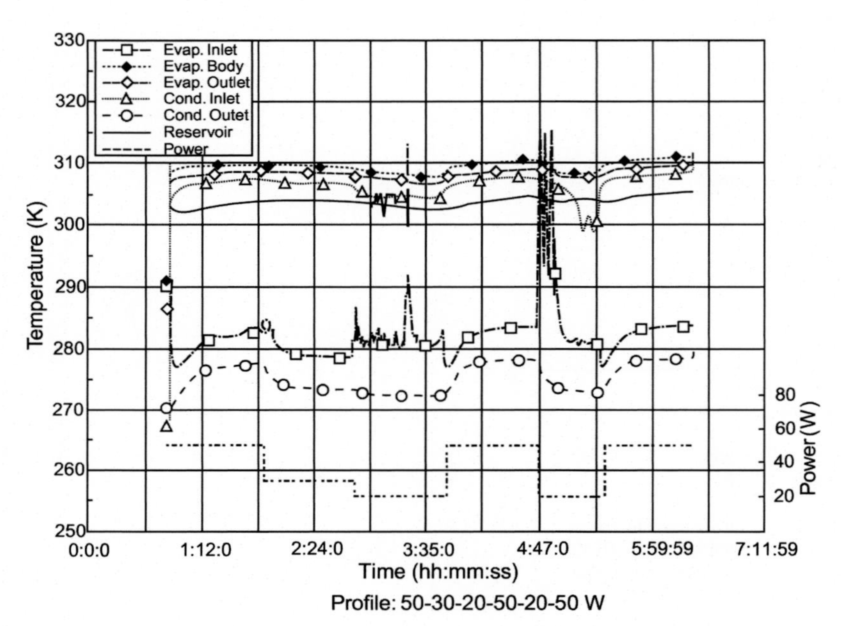


**FIGURE 2.31** Transient performance of capillary-pumped loop with acetone [56].

Whilst analysis of the steady-state performance of LHPs and CPLs is relatively straightforward and several examples of models are reported in the literature, for example, [57], the transient response is more difficult to analyse. Pouzet et al. [58] have developed a model that includes all the loop elements and physical processes in a CPL. The model reproduced the experimentally observed transient phenomena in a CPL and was able to explain both the oscillations observed during rapid changes of load. The model also explained the poor performance of CPLs after an abrupt decrease in load – a common phenomenon that is shown clearly in Fig. 2.31.

Whilst the original motivation for the development of LHP was for use in zero gravity environments, they are finding more use in terrestrial applications where orientation is important. Chen et al. [59] examined the performance of a miniature loop heat pipe at various orientations (as shown in Fig. 2.32) and horizontally. They found that the heat pipe had a similar resistance for all orientations except for the evaporator above the compensation chamber which performed less well. Start-up was smooth for most orientations, but there was a significant temperature overshoot for the two orientations 'condenser above evaporator' and 'evaporator above condenser'. Overall, they concluded that the device, intended for use in consumer electronics operating in the horizontal position, was satisfactory for the purpose and would

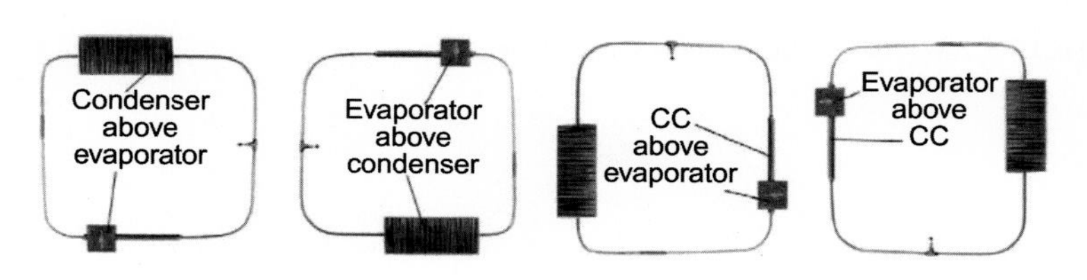


**FIGURE 2.32** Four vertical orientations as tested by Chen et al. [59].

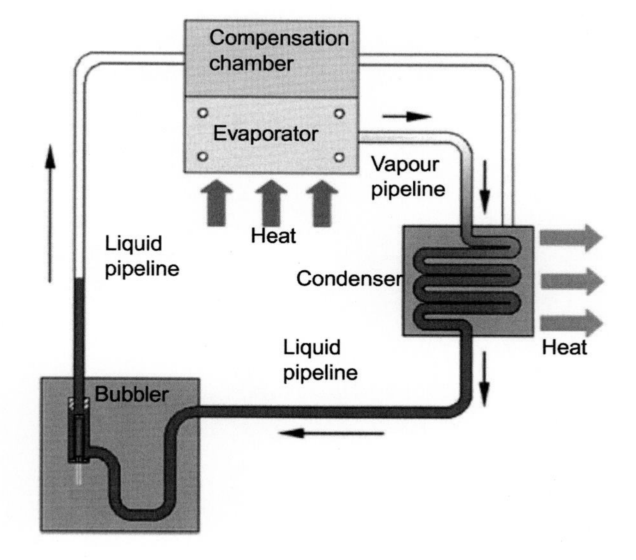


**FIGURE 2.33** Schematic of antigravity loop heat pipe incorporating bubble pump [61].

function at all likely orientations. It would not be expected that the gravitational head would measurably increase the thermal resistance of a miniature loop heat pipe; it is more significant with larger units: For example, Riehl [60] observed an additional thermal resistance when operating a loop heat pipe with the evaporator above the condenser (resistance 1 K/W) when compared to the condenser mounted above the evaporator (resistance 0.7 K/W orientation; the additional resistance was ascribed to gravitational head between condenser and evaporator. Deng et al. [61] have proposed the addition of a bubbler in the loop to function as a fluid pump as shown in Fig. 2.33 to mitigate the effect of gravity. Although this has been shown to work experimentally, it is questionable whether such additional complication, including the provision of a heat supply to the bubbler, is justified in most applications.

### 2.4.1 Thermosyphon loops

When working with the evaporator below the condenser in a gravitational field, it has been shown that the wick may be omitted from the classical heat pipe to give the two-phase thermosyphon. Similarly, if the condenser is above the evaporator in a loop, then this can operate satisfactorily with no wick. Thermosyphon loops have been investigated by several authors. Khodabandeh [62] has undertaken a detailed study of a loop suitable for cooling a radio base station and recommended suitable correlations for each component of heat transfer and pressure drop in the system (Fig. 2.34).

At steady state, liquid head in the downcomer must equal the sum of the pressure drops around the loop.

$$\rho_1 g h_L = \Delta P_{riser} + \Delta P_{condenser} + \Delta P_{downcomer} + \Delta P_{evaporator} \tag{2.8}$$

For a single-phase pressure drop in the downcomer and condenser, the equations given in Section 2.3.3 were employed. An additional term, $\Delta P_{LB}$, was added for each bend and change in section.

$$\Delta P_{LB} = \xi \frac{G^2}{2\rho_1} \tag{2.9}$$

where $\xi$ is an empirical constant for the geometry.

The flow in the evaporator and riser is two-phase, hence both void fraction and frictional pressure drop must be determined [63]. A range of void fraction correlations and frictional pressure drop calculations were tested, and it was

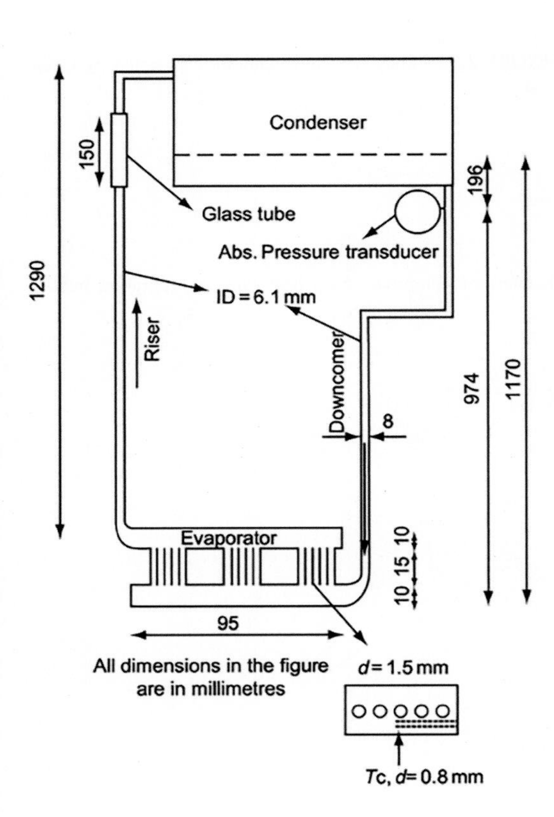


**FIGURE 2.34** Schematic of thermosyphon loop [62].

found that the gravitational pressure drop was best correlated by assuming homogeneous flow, whilst the Lockhart–Martinelli correlation [64] best fitted the frictional component of pressure drop.

The pressure drop in the short evaporator was principally due to the inertia term and was evaluated assuming homogeneous flow, as recommended by ref. [65],

$$\Delta P_{evaporator} = G^2 x (v_v - v_1) \tag{2.10}$$

The heat transfer in the evaporator [66] was well represented by an expression based on Coopers correlation, modified to account for the vapour mass fraction, $x$,

$$\alpha = C P_r^{0.394} \dot{q}^{0.54} \frac{1}{(1-x)^{0.65}} M^{-0.5} \tag{2.11}$$

with $C = 540$.

The thermal resistance at the condenser was dominated by the airside because of the high value of condensing heat transfer coefficient, thus the condensing coefficient was not evaluated.

The effect of noncondensable gas in an industrial scale (48 kW) loop thermosyphon heat exchanger has been investigated by Dube et al. [67]. As with all heat pipes, noncondensable gases may be present due to the release of gases dissolved in the working liquid when it boils produced by reactions between the working fluid and the material of the container walls and leakage of atmospheric gases inwards through the casing. Additionally, noncondensable gases may be introduced into the system as part of a control strategy [68]. These studies indicated that with an appropriately sized reservoir situated at the outlet of the condenser the adverse effects of noncondensable gases could be eliminated.

A useful review, if now slightly dated, of applications of what the authors call 'capillary heat loops' is given by Wang et al. [69]. The background centres around a project at the Canadian Space Agency to assist the understanding of CPLs, which led to a review of technology in space applications as well as in-house experiments and system modelling – the latter being based upon four different approaches.

## 2.5 Microheat pipes

Many microsystems proposed (and in some cases constructed) are directed at mimicking, or at least complementing one or more features of the living tissue, in particular the human body. The manufacture of replacement parts for our bodies, whether done with biological or nonbiological materials and structures, will necessitate excursions into fluid flow at the bottom end of the microfluidic scale (microfluidics being fluid flow at the microscale), encroaching on nanotechnology. It is therefore appropriate (and hopefully interesting for some readers) to 'set the scene' for microheat pipes – which operate mainly at the 'upper' end of the microfluidic size range – by briefly examining a feature of the human body which the vast majority of us have in abundance – the eccrine sweat gland [70,71].

An interesting aspect of sweat glands, and their behaviour during thermogenic, or heat-induced sweating is that they can be both active and passive. The secretion of liquid during active sweating involves the single-phase liquid flow along the sweat duct with expulsion and evaporation. The resting sweat gland, it is hypothesised, functions in thermoregulation by the two-phase heat transfer rather like a microheat pipe with a duct of typically 14 μ in diameter – perhaps in this decade we should refer to it as a 'nanoheat pipe'! (An average young man has 3 million sweat glands, of which more than 10% are on the head and the highest density is on the soles of the feet – 620/cm$^2$). Vapour exudation is illustrated in Fig. 2.35.

One may approach the fluid dynamics of these microchannels from a number of directions – conventional single- or two-phase flow theory or modified correlations from researchers such as Choi et al. [72] (sometimes limited to a single-phase flow), before considering mechanisms such as diffusion, transpiration streaming and osmosis – more commonly associated with water flows in plants and trees, and only the latter having been seriously considered in heat pipes.

Table 2.4 gives some of the characteristics of the porous structures which might be said to represent microfluidics in the 1960s – heat pipe wick structures. Interestingly, whilst capillary forces perhaps cannot dominate larger scale systems, they can have a positive or negative impact on microfluidic devices, and interface phenomena in general can become important players in microsystems where fluid flow is a feature. (The heat pipe specialist may well see opportunities for his or her expertise in the wider area of microfluidics – both in extending heat pipe models as dimensions reduce or in modelling other systems).

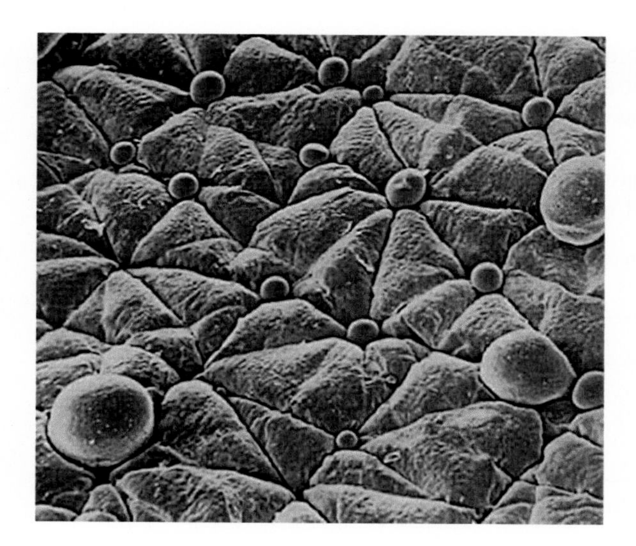

**FIGURE 2.35** Vapour bubbles shown at the exit of sweat glands by a scanning electron micrograph of forearm skin [69].

**TABLE 2.4 Characteristics of capillary structures.**

| Material | Porosity (%) | Effective pore radius (μm) | Permeability × $10^{13}$ (m$^2$) | Thermal conductivity (W/mK) |
|---|---|---|---|---|
| Nickel | 60–75 | 0.7–10 | 0.2–20 | 5–10 |
| Titanium | 55–70 | 3–10 | 4–18 | 0.6–1.5 |
| Copper | 55–75 | 3–15 | – | – |

Although reference was made in the 1970s, as mentioned above, to very small 'biological heat pipes' [70], the first published data on the fabrication of microheat pipes did not appear until 1984 [73]. In this paper, Cotter also proposed a definition of a microheat pipe as being one in which the mean curvature of the vapour–liquid interface is comparable in magnitude to the reciprocal of the hydraulic radius of the total flow channel.

This definition was later simplified by Peterson [74] so that a microheat pipe could be defined as one that satisfies the condition that

$$\frac{r_c}{r_h} \geq 1$$

Vasiliev et al. [75] define micro- and mini-heat pipes (HP) as follows:

| Micro HP | Mini HP | HP of 'usual' dimension |
|---|---|---|
| $r_c \leq d_v < l_c$ | $r_c < d_v \leq l_c$ | $r_c < l_c < d_v$ |

where $r_c$ is the effective liquid meniscus radius, $d_v$ the lesser dimension of vapour channel cross-section, $l_c$ the capillary constant of fluid:

$$l_c = \sqrt{\frac{\sigma}{g(\rho_1 - \rho_v)}}$$

Conventional small-diameter wicked heat pipes of circular (and noncircular) cross-section have been routinely manufactured for many years. Diameters of 2–3 mm still permit wicks of various types (e.g. mesh, sinter, grooves) to be employed. However, when internal diameters become of the order of 1 mm or less, the above condition applies.

The early microheat pipe configurations studied by Cotter include one having a cross-section in the form of an equilateral triangle, as illustrated in Fig. 2.36.

The length of each side of the triangle was only 0.2 mm. More recently, Itoh in Japan has produced a range of microheat pipes with different cross-sectional forms. Some of these are illustrated in Fig. 2.37, where it can be seen that the function of the conventional wick in the heat pipe is replaced by profiled corners that trap and carry the condensate to the evaporator [76].

Heat transport capabilities of microheat pipes are, of course, quite low. The data below, taken from ref. [76], are maximum thermal performances at different inclination angles of a selection of microheat pipes manufactured by Itoh. All use water as the working fluid, have copper walls, and operate with a vapour temperature of 50°C. Wall thickness is 0.15 mm.

Itoh later reported on the development of a microheat pipe having the dimensions shown in Fig. 2.38 with a thickness of only 0.1 mm, and a length of 25 mm, groove depth is 0.01 mm and minimum wall thickness 0.025 mm.

| External cross-section (mm) | Length (mm) | Heat transfer (W) | | |
|---|---|---|---|---|
| | | −90°C | 0°C | 90°C |
| 0.8 × 1.0 | 30 | 0.25 | 0.3 | 0.32 |
| 2.0 × 2.0 | 65 | 1.2 | 1.5 | 1.7 |
| 1.2 × 3.0 | 150 | 0.6 | 2.2 | 4.0 |
| 0.5 × 5.0 | 50 | 0.4 | 1.1 | 2.1 |

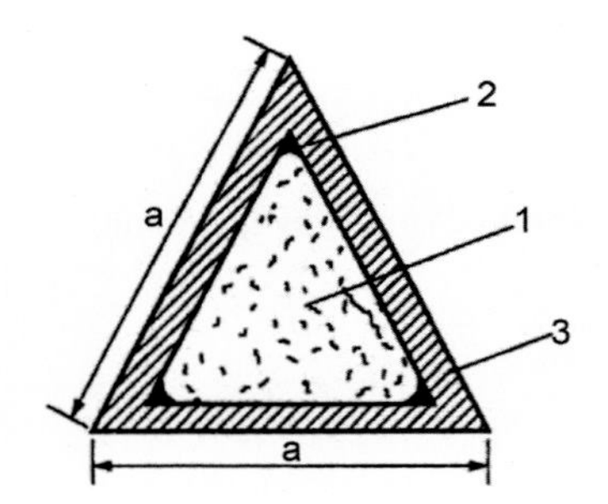


**FIGURE 2.36** Schematic diagram of Cotter's microheat pipe. 1, vapour; 2, liquid; 3, wall; $a = 0.2$ mm.

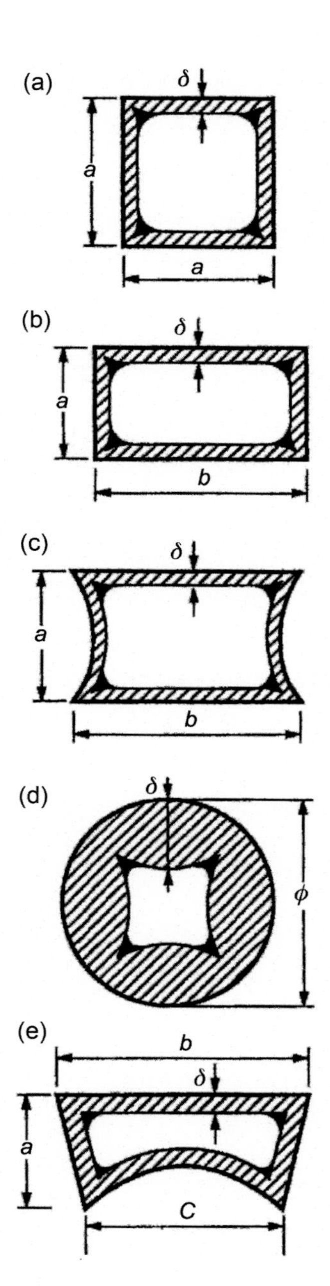


**FIGURE 2.37** Cross-sectional areas of Itoh's microheat pipes: (a) square; $a$ (0.3−1.5 mm) and $\delta$ (0.1−0.2 mm). (b) rectangular; $a$ (0.3−1.5 mm), $b$ (3.0−10.0 mm), and $\delta$ (0.1−0.2 mm). (c) modified rectangular; $a$ (0.5−1.2 mm), $b$ (3.0−5.0 mm), and $\delta$ (0.1−0.2 mm). (d) circular (outer) and square (inner); $\Phi$ (0.5−1.5 mm) and $\delta$ (0.1−0.2 mm). (e) tapered; $a = 0.6$ mm, $b = 2.0$ mm, $c = 1.3$ mm, and $\delta = 1.15$ mm.

Cotter pointed out that the cooling rates achievable using microheat pipes in, for example, electronics cooling is considerably inferior to that offered by forced convection systems. However, the other benefits attributed to heat pipes such as isothermalisation, passive operation and the ability to control temperatures in environments where heat loads vary, do suggest that microheat pipes have a potential role to play in a growing number of applications.

The field of electronics cooling was the first application area investigated. Peterson [74] showed a trapezoidal microheat pipe arrangement for the thermal control of ceramic chip carriers. Illustrated in Fig. 2.39 the heat pipes were 60 mm long and approximately 2 mm across the top surface. In this example, the condenser (not shown in the figure) was finned and located at one end of the assembly. Itoh has constructed units that can carry multiple integrated circuit packages (i.c. p.), with cooling fins interspaced between each i.c. p. Thus a microheat pipe with multiple evaporators and condensers can be formed.

The thermal management of electronics systems remains a prime candidate for microheat pipes. The Furukawa Company stressed the importance of thermal resistance (K/W), or the temperature drop per unit of heat transfer, in developing high-performance microheat pipes. It was claimed that internally grooved wick structures had an inherently

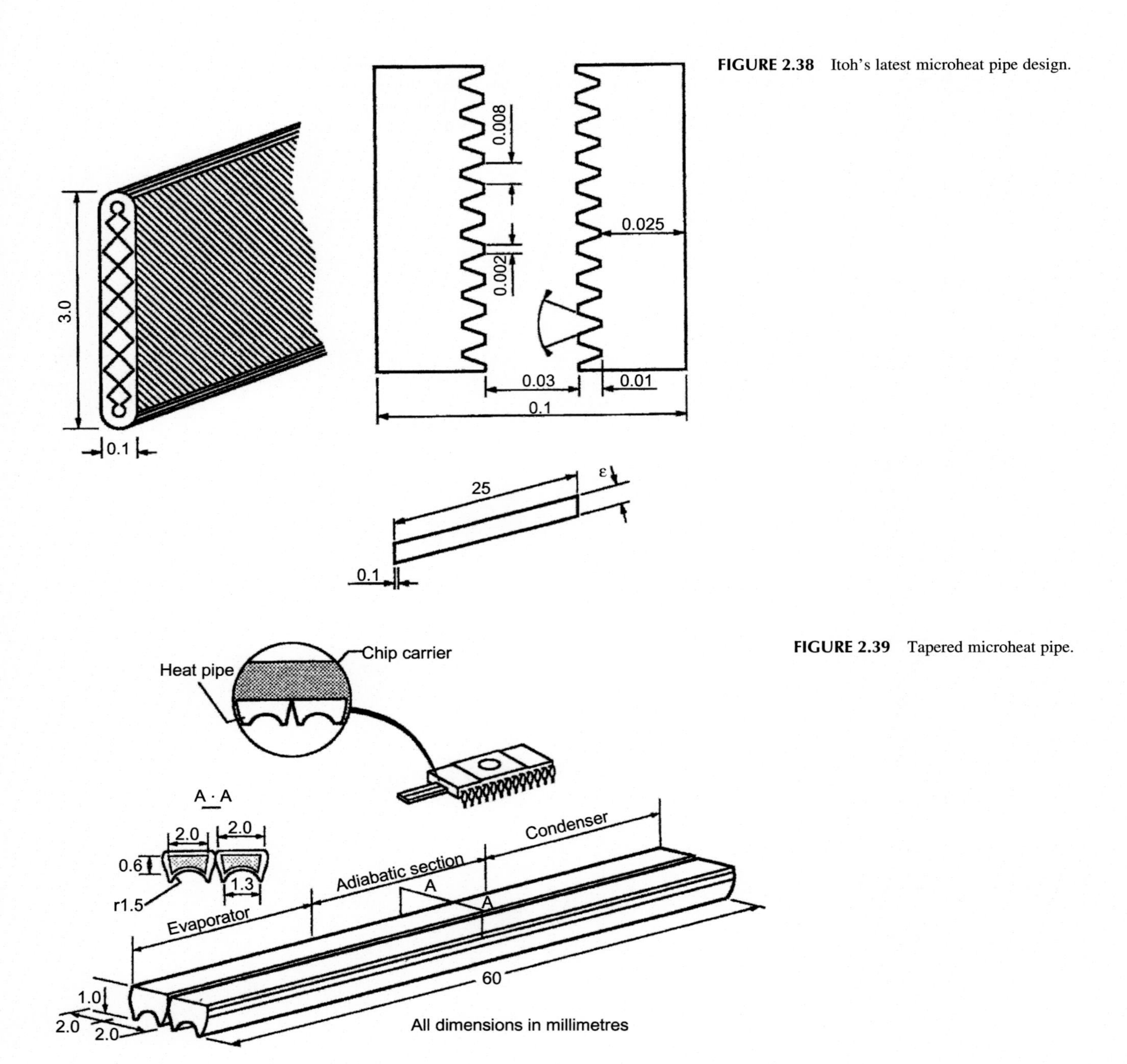


**FIGURE 2.38** Itoh's latest microheat pipe design.

**FIGURE 2.39** Tapered microheat pipe.

lower thermal resistance than sintered or mesh wicks. The high-performance units tested by Furukawa were grooved (pipes of 4 mm diameter – perhaps mini rather than micro by some definitions) and were able to transport up to about 25 W whilst having a thermal resistance of about 0.1 K/W. At 10 W heat transport, the resistance dropped to <0.05 K/W, whilst the conventional microheat pipe dried out at 15 W and had a consistently greater thermal resistance. In all cases, water was the working fluid and the high-performance grooved unit was operated successfully at a tilt of 30 degrees against gravity.

By bonding an array of wires to a plate, so that the line contact between the wire and the plate allows a liquid flow path to be formed generating capillary forces, a new form of microheat pipe can be made [77].

The analysis of microgrooved heat pipe structures, where grooves are of a polygonal form, has been detailed recently by Suman et al. [78] working in the United States. Structures of a similar form to those illustrated in Fig. 2.36 and Fig. 2.37 were rigorously analysed and a performance factor was obtained for heat pipes of triangular or rectangular shape. Here the microchannel dimensions were measured in the tens and hundreds of microns.

In Belarus, Aliakhnovich et al. [79] have recently reported on a pulsating microheat pipe where the wick can be discarded completely (see also Section 2.3). Capillary channels of 0.5 mm and greater were investigated. The minimum thermal resistance achieved (0.02 K/W) corresponded to the maximum performance of the microheat pipe array as determined by the capillary limit of 1.5 W. Water was the working fluid and the overall dimensions of the array were $78 \times 10 \times 2\ mm^3$. However, maximum heat fluxes were rather low.

Microheat pipes have also been used for cooling laser diodes, bringing about a temperature reduction of approximately 7°C in the chip which had been at 47°C without the heat pipe. The maximum power input to the diodes was of the order of 250 mW. In a number of similar applications, the microheat pipe, it is claimed, is preferable to thermoelectric cooling elements, in spite of the greater cooling capacity of the latter. The heat pipe is smaller, lighter and of lower cost.

Tang et al. [80] reviewed applications and developments in ultra-thin microheat pipes for cooling particularly in the field of portable electronic devices, and applications have also been briefly reported in the medical field, where local heating and/or cooling may be required as a treatment.

In addition, Han, Wang and Liang [81] also investigated the performance of a flat plate heat pipe sink with multiple sources.

## 2.6 Use of electrokinetic forces

Earlier editions of *Heat Pipes* dealt in some detail with a variety of electric field effects that could be used to assist heat pipe performance. These included electro-osmosis, electrohydrodynamics (EHD) and ultrasonics.

Within the heat-pipe-related two-phase systems (liquid—vapour) the influence of capillary action can be high, from a negative or positive viewpoint. Where the force is used to drive liquid, as in the microheat pipe discussed in Section 2.5, there is a wealth of literature examining optimum configurations of channel cross-section, pore sizes, etc. However, the impression is that in many microfluidic devices[1] capillary action is a disadvantage that needs to be overcome by external 'active' liquid transport mechanisms. Perhaps there is room for a more positive examination of passive liquid transport designs, but the use of electric fields and other more complicated procedures can provide the necessary control that may not always be available using passive methods alone. Some of these 'active' methods are described below.

### 2.6.1 Electrokinetics

Electrokinetics is the name given to the electrical phenomena that accompany the relative movement of a liquid and a solid. These phenomena are ascribed to the presence of a potential difference at the interface between any two phases at which movements occur. Thus if the potential is supposed to result from the existence of electrically charged layers of opposite sign at the interface, then the application of an electric field must result in the displacement of one layer with respect to the other. If the solid phase is fixed whilst the liquid is free to move, as in a porous material, the liquid will tend to flow through the pores as a consequence of the applied field. This movement is known as electro-osmosis.

Electro-osmosis using an alternating current has been used as the basis of microfluidic pumps allowing normal batteries to be used at the modest applied voltages. This helps to integrate such a pumping method with portable lab-on-a-chip devices. Cahill et al. in Zurich [82] have shown that a velocity of up to 100 μm/s could be achieved for applied potentials of less than 1 $V_{rms}$. The impact of the chemical state of the surfaces of the channel on electro-osmotic flow has also been investigated. However, of significance to reactions and other unit operations carried out in microfluidic devices, although not so important for heat pipes, is the ability to enhance mixing [83]. Heat pipe work is reported in [84—87]. For the reader interested in early work on osmotic heat pipes (without an applied electric field), see [88,89].

### 2.6.2 Electrohydrodynamics

For those willing to contemplate more complex systems, the EHD (or possibly MHD – magnetohydrodynamics; see Refs. [90,91]) can be used for fluid propulsion, mixing and separations as well as heat pipe flow enhancement. Qian and Bau [92] have tested an MHD stirrer, illustrated in Fig. 2.40. When a potential difference (PD) is applied across one or more pairs of electrodes, the current that results interacts with the magnetic field to induce Lorentz forces and

1. Microfluidics, or the flow of fluids at the microscale, is discussed in the context of microheat pipes in Sections 6.2 and 6.8.

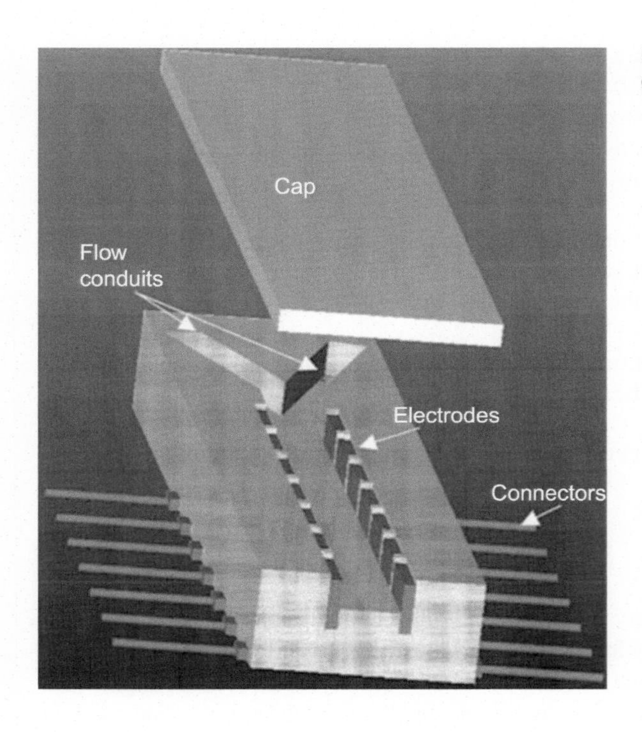


**FIGURE 2.40** The Y-shaped microchannels lead into a section with electrodes on both walls. The gap between adjacent electrodes is 0.5 mm [92].

fluid motion. The alternating application of the PD results in chaotic advection and mixing, but the authors point out that the system needs perfecting.

The work of Jones [93] involved replacing the conventional capillary structure in a heat pipe with electrodes that generated an EHD force.

The heat pipe is restricted to the use of insulating dielectric liquids as the working fluid, but as these tend to have a poor wicking capability and can be used in the vapour temperature range between 150°C and 350°C, and where suitable working fluids are difficult to find performance enhancement is useful. This particular type of unit is still attracting considerable interest.

The EHD heat pipe proposed by Jones consisted of a thin-walled tube of aluminium or some other good electrical conductor, with end caps made of an insulating material such as plexiglass. A thin ribbon electrode is stretched and fixed to the end caps in such a way that a small annulus is formed between it and the heat pipe wall over the complete length of the heat pipe. (This annulus is only confined to about 20% of the heat pipe circumference, and provision must be made for distributing the liquid around the evaporator by conventional means).

When a sufficiently high voltage is applied, the working fluid collects in the high electric field region between the electrode and the heat pipe wall, forming a type of artery as shown in Fig. 2.41.

Evaporation of the liquid causes an outward bulging of the liquid interface. This creates an inequality in the electromechanical surface forces acting normal to the liquid surface, causing a negative pressure gradient between condenser and evaporator. Thus a liquid flow is established between the two ends of the heat pipe.

Jones calculated that Dowtherm A could be pumped over a distance approaching 50 cm against gravity, much greater than achievable with conventional wicks. Applications of this technique could include temperature control and arterial priming.

The use of an insulating dielectric liquid somewhat limits the applicability of such heat pipes; but within this constraint, performances have shown 'significant improvement' over comparable capillary wicked heat pipes. Loehrke and Debs [94] tested a heat pipe with Freon 11 as the working fluid, and Loehrke and Sebits [95] extended this work to flat plate heat pipes, both systems using open grooves for liquid transport between evaporator and condenser. It is of particular interest to note that evaporator liquid supply was maintained even when nucleate boiling was taking place in the evaporator section, corresponding to the highest heat fluxes recorded.

Jones [93], in a study of EHD effects on film boiling, found that the pool boiling curve revealed the influence of electrostatic fields of varying intensity, and both the peak nucleate flux and minimum film flux increased as the applied voltage was raised, as shown in Fig. 2.42. Jones concluded that this was a surface hydrodynamic mechanism acting at the liquid–vapour interface.

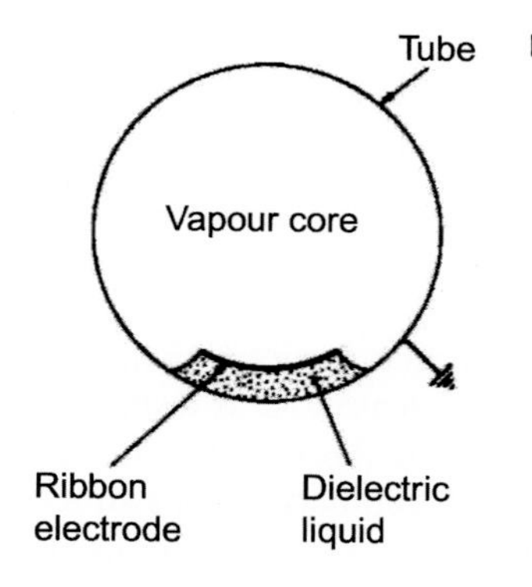


**FIGURE 2.41** Electrohydrodynamic liquid pump in a heat pipe.

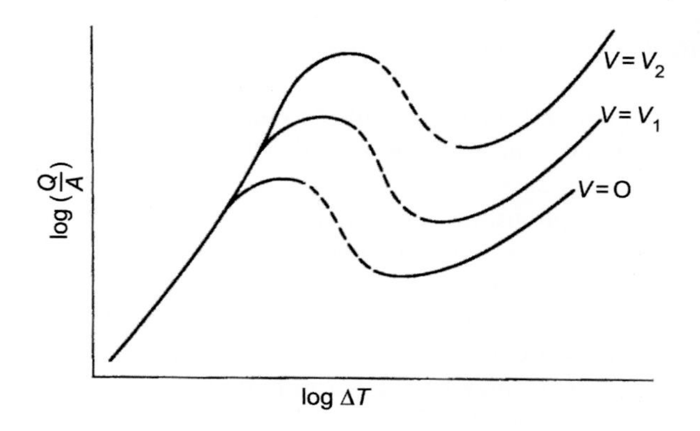


**FIGURE 2.42** Electrohydrodynamics – the effect of an electrostatic field on the pool boiling curve [93].

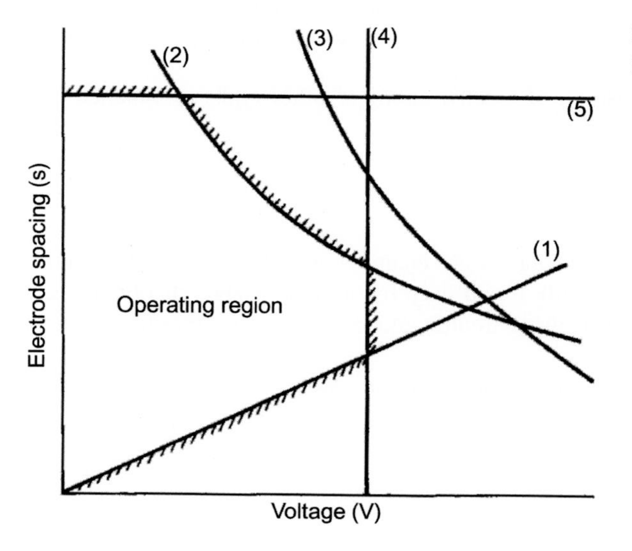


**FIGURE 2.43** Theoretical operational limits of an electrohydrodynamic heat pipe [90].

Because of the limitation in the range of working fluids available, it is unlikely that EHD heat pipes will compete with conventional or other types where water can be used as the working fluid. However, Jones and Perry [96] suggest that at vapour temperatures in excess of 170°C, where other dielectric fluids are available, advantages may result from the lower vapour pressures accruing to their use. To date, no experiments appear to have been carried out on EHD heat pipes over such a temperature range (e.g. 170°C–300°C). However, Kikuchi [97] has demonstrated that it is possible to plot an envelope showing the operating limits of an EHD heat pipe in an identical way to that for capillary-driven heat pipes. Illustrated in Fig. 2.43 the limits are as follows:

1. vapour breakdown limit
2. EHD wave speed limit

3. vapour sonic limit
4. voltage supply limit
5. size limit

The electrode spacing S has a strong influence on the vapour breakdown limit, and the size limit occurs because EHD liquid flow structures brought close together can interact with one another.

Kikuchi also constructed a flat plate heat pipe that utilised EHD forces [98].

A useful review of EHD heat pipes, based on the extensive research carried out in the former Soviet Union, was written by Bologa and Savin [99]. The basic use of EHD forces to enhance evaporation and condensation, as evaluated in many instances in shell-and-tube and other heat exchangers, can extend to liquid transport, as illustrated above. A further variation is the introduction of a pulsed electric field to accelerate the flow of liquid in capillary porous structures. Variations in either the steady state or the fluctuating EHD field can, depending upon the porous structure used, be employed to regulate the heat transfer properties of the heat pipe over a wide range.

Fig. 2.44 shows the effect of EHD on the temperature difference between the evaporator and the condenser of a heat pipe as the heat load is increased. It can be seen that EHD effects are effective in minimising the $\Delta T$, with, it is claimed, negligible additional power consumption.

Proposals have also been put forward in the former USSR and China [100] for using static or pulsed electric fields to enhance external forced convection heat transfer – 'Corona wind' cooling. The effect of these fields on Nusselt number, as a function of *Re* for applied voltages of up to 30 kV, is shown in Fig. 2.45.

In the field of heat pipes, as in many other heat and mass transfer applications, the use of EHD remains largely confined to the laboratory. Whether potential applications become realised in practice depends upon trends in the cost and reliability of the power sources and their ease of integration into the heat pipe environment. The area of microfluidics may be the one where practical use is realised, and this may arise from research in the United States [101] where a variant on EHD – the EHD conduction pump – is being investigated for two-phase loops.

As with other fluid transport systems used in heat pipes, the prefix 'micro' may be applied to EHD heat pipes. Yu et al. [102] have reported the active thermal control of an EHD-assisted microheat pipe array. The authors had earlier shown that EHD had led to a sixfold increase in the heat dissipation capability of a microheat pipe. The extension to active thermal control, as may be achieved with active feedback-controlled VCHPs discussed earlier in this chapter, may be useful in some applications.

However, the reader may like to consider that the superimposition of electric fields of other types to enhance the flow of liquids through a heat pipe could be used to effect similar active control.

### 2.6.3 Optomicrofluidics

Research at the Georgia Institute of Technology [103] is directed at using a light beam as an energy source for liquid manipulation. A dye injected into a larger droplet was mixed in the bulk liquid drop by, as with MHD, chaotic advection. In other experiments, the modulated light field can be used to drive liquids and droplets using thermocapillary

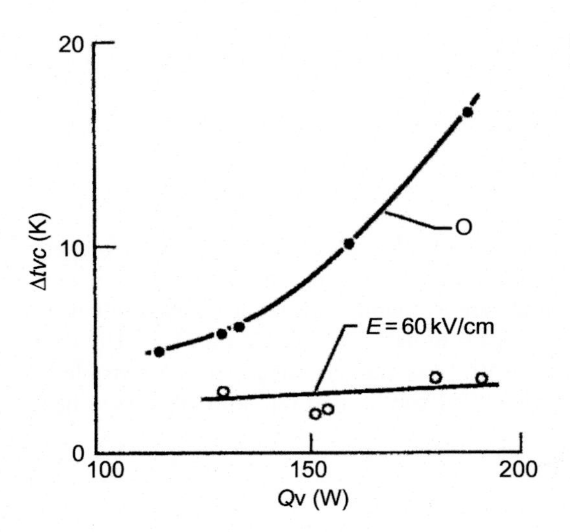


FIGURE 2.44 Temperature difference between evaporator and condenser – with and without electrohydrodynamics.

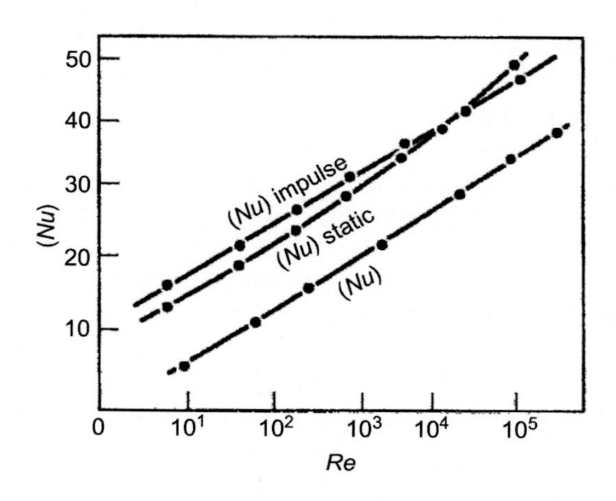


**FIGURE 2.45** Comparison of Nusselt numbers for no electric field (*Nu*) and static or impulse field.

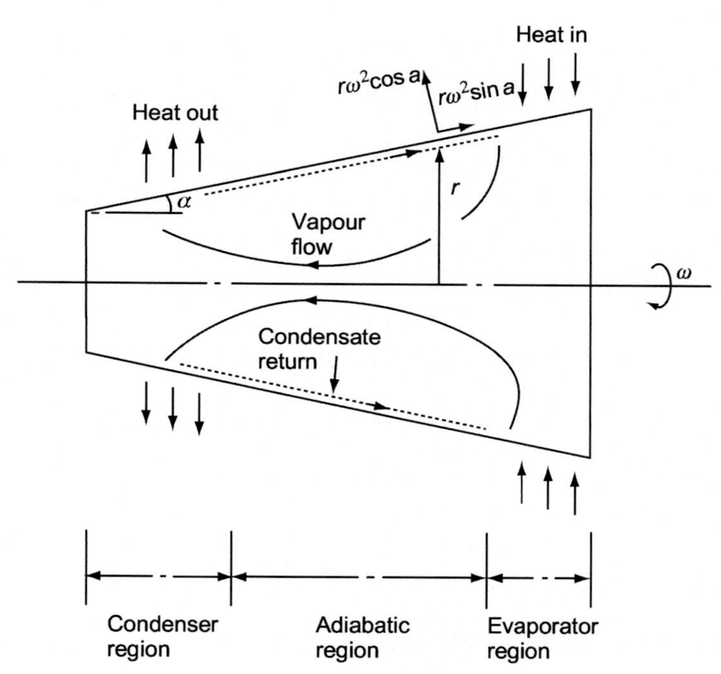


**FIGURE 2.46** Rotating heat pipe.

forces thus generated. (Of course, it may be ultimately shown that thermocapillary driving forces could be dominant in a system where previously conventional capillary forces had, perhaps to the detriment of performance, governed flow characteristics). Even optical tweezers are proposed in the patent literature for micromanipulation.

For those wishing to follow-up, the numerous techniques for influencing capillary forces, the paper by Le Berre et al. [104] on electrocapillary forces actuation in microfluidic elements is worthy of study. At the small scale, where particles might be involved in the fluid stream, the impact of electrophoresis and thermophoresis – the motion of a particle induced by a temperature gradient (from a hot to a cold region) should not be neglected.

## 2.7 Rotating heat pipes

In this section, unless otherwise stated, the rotating heat pipe is of the axial type, where the evaporator and condenser are separated in the direction of the axis of rotation, as in a turbine shaft. A radial rotating heat pipe would, for example, be that in a turbine blade, where the evaporator was radially displaced from the condenser.

The rotating heat pipe is a two-phase thermosyphon in which the condensate is returned to the evaporator by means of centrifugal force. The rotating heat pipe consists of a sealed hollow shaft, commonly having a slight internal taper along its axial length (although the taper is not strictly necessary but, as will be described later, can benefit performance) and containing a fixed amount of working fluid, Fig. 2.46.

The rotating heat pipe, like the conventional capillary heat pipe, is divided into three sections, the evaporator region, the adiabatic region and the condenser region. However, the rotation about the axis will cause a centrifugal acceleration $w^2r$ with a component $w^2r \sin \alpha$ along the wall of the pipe. The corresponding force will cause the condensed working fluid to flow along the wall back to the evaporator region.

The first reference to the rotating heat pipe was given in an article by Gray [105].

Centrifugal forces will significantly affect the heat and mass transfer effects in the rotating heat pipe and the three regions will be considered in turn.

Published work on evaporation from rotating boilers, Gray et al. [106], suggests that high centrifugal accelerations have smooth, stable interfaces between the liquid and the vapour phases. Using water at a pressure of 1 atm and centrifugal accelerations up to 400 g heat fluxes of up to 257 W/cm$^2$ were obtained. The boiling heat transfer coefficient was similar to that at 1 g, however the peak, or critical flux, increased with acceleration, Costello and Adams [107] derived a theoretical relationship which predicts that the peak heat flux increases as the one-fourth power of acceleration.

In the rotating condenser region, a high condensing coefficient is maintained due to the efficient removal of the condensate from the cooled liquid surface by centrifugal action. Ballback [108] carried out a Nusselt-type analysis but neglected vapour drag effects. Daniels and Jumaily [109] carried out a similar analysis but have taken account of the drag force between the axial vapour flow and the rotating liquid surface. They concluded that the vapour drag effect was small and could be neglected except at high heat fluxes. These workers also compared their theoretical predictions with measurements made on rotating heat pipes using Arcton 113, Arcton 21 and water as the working fluid. They stated that there is an optimum working fluid loading for a given heat pipe geometry, speed and heat flux. The experimental results appear to verify the theory over a range of heat flow and discrepancies can be explained by experimental factors. An interesting result [110] from this work is the establishment of a figure of merit $M$ for the working fluid where:

$$M' = \frac{\rho_1^2 L k_1^3}{\mu_1}$$

$\rho_1$ the liquid density, $L$ the latent heat of vapourisation, $k_1$ the liquid thermal conductivity and $\mu_1$ the liquid viscosity.

$M'$ is plotted against temperature for a number of working fluids in Fig. 2.47.

Normally the rotating heat pipe should have a thermal conductance comparable to or higher than that of the capillary heat pipe. The low equivalent conductance quoted by Daniels and Jumaily [109] may have been due to a combination of very low thermal conductivity of the liquid Arcton[2] and a relatively thick layer in the condenser.

In the adiabatic region, as in the same region of the capillary heat pipe, the vapour and liquid flows will be in opposite directions, with the vapour velocity much higher than the liquid.

### 2.7.1 Factors limiting the heat transfer capacity of the rotating heat pipe

The factors which will set a limit to the heat capacity of the rotating heat pipe will be sonic, entrainment, boiling and condensing limits (and noncondensable gases). The sonic limit and noncondensable gas effects are the same as for the capillary heat pipe. Entrainment will occur if the shear forces due to the counter flow vapour are sufficient to remove droplets and carry them to the condenser region. The radial centrifugal forces are important in inhibiting the formation of the ripples on the liquid condensate surface which precede droplet formation.

The effect of rotation on the boiling limit has already been referred to in Ref. [106,107] as has the condensing limit [109].

The Second International Heat Pipe Conference provided a forum for reports of several more theoretical and experimental studies on rotating heat pipes.

Marto [111] reported on recent work at the Naval Post-graduate School, Monterey, where studies on this type of heat pipe had been continuing since 1969. The developments at Monterey concern the derivation of a theoretical model for laminar film condensation in the rotating heat pipe taking into account vapour shear and vapour pressure drop. Marto found that these effects were small, but he recommended that the internal condensation resistance be of the same order as the condenser wall resistance and outside convection resistance. Thus it is desirable in rotating heat pipes to make the wall as thin as possible although in rotating electrical machines this is often inconsistent with structural requirements.

2. Arcton 21 and Arcton 113 are chlorofluorocarbons and would no longer be the viable working fluids as they have high ozone depletion potentials.

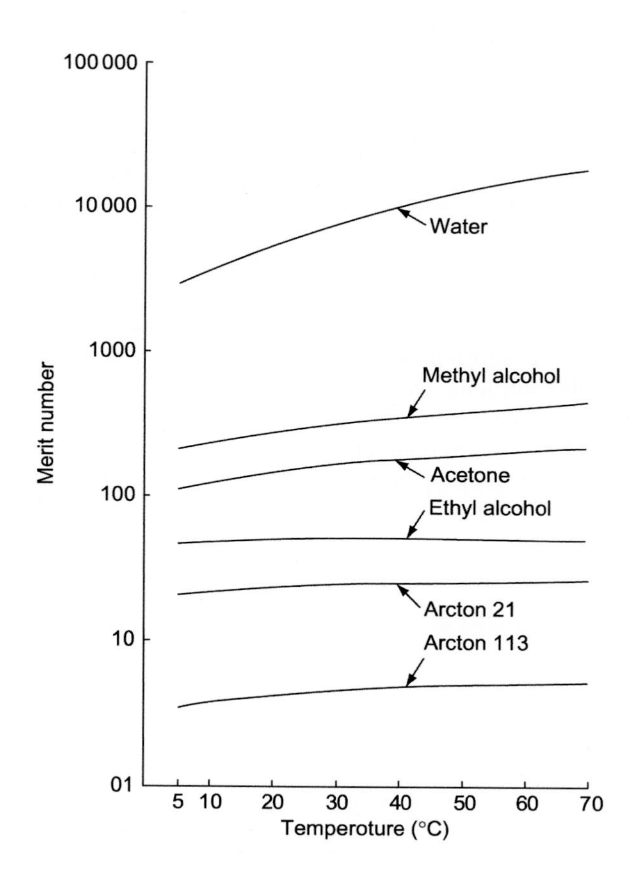


**FIGURE 2.47** Property group (figure of merit) versus temperature for various working fluids in a rotating heat pipe.

Vasiliev and Khrolenok [112] extended the analysis to the evaporator section, and recommended that

1. The most favourable mode of evaporator operation is fully developed nucleate boiling with minimal free convection effects because high heat transfer coefficients are obtained, independent of liquid film thickness and rotational speed.
2. For effective operation when boiling is not fully developed, the liquid film, preferably as thin as possible, should cover the whole evaporator surface.
3. Condenser heat flux is improved with increasing rotational speed. (For a 400 mm long × 70 mm internal diameter unit, with water as the working fluid, Vasiliev achieved condenser coefficients of 5000 W/m$^{2}$°C at 1500 rpm with an axial heat transport of 1600 W.)

In a useful review of the performance of rotating heat pipes, Marto and Weigel [113] reported their experiments on rotating heat pipes employing several forms of cylindrical condenser sections, including the use of cylinders having smooth walls, fins and helical corrugations. Water, ethanol and R113 were studied as working fluids, and heat transfer coefficients in the condenser section were measured as a function of rotational speed (varied between 700 and 2800 rpm).

The results showed that in a rotating heat pipe it was not necessary for the condenser section to be tapered, and cheap smooth wall condensers, which are slightly inferior in heat transport capability, could be used. However, much improved performance could be achieved with finned condensers, one particular unit produced by Noranda and employing spiralled fins giving enhancements of 200%–450%. The authors proposed that the surface acted as pump impellers to force the condensate back to the evaporator section.

Recent work on rotating heat pipes at speeds of up to 4000 rpm suggests that the effect of a taper in the condenser section can be positive [114]. In a heat pipe where tests were carried out with the fluid inventory (water in a copper pipe) varying between 5% and 30% of the internal volume, one set of experimental data showed that for a given temperature difference and fluid inventory, the heat transfer in the tapered heat pipe was slightly over 600 W, whilst in the cylindrical unit it was just over 200 W. A reduction in fluid inventory aided the cylindrical unit more (rising to 450 W) than the tapered unit, which stayed about the same. The differences are attributed to the reduction in the thermal resistance across the film.

The reader may find the review paper by Peterson and Win [115] useful as a summary of rotating heat pipe theory and performance.

## 2.7.2 Applications of rotating heat pipes

The rotating heat pipe is obviously applicable to rotating shafts having energy-dissipating loads, for example the rotors of electrical machinery, rotary cutting tools, heavily loaded bearings and the rollers for presses. Polasek [116] reported experiments on cooling on a.c. motor incorporating a rotating heat pipe in the hollow shaft, Fig. 2.48. He reported that the power output can be increased by 15% with the heat pipe, without any rise in winding temperature. Gray [105] suggested that the use of a rotating heat pipe in an air-conditioning system, Groll et al. [117] reported the use of a rotating heat pipe for temperature flattening of a rotating drum. The drum was used for stretching plastic fibres and rotated at 4000–6000 rpm, being maintained at a temperature of 250°C; Groll selected diphenyl as the working fluid.

As well as the use of 'conventional' rotating heat pipes to cool electric motors (as illustrated in Fig. 2.49) where the central shaft is in the form of a heat pipe, other geometries have been adopted. Groll and his colleagues at IKE, Stuttgart [118], arranged a number of copper/water heat pipes, each 700 mm long and 25 mm outside diameter, around the rotor, as shown in the end view in Fig. 2.50. It was found that adequate pumping capability was provided by the axial component of the centrifugal force, but when the rotor was stationary, circumferential grooves were needed to ensure that the heat pipes functioned. Each heat pipe was capable of transferring in an excess of 500 W, and rotor speeds were up to 6000 rpm.

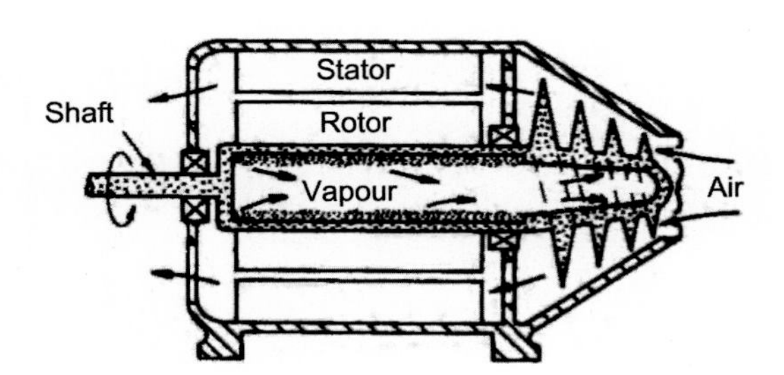


**FIGURE 2.48** Application of rotating heat pipes to cooling of motor rotors.

**FIGURE 2.49** Electric motors employing rotating heat pipes. *Courtesy: Furukawa Electric Co. Ltd, Tokyo.*

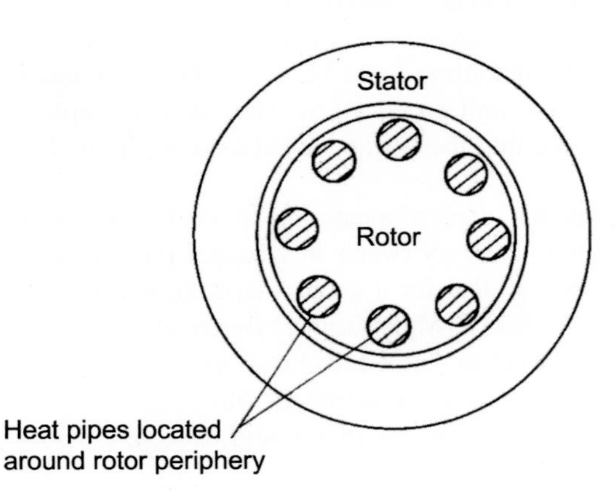


**FIGURE 2.50** Location of heat pipes outside rotor centreline.

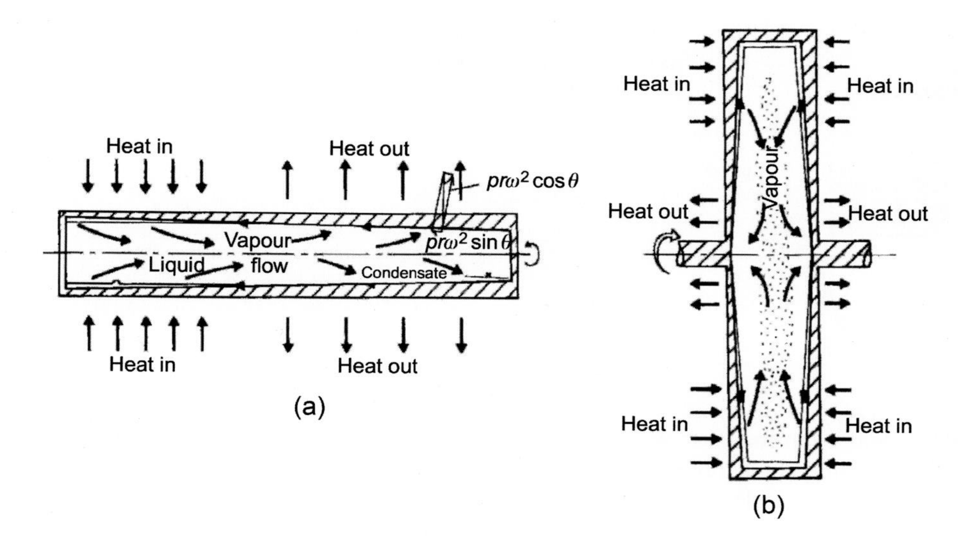


**FIGURE 2.51** The disc-shaped rotating heat pipe: (a) axial flow type and (b) radial flow type [119].

The disc-shaped rotating heat pipe proposed by Maezawa et al. [119] also utilises centrifugal forces for condensate return, but in this case the unit resembles a wheel or disc on a shaft rather than being an integral part of the shaft itself. The disc-shaped rotating heat pipe is illustrated in Fig. 2.51, where it is compared to a 'conventional' rotating heat pipe. Vapour and liquid flow is predominantly radial – as opposed to the axial flow in the more common form. Using this type of heat pipe work has been carried out on cooling the brakes of heavy road vehicles and the experiments have successfully demonstrated that brake temperatures can be significantly reduced. Ethanol was used as the working fluid.

The disc heat pipe has its equivalent in an interesting area of chemical engineering – that of intensified process technology. Protensive, a UK-based company, has employed enhanced two-phase heat transfer on spinning discs, rotating at 500–10,000 rpm, for the thermal control of chemical reactions on the 'spinning disc reactor' and for stripping and concentration of chemicals [120]. The equipment for high-capacity evaporative stripping, illustrated in Fig. 2.52, uses the spinning disc to generate a thin film of liquid organic solvent that is readily evaporated via heat input to the spinning disc. In variants, the disc itself may be hollow and act as a vapour chamber. Protensive quoted an example of toluene evaporation on a 1-m-diameter disc where 1.5 tonne/h is evaporated with a temperature driving force of 20°C. An interesting aspect of the enhancement is the presence of surface waves – perhaps worth examining in disc and other rotating heat pipe concepts – that allow film coefficients of 20–50 kW/m$^2$/K for low viscosity fluids (of the types used in heat pipes).

Most applications of axial rotating heat pipes have been for operation at modest rotational speeds, typically up to 3000 rpm. However, the demands for high-speed rotating motors, and systems where generators or motors may be directly connected to high-speed prime movers such as gas turbines, have spurred interest in higher rotational speeds, as high as 60,000 rpm.

Amongst the first reports on work at very high speeds was that of Ponnappan et al. [121] presented at the 10th International Heat Pipe Conference. The ultimate use was to be in a generator directly coupled to an aircraft gas turbine. With power densities of up to 11 kW/kg being predicted and rotor tip velocities of over 300 m/s, the machine efficiency was such that about 20% of the rated power capacity was dissipated as a thermal energy in the generator. In the specific unit for which the rotating heat pipe was investigated, the switched reluctance generator heat rate was over 1.9 kW, and the rotor temperature needed to be maintained below 388°C, when rotating at 40,000 rpm. The shaft size was 12.3 cm in length and 2.54 cm in diameter.

The heat pipe tested had a tapered condenser section, leading into an untapered adiabatic and condenser sections. Water was used as the working fluid with 316 stainless steel, and with a total heat pipe internal volume of 57.44 cm$^3$, the water fill was 0.0105 kg, representing 18.3% of the internal volume at ambient temperature. Design capacity was 2 kW, and the maximum operating temperature was 250°C.

The paper presents full performance data, and early data on modelling. However, with a heat input of 1.3 kW, at 30,000 rpm,[3] the steady-state temperature distribution with jet oil cooling of the condenser indicated that the temperature difference between extreme ends of the evaporator and condenser was 20°C at a vapour temperature of about 200°C. Conventional rotating heat pipe theory gave unsatisfactory agreement, but the authors point out, interestingly,

3. If directly coupled to the gas turbine, rotational speeds might reach 60,000 rpm.

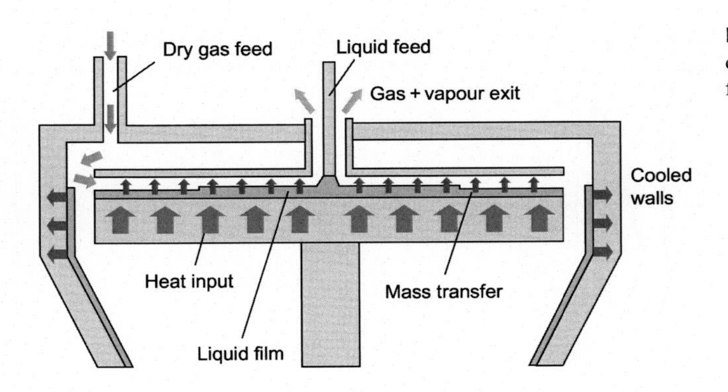


**FIGURE 2.52** The spinning disc reactor, with liquid introduced to the centre of the disc, and rotation driving it to the periphery, mirrors the fluid movement within disc heat pipes. *Courtesy: Protensive Ltd.*

that at centrifugal accelerations of 1000–9000 G at the inner wall, the impact of this on hydrodynamics and heat transfer is not known. Further data from Ponnappan and coworkers [122] detail experiments with water and the application again is in the important and upcoming area of more-electric (or all-electric) aircraft.

It was proposed by Song et al. [123] that the discrepancies between theory and experiment were due to the large fluid loading in the heat pipe, affecting film thickness in the condenser.

The R&D Centre at Hiroshima Machine Tool Works in Japan successfully operated a high-speed rotating heat pipe (24,000 rpm) for cooling the inner-race bearings on a spindle motor. The 44-mm outside diameter heat pipe was able to decrease the bearing temperature from 70°C to 45°C, thus avoiding a seizure problem [124].

### 2.7.3 Microrotating heat pipes

The ability to miniaturise systems, or to construct micro versions of equipment, is increasingly necessary as the demand for increased power out of ever smaller unit operations, be they static or dynamic, has to be met. As seen in Section 2.5 in this chapter, the heat pipe has had to develop in its micro form in common with many other thermal control technologies.

Modelling of the heat transfer in small rotating heat pipes has been carried out by Lin et al. [125]. In this analysis, the rotating unit had axial internal grooves of triangular cross-section running the length of the internal surface. The unit used ammonia working fluid in the simulation and had an internal diameter of 4 mm and a rotational speed of 3000 rpm. Interestingly, it was found that the influence of rotation on the heat transfer in the microregion of evaporation could be neglected.

Ling et al. [126] carried out experimental work on radially rotating miniature high-temperature heat pipes; the radial rotating form being, for example, suitable for cooling gas turbine blades. In such environments, as turbine inlet temperatures are consistently on an increasing trend, the use of such heat pipes becomes important if the blade life is to be maintained in increasingly hostile conditions. It is sometimes found that conventional air cooling methods are no longer sufficient to allow the blades to perform adequately.

The heat pipes tested had diameters of 2 and 1.5 mm and lengths of 80 mm. The construction material was s/s 304 W, and the working fluid was sodium. The operating temperature was up to 900°C. Heat transfer capabilities were up to 280 W in the 2 mm diameter unit, and slightly less in the smaller heat pipe. The test facility was capable of a maximum speed of 3600 rpm, quite modest in gas turbine terms. The work demonstrated the adverse effect of noncondensible gases in high-temperature rotating heat pipes (in common with its effect on other types), and the experimental data emphasised the need for compatibility and high-quality manufacturing procedures, should heat pipes 'fly' in such applications.

Cao [127] updates work reported over many years on miniature rotating heat pipes in thermal control of turbine blades and discs and suggests that lifetime is proven over a period of 10 years. Whether we shall see these used in commercial gas turbines is debatable as materials used for hot-end blades are developing in such a manner that turbine inlet temperatures not envisaged when this research started are now feasible. Finally, the 'radial rotating heat pipe', albeit with a rather low rotational speed, has been proposed for one application where centripetal forces are unlikely to contribute greatly to performance! Patent GB 2409716 [128] describes heat pipes used in an automotive steering wheel connecting the hub to the rim. The effect of these, depending upon the function required, is to heat or cool the rim to maintain a comfortable feel for the driver!

## 2.8 Miscellaneous types

### 2.8.1 The sorption heat pipe

As mentioned in the Introduction, there are instances where heat pipes alone may not satisfy the needs of high heat flux control. One approach is to combine the heat pipe with another heat and/or mass transfer phenomenon to improve the overall system capability. A device that combines heat pipes and sorption (in this case adsorption) is the SHP [129].

Proposed for terrestrial and space applications, the basic unit is illustrated in Fig. 2.53, and in more functional form in Fig. 2.54.

Solid adsorption cooling systems are well known in the air-conditioning field (competing, so far unsuccessfully, with the more familiar absorption cycle units), where ammonia gas and activated carbon are a typical gas–solid pair. In the system proposed here, it is therefore logical to select ammonia as the heat pipe working fluid as it is readily adsorbed and desorbed into and from the sorption structure shown in Fig. 2.54 which uses activated carbon as the sorbent bed.

The cycle is more complex than for a simple heat pipe and involves the following stages:

1. Desorption by applying heat to the sorption structure (2) that also has heaters embedded in it.
2. Condensation of the desorbed fluid (ammonia) in (3).
3. Excess ammonia liquid is filtered through the porous valve (5) into the evaporator (6).
4. Liquid in (3) that saturates the wick (4) returns via the wick to the bed (2), enhancing heat and mass transfer there.
5. When desorption is complete the heaters in (2) are turned off, the working fluid collects in (6) and the pressure in the bed (2) decreases.

In phase 2 (sorption cycles of this type are cyclic), the following processes occur:

1. Valve (5) is opened allowing the vapour pressure in the heat pipe to equalise.
2. During liquid evaporation the air in the cold box (8) cools down.
3. Valve (5) is closed once the sorbent bed is saturated with ammonia.
4. The bed (2) starts to cool down, aided by the heat pipe condenser (3).

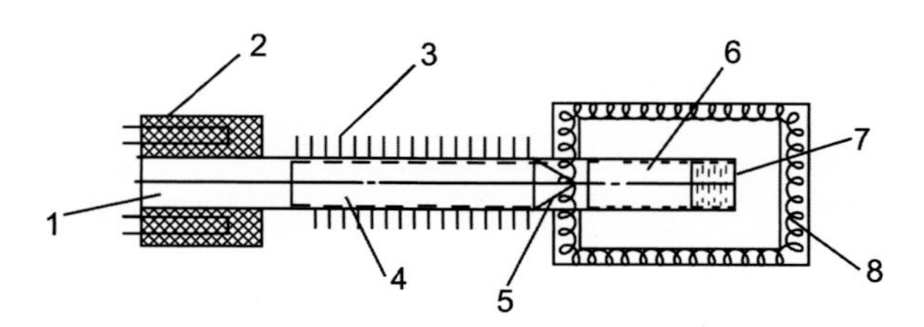


**FIGURE 2.53** Sorption heat pipe showing (1) vapour channel; (2) porous sorption structure; (3) finned heat pipe evaporator/condenser; (4) heat pipe wick; (5) porous valve; (6) heat pipe low-temperature evaporator (wicked); (7) working fluid accumulated in the evaporator; (8) cold box [129].

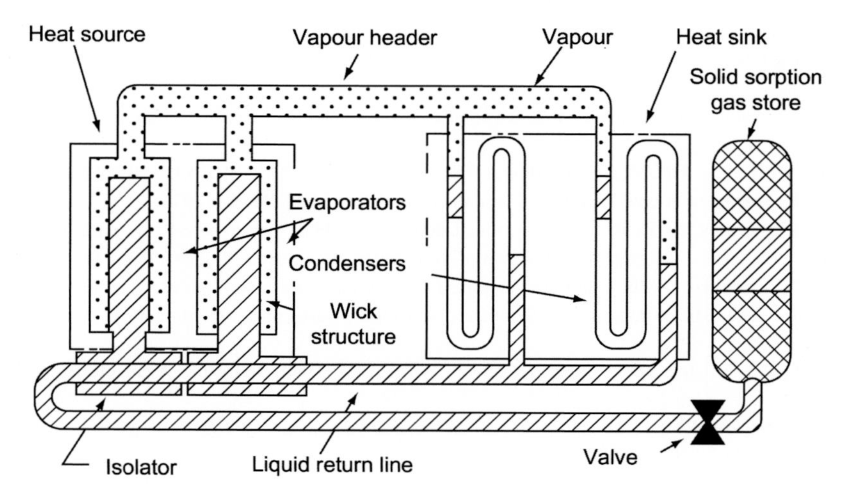


**FIGURE 2.54** Schematic of the sorption heat pipe [129].

It is claimed that the system is superior to an LHP with the same working fluid and evaporator dimensions and that the unit can be considered for sensor cooling for space uses. In particular, the integration of the sorption cooler with an LHP is seen as beneficial, and measurements with heat fluxes of 100–200 W/cm$^2$ give average evaporator thermal resistances of 0.07–0.08 K/W.

### 2.8.2 Magnetic fluid heat pipes

There has been interest for many years in the use of magnetic fluids in a range of engineering applications, extending also to heat transfer. The motion of magnetic fluids can be influenced by applied magnetic fields and in heat pipes, it has been proposed that this could enhance performance and possibly allow a new control mechanism.

Work in Japan by Jeyadevan et al. [130] using a citric ion-stabilised magnetic fluid showed that the application of a magnetic field in such a case could enhance heat transfer by up to 30%. Over a temperature range of typically 110°C–130°C, the heat transport capability was 10% greater than that of a comparable water heat pipe. No noncondensable gas was formed during the operation.

As described earlier in this chapter regarding a number of other types of heat pipe, the need for external energy inputs makes an essentially passive device an active one. Thus the reasons for employing magnetic fields need, in the opinion of the authors, to be strong before such a step is further investigated.

## References

[1] T. Wyatt. A controllable heat pipe experiment for the SE-4 satellite. JHU Tech. Memo APL-SDO-1134, John Hopkins University, Appl. Phys. Lab., March 1965, AD 695433.

[2] R. Kosson, R. Hembach, F. Edelstein, J. Loose. Development of a high capacity variable conductance heat pipe. 8th Thermophysics Conference, AIAA Paper 1973, p. 728.

[3] D.B. Marcus, G.L. Fleischman, Steady state and transient performance of hot reservoir gas controlled heat pipes, ASME Paper 70-HT/SpT-11, 1970.

[4] R Turner. The "constant temperature" heat pipe – A unique device for the thermal control of spacecraft components. AIAA with 4th Thermophysics Conference Paper, p. 69–632. June 1969 (RCA).

[5] J. Rogovin, B. Swerdling. Heat pipe applications to space vehicles. AIAA Paper 1971:71–421.

[6] W.B. Bienert, P.J. Brennan. Transient performance of electrical feedback-controlled variable-conductance heat pipes. LIFE SUPPORT AND ENVIRONMENTAL CONTROL CONFERENCE ASME Paper 71-Av-27, 1971.

[7] W.B. Bienert, et al. TECHNICAL SUMMARY REPORT FOR STUDY TO EVALUATE THE FEASIBILI!TY OF A FEEDBACK CONTROLLED VARIABLE CONDUCTANCE HEAT PIPE. Contract NAS2-5772, Tech. Summary Report DTM-70-4, Dynatherm. September 1970.

[8] W. Bienert, P.J. Brennan, T.P. Kirkpatrick. Feedback controlled variable conductance heat pipe. AIAA Paper 71-421, 1971.

[9] C.A. Depew, W.J. Sauerbrey, B.A. Benson. Construction and testing of a gas-loaded passive control variable conductance heat pipe. AIAA Paper 1973:73–727.

[10] P. Marcarino, A. Merlone, Gas-controlled heat-pipes for accurate temperature measurements, Appl Therm Eng 23 (2003) 1145–1152. Available from: https://doi.org/10.1016/S1359-4311(03)00045-0.

[11] P. Marcarino, A. Merlone, R. Dematteis. Determination of the mercury vapour P–T relation using a heat pipe. 8th International Symposium on Temperature and Thermal Measurements in Industry and Science, VDE Verlag, Berlin, pp. 1203-1208, 2002.

[12] K. Ohashi, H. Hayakawa, M. Yamada, T. Hayashi, T. Ishii, Preliminary study on the application of the heat pipe to the passive decay heat removal system of the modular HTR, Progress in Nuclear Energy 32 (1998) 587–594. Available from: https://doi.org/10.1016/S0149-1970(97)00047-4.

[13] K. Watanabe, A. Kimura, K. Kawabata, T. Yanagida, M. Yamauchi, Development of a variable-conductance heat-pipe for a sodium–sulphur (NAS) battery, Furukawa Rev (20)(2001) 71–76.

[14] J.P. Kirkpatrick. VARIABLE CONDUCTANCE HEAT PIPES – FROM THE LABORATORY TO SPACE VARIABLE CONDUCTANCE HEAT PIPES FROM THE LABORATORY TO SPACE. First International Heat Pipe Conference, Stuttgart, 15-17 October 1973.

[15] A. Chernenko, V. Kostenko, V. Loznikov, N. Semena, S. Konev, B. Rybkin, et al., Optimal cooling of HPGe spectrometers for space-born experiments, Nucl. Instrum. Methods Phys. Res. A 442 (2000) 404–407. Available from: https://doi.org/10.1016/S0168-9002(99)01262-0.

[16] A. Pashkin, I. Prokopenko., B. Rybkin, A. Chernenko, V. Kostenko, I. Mitrofanov. Development of cryogenic heat pipe diodes for Germanium gamma-spectrometer thermal control system. Proceedings of IV Minsk International Seminar: 'Heat Pipes, Heat Pumps, Refrigerators', Miksk, Belarus, 4-7 September 2000.

[17] C.I. Ezekwe, Performance of a heat pipe assisted night sky radiative cooler, Energy Convers. Manag. 30 (1990) 403–408. Available from: https://doi.org/10.1016/0196-8904(90)90041-V.

[18] S. Varga, A.C. Oliveira, C.F. Afonso, Characterisation of thermal diode panels for use in the cooling season in buildings, Energy Build 34 (2002) 227–235. Available from: https://doi.org/10.1016/S0378-7788(01)00090-1.

[19] A. Bahr, H, Piwecki. Passive solar heating with heat storage in the outside walls. European Commission Report, EUR 7077EN, JRC Ispra, Italy, 1981.

[20] T. Ochi, T. Ogushi, R. Aoki, Development of a heat pipe thermal diode and its heat transport performance, JSME Int J 39 (2) (1996) 419–425.
[21] P. Hurley. 2001. SYSTEM FOR TRANSFERRING THERMAL ENERGY. US Patent Application number WO 01/04549 A1. Applicant: Solar Dynamics Inc. Florida, USA.
[22] O. Brost, K.P. Schubert. Development of alkali-metal heat pipes as thermal switches. Micromechanical Sensors, Actuators and Systems, vol. 32, ASME DSC, pp. 123-124, 1991.
[23] B.N.F. Eddleston, K. Hecks. Application of heat pipes to the thermal control of advanced communications spacecraft. 1st International Heat Pipes Conference, Paper 9-4, Stuttgart, Oct 1973.
[24] D.A. Wolf. Flexible Heat Pipe Switch. Final Rep NASA Contract No NASS–255689, Mod 1981; 4.
[25] G.P. Peterson. Analytical development and computer modelling of a bellows type heat pipe for the cooling of electronic components. ASME Winter Annual Meeting Paper, 86-WA/HT-89, Ansheim, CA, 1986.
[26] F.C. Prenger, W.F. Stewart, J.E. Runyan. DEVELOPMENT OF A CRYOGENIC HEAT PIPE. Proceedings of the International Engineering Conference and International Cryogenic Materials Conference, Albuquerque, NM, 12-16 July 1993.
[27] H. Akachi. US Patent No. 5219020, 1993.
[28] H. Akachi. US Patent No. 5490558, 1996.
[29] Akachi. US Patent No. 4921041, 1990.
[30] F. Polasek, L. Rossi. Thermal control of electronic equipment and two-phase thermosyphons. 11th IHPC, 1999.
[31] P. Charoensawan, S. Khandekar, M. Groll, P. Terdtoon, Closed loop pulsating heat pipes: Part A: parametric experimental investigations, Appl. Therm. Eng 23 (2003) 2009–2020. Available from: https://doi.org/10.1016/S1359-4311(03)00159-5.
[32] M. Vogel, G. Xu, Low profile heat sink cooling technologies for next generation CPU thermal designs, Electronic Cooling Online 11 (1) (2005).
[33] S. Duminy. Experimental investigation of pulsating heat pipes (Diploma Thesis). Institute of Nuclear Engineering and Energy Systems (IKE), University of Stuttgart, Germany, 1998.
[34] S. Khandekar, P.K. Panigrahi, F. Lefèvre, J. Bonjour, Local hydrodynamics of flow in a pulsating heat pipe: a review, Frontiers in Heat Pipes 1 (2010) 023003.
[35] S. Khandekar, P. Charoensawan, M. Groll, P. Terdtoon, Closed loop pulsating heat pipes Part B: visualization and semi-empirical modeling, Appl. Therm. Eng. 23 (2003) 2021–2033. Available from: https://doi.org/10.1016/S1359-4311(03)00168-6.
[36] H. Akachi, PF and SP. Pulsating Heat Pipes. Proceedings of the Fifth International Heat Pipe Symposium, Melbourne, Australia, 1996. ISBN 0-08-042842-8, pp. 208-217.
[37] K. Cornwell, P.A. Kew. Boiling in small parallel channels. Proceedings of the CEC Conference on Energy Efficiency. Process Technology, Elsevier Applied Sciences, Paper 22, 1992: 624–38.
[38] S. Khandekar, N. Dollinger, M. Groll, Understanding operational regimes of closed loop pulsating heat pipes: an experimental study, Appl. Therm. Eng. 23 (2003) 707–719. Available from: https://doi.org/10.1016/S1359-4311(02)00237-5.
[39] X.M. Zhang, J.L. Xu, Z.Q. Zhou, Experimental study of a pulsating heat pipe using FC-72, ethanol, and water as working fluids| Request PDF Experimental Heat Transfer 17 (2004) 47–67.
[40] P. Sakulchangsatjatai, P. Terdtoon, T. Wongratanaphisan, P. Kamonpet, M. Murakami, Operation modeling of closed-end and closed-loop oscillating heat pipes at normal operating condition, Appl. Therm. Eng. 24 (2004) 995–1008. Available from: https://doi.org/10.1016/J.APPLTHERMALENG.2003.11.006.
[41] T. Katpradit, T. Wongratanaphisan, P. Terdtoon, P. Kamonpet, A. Polchai, A. Akbarzadeh, Correlation to predict heat transfer characteristics of a closed end oscillating heat pipe at critical state, Appl. Therm. Eng. 25 (2005) 2138–2151. Available from: https://doi.org/10.1016/J.APPLTHERMALENG.2005.01.009.
[42] H. Yang, S. Khandekar, M. Groll, Operational limit of closed loop pulsating heat pipes, Appl. Therm. Eng. 28 (2008) 49–59. Available from: https://doi.org/10.1016/J.APPLTHERMALENG.2007.01.033.
[43] Z.H. Liu, Y.Y. Li, A new frontier of nanofluid research – Application of nanofluids in heat pipes, Int. J. Heat Mass Transf. 55 (2012) 6786–6797. Available from: https://doi.org/10.1016/J.IJHEATMASSTRANSFER.2012.06.086.
[44] Y. Ji, H. Chen, Y.J. Kim, Q. Yu, X. Ma, H.B. Ma, Hydrophobic surface effect on heat transfer performance in an oscillating heat pipe, J. Heat Transfer 134 (2012) 1–4.
[45] G. Karimi, J.R. Culham. Review and assessment of pulsating heat pipe mechanism for high heat flux electronic cooling. Inter Society Conference on Thermal Phenomena, 2004.
[46] J. Clement, X. Wang, Experimental investigation of pulsating heat pipe performance with regard to fuel cell cooling application, Appl. Therm. Eng 50 (2) (2013) 68–74. Available from: https://doi.org/10.1016/J.APPLTHERMALENG.2012.06.017.
[47] F.J. Stenger. Experimental feasibility study of water-filled capillary-pumped heat-transfer loops - NASA Technical Reports Server (NTRS). OH: NASA TM X-1310, NASA, Washington, DC, 1966.
[48] Y.F. Gerasimov, Y.F. Maidanik, G.T. Shchegolev, G.A. Filippov, L.G. Starikov, V.M. Kiseev, et al., Low-temperature heat pipes with separate channels for vapor and liquid, Journal of Engineering Physics 28 (1975) 957–960. Available from: https://doi.org/10.1007/BF00867371.
[49] A. Delil. Research Issues on Two-Phase Loops for Space Applications | Semantic Scholar. National Aerospace Laboratory Report NLR-TP-2000-703, 2000.
[50] Y.F. Maydanik, Loop heat pipes, Appl. Therm. Eng. 25 (2005) 635–657. Available from: https://doi.org/10.1016/J.APPLTHERMALENG.2004.07.010.
[51] G. Anderson, Y. Eastman, D. Sarraf, J. Zuo. 2005. HYBRID LOOP COOLING OF HIGH POWERED DEVICES. US Patent application Number US 6,948,556 B1.

[52] M. Nikitkin, B. Cullimore. CPL and LHP Technologies: What are the Differences, What are the Similarities? SAE Technical Papers 981587, pp 400408, 1998. https://doi.org/10.4271/981587.

[53] I. Muraoka, F.M. Ramos, V.v Vlassov, Analysis of the operational characteristics and limits of a loop heat pipe with porous element in the condenser, Int. J. Heat Mass Transf. 44 (2001) 2287–2297. Available from: https://doi.org/10.1016/S0017-9310(00)00259-3.

[54] S. Launay, V. Sartre, J. Bonjour, Parametric analysis of loop heat pipe operation: a literature review, International Journal of Thermal Sciences 46 (2007) 621–636.

[55] A. Ambirajan, A.A. Adoni, J.S. Vaidya, A.A. Rajendran, D. Kumar, P. Dutta, Loop heat pipes: a review of fundamentals, operation, and design, Heat Transfer Engineering 33 (2012) 387–405.

[56] E. Bazzo, R.R. Riehl, Operation characteristics of a small-scale capillary pumped loop, Appl. Therm. Eng. 23 (2003) 687–705. Available from: https://doi.org/10.1016/S1359-4311(03)00017-6.

[57] M. Hamdan. et al. Loop heat pipe (LHP) development by utilizing coherent porous silicon (CPS) wicks. Inter Society Conference on Thermal Phenomena, IEEE, 2002.

[58] E. Pouzet, J.L. Joly, V. Platel, J.Y. Grandpeix, C. Butto, Dynamic response of a capillary pumped loop subjected to various heat load transients, Int. J. Heat Mass Transf. 47 (2004) 2293–2316. Available from: https://doi.org/10.1016/J.IJHEATMASSTRANSFER.2003.11.003.

[59] Y. Chen, M. Groll, R. Mertz, Y.F. Maydanik, S.v Vershinin, Steady-state and transient performance of a miniature loop heat pipeInternational Conference on Nanochannels, Microchannels, and Minichannels Int. J. Therm. Sci. 45 (2006) 1084–1090.

[60] R.R. Riehl. Comparing the behavior of a loop heat pipe with different elevations of the capillary evaporator. International Conference On Environmental Systems, Colorado Springs, Citeseer. SAE 2004.-01-2510.

[61] W. Deng, Z. Xie, Y. Tang, R. Zhou, Experimental investigation on anti-gravity loop heat pipe based on bubbling mode, Exp. Therm. Fluid Sci. 41 (2012) 4–11. Available from: https://doi.org/10.1016/J.EXPTHERMFLUSCI.2012.01.030.

[62] R. Khodabandeh, Heat transfer and pressure drop in a thermosyphon loop for cooling of electronic components, Int. J. Refrig. 28 (5) (2005) 725–734.

[63] R. Khodabandeh, Pressure drop in riser and evaporator in an advanced two-phase thermosyphon loop, International Journal of Refrigeration 28 (2005) 725–734. Available from: https://doi.org/10.1016/J.IJREFRIG.2004.12.003.

[64] R.W. Lockhart, R.C. Martinelli, Proposed correlations for isothermal two-phase two-component flow in pipes, Berkely. Chem. Eng. Prog. 45 (1949) (1949) 39–48.

[65] M.B. Bowers, I. Mudawar, Two-phase electronic cooling using mini-channel and micro-channel heat sinks: Part 2—flow rate and pressure drop constraints, Journal of Electronic Packaging, Transactions of the ASME 116 (1994) 298–305. Available from: https://doi.org/10.1115/1.2905701.

[66] R. Khodabandeh, Heat transfer in the evaporator of an advanced two-phase thermosyphon loop, International Journal of Refrigeration 28 (2005) 190–202. Available from: https://doi.org/10.1016/J.IJREFRIG.2004.10.006.

[67] V. Dube, A. Akbarzadeh, J. Andrews, The effects of non-condensable gases on the performance of loop thermosyphon heat exchangers, Appl Therm Eng 24 (2004) 2439–2451. Available from: https://doi.org/10.1016/J.APPLTHERMALENG.2004.02.013.

[68] V. Dube, The Development and Application of a Loop Thermosyphon Heat Exchanger for Industrial Waste Heat Recovery and Determination of the Influence of Non- Condensable Gases on its Performance (Ph.D. Thesis), - Vipin Dube - Google Books. Melbourne: RMIT University of Melbourne, Australia, 2003.

[69] G. Wang, D. Mishkinis, D. Nikanpour, Capillary heat loop technology: Space applications and recent Canadian activities, Appl. Therm. Eng. 28 (2008) 284–303. Available from: https://doi.org/10.1016/J.APPLTHERMALENG.2006.02.027.

[70] F.A.J. Thiele, P.D. Mier, Reay D.A., Heat transfer across the skin: the role of the resting sweat gland, in: Proceedings of the Congress on Thermography, Amsterdam, June 1974.

[71] F.A.J. Thiele, D.A. Reay, J.W.H. Mali, G.J. de Jongh, A possible contribution to heat transfer through human skin by the eccrine (atrichial) sweat gland, in: J. Jadassohn (Ed.), Handbuch Der Haut Und Geschlechtskrankheiten: Normale Und Pathologische Physiologic Der Haut III, Springer-Verlag, Berlin, 1981, pp. 123–203.

[72] S.B, Choi. Fluid flow and heat transfer in microtubes. Micromechanical Sensors, Actuators, and Systems, ASME DSC 1991:123–34.

[73] T.P, Cotter. Principles and prospects for micro heat pipes (Conference) | OSTI.GOV, Los Alamos: Proceedings of Fifth International Heat Pipe Conference, vol. 1, Tsukuba, pp. 328-335, 1984.

[74] G.P, Peterson. Investigation of miniature heat pipes, Final Report, Wright Patterson AFB, Contract F33615-86-C-2733, Task 9, 1988.

[75] L.L, Vasiliev. et al. Copper sintered powder wick structures of miniature heat pipes. VI Minsk International Seminar 'Heat Pipes, Heat Pumps, Refrigerators,' Minsk, Belarus: 12-15 September 2005.

[76] A. Itoh, F. Polasek. Development and application of micro heat pipes. Proceedings of 7th International Heat Pipe Conference, Minsk: 1990. Hemisphere, New York, NY, 1990.

[77] S. Launay, V. Sartre, M.B.H. Mantelli, K.V. de Paiva, M. Lallemand, Investigation of a wire plate micro heat pipe array, International Journal of Thermal Sciences 43 (2004) 499–507. Available from: https://doi.org/10.1016/J.IJTHERMALSCI.2003.10.006.

[78] B. Suman, P. Kumar, An analytical model for fluid flow and heat transfer in a micro-heat pipe of polygonal shape, Int. J. Heat Mass Transf. 48 (2005) 4498–4509. Available from: https://doi.org/10.1016/J.IJHEATMASSTRANSFER.2005.05.001.

[79] A. Aliakhnovich. et al. Investigation of a heat transfer device for electronics cooling. I Minsk Int. Seminar 'Heat Pipes, Heat Pumps, Refrigerators,' Minsk, Belarus: 12-15 September 2005.

[80] H. Tang, Y. Tang, Z. Wan, J. Li, L. Lu, Y. Li, K. Tang, Review of applications and developments of ultra-thin micro heat pipes for electronic cooling, Journal of Applied Energy 223 (2018) 383–400.

[81] X. Han, Y. Wang, Q. Liang, Investigation of the thermal performance of a novel flat heat pipe sink with multiple heat sources, International Communications in Heat and Mass Transfer 94 (2018) 71–76.

[82] B.P. Cahill, L.J. Heyderman, J. Gobrecht, A. Stemmer, Electro-osmotic pumping on application of phase-shifted signals to interdigitated electrodes, Sens. Actuators B Chem 110 (2005) 157–163. Available from: https://doi.org/10.1016/J.SNB.2005.01.006.

[83] H.Y. Wu, C.H. Liu, A novel electrokinetic micromixer, Sens. Actuators A Phys 118 (2005) 107–115. Available from: https://doi.org/10.1016/J.SNA.2004.06.032.

[84] L. Dresner, Electrokinetic phenomena in charged microcapillaries, Journal of Physical Chemistry 67 (1963) 1635–1641. Available from: https://doi.org/10.1021/J100802A015/ASSET/J100802A015.FP.PNG_V03.

[85] D. Burgreen, F.R. Nakache, Electrokinetic flow in ultrafine capillary slits, Journal of Physical Chemistry 68 (1964) 1084–1091. Available from: https://doi.org/10.1021/J100787A019/ASSET/J100787A019.FP.PNG_V03.

[86] M. Abu-Romia. Possible application of electro-osmotic flow pumping in heat pipes. 6th Thermophysics Conference. AIAA Paper 71423, 1971. https://doi.org/10.2514/6.1971-423.

[87] J.H. Cosgrove, J.K. Ferrell, A. Carnesale, Operating characteristics of capillarity-limited heat pipes, Journal of Nuclear Energy 21 (1967) 547–558.

[88] C.P. Minning. et al. Development of an osmotic heat pipe. AIAA Paper 78-442. Proceedings of III International Heat Pipe Conference, Palo Alto: AIAA Report CP784; 1978, p. 78–442.

[89] C.P. Minning, A. Basiulis, Application of osmotic heat pipes to thermal-electric power generation systemsProceedings IV International Heat Pipe Conference in: D.A. Reay (Ed.), Advances in Heat Pipe Technology, Pergamon, Oxford, 1981.

[90] T.B. Jones, Electrohydrodynamic heat pipes, Int. J. Heat Mass Transf. 16 (1973) 1045–1048. Available from: https://doi.org/10.1016/0017-9310(73)90043-4.

[91] J.E. Bryan, J. Seyed-Yagoobi, Heat transport enhancement of monogroove heat pipe with electrohydrodynamic pumping, J. Thermophysics Heat Trans. 11 (1997) 454–460. Available from: https://doi.org/10.2514/2.6261.

[92] S. Qian, H.H. Bau, Magneto-hydrodynamic Stirrer for Stationary and Moving Fluids, Sens Actuators B Chem 106 (2005) 859–870. Available from: https://doi.org/10.1016/j.snb.2004.07.011.

[93] T.B. Jones. Electro-hydrodynamic effects on minimum film boiling, Report PB-252 320, Colorado State University, 1976.

[94] R. Loehrke, R. Debs. Measurements of the performance of an electrohydrodynamic heat pipe. 10th Thermophysics Conference, AIAA Paper 75-659 1975, p. 659.

[95] R.I. Loehrke. and SDR. Flat plate electro-hydrodynamic heat pipe experiments. Proceedings of 2nd International Heat Pipe Conference, Bologna, ESA Report SP 112, 1976.

[96] T.B. Jones, M.P. Perry, Electrohydrodynamic heat pipe experiments, J. Appl. Phys. 45 (1974) 21–29. Available from: https://doi.org/10.1063/1.1663557.

[97] K. Kikuchi, Study of EHD heat pipe, Technocrat 10 (11) (1977) 3841.

[98] K. Kikuchi, et al., Large scale EHD heat pipe experimentsProceedings of Fourth International Heat Pipe Conference, London in: D.A. Reay (Ed.), Advanced in Heat Pipe Technology, Pergamon, Oxford, 1981.

[99] M.K. Bologa. and SIK. Electro-hydrodynamic heat pipes. Proceedings of 7th International Heat Pipe Conference, Minsk, May 1990. Hemisphere, New York, NY, 1991.

[100] L. Kui. The enhancing heat transfer of heat pipes by the electric field. Proceedings of 7th International Heat Pipe Conference, 1990. Hemisphere, New York, NY, 1991.

[101] Y. Feng, Y. Wang, D. Huang, J. Seyed-Yagoobi. Refrigerant flow controlled/driven with electrohydrodynamic conduction pump, in: Proceedings of Sixth Minsk International Seminar 'Heat Pipes, Heat Pumps, Refrigerators', Minsk, Belarus, 1215 September 2005.

[102] Z. Yu, K.P. Hallinan, R.A. Kashani, Temperature control of electrohydrodynamic micro heat pipes, Exp. Therm. Fluid Sci. 27 (2003) 867–875. Available from: https://doi.org/10.1016/S0894-1777(03)00059-1.

[103] N. Garnier, R.O. Grigoriev, M.F. Schatz, Optical Manipulation of Microscale Fluid Flow, Phys. Rev. Lett (2003) 91. Available from: https://doi.org/10.1103/PHYSREVLETT.91.054501.

[104] M.le Berre, Y. Chen, C. Crozatier, Z.L. Zhang, Electrocapillary force actuation of microfluidic elements, Microelectron Eng 7879 (2005) 93–99. Available from: https://doi.org/10.1016/J.MEE.2004.12.014.

[105] V.H. Gray. The rotating heat pipe-A wickless, hollow shaft for transferring high heat fluxes. AMERICAN SOCIETY OF MECHANICAL ENGINEERS AND AMERICAN INST. OF CHEMICAL ENGINEERS, HEAT TRANSFER CONFERENCE, AMERICAN SOCIETY OF MECHANICAL ENGINEERS. Paper No. 69-HT-19, 1969.

[106] V.H. Gray, P.J. Marto, A.W. Joslyn. Boiling heat transfer coefficients: interface behaviour and vapour quality in rotating boiler operation to 475 g. NASA TN D- 4136, March 1968.

[107] C.P. Costello. and AJM. Burnout fluxes in pool boiling at high accelerations. Mechanical Engineering Department, University of Washington, Washington, DC, 1960.

[108] L.J. Baliback. The operation of a rotating wickless heat pipe. (M. Sc. Thesis), United States Naval Postgraduate School, Monterey, CA, 1969.

[109] T.C.. Daniels, F.K Al-Jumaily. Theoretical and experimental analysis of a rotating wickless heat pipe. Proceedings of the 1st International Heat Pipe Conference, Stuttgart, 1973.

[110] F.K. Al-Jumaily. An investigation of the factors affecting the performance of a rotating heat pipe (Ph.D. Thesis). University of Wales, December 1973.

[111] P.J. Marto. Performance characteristics of rotating wickless heat pipes. Heat Pipes, Proceedings of Second International Heat Pipe Conference, Bologna, ESA Report SP 112, 1976.

[112] L.L. Vasiliev, V.V. Khrolenok. Centrifugal coaxial heat pipes. Proceedings of 2nd International Heat pipe Conference, Bologna 1969. ESA Report SP 112, 1976.

[113] R. Marto, H. Weigel, The development of economical rotating heat pipes, in: D.A. Reay (Ed.), Proceedings of Fourth International Heat Pipe Conference, Advances in Heat Pipe Technology, Pergamon Press, Oxford, 1981.

[114] F. Song, D. Ewing, C.Y. Ching, Experimental investigation on the heat transfer characteristics of axial rotating heat pipes, Int. J. Heat Mass Transf. 47 (2004) 4721–4731. Available from: https://doi.org/10.1016/J.IJHEATMASSTRANSFER.2004.06.001.

[115] G.P. Peterson, D. Win. A review of rotating and revolving heat pipes. ASME Paper 91-HT-24, New York: 1991.

[116] F. Polasek. Cooling of a.c. motor by heat pipes, Proceedings of First International Heat Pipe Conference, Stuttgart, October 1973.

[117] M. Groll, G. Kraus, H. Kreel, P. Zimmerman. Industrial applications of low temperature heat pipes. Proceedings of 1st International Heat Pipe Conference, Stuttgart: October 1973.

[118] M. Groll, H. Kraehling, W.D. Muenzel, Heat Pipes for Cooling of an Electric Motor, J. Energy 2 (1978) 363–367. Available from: https://doi.org/10.2514/3.62387 (United States).

[119] S. Maezawa, Y. Suzuki, A. Tsuchida, Heat transfer characteristics of disk-shaped rotating, wickless heat pipeProceedings of IV International Heat Pipe Conferencein: D.A. Reay (Ed.), Advances in heat pipe technology, Pergamon Press, Oxford, 1981. Available from: https://doi.org/10.1016/B978-0-08-027284-9.50068-1.

[120] Protensive Ltd n.d. http://www.protensive.co.uk/pages/technologies/category/categoryid = heat.

[121] R. Ponnappan, Q. He, J. Baker, J.G. Myers, J.E. Leland. High speed rotating heat pipe: analysis and test results. Proceedings of 10th International Heat Pipe Conference, Stuttgart, 21-25 September 1997.

[122] R. Ponnappan, Q. He, T.E. Leland, Test Results of Water and Methanol High-Speed Rotating Heat Pipes, J. Thermophys. Heat Transf. 12 (1998) 391–397. Available from: https://doi.org/10.2514/2.6350.

[123] F. Song, D. Ewing, C.Y. Ching, Fluid flow and heat transfer model for high-speed rotating heat pipes, Int. J. Heat Mass Transf. 46 (2003) 4393–4401. Available from: https://doi.org/10.1016/S0017-9310(03)00292-8.

[124] R. Hashimoto, H. Itani, K. Mizuta, K. Kura, Y. Takahashi, Heat Transport Performance of Rotating Heat Pipes Installed in High Speed Spindle, MITSUBISHI JUKO GIHO 32 (1995) 366–369.

[125] L. Lin, A. Faghri, Heat transfer in micro region of a rotating miniature heat pipe, Int .J. Heat Mass Transf. 42 (1999) 1363–1369. Available from: https://doi.org/10.1016/S0017-9310(98)00270-1.

[126] J. Ling, Y. Cao, Closed-form analytical solutions for radially rotating miniature high-temperature heat pipes including non-condensable gas effects, Int. J. Heat Mass Transf. 43 (2000) 3661–3671. Available from: https://doi.org/10.1016/S0017-9310(99)00339-7.

[127] Y. Cao, Miniature High-Temperature Rotating Heat Pipes and Their Applications in Gas Turbine Cooling, Frontiers in Heat Pipes (FHP) 1 (2010).

[128] GB Patent 2409716, Assigned to Autoliv Development (Sweden)., A steering wheel with heat pipes and a plastic thermally insulating hub, Published on 6 July 2005.

[129] L. Vasiliev, L. Vasiliev, Sorption heat pipe—a new thermal control device for space and ground application, Int. J. Heat Mass Transf. 48 (2005) 2464–2472. Available from: https://doi.org/10.1016/J.IJHEATMASSTRANSFER.2005.01.001.

[130] B. Jeyadevan, H. Koganezawa, K. Nakatsuka, Performance evaluation of citric ion-stabilized magnetic fluid heat pipe, J. Magn. Magn. Mater 289 (2005) 253–256. Available from: https://doi.org/10.1016/J.JMMM.2004.11.072.

Chapter 3

# Heat pipe materials, manufacturing and testing

This chapter will discuss the main components of a heat pipe and the materials used in its construction. Since the previous edition of *Heat Pipes* was written, the materials and components of heat pipes have remained essentially the same, with perhaps the exception of the addition of nanoparticles to working fluids and an interest in magnesium as a wall/wick material. Nevertheless, because of the time gap, life tests have now been extended for a further period and consequently some working fluids have lost their attractiveness. This may be dictated by health and safety considerations or by environmental pressures – for example, the use of chlorofluorocarbons is now banned, and in some European countries, hydrofluorocarbons (HFCs) are being phased out in favour of fluids that contribute less to global warming.

The temperature range affected by these trends is principally between $-50°C$ and $+100°C$, and this affects products in the areas of electronics thermal control, domestic and heat recovery areas, as well as (although less important in the context of global warming) spacecraft.

The issues of compatibility and the results of life tests on heat pipes and thermosyphons remain critical aspects of heat pipe design and manufacture. In particular, the generation of noncondensable gases that adversely affects the performance of heat pipes in either short-term or long-term applications must be taken particularly seriously in the emerging technology of microheat pipes and arrays of such units. It is also encouraging to note that the reporting of extended life tests on heat pipes (including loop thermosyphons) has not abated and data from the 16th International Heat Pipe Conference for sodium, naphthalene and a selection of high-temperature organic fluids were reported. Even $CO_2$ has been examined in its supercritical form as a working fluid.

An aspect of heat pipes that has always been of interest to researchers is the compatibility of water with steel (mild or stainless variants). The superior properties of water and the low cost of some steels, together with their strength, make the combination of water–steel an attractive if sometimes elusive proposition and substantial discussion on this combination is given later within this chapter.

A considerable quantity of data on heat pipe life tests was accumulated in the 1960s–1980s. To many, this was the most active period of heat pipe research and development, and heat pipes are now routine components for both terrestrial and space applications. It is therefore important to retain much of the early life test data (often now discarded as paper copies of reports from company libraries etc., are dispensed with). So fortunately, apart from examples where the fluids have been discarded due to environmental or other considerations (e.g. refrigerants such as R11 and R113), some historical data are retained within this 7th Edition of *Heat Pipes* – and rightly so because the majority of the fluids, materials and operating conditions remains the same so the life test data remain valid and should not just be of archival value to those entering the heat pipe field. These data, and some additional results, are given in Section 3.5.1.

Notwithstanding, and as explained within this chapter, once designs have been finalised the importance of life tests at each laboratory should not be underestimated; the three basic components of a heat pipe being:

1. the working fluid
2. the wick or capillary structure
3. the container.

Inevitably in the selection of a suitable combination of the above, a number of conflicting factors may arise, and the principal bases for the final selection are discussed below.

**Heat Pipes. DOI: https://doi.org/10.1016/B978-0-12-823464-8.00001-2**

## 3.1 The working fluid

The first consideration in the identification of a suitable working fluid is the operating vapour temperature range and a selection of fluids is shown in Table 3.1. Within the approximate temperature band, several possible working fluids may exist and a variety of characteristics must be examined to determine the most acceptable of these fluids for the application being considered. The prime requirements are as follows:

1. Compatibility with wick and wall materials
2. Good thermal stability
3. Wettability of wick and wall materials
4. Vapour pressures not too high or low over the operating temperature range
5. High latent heat
6. High thermal conductivity
7. Low liquid and vapour viscosities
8. High surface tension
9. Acceptable freezing or pour point.

**TABLE 3.1 Heat pipe working fluids.**

| Medium | Melting point (°C) | Boiling point at atmos. press. (°C) | Useful range (°C) |
|---|---|---|---|
| Helium | −271 | −261 | −271 to −269 |
| Nitrogen | −210 | −196 | −203 to −160 |
| Freon R410A | −155 | −48.5 | −100 to 35 |
| Freon R408A | NA | −44.4 | −82 to 48 |
| Freon R134a | −103.3 | −27 | −75 to 75 |
| Ammonia | −78 | −33 | −60 to 100 |
| R1234yf | −150.4 | −29.5 | −60 to 65 |
| Pentane | −130 | 28 | −20 to 120 |
| Acetone | −95 | 57 | 0 to 120 |
| Methanol | −98 | 64 | 10 to 130 |
| Flutec PP2[a] | −50 | 76 | 10 to 160 |
| Flutec PP9[a] | −70 | 160 | 0 to 225 |
| Ethanol | −112 | 78 | 0 to 130 |
| Heptane | −90 | 98 | 0 to 150 |
| R1336mzz | −90.5 | 33.15 | −20 to 160 |
| Water | 0 | 100 | 30 to 200 |
| Toluene | −95 | 110 | 50 to 200 |
| Thermex[b] | 12 | 257 | 150 to 350 |
| Mercury | −39 | 361 | 250 to 650 |
| Caesium | 29 | 670 | 450 to 900 |
| Potassium | 62 | 774 | 500 to 1000 |
| Sodium | 98 | 892 | 600 to 1200 |
| Lithium | 179 | 1340 | 1000 to 1800 |
| Silver | 960 | 2212 | 1800 to 2300 |

*Note:* (The useful operating temperature range is indicative only.) Full properties of most of the above are given in Appendix 1.
[a]*Included for cases where electrical insulation is a requirement.*
[b]*Also known as Dowtherm A, a eutectic mixture of diphenyl ether and diphenyl.*

The selection of the working fluid must also be based on thermodynamic considerations which are themselves concerned with the various limitations to heat flow occurring within the heat pipe. These are also discussed in Chapter 4 and are the viscous, sonic, capillary, entrainment and nucleate boiling limitations.

Many of the problems associated with long-life heat pipe operation are a direct consequence of material incompatibility. This involves all three components of the heat pipe and is fully discussed later. However, one aspect peculiar to the working fluid is the possibility of thermal degradation. With certain organic fluids it is necessary to keep the film temperature below a specific value to prevent the fluid from breaking down into different compounds. Hence, a good thermal stability is therefore a necessary feature of the working fluid over its likely operating temperature range.

The surface of a liquid behaves like a stretched skin except that the tension in the liquid surface is independent of surface area, and so over the entire surface area of a liquid, there is a pull due to the attraction of the molecules tending to prevent their escape. This surface tension varies with temperature and pressure although frequently the variation with pressure is relatively small. Additionally, the effective value of surface tension may be considerably altered by the accumulation of foreign matter at the liquid/vapour, liquid/liquid or solid surfaces. Prediction of surface tension is more fully discussed within Chapter 4 but in heat pipe design a high value of surface tension is desirable in order to enable the heat pipe to operate against gravity and to generate a high capillary driving force.

In addition to high surface tension, it is also necessary for the working fluid to wet the wick and container material, that is, the contact angle must be zero (or at least very small). Despite suggestions that additives can improve the performance of heat pipes – for example, by the addition of small amounts of long-chain alcohol to water heat pipes [1] – such a practice is not generally recommended. Those designing and assembling heat pipes for testing should therefore not be tempted to consider additives of which, it may be claimed, improve the 'wettability' of surfaces because in two-phase systems additives tend to get left behind when phase change occurs! (However, the addition of a fluid such as ethylene glycol to a water thermosyphon can be shown to have some benefits, not least as an 'antifreeze solution' – see later.)

The vapour pressure over the operating temperature range must be sufficiently high to avoid undue vapour velocities which tend to set up a large temperature gradient, entrain the refluxing condensate in the countercurrent flow, or cause flow instabilities associated with compressibility. Conversely, the pressure must not be too high because this will then necessitate a thick-walled container. A high latent heat of vapourisation is also desirable to transfer large amounts of heat with the minimum fluid flow and so help maintain a low-pressure drop within the heat pipe. In addition, the thermal conductivity of the working fluid should (preferably) also be high to minimise the radial temperature gradient and to reduce the possibility of nucleate boiling at the wick/wall interface.

In practice, the resistance to fluid flow will be minimised by choosing fluids with low values of vapour and liquid viscosity, and a convenient approach for quickly comparing working fluids is provided by comparing their Merit number. Introduced more fully in Chapter 4, this is defined as $\sigma_1 L\rho_1/\mu_1$ where $\sigma_1$ is the surface tension, $L$ the latent heat of vaporisation, $\rho_1$ the liquid density and $\mu_1$ the liquid viscosity. Fig. 3.1 gives the Merit number at the boiling point for working fluids, covering temperature ranges between 200 and 1750K, and compared with all organic fluids (such as acetone and alcohols), one obvious feature is the marked superiority of water with its high latent heat and surface tension. The final fluid selected is, of course, also based on cost, availability, compatibility and the other factors noted above. Graphical and tabulated data on Merit numbers for a range of common working fluids are presented in [2]. (Note this reference also presents thermosyphon figures of merit.)

A high Merit number is not the only criterion for the selection of the working fluid and other factors may, for a particular situation, be of greater importance. For example, on a cost basis, potassium might be chosen rather than caesium or rubidium which are one hundred times more expensive. It is also worth noting that although over the temperature range 1200–1800K lithium has a higher Merit number than most metals (including sodium), its use requires a container made from an expensive lithium-resistant alloy whereas sodium can be contained using stainless steel. It may therefore be cheaper and more convenient to accept the lower performance of a heat pipe made from sodium/stainless steel.

Practical working fluids used in heat pipes range from helium at 4K up to lithium at 2300K, and Fig. 3.1 shows the superiority of water over the range 350–500K where the alternative organic fluids tend to have considerably lower Merit numbers. At slightly lower temperatures, 270–350K, ammonia is a desirable fluid although it requires careful handling to retain high purity, and acetone and the alcohols are alternatives but with lower vapour pressures. These fluids are commonly used in heat pipes for space applications. Water and methanol, both being compatible with copper, are often used for cooling electronic equipment.

Where HFCs are acceptable, R134A and R407C have been investigated as working fluids for heat pipes/thermosyphons in the context of solar collectors [3]. In a comparison with a third fluid, R22 (now being phased out), it was

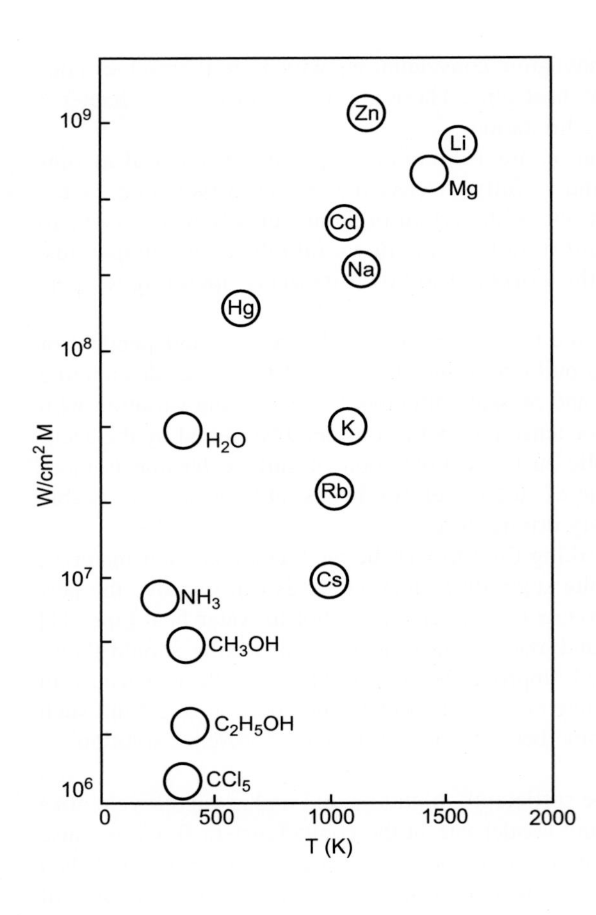


**FIGURE 3.1** Merit number for selected working fluids for their boiling point. *Courtesy: Philips Technical Review.*

found that R407C gave a superior performance to the other two fluids, but it is still unclear why water was not selected. Some of the 'new' HFCs have boiling ranges, and care should be exercised in their use for heat pipes (see also Chapter 6 where renewable energy uses of heat pipes are discussed).

Another alternative to CFCs is PFCs (perfluorocarbons). These generically have the formula $F_nC_{2n+2}$ and are made by a number of companies (see Flutec PP fluids in Table 3.1 as examples). Their merit is that they are dielectric fluids and can be used for direct cooling of electronics systems (as well as other uses). Although they have not received a great deal of attention for heat pipe working fluids – they were promoted as relatively benign alternatives to chlorofluorocarbons many years ago as they do not suffer from having a high global warming potential (GWP) – recent work has looked at their performance in small thermosyphons, with the possibility for their use in electronics cooling [4]. The 3 M products Fluorinert FC-84, FC-77 and FC-328 were tested in 6 mm internal diameter thermosyphons, but unless the dielectric properties were essential, water would be the fluid of choice – dictated principally by the abysmally low latent heats of evaporation of PFCs in general.

Work by Robert Dobson in South Africa [5] introduced the use of supercritical $CO_2$ as a working fluid, albeit in closed-loop thermosyphons. The particular paper discussing its use presents a methodology for evaluation of the transient performance of such systems and does not contain experimental verification, but the analysis indicates that using $CO_2$ as the working fluid in a natural circulation heat transfer loop would result in 95% more heat transfer than in a similar water-based loop.

For temperatures over 500K and up to 650K, high-temperature organic heat transfer fluids, normally offered for single-phase duties, may be used. These are available from companies such as The Dow Chemical Company and are eutectics of diphenyl and diphenyl oxide. The boiling point is around 260°C at atmospheric pressure. Unfortunately, they have low surface tension and a poor latent heat of vapourisation. Plus, as with many other organic compounds, diphenyls are readily broken down when film temperatures exceed a critical value. However, unlike many other fluids having similar operating temperature ranges, these eutectic mixtures have a specific boiling point, rather than a boiling range. Other fluids such as silicons are being studied for use at above 600K.

**TABLE 3.2 Compatibility tests with organic working fluids.**

| Test duration (year) | Working fluids | Structural materials (operating temperatures) | | |
|---|---|---|---|---|
| | | Compatible | Fairly compatible | Incompatible |
| 4.5–5 | *N*-Octane | • ST35 (230°C) | • X10CrNiT189<br>• (200°C, 250°C) | |
| | Diphenyl | | • ST35 (270°C)<br>• X10CRNITI189 (300°C) | • ST35 (300°C)<br>• X10CRNITI189 (350°C) |
| | Diphenyl Oxide | • ST35 (220°C) | | |
| 3 | Toluene | • ST35 (250°C)<br>• 13CRMO44 (250°C)<br>• X2CRNIMO1812 (280°C) | | |
| | Naphthalene | • ST35 (270°C) | • 13CRMO44 (270°C) | |
| 1 | Diphenyl | • Ti99.4 (270°C) | | • CUNI10FE (250°C) |
| | Diphenyl | • 13CRMO44 (250°C)<br>• X2CRNIMO1812 (250°C) | | • 13CRMO44 (400°C)<br>• X2CRNIMO1812 (400°C) |
| | QM | | | • 13CRMO44 (320°C,400°C)<br>• X2CRNIMO1812 (350°C, 400°C) |
| | OMD | | | • 13CRMO44 (350°C, 400°C)<br>• X2CRNIMO1812(350°C, 400°C) |
| | Toluene | • Ti99.4 (250°C)<br>• CUNI10FE (280°C) | | |
| | Naphthalene | • X2CRNIMO1812 (320°C)<br>• Ti99.4 (300°C) | • CUNI10FE (320°C) | |

One of the most comprehensive sets of compatibility tests was that carried out at Institut für Kernenergetik und Energiesysteme (IKE), Stuttgart [6,7], on a number of organic working fluids. These concentrated on thermosyphons made using a range of boiler and austenitic steels and the data obtained are summarised in Table 3.2. It can be seen from the test results that organic fluids operating at temperatures well in excess of 300°C tend to be unsuitable for long-term use in heat pipes.

It was pointed out by the research workers [8] that diphenyl and naphthalene, the two working fluids most suitable for operation in this temperature regime, can both suffer from decomposition caused by overheating at the evaporator section. This leads to the generation of noncondensable gas that can be vented via, for example, a valve.

The use of naphthalene was also reported in thermosyphons by Chinese researchers [9]. Vapour temperatures in excess of 250°C were achieved in the experiments but no degradation/compatibility data were given. Although not covering compatibility, the use of naphthalene in a stainless steel loop thermosyphon was reported by academics in Brazil and although the overall system performance had limitations, these were not influenced by incompatibilities [10].

More recently [11], the temperature range 400–700K has received attention from workers at the NASA Glenn Research Centre, as a result of which there is a suggestion that metallic halides might be used as working fluids within this temperature range. The halides are typically compounds of lithium, sodium, potassium, rubidium and copper, with fluorine, iodine, bromine and iodine [12]. The suggestion (in much of the literature) is that these will be reactive, and it should be pointed out that mixtures of iodine and sulphur were investigated as heat pipe fluids by the UK Atomic Energy Authority, some 25–30 years ago.

The NASA Glenn Research Centre has, together with a heat pipe manufacturer – Advanced Cooling Technologies, Inc. (see Appendix 4) – updated, in the context of both conventional and loop heat pipes, its life test report in 2011 [13], labelling in this case, the *intermediate* temperature range as 450–750K. Water was shown to be compatible with titanium and monel for temperatures up to 550K (277°C), based upon life tests over 54,000 h. The work on halides mentioned earlier has been continued, and it was concluded that long-term tests (at the time of reporting) had reached 50,000 h and exhibited compatibility between titanium and TiBr at 653K. Additionally, $AlBr_3$ with Hastelloy C-22 was deemed 'viable' at 673K, but with some slight corrosion occurring. As an alternative, titanium with $TiBr_4$ was proposed based on the same life test period. It should be noted that the interest in these latter nonorganic fluids is prompted by the potential use of heat pipes within nuclear reactors where radioactivity can adversely affect organic fluids.

At higher temperatures, we now enter the regime of liquid metals. Mercury has a useful operating temperature range of about 500–950K and has attractive thermodynamic properties. It is also liquid at room temperature which facilitates handling, filling and the start-up of a heat pipe.

Apart from its toxicity the main drawback to the use of mercury as a working fluid in heat pipes, as opposed to thermal syphons, is the difficulty encountered in wetting the wick and wall of the container. There are few papers specifically devoted to this topic, but Deverall [14] at Los Alamos and Reay [15] have both reported work on mercury wetting.

Japanese work on mercury heat pipes using type 316 L stainless steel as the container material showed that good thermal performance could be achieved once full start-up had been achieved, but compatibility of materials proved a problem owing to corrosion [16]. Such problems were not reported in the series of experiments, including a 1-month 'life test' by Macarino and colleagues at Istituto di Metrologia "G Colonnetti" (IMGC) in Turin, Italy, who used a unit supplied by the Joint Research Centre, Ispra. This was one of a series of gas-controlled, stainless steel heat pipes for accurate temperature measurements and operated at a vapour temperature of around 350°C [17].

Bienert [18], in proposing mercury/stainless steel heat pipes for solar energy concentrators, used Deverall's technique for wetting the wick in the evaporator section of the heat pipe and achieved sufficient wetting for gravity-assisted operation. He argued that nonwetting in the condenser region of the heat pipe should enhance dropwise condensation which would result in higher film coefficients than those obtainable with film condensation. In addition, work at Los Alamos has suggested that magnesium can be used to promote mercury wetting [19]. Moving even higher up the vapour temperature range, caesium, potassium and sodium are acceptable working fluids and their properties relevant to heat pipes are well documented (see Appendix 1). Above 1400K, lithium is generally the first choice as a working fluid but silver has also been used [20]. Working on applications of liquid-metal heat pipes for nuclear and space-related uses, Tournier and others at the University of New Mexico [21] suggest that lithium is the best choice of working fluid at temperatures above 1200K. For those interested in the start-up of liquid metal, in particular lithium heat pipes, the research at the University of New Mexico is well worth studying.

### 3.1.1 Nanofluids

Introduced in this chapter: The literature on nanofluids (fluids to which nanoparticles have been added) is growing at a rapid rate, but this is not a new phenomenon as far as heat pipes are concerned (see e.g. Tsai et al. in 2004 [22]). Heat pipes and thermosyphons have not escaped the attention of this trend, but a number of aspects must be considered when addressing the use of such particles, including, but not limited to, the following:

1. What is their role?
2. What size/shape of nanoparticle is needed and what is available?
3. How much do they cost (some have used gold particles)?
4. Will they adversely affect any aspect of heat pipe performance?
5. Are they compatible with the working fluid?
6. Can they be recovered safely at the end of the heat pipe life?
7. Is there adequate data from earlier tests on nanoparticles in heat pipes?

The usual reason for including such particles in a liquid is to enhance heat transfer, most commonly by increasing the liquid thermal conductivity or by reducing the size of nucleation bubbles/increasing their number.

This seemed to be the outcome of the work reported in [22]. Gold nanoparticles of different sizes were prepared by reduction of HAuCl4 with solutions of tetrachloroaurate, trisodium citrate and tannic acid. In the present study, the size of the gold nanoparticle was adjusted by changing the amounts of tetrachloroaurate, trisodium citrate and tannic acid, and the heat pipe using the nanofluid had a length of 170 mm and an outer diameter of 6 mm. The thermal resistance of the heat pipes ranged from 0.17 to 0.215°C/W with different nanoparticle solutions. These values indicated that the thermal resistance of the heat pipes with a nanoparticle solution was up to 37% lower than that with water alone. The authors concluded that '*as a result, the higher thermal performances of the new coolant (really the working fluid!) have proved its potential as a substitute for conventional water in a vertical circular meshed heat pipe*'.

As one would expect, whilst most of the improvement in thermal resistance took place in the evaporator – and here the authors attributed the improvement largely to the reduced size of nucleation bubbles – the authors also claimed small reductions in the condenser. This latter improvement was attributed to improved liquid thermal conductivity, but one must query how solid particles reached the condenser by way of the vapour space, unless the entrainment limit was exceeded.

Nanofluids have been proposed for use in heat pipes cooling Pentium chips [23] where aluminium oxide – a very common nanoparticle – was investigated together with titanium dioxide, and several fluids were tested using nanoparticle concentrations of 1%–5% by volume of the fluid. It was found that the screen mesh was coated by the particles and it was proposed that this promoted a good capillary structure for wicking.

This was born out from the research by Brusly Solomon et al. [24] in a comprehensive assessment of the performance of wicked heat pipes using mesh already coated with nanoparticles. A 40% thermal resistance reduction at the evaporator was noted, and the results of wick coating are illustrated in Fig. 3.2.

Leong and colleagues in Malaysia investigated nanofluids in a thermosyphon air preheater [25]. It was found, as shown in Fig. 3.3, that the impact on the performance of the thermosyphon was influenced much more by the air velocity across the thermosyphons than the amount of nanoparticles included in the fluid. Results were similar in trend for both alumina and titanium dioxide nanoparticles, from which Leong et al. therefore suggested that the role of enhanced liquid thermal conductivity due to nanoparticles was minor.

## 3.2 The wick or capillary structure

The selection of the wick for a heat pipe depends on many factors, several of which are closely linked to the properties of the working fluid. Obviously, the prime purpose of the wick is to generate capillary pressure to transport the working fluid from the condenser to the evaporator, but it must also be able to distribute the liquid around the evaporator section to any areas where heat is likely to be received by the heat pipe. Often these two functions require wicks of different forms, particularly where the condensate has to be returned over a distance of, say, 1 m, in zero gravity. But where a wick is retained in a 'gravity-assisted' heat pipe, its roles may be to enhance heat transfer and to circumferentially distribute liquid.

Chapter 4 confirms that the maximum capillary head generated by a wick increases with decreasing pore size, but wick permeability (another desirable feature) increases with increasing pore size. So, for homogeneous wicks, there will be an optimum pore size and a necessary compromise and there are three main types in this context. Low-performance wicks in horizontal and gravity-assisted heat pipes should permit maximum liquid flow rate by having a comparatively large pore size, for example with 100 or 150 mesh, but where pumping capability is required against gravity then small pores are needed. Whereas in space, the constraints on size and the needed general high-power capability necessitate the use of nonhomogeneous or arterial wicks aided by small pore structures for axial liquid flow. Thickness is another feature of the wick which must be optimised as the heat transport capability of a heat pipe is raised by increasing the wick thickness. Unfortunately, this also increases its radial thermal resistance which would work against this increased capability and so lower the allowable maximum evaporator heat flux. Additionally, the overall thermal resistance at the evaporator also depends on the conductivity of the working fluid in the wick.

Table 3.3 gives measured values of evaporator heat fluxes for various wick/working fluid combinations. Other necessary properties of the wick are (1) compatibility with the working fluid and its wettability; (2) it should be easily formed to mould into the wall shape of the heat pipe and should preferably be of a form that enables repeatable performance to be obtained; (3) it should be cheap.

A guide to the relative cost of wicks may be given by noting that mass-produced heat pipes for electronics applications use sintered copper powder or woven wire mesh, but where aluminium gravity-assisted units can be used (as in some solar collectors), extruded grooves are employed. Microheat pipes may also not need a separate 'wick' as the corners of the pipe (where noncircular) may generate the necessary capillary action.

### 3.2.1 Homogeneous structures

Of the wick forms available, meshes and twills are the most common and these are manufactured in a range of pore sizes and materials – the latter including stainless steel, nickel, copper and aluminium. Table 3.4 shows measured values of pore size and permeabilities for a variety of meshes and twills. Homogeneous wicks fabricated using metal foams, and more particularly felts, are becoming increasingly useful, and by modifying the pressure on the felt during

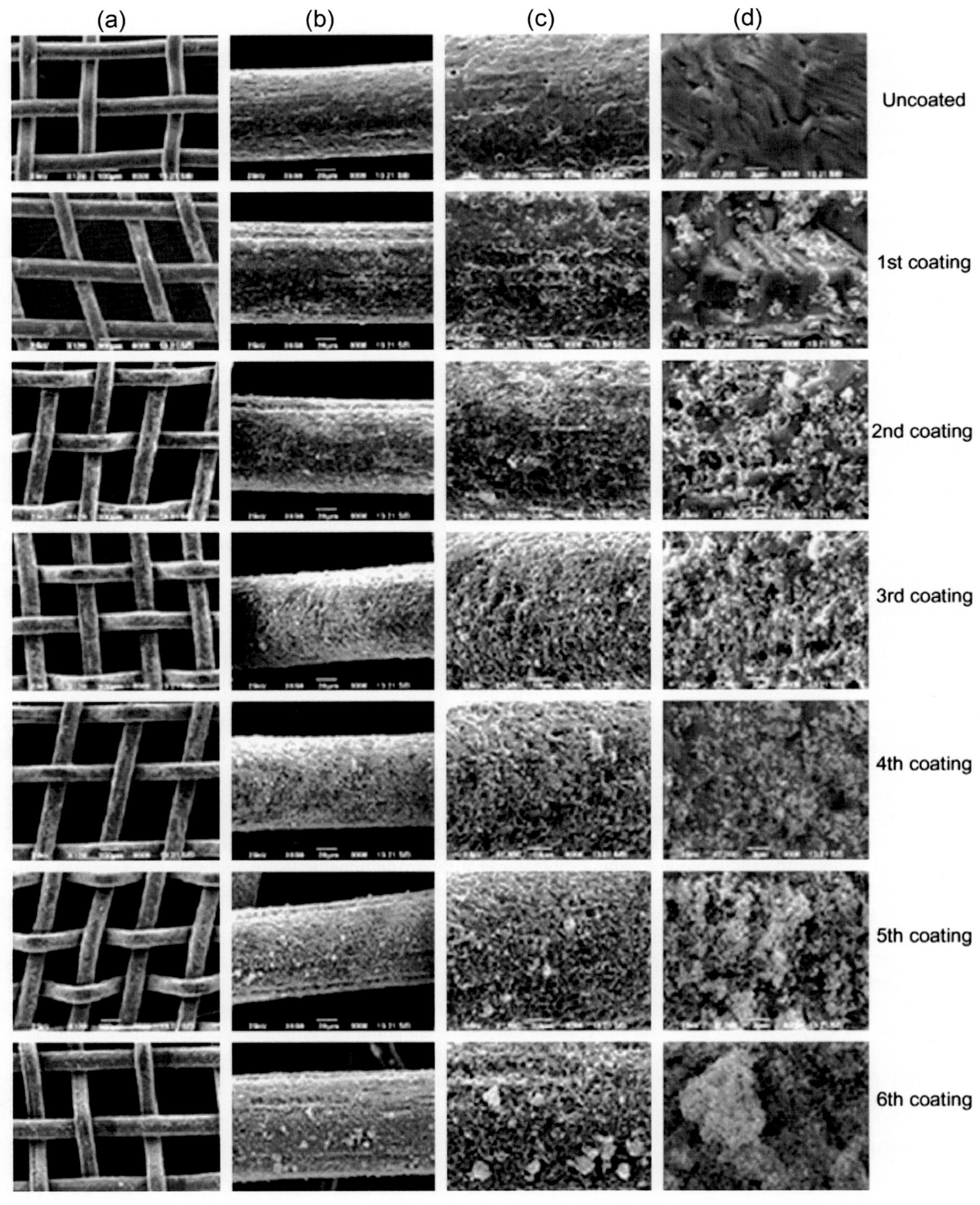


**FIGURE 3.2** Depositions of nanoparticles on the wick surface at different magnifications (a) 120 ×, (b) 550 ×, (c) 1500 × and (d) 7000 × [24].

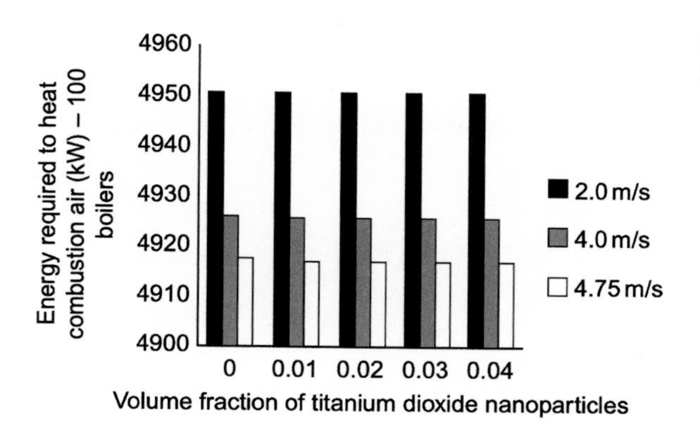


**FIGURE 3.3** Effect of titanium dioxide nanoparticles' volume fraction and hot air velocity to energy performance of thermosyphon heat exchanger [25].

**TABLE 3.3** Measured radial evaporator heat fluxes in heat pipes (these are not necessarily limiting values).

| Working fluid | Wick | Vapour temp. (°C) | Rad. flux (W/cm$^2$) |
|---|---|---|---|
| Helium [26] | s/s mesh | −269 | 0.09 |
| Nitrogen [26] | s/s mesh | −163 | 1.0 |
| Ammonia [27] | Various | 20–40 | 5–15 |
| Ethanol [28] | 4 × 100 mesh s/s | 90 | 1.1 |
| Methanol [29] | Nickel foam | 25–30 | 0.03–0.4 |
| Methanol [29] | Nickel foam | 30 | 0.24–2.6 |
| Methanol [29] | 1 × 200 mesh (horiz.) | 25 | 0.09 |
| Methanol [29] | 1 × 200 mesh (−2.5 cm head) | 25 | 0.03 |
| Water [27] | Various | 140–180 | 25–100 |
| Water [28] | Mesh | 90 | 6.3 |
| Water [28] | 100 mesh s/s | 90 | 4.5 |
| Water [29] | Nickel felt | 90 | 6.5 |
| Water [30] | Sintered copper | 60 | 8.2 |
| Mercury [26] | s/s mesh | 360 | 180 |
| Potassium [26] | s/s mesh | 750 | 180 |
| Potassium [27] | Various | 700–750 | 150–250 |
| Sodium [26] | s/s mesh | 760 | 230 |
| Sodium [27] | Various | 850–950 | 200–400 |
| Sodium [31] | 3 × 65 mesh s/s | 925 | 214 |
| Sodium [32] | 508 × 3600 mesh s/s twill | 775 | 1250 |
| Lithium [26] | Niobium 1% zirconium | 1250 | 205 |
| Lithium [33] | Niobium 1% zirconium | 1500 | 115 |
| Lithium [27] | SGS-tantalum | 1600 | 120 |
| Lithium [34] | W-26 Re grooves | 1600 | 120 |
| Lithium [34] | W-26 Re grooves | 1700 | 120 |
| Silver [26] | Tantalum 5% tungsten | – | 410 |
| Silver [34] | W-26 Re grooves | 2000 | 155 |

**TABLE 3.4 Wick pore size and permeability data.**

| Material and mesh size | Capillary height[a] (cm) | Pore radius (cm) | Permeability ($m^2$) | Porosity (%) |
|---|---|---|---|---|
| Glass fibre [35] | 25.4 | – | $0.061 \times 10^{-11}$ | – |
| Refrasil | | | | |
| Sleeving [35] | 22.0 | – | $0.104 \times 10^{-10}$ | – |
| Refrasil (bulk) [36] | – | – | $0.18 \times 10^{-10}$ | – |
| Refrasil (batt) [36] | – | – | $1.00 \times 10^{-10}$ | – |
| Monel beads [37] | | | | |
| 30–40 | 14.6 | 0.052[b] | $4.15 \times 10^{-10}$ | 40 |
| 70–80 | 39.5 | 0.019[b] | $0.78 \times 10^{-10}$ | 40 |
| 100–140 | 64.6 | 0.013[b] | $0.33 \times 10^{-10}$ | 40 |
| 140–200 | 75.0 | 0.009 | $0.11 \times 10^{-10}$ | 40 |
| Felt metal [38] | | | | |
| FM1006 | 10.0 | 0.004 | $1.55 \times 10^{-10}$ | – |
| FM1205 | – | 0.008 | $2.54 \times 10^{-10}$ | – |
| Nickel powder [35] | | | | |
| 200 *n* | 24.6 | 0.038 | $0.027 \times 10^{-10}$ | – |
| 500/a | >40.0 | 0.004 | $0.081 \times 10^{-11}$ | – |
| Nickel fibre [35] | | | | |
| 0.01 mm dia. | >40.0 | 0.001 | $0.015 \times 10^{-11}$ | 68.9 |
| Nickel felt [39] | – | 0.017 | $6.0 \times 10^{-10}$ | 89 |
| Nickel foam [39] | | | | |
| Ampornik 220.5 | – | 0.023 | $3.8 \times 10^{-9}$ | 96 |
| Copper foam [39] | | | | |
| Amporcop 220.5 | – | 0.021 | $1.9 \times 10^{-9}$ | 91 |
| Copper powder | | | | |
| (sintered) [38] | 156.8 | 0.0009 | $1.74 \times 10^{-12}$ | 52 |
| (sintered) [40] | | | | |
| 45 – 56/*x* | – | 0.0009 | – | 28.7 |
| 100 – 145 *ix* | – | 0.0021 | – | 30.5 |
| 150 – 200 *ix* | – | 0.0037 | – | 35 |
| Nickel 50 [35] | 4.8 | – | – | 62.5 |
| 50 [41] | – | 0.0305 | $6.635 \times 10^{-10}$ | – |
| Copper 60 [38] | 3.0 | – | $8.4 \times 10^{-10}$ | – |
| Nickel 60 [40] | – | 0.009 | – | – |
| 100 [41] | – | 0.0131 | $1.523 \times 10^{-10}$ | – |
| 100 [42] | – | – | $2.48 \times 10^{-10}$ | – |
| 120 [38] | 5.4 | – | $6.00 \times 10^{-10}$ | – |
| 120[c] [38] | 7.9 | 0.019 | $3.50 \times 10^{-10}$ | – |
| 2[d] × 120 [43] | – | – | $1.35 \times 10^{-10}$ | – |
| 120 [44] | – | – | $1.35 \times 10^{-10}$ | – |
| S/s 180 (22°C) [45] | 8.0 | – | $0.5 \times 10^{-10}$ | – |

*(Continued)*

**TABLE 3.4 (Continued)**

| Material and mesh size | Capillary height[a] (cm) | Pore radius (cm) | Permeability ($m^2$) | Porosity (%) |
|---|---|---|---|---|
| 2x180 (22°C) [45] | 9.0 | – | $0.65 \times 10^{-10}$ | – |
| 200 [40] | – | 0.0061 | $0.771 \times 10^{-10}$ | – |
| 200 [38] | – | – | $0.520 \times 10^{-10}$ | – |
| Nickel 200 [35] | 23.4 | 0.004 | $0.62 \times 10^{-10}$ | 68.9 |
| 2 × 200 [43] | – | – | $0.81 \times 10^{-10}$ | – |
| Phosp./bronze | – | 0.003 | $0.46 \times 10^{-10}$ | 67 |
| 200 [46] | | | | |
| Titanium 2 × 200 [40] | – | 0.0015 | – | 67 |
| 4 × 200 [40] | – | 0.0015 | – | 68.4 |
| 250 [42] | – | – | $0.302 \times 10^{-10}$ | – |
| Nickel[c] 2 × 250 [40] | – | 0.002 | – | 66.4 |
| 4 × 250 [40] | – | 0.002 | – | 66.5 |
| 325 [40] | – | 0.0032 | – | – |
| Phosp/bronze [44] | – | 0.0021 | $0.296 \times 10^{-10}$ | 67 |
| S/s (twill) 80[e] [47] | – | 0.013 | $2.57 \times 10^{-10}$ | – |
| 90[e] [47] | – | 0.011 | $1.28 \times 10^{-10}$ | – |
| 120[e] [47] | – | 0.008 | $0.79 \times 10^{-10}$ | – |
| 250 [43] | – | 0.0051 | – | – |
| 270 [43] | – | 0.0041 | – | – |
| 400 [43] | – | 0.0029 | – | – |
| 450 [47] | – | 0.0029 | | |

[a]*Obtained with water unless stated otherwise.*
[b]*Particle diameter.*
[c]*Oxidised.*
[d]*Number of layers.*
[e]*Permeability measured in direction of warp.*

assembly varying pore sizes can be produced. Additionally, by incorporating removable metal mandrels, an arterial structure can also be moulded in the felt.

Foams (see http://www.Porvair.com for data on a variety of foams) and felts are growing in importance for a variety of heat transfer duties because (like foams), as well as incorporating arteries in felts, their structures can be 'graded' and so allow the structure to be designed for 'local' conditions that may differ from those in other parts of a heat pipe wick (e.g. radially or axially). The research at Auburn University in the United States [48] was directed at examining whether the heat transfer limits of heat pipes could be increased by gradation of the pore structure. In particular, they investigated whether variables could include porosity, fibre diameter, pore diameter and pore diameter distribution. The layouts selected by the Auburn research team – which resemble composite mesh wicks in concept – were unsuccessful due to vapour being unable to escape into the vapour space, however the concept, with different variants, has merit.

Fibrous materials have been widely used in heat pipes and generally have small pore sizes. The main disadvantage is that ceramic fibres have little stiffness and usually require a continuous support for example by a metal mesh. Hence, whilst the fibre itself may be chemically compatible with the working fluids the supporting materials may cause problems. Semena and Nishchik discuss metallic fibre (and sintered) wick properties in reference [49] and recently Kostornov et al. [50] concentrating upon evaporator behaviour studied the performance of wicks made from copper and stainless steel fibres. In this work, capillary structure porosities were examined over a wide range (40%–90%), with wick thicknesses from 0.2 to 10 mm, and water and acetone were amongst the fluids tested.

Interest has also been shown in using carbon fibre as a wick material because carbon fibre filaments have many fine longitudinal grooves on their surface and so have high capillary pressures. In addition, they are, of course, chemically stable. A number of heat pipes have been successfully constructed using carbon fibre wicks including one having a length of 100 m. This demonstrated a heat transport capability three times that of one having a metal mesh wick [51]. In addition, the use of carbon fibre-reinforced structures as wall material also cannot be ruled out for future aerospace applications. Other designs incorporating this wick material have been reported [52,53].

Sintered powers are available in a spherical form for a number of materials from which fine pore structures may be made, possibly incorporating larger arteries for added liquid flow capability. Leaching has been used to produce fine longitudinal channels, and grooved walls in copper and aluminium heat pipes have been applied for heat pipes in zero-gravity environments (in general, grooves alone are unable to support significant capillary heads in earth gravity applications and entrainment may limit the axial heat flow although covering the grooves with a mesh prevents this) (see Fig. 3.4).

'Biporous' is a relatively new term that is being used to describe heat pipe wicks in which sintered structures of differing porosity might be used. These are similar to composite wicks and sintered structures where a low-pressure drop flow path may parallel a high capillary driving force; mono- and biporous wicks are shown in Fig. 3.5 [54].

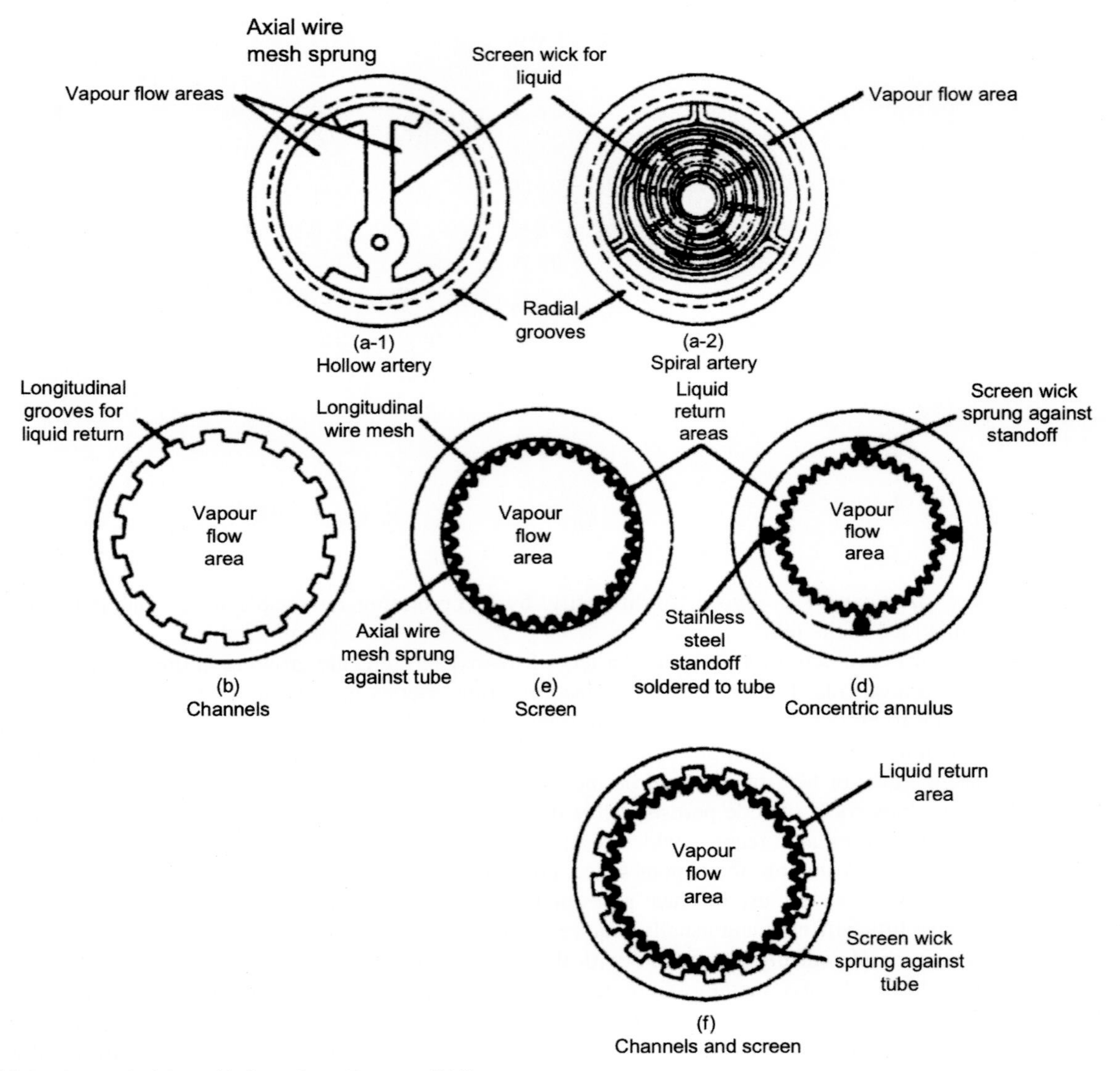


**FIGURE 3.4** Forms of wick used in heat pipes. *Courtesy: NASA.*

**FIGURE 3.5** Sintered glass wick samples showing in the upper picture a monoporous wick and in the lower picture the biporous unit: (a) monoporus wick (*d*: 275 μm) and (b) biporous wick ($d = 125$ μm, $D = 675$ μm) [54].

In the past, polymers have been proposed for use as heat pipe wall and wick materials. For example, flexible heat pipes were illustrated in early editions of *Heat Pipes* which included a polymer section to depict this flexibility. Today, the specific porosity/pore size requirements used in the wicks of some modern loop heat pipes (LHPs) have encouraged users to examine a variety of porous structures, including ceramics and polyethylene as well as porous nickel. As stated by Figus and colleagues at Astrium société par actions simplifiée (SAS) in France [55], high-performance porous media are necessary to achieve high evaporator heat fluxes (greater than 10,000 $W/m^2K$) and early polymer wicks had very small pore sizes but low permeabilities ($10^{-4}$ $m^2$ – orders of magnitude lower than the examples of those wicks listed in Table 3.4). For example, PTFE has also been used by Astrium in LHP evaporators. (Note the wicks in Table 3.4 are principally for 'conventional' heat pipes, not LHPs.)

In December 2001 during a flight on the Shuttle spacecraft (STS-108) [56,57], the capillary pumped loop (CPL) successfully demonstrated that a CPL with two parallel evaporators, using polyethylene wicks, could start-up and operate for substantial periods in high-power (1.5 kW) and low-power (100 W) modes. Additionally, over the preceding 2-year fully charged storage period, no gas generation had been observed.

## 3.2.2 Arterial wicks

Arterial wicks are necessary in high-performance heat pipes for spacecraft where temperature gradients in the heat pipe have to be minimised to counter the adverse effect of what are (generally) low thermal conductivity working fluids. An arterial wick developed by one of the authors at International Research and Development (IRD) is shown in Fig. 3.6 and in this example, with a wall material of aluminium and acetone as the working fluid, the bore of the heat pipe was only 5.25 mm. Developed for the European Space Agency (ESA), it was designed to transport 15 W over a distance of 1 m with an overall temperature drop not exceeding 6°C.

The aim of this wick system was to obtain liquid transport along the pipe with the minimum pressure drop, and the necessary driving force was achieved by covering the six arteries with a fine screen.

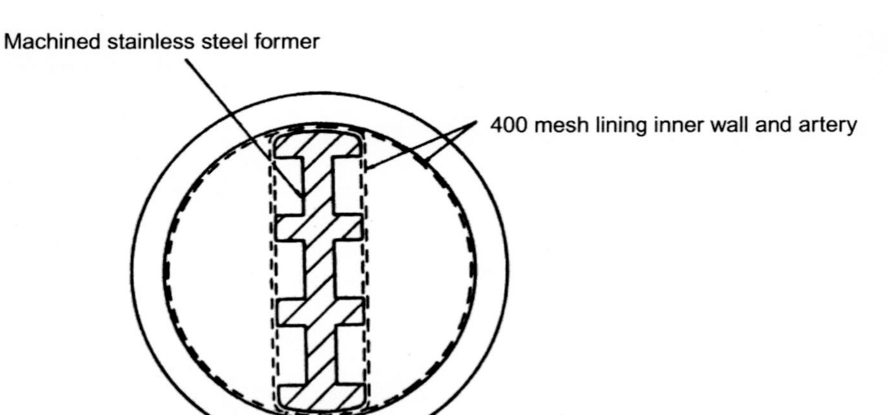


**FIGURE 3.6** Arterial wick developed at IRD.

In addition, to achieve the full heat transport potential of an arterial wick, the artery must be completely shut off from the vapour space; consequently, as the maximum capillary driving force is determined by the pore size of the screen, a high degree of quality control was required during the manufacturing process to ensure the artery was successfully closed and the screen undamaged.

A further consideration in the design of arterial heat pipes is that of a vapour or a gas blockage of the arteries. If a vapour or gas bubble forms within (or is vented into) the artery, then its transport capability is seriously reduced. Indeed, if a bubble completely blocks the artery, then the heat transport capability is solely dependent on the effective capillary radius of the artery – that is, there is an effective state of open artery. In which case, in order (following this condition) that the artery will reprime, the heat load must be reduced to a value below the maximum capability associated with the open artery.

The implications of wick design and working fluid properties in arterial heat pipes are therefore as follows:

1. The working fluid must be thoroughly degassed prior to filling to minimise the risk of noncondensable gases blocking the artery.
2. The artery must not be in contact with the wall to prevent nucleation within it.
3. A number of redundant arteries should be provided to allow for some degree of failure.
4. Successful priming (i.e. refilling) of the artery, if applied to spacecraft, must be demonstrated in a one 'g' environment – it being expected that priming in a zero 'g' environment will be easier.

As mentioned earlier, arterial heat pipes have been developed principally to meet the demands of spacecraft thermal control and these demands have increased rapidly over the last decade. So that whilst mechanical pumps can be used in two-phase 'pumped loops', the passive attraction of the heat pipe has spurred the development of several derivatives of 'conventional' arterial heat pipes.

One such derivative is the monogroove heat pipe, shown in Fig. 3.7. This heat pipe comprises two axial channels, one for the vapour and one for the liquid flow. A narrow axial slot between these two channels creates a high capillary pressure difference and, in this particular design, circumferential grooves are used for liquid distribution and to maximise the evaporation and condensation heat transfer coefficient [58,59]. The principal advantage of this design being the separation of the liquid and vapour flows thereby eliminating the entrainment or countercurrent flow limitation.

Not all wicks are associated solely with heat pipes. Research at GE Global Research Centre in the United States [60] is examining the wicking capability of a number of cryogenic fluids that are used in spacecraft for thermal management and as propellants. These include liquid nitrogen, helium, hydrogen and oxygen. (The authors point out that the next generation of cryogen propellants [liquid oxygen and hydrogen] is currently planned by NASA for future space missions.) In some applications, the capillary structure is used to supply liquid propellant.

Using a commercially available stainless steel mesh with a pore size of about 5 μm, the research team indicated that a wicking height of approximately 60 mm was achieved using liquid nitrogen. The next challenge they are addressing is the construction of wicking structures with nanosized pores to demonstrate helium wicking.

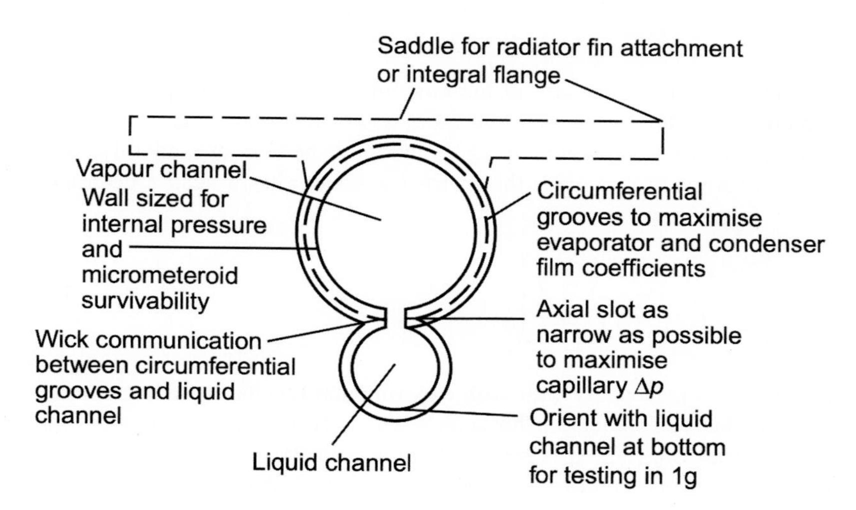


**FIGURE 3.7** Monogroove heat pipe configuration.

## 3.3 Thermal resistance of saturated wicks

One feature mentioned in discussions of the desirable properties for both the wick and the working fluid is their thermal conductivity. This is because conductivity is an important factor in determining the allowable wick thickness and the available expressions for predicting the thermal conductivities for several types of saturated wick are discussed below.

### 3.3.1 Meshes

Gorring and Churchill [61] present solutions to determine the thermal conductivity of heterogeneous materials that are divided into three categories: dispersions, packed beds and continuous pairs. But no satisfactory solution for a mesh is given because a mesh is a limiting case of dispersion, that is, the particles are in contact but not tightly packed. However, since the conductivity of dispersions is less than that of packed beds, an estimate of mesh conductivity can be made using Rayleigh's expression for the effective conductivity of a dispersion consisting of a square array of uniform cylinders:

$$k_w = \left(\frac{\beta - \varepsilon}{\beta + \varepsilon}\right) k_1 \tag{3.1}$$

where:

$$\beta = \left(1 + \frac{k_s}{k_1}\right) \Big/ \left(1 - \frac{k_s}{k_1}\right) \tag{3.2}$$

and where $k_s$ is the thermal conductivity of the solid phase, $k_1$ is the thermal conductivity of the liquid phase, and $\varepsilon$ is the volume fraction of the solid phase.

Recent work studying the effect of the number of layers of mesh in a wick on the overall heat pipe performance has thrown some doubt on the validity of conduction heat transfer models for wicks [62]. In this, most of the comparisons were made with the model of Kar and Dybbs [63] and no reference is made to the work of Goring and Churchill [61] and the conductivity calculations derived as a basis of their work. The experiments were limited to copper–water heat pipes with copper mesh as the wick. The mesh was made of 0.109 mm diameter wire, with 3.94 strands per mm. The effect of fluid inventory (too little or too much fluid to saturate the wick) was investigated, and the effect of mesh layers was investigated using heat pipes with one, two and six layers of wick. In the latter three units, the exact amount of fluid to fully saturate the wick was added – no more and no less.

It was found with fully saturated wicks that the number of layers of wick had little influence on the effective thermal resistance. The difference in thermal resistance between a single mesh layer unit and one with six layers was only 40%, much less than would be predicted by conduction models. However, no alternatives to the conduction models were proposed.

### 3.3.2 Sintered wicks

The exact geometric configuration of a sintered wick is unknown because of the random dispersion of the particles and the varying degree of deformation and fusion which occurs during the sintering process. For this reason, it is suggested that the sintered wick is best represented by a continuous solid phase containing a random dispersion of randomly sized spheres of liquid, and Maxwell [64] has derived an expression that gives the thermal conductivity of such a heterogeneous material as

$$k_w = k_s \left[ \frac{2 + k_1/k_s - 2\varepsilon\left(\frac{1-k_1}{k_s}\right)}{2 + k_1/k_s + \varepsilon\left(\frac{1-k_1}{k_s}\right)} \right] \tag{3.3}$$

Gorring and Churchill show that this expression agrees reasonably well with experimental results; in addition, work in Japan [65] has also confirmed the applicability of Maxwell's equation for screen wicks. However, it is stressed that when the effective thermal conductivity is predicted from Eq. (3.3) it is necessary to accurately estimate the porosity. In order to do so, the intermeshing between the screen layers should be taken into account, but this can be done by accurately measuring the overall thickness of the wick.

More recently, Xu et al. [66] have reported on the thermal conductivity of sintered wicks used in LHPs. As with conventional heat pipes, the wick thermal conductivity in the evaporator is critical to optimum performance, and the work by these researchers concentrated on sintered nickel wicks with controlled porosity and pore characteristics. The porosities were between 70% and 75% with a mean pore size within a range of 10–80 μm. Thermal conductivities varied from 2 to 4 W/mK, and the thermal conductivity was influenced not only by the porosity but also by the pore size distribution. The conclusion was that for the same porosity, wicks with a smaller pore size and more uniform pore size distribution showed much lower thermal conductivity.

### 3.3.3 Grooved wicks

The radial thermal resistance of grooves will be radically different for the evaporator and condenser sections and this occurs because of their different mechanisms of heat transfer. In the evaporator the land or fin tip plays no active part in the heat transfer process – the probable heat flow path is conduction via the fin, conduction across a liquid film at the meniscus and evaporation at the liquid–vapour interface. Whilst in the condenser section, the grooves will be flooded and so the fin tip plays an active role in the heat transfer process – the build-up of a liquid film at the fin tip providing the major resistance to heat flow and the thickness of this (liquid film) being a function of the condensation rate and the wetting characteristics of the working fluid.

Since the mechanism in the evaporator section is less complex and should provide the greatest resistance, we will concentrate on the analysis of that region.

Joy [67] and Eggers and Serkiz [68] propose identical models that each assume one-dimensional heat conduction along the fin and one-dimensional conduction near the fin tip across the liquid to the liquid/vapour interface where evaporation occurs. In the liquid, the average heat flow length is taken to be a quarter of the channel width, and the heat flow area is the channel half-width times the input length.

Thus

$$\frac{\Delta T}{Q} = \frac{a}{k_s N f l_e} + \frac{1}{4 k_1 l_e N} + \frac{1}{h_e \pi b N l_e} \tag{3.4}$$

where $N$ is the number of channels, $a$ the channel depth, $b$ the channel half-width and $f$ the fin thickness.

Kosowski and Kosson [69] have made measurements of the maximum heat transport capability and radial thermal resistance of an aluminium grooved heat pipe using Freon 21 and Freon 113[1] and ammonia as the working fluids.

1. Note that R21 is not now recommended for use because of its toxicity. Other refrigerants, in particular those classed as chlorofluorocarbons (CFCs) and including R113 are banned or limited in their use because of their ozone layer depletion effects. 'New' refrigerants may be substituted as appropriate, for example R141b has been used in heat pipes.

The relevant dimensions of their heat pipes were as follows:

$N = 30$
$a = 0.89$ mm
$2b = 0.76$ mm
Pipe outside diametre = 12.7 mm
Heat pipe no.1:
$l_e = 304.8$ mm,
$l_c = 477.6$ mm
Heat pipe no. 2:
$l_e = 317.5$ mm,
$l_c = 503$ mm

The following heat transfer coefficients (based on the outside area) were measured:

| Fluid | $h_c$ W/m$^2$ °C | $h_c$ W/m$^2$ °C |
|---|---|---|
| Freon 21 (heat pipe no. 1) | 1134 | 1700 |
| Freon 113 (heat pipe no. 2) | 652 | 1134 |
| Ammonia (heat pipe no. 3) | 2268 | 2840 |

Converting the heat transfer coefficients into a thermal resistance:

| Fluid | R°C/W (evap.) | R°C/W (cond.) |
|---|---|---|
| Freon 21 | 0.0735 | 0.031 |
| Freon 113 | 0.122 | 0.044 |
| Ammonia | 0.035 | 0.0175 |

The contribution due to fin conduction is 0.0018°C/W, ($l = 0.25$ mm, $k_w = 200$ W/m°C) and is negligible. This bears out Kosowski and Kosson's observation that the percentage fill has little effect on the thermal resistance.

The evaporation term is also small, and the most significant contribution to the resistance is the liquid conduction term. Comparing the theory and experiment results suggest that the theory overpredicts the conduction resistance by a considerable amount (50%–300%). It would therefore be more accurate to use the integrated mean heat flow length

$$\left(1 - \frac{\pi}{4}\right)b \tag{3.5}$$

Rather than $\frac{b}{2}$ such that:

$$\frac{\Delta T}{Q} = \frac{\left(1 - \frac{\pi}{4}\right)}{2k_1 l_e N} \tag{3.6}$$

Then, knowing the duty and allowable temperature drop into the vapour space, the number of grooves can be calculated for various geometries and working fluids.

In the heat pipe condenser section, or when the channels are mesh-covered, the fin tip plays an active part in the heat transfer process and the channels are filled. In this case, the parallel conduction equation is used:

$$k_w = k_s\left\{1 - \varepsilon\left(1 - \frac{k_1}{k_s}\right)\right\} \tag{3.7}$$

where $\varepsilon$, the liquid void fraction, is given by:

$$\varepsilon = \frac{2b}{2b + f} \tag{3.8}$$

and where $f$ is the fin thickness

### 3.3.4 Concentric annulus

In this case, the capillary action is derived from a thin annulus containing the working fluid. Thus

$$k_w = k_1 \tag{3.9}$$

This case may also be used to analyse the effects of loose-fitting mesh and sintered wicks.

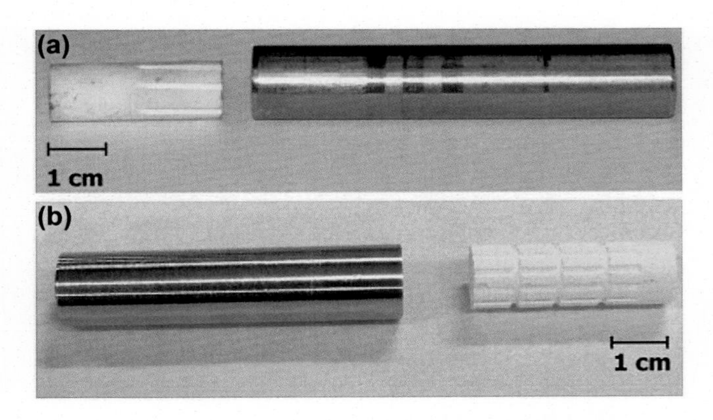


**FIGURE 3.8** (a) Ceramic wick with vapour channels (grooves) used in the LHP and (b) CPL [73].

### 3.3.5 Sintered metal fibres

Data tabulated by the Engineering Sciences Data Unit [70] on a variety of the more common wicks used in heat pipes include expressions for thermal conductivity. (Also included are expressions for porosity, minimum capillary radius, permeability and equivalent diameter.) The following equation is given for the thermal conductivity of sintered metal fibre wicks:

$$k_w = \varepsilon^2 k_1 + (1-\varepsilon)^2 k_s + \frac{4\varepsilon(1-\varepsilon)k_1 k_s}{k_1 + k_s} \tag{3.10}$$

Values obtained using this equation may be compared with data obtained experimentally by Semena and his coworkers [49,71]. A more detailed discussion of metal felt wicks, with a presentation of data on other wick properties, is given by Acton [72].

### 3.3.6 Ceramic wick structures

Like the nickel sintered wick mentioned above, the type of geometry of a wick in a LHP or CPL necessitates slightly different approaches – a typical LHP evaporator is shown in Chapter 2 – and the emphasis on thermal conductivity is obvious in this illustration. Santos et al. [73] in Brazil studied the use of ceramic wicks in both LHPs and capillary pumped loops, and the ceramic wicks used were sintered from alumina and mullite powders, giving porosities of 50%, a 1–3 μm pore size distribution and a permeability of $35 \times 10^{-15}$ m$^2$. Grooves that act as vapour channels are visible in Fig. 3.8, and it was found that with practice a larger number of grooves could be machined in the units as long as the wick was cleaned at 800°C for 1 h afterwards to remove any oil contamination.

## 3.4 The container

The function of the container is to isolate the working fluid from the outside environment. It has, therefore, to be leak-proof in order to maintain the pressure differential across its walls and to enable the transfer of heat to take place into and from the working fluid.

Selection of the container material depends on several factors, and these are:

1. compatibility (both with working fluid and the external environment)
2. strength-to-weight ratio
3. thermal conductivity
4. ease of fabrication, including weldability, machineability and ductility
5. porosity
6. wettability,

and most of these are self-explanatory.

A high strength-to-weight ratio is more important in spacecraft applications and the material should also be nonporous to prevent the diffusion of gas into the heat pipe. A high thermal conductivity ensures minimum temperature drop between the heat source and the wick.

Recently work has been done on the potential of magnesium as a light-weight heat pipe structural material [74], where increasingly demanding electronics cooling applications (both in space and for items such as graphics cards) require increased heat pipe duties, more compact systems and lower weights. Compatibility data cited in [74] suggests that, subject to proper cleaning procedures, both pure ammonia and acetone were compatible with magnesium alloy.

The thermal conductivity of some wall materials is given in Appendix 2.

## 3.5 Compatibility

Compatibility has already been discussed in relation to the selection of the working fluid, the wick and containment vessel of the heat pipe. However, this feature is of prime importance and warrants particular attention here.

The two major effects from incompatibility are corrosion and the generation of noncondensable gas. If the wall or wick material is soluble in the working fluid, mass transfer is likely to occur between the condenser and the evaporator with solid material being deposited in the latter. This will result either in local hot spots or blocking of the pores of the wick. Noncondensable gas generation is probably the most common indication of heat pipe failure which gradually becomes blocked because the noncondensables tend to accumulate in the heat pipe condenser section. It is easy to identify because of the sharp temperature drop that exists at the gas/vapour interface.

Some compatibility data is of course available from general scientific publications and from trade literature on chemicals and materials. However, it has become common practice to carry out life tests on representative heat pipes, the main aim of these being to estimate long-term materials compatibility under heat pipe operating conditions. At the termination of life tests, gas analyses and metallurgical examinations as well as chemical analysis of the working fluid may be carried out (see also Section 3.9.2).

Many laboratories have carried out life tests and a vast quantity of data has been published. However, it is important to remember that whilst life test data obtained by one laboratory may indicate satisfactory compatibility, different assembly procedures at another laboratory, involving, for example, a nonstandard materials treatment process, may result in a different corrosion or gas generation characteristic. Thus it is important to obtain compatibility data whenever procedural changes to cleaning or pipe assembly are made.

Stainless steel is a suitable container and wick material for use with working fluids such as acetone, ammonia and liquid metals from the point of view of compatibility, but its low thermal conductivity is a disadvantage and copper and aluminium are used where this feature is important. The former is particularly attractive for mass-produced units using water as the working fluid. Plastic has been used as the container material, and at very high temperatures ceramics and refractory metals such as tantalum have been given serious consideration. In order to introduce a degree of flexibility in the heat pipe wall, stainless steel bellows have been used, and in cases where electrical insulation is important, a ceramic or glass-to-metal seal has been incorporated. This must of course be used in conjunction with electrically nonconducting wicks and working fluids.

### 3.5.1 Historic compatibility data

Long-term life tests on cryogenic heat pipes started a little later than those for higher temperature units. Nonetheless, there are comprehensive data from ESA sources [75] on stainless steel (container was type 304 L and wick type 316) heat pipes using as working fluids – methane, ethane, nitrogen or oxygen, arising out of tests extending over a period of up to 13 years.

The test units were 1 m in length and of either 3.2 or 6.35 mm outside diameter. Heat transport capability was up to 5 W m (meaning that the pipe transport 5 W over 1 m, or, for example, 10 W over 0.5 m), and vapour temperatures were 70–270K. Tests were completed in the mid-1990s, and the main outcomes were as follows:

- All pipes retained maximum heat transport capability.
- All pipes maintained maximum tilt capability (capillary pumping demonstration).
- The evaporator heat transfer coefficient remained constant.
- No incompatibility or corrosion was evident in the oxygen and nitrogen pipes.
- Slight incompatibility, resulting in noncondensable gas extending over 1% of the heat pipe length and therefore affecting condenser efficiency, was noted with the ethane and methane units.

Unlike the oxygen and nitrogen tungsten inert gas (TIG)-welded pipes, the ethane and methane units were hard-brazed, and the implication is that the gas generation was attributed to this.

A comprehensive review of material combinations in the intermediate temperature range has been carried out by Basiulis and Filler [76] and is summarised below. Results are given in the paper over a wider range of organic fluids (most produced by Dow Chemicals) than given in Table 3.5.

Tests in excess of 8000 h with ammonia/aluminium were reported, but only 1008 h had been achieved at the time of data compilation for aluminium/acetone. No temperatures were specified by Basiulis for these tests; other workers have exceeded 16,000 h with the latter combination.

Later, the life test work at IKE, Stuttgart was published [77], involving tests on about 40 heat pipes. The tests indicated that copper/water heat pipes could be operated without degradation over long periods (over 20,000 h), but severe gas generation was observed with stainless steel/water heat pipes. IKE also had some reservations concerning acetone with copper and stainless steel; whilst compatible, it was stressed that proper care had to be given to the purity of both the acetone and the metal. The same reservation also applied to the use of methanol.

Exhaustive tests on stainless steel/water heat pipes were also carried out at Ispra [78], where experiments were conducted with vapour temperatures as high as 250°C.

It was found that neither the variation of the fabrication parameters nor the addition of a large percentage of oxygen to the gas plug resulted in a drastic reduction of the hydrogen generation at 250°C. Hydrogen was generated within 2 h of start-up in some cases. The stainless steel used was type 316 and such procedures as passivation and outgassing were ineffective in arresting generation. However, it was found that the complimentary formation of an oxide layer on the steel did inhibit further hydrogen generation.

Gerrels and Larson [79], as part of a study of heat pipes for satellites, also carried out comprehensive life tests to determine the compatibility of a wide range of fluids with aluminium (6062 alloy) and stainless steel (type 321). The fluids used included ammonia which was found to be acceptable. It is important, however, to ensure that the water content of the ammonia is very low, only a few parts per million concentration being acceptable with aluminium and stainless steel.

The main conclusions concerning compatibility made by Gerrals and Larson are given below, with data being obtained for the following fluids:

| | |
|---|---|
| *n*-Pentane | CP-32 (Monsanto experimental fluid) |
| *n*-Heptane | CP-34 (Monsanto experimental fluid) |
| Benzene | Ethyl alcohol |
| Toluene | Methyl alcohol |
| Water (with stainless steel) | Ammonia |
| | *n*-butane |

The stainless steel used with the water was type 321.

All life tests were carried out in gravity-refluxing containers, with heat being removed by forced air convection and being introduced by immersion of the evaporator sections in a temperature-controlled oil bath.

Preparation of the aluminium alloy was as follows: initial soak in a hot alkaline cleaner followed by deoxidation in a solution of 112 g sodium sulphate and 150 mL concentrated nitric acid in 850 mL water for 20 min at 60°C. In addition, the aluminium was either machined or abraded in the area of the welds. A mesh wick of commercially pure

**TABLE 3.5 Compatibility data (low-temperature working fluids).**

| Wick material | Working fluids | | | | | |
|---|---|---|---|---|---|---|
| | Water | Acetone | Ammonia | Methanol | Dow-A | Dow-E |
| Copper | RU | RU | NU | RU | RU | RU |
| Aluminium | GNC | RL | RU | NR | UK | NR |
| Stainless steel | GNT | PC | RU | GNT | RU | RU |
| Nickel | PC | PC | RU | RL | RU | RL |
| Refrasil fibre | RU | RU | RU | RU | RU | RU |

*RU*, recommended by past successful usage; *RL*, recommended by literature; *PC*, probably compatible; *NR*, not recommended; *UK*, unknown; *GNC*, generation of gas at all temperatures; *GNT*, generation of gas at elevated temperatures when oxide present.

aluminium was inserted into the heat pipes. The capsules were TIG welded under helium in a vacuum-purged inert gas welding chamber. Leak detection followed welding, and the capsules were also pressure tested to 70 bar. A leak check also followed the pressure test.

The type 321 stainless steel container was cleaned before fabrication by soaking in hot alkaline cleaner and pickling for 15 min at 58°C in a solution of 15% by volume of concentrated nitric acid, 5% by volume of concentrated hydrochloric acid and 80% water. In addition, the stainless steel was passivated by soaking for 15 min at 65°C in a 15% solution of nitric acid. Type 316 stainless steel was used as the wick. The capsule was TIG welded in air with argon purging.[2] In both series of US tests, mixtures of materials were used, but this is not good practice in life tests as any degradation may not be identified as being caused by the particular single material. The boiling-off technique was used to purge the test capsules of air.

In the case of methyl alcohol, reaction was noted during the filling procedure, and a full life test was obviously not worth proceeding with.

Sealing of the capsules was by a pinch-off, followed by immersion of the pinched end in an epoxy resin for final protection.

The following results were obtained:

*n*-Pentane: Tested for 750 h at 150°C. Short-term instabilities were noted with random fluctuation in temperature of 0.2°C. On examination of the capsule, very light brownish areas of discolouration were noted on the interior wall but the screen wick appeared clean. No corrosion evidence was found. The liquid removed from the capsule was found to be slightly brown in colour.

*n*-Heptane: Tested for 600 h at 160°C. Slight interior resistances were noted after 465 h, but on opening the capsule at the end of the tests, the interior, including the screen, was clear, and the working fluid was clean.

Benzene: Tested for 750 h at 150°C plus vapour pressure 6.7 bar. Very slight local areas of discolouration were found on the wall, the wick was clean, there was no evidence of corrosion, and the liquid was clear. It was concluded that benzene was very stable with the chosen aluminium alloy.

Toluene: Test run for 600 h at 160°C. A gradual decrease in condenser section temperature was noted over the first 200 h of testing but no change was noted following this. On opening the capsule, slight discolouration was noted locally on the container wall. This seemed to be a surface deposit with no signs of attack on the aluminium. The screen material was clean and the working fluid was clear at the end of the test.

Water: (stainless steel) Tested at 150°C plus for 750 h. Vapour pressure 6.7 bar. Large concentration of hydrogen was found when analysis of the test pipe was performed. this was attributed in part to poor purging procedure as there was discolouration in the area of the welds, and the authors suggest that oxidation of the surfaces had occurred. A brown precipitate was also found in the test heat pipe.

CP-32: Tested for 550 h at 158°C. A brownish deposit was found locally on interior surfaces. Screen was clean, but the working fluid was darkened.

CP-34: Tested for 550 h at 158°C. Gas generation was noted. Also, extensive local discolouration on the capsule wall near the liquid surface. No discolouration on the screen. The fluid was considerably darkened.

Ammonia: Tested at 70°C for 500 h. Some discolouration of the wall and mesh was found following the tests. This was attributed to some nonvolatile impurity in the ammonia which could have been introduced when the capsule was filled. In particular, some of the lubricant on the valve at the filling position could have entered with the working fluid. (This was the only test pipe on which the filling was performed through a valve.)

*n*-Butane: Tested for 500 h at 68°C. It was considered that there could have been noncondensable gas generation, but the fall-off in performance was attributed to some initial impurity in the *n*-butane prior to filling. The authors felt that this impurity could be isobutane. Further tests on purer *n*-butane gave better results, but the impurity was not completely removed.

Gerrals and Larson argued thus (concerning the viability of their life tests): 'It should be emphasised that the present tests were planned to investigate the compatibility of a particular working fluid-material combination for long-term (5-year) use in a vapour chamber radiator under specified conditions. The reference conditions call for a steady-state temperature of 143°C for the primary radiator fluid at the inlet to the radiator and a 160°C short-term peak temperature. The actual temperature to which the vapour chamber working fluid is exposed must be somewhat less than the primary radiator fluid temperature because some temperature drop occurs from the

2. In both series of US tests, mixtures of materials were used. This is not good practice in life tests as any degradation may not be identified as being caused by a particular single material.

primary radiator fluid to the evaporative surface within the vapour chamber. It is estimated, in these capsule tests, that the high-temperature fluids were exposed to temperatures at least 10°C higher than the peak temperature, and at least 20°C greater than the long-term steady-state temperatures the fluids would experience in the actual radiator. Although the time of operation of these capsule tests is only about 1% of the planned radiator lifetime, the conditions of exposure were much more severe. It, therefore, seems reasonable to assume that if the fluid-material combination completed the capsule tests with no adverse effects, it is a likely candidate for a radiator with a 5-year life'.

On the basis of the above tests, Gerrels and Larson selected the following working fluids:

- 6061 Aluminium at temperatures not exceeding 150°C:
  - Benzene
  - *n*-Heptane
  - *n*-Pentane
- 6061 Aluminium at temperatures not exceeding 65°C:
  - Ammonia *n*-Butane
- The following fluids were felt not to be suitable:
  - Water (in type 321 stainless steel)
  - CP-32 (in 6061 aluminium)
  - CP-34 (in 6061 aluminium)
  - Methyl alcohol (in 6061 aluminium)
  - Toluene (in 6061 aluminium).

Gerrels and Larson point out that the Los Alamos Laboratory obtained heat pipe lives in excess of 3000 h without degradation using a combination of water and type 347 stainless steel.

Other data [80] suggest that alcohols in general are not suitable with aluminium.

Summarising the data of Gerrels and Larson, ammonia was recommended as the best working fluid for vapour chambers operating below 65°C and *n*-pentane the best for operation above this temperature, assuming that aluminium is the container material.

At the other end of the temperature scale, long lives have been reported [19] for heat pipes with lithium or silver as the working fluid. With a tungsten rhenium (W-26 Re) container, a life of many years was forecast with lithium as the working fluid operating at 1600°C. At 1700°C, significant corrosion was observed after 1 year, whilst at 1800°C the life was as short as 1 month. W-26 Re/silver heat pipes were considered capable of operating at 2000°C for 10,900 h. Some other results are presented in Table 3.6 updated using information summarised in [81].

Sodium is the working fluid of choice for many high-temperature applications, and the 10-year sodium heat pipe life text carried out by Thermacore Inc (see Appendix 4) in conjunction with Sest Inc. (a company that has dynamically modelled the Stirling Converter and is adept at life predictions) gave most encouraging results [82]. One application for which such heat pipes are proposed (as when liquid-metal heat pipes were first examined over 50 years ago) is for thermal control in space nuclear power plants which are believed to be making a come-back. The Thermacore study revolved around a Stirling-cycle-based Space Power Converter heat pipe using Inconel 718 as the container material and a stainless steel wick. Examination after 10 years showed that a 30 μm thick layer of the container's inner wall had been depleted of nickel – the result of its long-term operation at 700°C – but it was concluded that this did not seem to affect the operation of the heat pipe.

The work also involved variable conductance sodium heat pipes (see Section 3.8.11 for a discussion of the variable conductance heat pipe [VCHP]) using Haynes 230 alloy as the container material [83].

Confirmation on the importance of the purity of the working fluid in contributing to a satisfactory life with lithium heat pipes is given by the work at the Commissariat à l'Energie Atomique in Grenoble [84]. Who, in pointing out that failures in liquid-metal heat pipes most commonly arise due to impurity-driven corrosion mechanisms, suggest that a rigorous purifying procedure for the lithium is carried out. This involves forced circulation on a cold and on a hot trap filled with Ti–Zr alloy as a getter to lower the oxygen and nitrogen content, respectively. The lithium is then distilled under vacuum before filling the heat pipe.

Data on extended tests on units using lead at more modest temperatures (around 1340°C) have been obtained over 4800 h at the Los Alamos Laboratory [85]. Using one heat pipe fabricated from a molybdenum tube with a Ta–10% W wall and a W wick, no degradation in performance was noted over the test period.

One subject of much argument is the method of conducting life tests and their validity when extrapolating likely performance over a period of several years. For example, on satellites, where remedial action in the event of failure is

**TABLE 3.6 Compatibility data (life tests on high-temperature heat pipes).**

| Working | Material | | Vapour temp. (°C) | Duration (h) |
|---|---|---|---|---|
| | Wall | Wick | | |
| Caesium | Ti | | 400 | >2000 |
| | Nb + 1%Zr | | 1000 | 8700 |
| Potassium | Ni | | 600 | >6000 |
| | Ni | | 600 | 16,000 |
| | Ni | | 600 | 24,500 |
| | 304, 347 ss | | 510, 650 | 6100 |
| Sodium | Hastelloy X | | 715 | >8000 |
| | Hastelloy X | | 715 | >33,000 |
| | 316 ss | | 771 | >4000 |
| | Nb + 1%Zr | | 850 | >10,000 |
| | Nb + 1%Zr | | 1100 | 1000 |
| | 304, 347 ss | | 650–800 | 7100 |
| Bismuth | Ta | | 1600 | 39 |
| | W | | 1600 | 118 |
| Lithium | Nb + 1%Zr | | 1100 | 4300 |
| | Nb + 1%Zr | | 1500 | >1000 |
| | Nb + 1%Zr | | 1600 | 132 |
| | Ta | | 1600 | 17 |
| | W | | 1600 | 1000 |
| | SGS-Ta | | 1600 | 1000 |
| | TMZ | | 1500 | 9000 |
| | W + 26%Re | | 830–1000 | 7700 |
| Lead | Nb + 1%Zr | | 1600 | 19 |
| | SGS-Ta | | 1600 | 1000 |
| | W | | 1600 | 1000 |
| | Ta | | 1600 | >280 |
| Silver | Ta | | 1900 | 100 |
| | W | | 1900 | 335 |
| | W | | 1900 | 1000 |
| | Re | W | 2000 | 300 |

For European Space Agency requirements on duration also see Section 3.9.

difficult (if not impossible) to implement, a life of 7 years is a standard minimum requirement.[3] It is therefore necessary to accelerate the life tests so that reliability over a longer period can be predicted with a high degree of accuracy.

Life tests on heat pipes are commonly regarded as being primarily concerned with the identification of any incompatibilities that may occur between the working fluid and the wick and the wall materials, and the ultimate life test

3. ESA requirement.

would be in the form of a long-term performance test under likely operating conditions. However, if this is carried out, it is difficult to accelerate the life test by increasing, say, the evaporator heat flux as any significant increase is likely to cause dry-out because then the pipe will be operating well in excess of its probable design capabilities. Therefore any accelerated life test that involves heat flux increases of the order of (say) four over that required under normal operating conditions must be carried out in the reflux mode, with regular performance tests to ensure that the design capability is still being obtained.

An alternative possibility as a way of accelerating any degradation processes is to raise the operating temperature of the heat pipe. One drawback of this method is the effect that increased temperature may have on the stability of the working fluid itself. Acetone cracking, for example, might be a factor where oxides of metals are present, resulting in the formation of diacetone alcohol that has a much higher boiling point than pure acetone.

Obviously, there are many factors to be taken into account when preparing a life test programme, including such questions such as the desirability of heat pipes with valves or completely sealed units as used in practice. This topic is of major importance and life test procedures are discussed more fully in Section 3.9.

One of the most comprehensive life test programmes was that carried out by Hughes Aircraft Co. [86], for which a summary of recommendations based on these tests together with the additional experience gained with some of the material combinations discussed above are given in Table 3.7.

The lack of support given to a nickel/water combination is based on an Arrhenius-type accelerated life test carried out by Anderson [87]. Work carried out on nickel wicks in water heat pipes (generally with a copper wall) has, in the authors' experience, not created compatibility problems and this area warrants further investigation. With regards to water/stainless steel combinations – for some years the subject of considerable study and controversy – the Hughes work suggests that type 347 stainless steel is acceptable as a container with water. Tests had been progressing since December 1973 on a 347 stainless steel container and copper wicked water heat pipe operating at 165°C with no trace

**TABLE 3.7 Hughes aircraft compatibility recommendations.**

| | Recommended | Not recommended |
|---|---|---|
| Ammonia | • Aluminium<br>• Carbon steel<br>• Nickel<br>• Stainless steel | • Copper |
| Acetone | • Copper<br>• Silica<br>• Aluminium[a]<br>• Stainless steel[a] | |
| Methanol | • Copper<br>• Stainless steel<br>• Silica | • Aluminium |
| Water | • Copper<br>• Monel<br>• 347 Stainless steel[b] | • Stainless steel<br>• Aluminium<br>• Silica<br>• Inconel<br>• Nickel<br>• Carbon Steel |
| Dowtherm A | • Copper<br>• Silica<br>• Stainless steel[c] | |
| Potassium | • Stainless steel | • Titanium |
| Sodium | • Stainless steel<br>• Inconel | • Titanium |

*Note*: Type 347 stainless steel as specified in AISI codes does not contain tantalum. AISI type 348, which is otherwise identical except for a small tantalum content, should be used in the United Kingdom (Authors). Work on water in steel heat pipes is also discussed later in this chapter.

[a] *The use of acetone with aluminium and/or stainless steel presented problems to the authors, but others have had good results with these materials. The problem may be temperature related – use with caution.*

[b] *Recommended with reservations.*

[c] *This combination should be used only where some noncondensable gas in the heat pipe is tolerable, particularly at higher temperatures.*

of gas generation. (Type 347 stainless steel contains no titanium but does contain niobium.) Surprisingly, a type 347 wick caused rapid gas generation. The use of Dowtherm A (equivalent to Thermex, manufactured by ICI) is recommended for moderate temperatures only, breakdown of the fluid progressively occurring above about 160°C. With careful materials preparation Thermex appears compatible with mild steel but the Hughes data are limited somewhat by the low operating temperature conditions.

Hughes emphasised the need to carry out rigorous and correct cleaning procedures but also stressed that the removal of cleaning agents and solvents prior to filling with working fluid is equally important.

### 3.5.2 Compatibility of water and steel – a discussion

Water is an ideal working fluid for heat pipes because of its high latent heat (thus requiring a relatively low inventory), its low cost, and its high 'figure of merit'. As shown already in this chapter, water is compatible with a number of container materials the most popular being copper. However, ever since heat pipes were first conceived, experimenters have experienced difficulties in operating a water/steel (be it mild, boiler or stainless steel) heat pipe without the generation of hydrogen in the container. The generation always manifests itself as a cold plug of gas at the condenser section of the heat pipe – blocking off its surface for heat rejection – resulting in a sharply defined interface between the water vapour and the noncondensable gas and so making the presence readily identifiable. The effect is similar to that when air is present in a conventional domestic radiator system but of course the reasons may be different.

The earlier description in Chapter 1 of the Perkins Tube and its derivatives, which used iron or steel with water as the working fluid, shows that the fluid/wall combination has demonstrated a significant life although gas generation did occur.

#### *3.5.2.1 The mechanism of hydrogen generation and protective layer formation*

The reaction responsible for the generation of hydrogen is

$$Fe + 2H_2O \rightarrow Fe(OH)_2 + H_2$$

Corrosion of steel is negligible in the absence of oxygen, therefore in a closed system such as a steel heat pipe with water as the working fluid, the corrosion only takes place until all the free oxygen is consumed. When oxygen is deficient, low alloy steels develop a hydrated magnetite layer in neutral solutions by decomposition of the ferrous hydroxide. Further dehydration, or a chemical reaction of iron with water, occurs as follows:

$$3Fe(OH)_2 \rightarrow Fe_3O_4 + 2H_2O + H_2$$

The conversion of $Fe(OH)_2$ leads to a protective layer of $Fe_3O_4$ (especially with mild/carbon steels) and in high-temperature water, and it is this $Fe_4O_3$ layer that is responsible for the good corrosion resistance of boiler steels in power generation plant boilers.

A number of different methods have been developed to obtain nonporous and adhesive magnetite layers, and these are discussed below.

In addition, several other observations have been made concerning factors which influence the long-term compatibility of steel with water, and these are:

- The pH value of the water should be greater than 9 (some researchers quote a range of 6–11 as being satisfactory – see below for a discussion). This value can be adjusted by conditioning of the water. The high pH value has also been reported to favour the growth of effective passive layers.
- The water must be fully degassed and desalted (or demineralised) before conditioning.
- The protective layer can be destroyed if the metal is locally overheated above 570°C (thus the passivation reaction is best carried out once the container has been welded or subjected to any other high-temperature procedures during assembly).

#### *3.5.2.2 Work specifically related to passivation of mild steel*

There are a number of potential/active heat pipe applications where the advantages of using water with mild steels is attractive. For example, in domestic radiators attempts have been made to construct two-phase units with more rapid responses than those of a single-phase heating system. Japanese and Korean companies have marketed such products and research in the United Kingdom has also investigated such combinations.

Work (some years ago) in the then Czechoslovakia [88] discusses three approaches to the prevention of hydrogen generation:

- Preparation of heat pipes with an inhibitor added to the working fluid.
- Preparation of pipes provided with a protective oxide layer.
- Preparation of heat pipes involving a protective layer in conjunction with an inhibitor.

All the tests were carried out with the condenser above the evaporator, and with the evaporator in a sand bath heated to 200°C. The test period was 6000 h.

Measurements were made of the temperature difference along the pipes (higher values are poor), and the gases inside were analysed for hydrogen. All results are compared in Table 3.8. A steel heat pipe without any treatment was tested to form the basis for comparison – pipe (X). As can be seen from the table, this exhibited a high-temperature drop along its length, indicating gas generation.

### 3.5.2.3 Use of an inhibitor

The inhibitor selected was an anodic type based on chromate to minimise oxidation. $K_2CrO_4$ was selected and used in a concentration of 5 g for 1 l of water, with a pH of 7.87. The pipe internal surface had been previously degassed. The inhibitor acts to decelerate the anodic partial reaction ($Fe \rightarrow Fe^{2+} + 2e$). The chromate anion is first adsorbed on the active sites of the metal surface and the film formed is chromium III oxide ($Cr_2O_3$), or chromium III hydroxide-chromate [$Cr(OH)CrO_4$]. The passivating concentration varies from 0.1% to 1.0%.

The pipe (A) with an inhibitor and no protective oxide layer exhibited a temperature difference along its length of 6°C and traces of hydrogen were found on analysis of the gases in the heat pipe.

### 3.5.2.4 Production of a protective layer

As discussed above, the natural ageing process of mild steel leads to a protective layer of magnetite being formed which reduces hydrogen generation rates. It was found in tests on naturally passivated steel in the laboratory that whilst the performance stabilised after 2000 h of operation, there was, in the intervening period, sufficient generation of hydrogen to significantly reduce the performance (again evident in an increased temperature drop along the length). Therefore it was decided to accelerate the passivation process prior to filling and sealing the heat pipe by oxidising the pipe surface with superheated steam vapour, the efficiency of this process being determined by the thickness and porosity of the magnetite layer formed. Optimum thickness of the layer was found to be 3–5 μm. Both of these properties are functions of temperature and the duration of oxidation, but this can also be positively influenced by adding an oxidation catalyst, ammonium molybdate $(NH_4)_2MoO_4$ to the water vapour. This leads to molybdenum trioxide being deposited on the surfaces being treated. Note that the stability of magnetite reduces above 570°C and this temperature should not be exceeded during the process (as mentioned above).

The heat pipe tested (*b*) with vapour oxidation carried out at 550°C had, at the end of the test period, a working fluid with a pH of 8.01 and had a temperature difference along its length of 10°C, larger than that with the inhibitor alone. Only a trace of hydrogen was found on analysis.

**TABLE 3.8 Heat pipe specification and analytical results after 6000 h.**

| Code | Surface treatment | Working fluid | pH | Inhibitor concentration | Temperature difference (°C) | Amount of $H_2$ |
|---|---|---|---|---|---|---|
| X | None | Water | 8.32 | None | 33.2 | Maximum |
| A | None | +Inhibitor | 10.52 | 0.04 | 6 | Trace |
| b | Oxidation + vapour at 550 (°C) | Water | 8.01 | None | 10 | Trace |
| $c_1$ | Oxidation + vapour at 550 (°C) | +Inhibitor | 9.05 | 0.03 | 3 | None |
| $c_2$ | Oxidation + vapour + catalyst at 550 (°C) | +Inhibitor | 7.01 | 0.45 | 13 | Trace |

### 3.5.2.5 Pipes with both inhibitor and oxide layer

A combination of the magnetite layer and an inhibitor introduced into the working fluid was then tested, and it was necessary for the pH of the inhibitor solution to be maintained in the range $6 < pH < 11$ so that the inhibitor did not break down the magnetite layer.

The optimum performance was achieved with pipe $c_1$ which was passivated using superheated steam without the addition of the oxidation catalyst and also contained an inhibitor. The catalyst led to inferior performance in the pipe $c_2$ which again had an inhibitor present. The poor performance of this latter heat pipe is attributed to faults in the protective layer due to exceeding the limiting layer thickness and by the dissolution of $MoO_3$ in the aqueous solution of potassium chromate.

The temperature drop in pipe $c_1$ was 3°C, and there was no indication of hydrogen formation. An analysis of the working fluid showed that the pH was strongly shifted towards the alkaline region after the 6000 h test.

(Note that the temperature difference was measured along the condenser section of the heat pipe.)

In a further communication [89], the Czech research team reported on an extension of the tests to 18,000 h. They concluded, on the basis of the very small (1°C) increase in temperature difference for pipes (*a*) and (*q*) measured over the additional 12,000 h, that the inhibitor stabilises the performance of water/steel heat pipes, but a superior performance is achievable with the combination of inhibitor and oxidation. However, this latter process is substantially more costly.

The use of inhibitors tested for up to 35,000 h has been reported from Ukraine [90]. The steel used was Steel 10 GOST 8733-87, and the best results were achieved with only a minor increase in thermal resistance using a chromate-based inhibitor. Although the vapour temperature was only 90°C, the authors recommended that accelerated life tests should be done for at least 35,000 h at 200°C–250°C.

Work in China [91] on long-term compatibility of steel/water units, with a carbon content slightly higher than the steel used in the RRR (0.123%), gave similar results to those of Novotna et al. [88]. The Chinese tests were carried out over a period of 29,160 h, with temperature drops along the heat pipes with passivation (with or without an inhibitor) never exceeding 6°C, and frequently being substantially less. Since then, carbon steel/water heat pipes have become a regular feature of heat recovery plant in China, with one company producing over 1000 heat exchangers [92].

Reported in 2003, the extensive activities in China on steel/water heat pipes have continued [93]. Tests in heat recovery duties where exhaust gases are up to 700°C have been carried out safely. Typified by the data in [94], the National Technological Supervision Administration of China has, with others, formulated standards for the technical specification of carbon steel–water heat pipe heat exchangers and boiler heat pipe economisers.

Further work in France, following an almost identical path to that in Czechoslovakia, confirms the above results. The authors [95] concluded that: 'In accordance with other studies, the addition of chromates in a neutral solution of distilled water and the formation of a magnetite layer produce sufficient conditions to prevent corrosion and hydrogen release in water/mild steel heat pipes at any operating temperature'.

The importance of deoxygenation of the water, even in systems that are not sealed to such a high integrity as heat pipes, is demonstrated by work on district heating pipeline corrosion [96]. Effective passivation was achieved in conjunction with deaerated water with an oxygen content of 40 (μg/kg), whereas if normal water (with an oxygen content of 4 mg/kg was used) pitting on the steel surfaces appeared.

### 3.5.2.6 Comments on the water–steel data

The above data suggests, that with the correct treatment of the water and the internal wall of the heat pipe/thermosyphon, steel and water are compatible in terms of the rate of noncondensable gas generation. However, there is a cost associated with the prevention of gas generation, both in treating the water and in achieving a suitable magnetite (or other material/form) layer on the container wall. Because if any subsequent manufacturing process (e.g. welding, heat treatment, enamel coating) involves a high-temperature operation downstream of passivation, this may need to be rescheduled – that is, it is always important when considering a heat pipe procedure that involves changes to internal surface structures to remember that subsequent manufacturing processes may degrade such procedures. Similarly, the heat pipe manufacturer may have little or no control over the system installer.

## 3.6 How about water and aluminium?

It is interesting to observe that there is, on the other hand, little or no data on the compatibility of aluminium and water as a heat pipe combination, and few attempts have been made by the heat pipe fraternity to overcome the perceived incompatibility using this combination.

The Ukrainian research mentioned in Ref. [90] did look at aluminium–water thermosyphons, and using alloy 6060, it was found that some corrosion could be slowed by selection of optimum pH conditions (5–6.5) and passivation method. However, long-term compatibility was not demonstrated.

Geiger and Quataert [97] have carried out corrosion tests on heat pipes using tungsten as the wall material and silver (Ag), gold (Au), copper (Cu), gallium (Ga), germanium (Ge), indium (In) and tin (Sn) as working fluids. The results from these tests carried out at temperatures of up to 1650°C enable Table 3.6 to be extended above 2000°C, albeit for heat pipes having comparatively short lives. Of the above combinations, tungsten and silver proved the most satisfactory, giving a life of 25 h at 2400°C with a possible extension if improved quality tungsten could be used.

## 3.7 Heat pipe start-up procedure

Heat pipe start-up behaviour is difficult to predict and may vary considerably depending upon many factors. The effects of working fluid, wick behaviour, and configuration on start-up performance have been studied qualitatively and a general description of the start-up procedure has been obtained [98].

During start-up, vapour must flow at a relatively high velocity to transfer heat from the evaporator to the condenser, and the pressure drop through the centre channel will be large. Since the axial temperature gradient in a heat pipe is determined by the vapour pressure drop, the initial temperature of the evaporator will be much higher than that of the condenser and will, of course, depend on the working fluid used. If the heat input is large enough a temperature front will gradually move towards the condenser section, and during normal heat pipe start-up, the temperature of the evaporator will increase by a few degrees until the front reached the end of the condenser. At this point the condenser temperature will increase until the pipe structure becomes almost isothermal (when lithium or sodium are used as working fluids, this process occurs at temperature levels where the heat pipe becomes red hot, and the near isothermal behaviour is visible).

Heat pipes with screen-covered channels behave normally during start-up as long as heat is not added too quickly, but Kemme found that heat pipes with open channels did not exhibit straightforward start-up behaviour. Very large temperature gradients were measured and the isothermal state was reached in a peculiar manner. When the heat was first added, the evaporator temperature levelled out at 525°C (sodium being the working fluid) and the front, with a temperature of 490°C, extended only a short distance into the condenser section. In order to achieve a near isothermal condition more heat was added, but the temperature of the evaporator did not increase uniformly, a temperature of 800°C being reached at the end of the evaporator farthest from the condenser. Most of the evaporator remained at 525°C and a sharp gradient existed between these two temperature regions.

Enough heat was added so that the 490°C front eventually reached the end of the condenser but before this occurred, temperatures in excess of 800°C were observed over a considerable portion of the evaporator. Once the condenser became almost isothermal, its temperature rapidly increased and the very hot evaporator region quickly cooled in a pattern which suggested that liquid return flow was in fact taking place. From this point, the heat pipe behaved normally. In some instances during start-up, when the vapour density is low and its velocity high, the liquid can be prevented from returning to the evaporator, and this is more likely to occur when open return channels are used for liquid transfer than when porous media are used.

Further work by van Andel [99] on heat pipe start-up has enabled some quantitative relationships to be obtained which assist in ensuring that satisfactory start-up can occur. This is based on the criterion that burnout does not occur, that is, the saturation pressure in the heated zones should not exceed the maximum capillary force. If burnout is allowed to occur, drying of the wick results inhibiting the return flow of liquid.

A relationship that gives the maximum allowable heat input rate during the start-up condition is:

$$Q_{\max} = 0.4\pi r_c^2 \times 0.73L\left(P_E \rho_E\right)^{\frac{1}{2}} \tag{3.11}$$

where $r_c$ is the vapour channel radius, $L$ is the latent heat of vaporisation, and $P_E$ and $\rho_E$ are the vapour pressure and vapour density in the evaporator section respectively.

It is important to meet the start-up criteria when a heat pipe is used in an application that may involve numerous starting and stopping actions – for example, in cooling a piece of electronic equipment or cooling brakes. One way in which the problem can be overcome is to use an extra heat source connected to a small branch heat pipe when the primary role of cooling is required, thus reducing the number of start-up operations. The start-up time of gas-buffered heat pipes is quicker. Busse [100] has made a significant contribution to the analysis of the performance of heat pipes, showing that before sonic choking occurs, a viscous limitation (that can lie well below the sonic limit) can be met. This is described in detail in Chapter 4. Where it is required to calculate the transient behaviour of heat pipes during start-up and in later transient operation, time constants and other data may be calculated using equations presented in Ref. [81].

## 3.8 Heat pipe manufacture and testing

The manufacture of conventional capillary driven heat pipes involves a number of comparatively simple operations particularly when the unit is designed for operation at temperatures of the order of, say, 50°C−200°C. It embraces skills such as welding, machining, chemical cleaning and nondestructive testing and can be carried out following a relatively small outlay on capital equipment − the most expensive item of which is likely to be the leak detection equipment. (Note that many procedures described are equally applicable to thermosyphons). Tubular heat pipes are generally simpler to manufacture than flat plate forms, and the tubular unit is the heat pipe (or thermosyphon) of choice for those starting to examine their performance in the laboratory.

With all heat pipes, cleanliness is of prime importance to ensure that no incompatibilities exist (assuming that the materials selected for the wick, wall and working fluid are themselves compatible) and to make certain that the wick and wall will be wetted by the working fluid. As well as affecting the life of the heat pipe, negligence in assembly procedures can also lead to inferior performance, due, for example, to poor wetting. Atmospheric contaminants, in addition to those likely to be present in the raw working fluid, must be avoided. Above all, the heat pipe must be leak-tight to a very high degree. This can involve outgassing of the metal used for the heat pipe wall, end caps etc., although this is not essential for simple low-temperature operations. Quality control cannot be overemphasised in heat pipe manufacture, and in the following discussion of assembly methods, this will be frequently stressed.

A substantial part of this chapter is allocated to a review of life test procedures for heat pipes. The life of a heat pipe often requires careful assessment in view of the many factors that can affect long-term performance, and most establishments seriously involved in heat pipe design and manufacture have extensive life test programmes in progress. As discussed later, data available from the literature can indicate satisfactory wall/wick/working fluid combinations, but the assembly procedures used differ from one manufacturer to another, and this may introduce an unknown factor that will necessitate investigation. The outcomes of life tests using a wide variety of fluids are given in this chapter.

Measuring the performance of heat pipes is also a necessary part of the work leading to an acceptable product, and the interpretation of the results may prove difficult. Test procedures for heat pipes destined for use in orbiting satellites have their own special requirements brought about by the need to predict performance in zero gravity by testing in earth gravity. There are numerous examples in the literature of measured temperature profiles along heat pipes that show much greater temperature drops than one would anticipate if the operation was within limits and no noncondensable gases were present. It is then necessary to examine where temperature measurements were made and the thermal resistances between the inside of the condenser (e.g.) and the heat sink in order to fully understand the possibly high overall resistance.

Whilst the vast majority of heat pipes manufactured today are conventional wicked circular (or near circular) cross-section units, there is an increasing trend towards miniaturisation and the use of microgroove-type structures as wicks. The implications of these trends for manufacturing procedures are highlighted in appropriate parts of this chapter.

Although the manufacture of special types such as loop, oscillating or microheat pipes are outside the scope of this book there are, again, features unique to these types that require close attention during manufacture and assembly, and these are identified and referenced where it is believed appropriate.

The use of 3D printing, of which selective laser remelting is one form of construction, is new to the heat pipe manufacturing field and is described later − Section 3.8.15. Imagine being able to construct a heat pipe, together with a composite wick, as a single assembly from a CAD drawing via a 3D printing machine.

### 3.8.1 Manufacture and assembly

#### *3.8.1.1 Container materials*

The heat pipe container, including the end caps and filling tube, is selected on the basis of several properties of the material used and these are listed within this chapter (unless stated otherwise, the discussion in this chapter assumes that the heat pipes are tubular in geometry). However, the practical implications of this selection are numerous.

Of the many materials available for the container, three are by far the most common in use, namely copper, aluminium and stainless steel. Copper is eminently satisfactory for heat pipes operating between 0°C and 200°C in applications such as electronics cooling. Whilst a commercially pure copper tube is suitable, the oxygen-free high-conductivity type is preferable, and like aluminium and stainless steel, the material is readily available and can be obtained in a wide variety of diameters and wall thicknesses in its tubular form.

Aluminium is less common as a material for commercially available heat pipes but has received a great deal of attention in aerospace applications because of its obvious weight advantages. It is generally used in alloy form, typically

6061-T6, the nearest British equivalent being aluminium alloy HT30. Again, this is readily available and can be drawn to suit (by the heat pipe manufacturer), or extruded to incorporate, for example, a grooved wick.

Stainless steel unfortunately cannot generally be used as a container material with water where a long life is required owing to gas generation problems. However, it is perfectly acceptable with many other working fluids and is, in many cases, the only suitable container for example with liquid metals such as mercury, sodium and potassium. Types of stainless steel regularly used for heat pipes include 302, 316 and 321. (Comments on compatibility with water are made in Table 3.2.) Mild steel may be used with organic fluids and again reference to Table 3.2 can be made for a discussion on its possible compatibility with water.

In the assembly of heat pipes, provision must be made for a filling, and the most common procedure involves the use of an end cap with a small diameter tube attached to it, as shown in Fig. 3.9. The other end of the heat pipe contains a blank end cap. End cap and filling tube materials are generally identical to those of the heat pipe case, although for convenience a copper extension may be added to a stainless steel filling tube for cold welding (see Section 3.8.9). It may also be desirable to add a valve to the filling tube where, for example, the gas analysis may be carried out following life tests (see Section 3.9), but the valve material must, of course, be compatible with the working fluid.

If the heat pipe is to operate at high vapour pressures, a pressure test should be carried out to check the integrity of the vessel.

There have been a number of heat pipes and thermosyphons constructed using polymers. For example, polymer thermal diodes have been used to transfer solar thermal energy through the walls of buildings, and flexible sections in some early tubular heat pipes were formed from a polymer. However, the sealing of polymers has been a problem.

A group in Taiwan, one of several working on polymer heat pipes, has employed polyethylene-terephthalate (PET) plastic film in a flat plate heat pipe that used methanol as the working fluid with a copper mesh wick [101]. The PET sheets can be bonded together using hot lamination, but as with flat plate heat pipes in general (but more so with polymer ones), care has to be taken to ensure that the vapour pressure does not become too great. In the case of the PET unit, it was found that delamination could occur at higher pressures, so a limit of just over 14 kPa was put on the internal pressure. The group concluded that the device was a viable alternative (and a low-cost one) in some electronics cooling applications.

### 3.8.2 Wick materials and form

The number and form of materials that have been tested as wicks in heat pipes is very large and reference has already been made to some of these in the analysis of the liquid pressure drop, presented in Chapter 4, and in the discussion on selection criteria in Table 3.4.

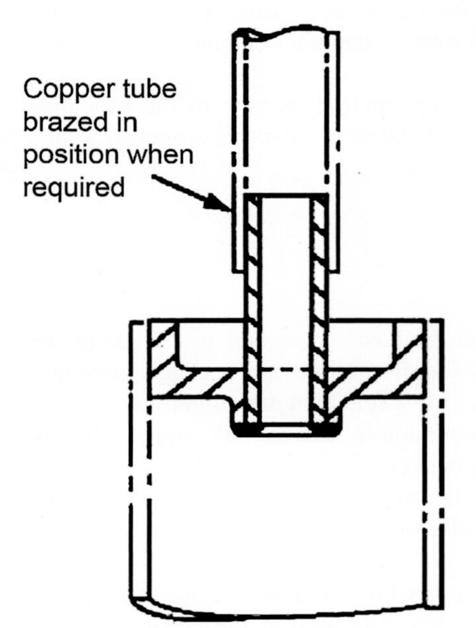


**FIGURE 3.9** End cap and filling tube.

### 3.8.2.1 Wire mesh

The most common form of wick is a woven wire mesh or twill which can be made from many metals. Stainless steel, monel and copper are woven to produce meshes having very small pore sizes (see Table 3.4) and 400 mesh stainless steel is available 'off the shelf' from several manufacturers. Aluminium is available, but because of difficulties in producing and weaving fine aluminium wires, the requirements of small pore wicks cannot be met. Stainless steel is the easiest material to handle in mesh form. It can be rolled and retains its shape well particularly when a coarse mesh is used. The inherent springiness in coarse meshes assists in retaining the wick against the heat pipe wall, in some cases obviating the need for any other form of wick location. In heat pipes where a 400 mesh is used, a coarse 100 mesh layer located at the inner radius can hold the finer mesh in shape. Stainless steel can also be diffusion bonded, giving strong permanent wick structures attached to the heat pipe wall. The diffusion bonding of stainless steel is best carried out in a vacuum furnace at a temperature of 1150°C–1200°C.

The spot welding of wicks is a convenient technique for preserving shape or for attaching the wick to the wall in cases where the heat pipe diameter is sufficiently large to permit the insertion of an electrode. Failing this, a coil spring can be used.

It is important to ensure (whatever the wick form) that it is in very close contact with the heat pipe wall, particularly at the evaporator section; otherwise, local hot spots will occur. With mesh the best way of making certain that this is the case is to diffusion bond the assembly.

The manufacture of heat pipes for thermal control of the chips in laptop computers and the like conventionally involves wicked copper heat pipes. Here copper or nickel mesh may be employed instead of stainless steel, depending upon the choice of the working fluid.

### 3.8.2.2 Sintering

A similar structure having an intimate contact with the heat pipe wall is a sintered wick. Sintering is often used to produce metallic filters and many components of machines are now produced by this process as opposed to die casting or moulding.

The process involves bonding together a large number of particles in the form of a packed metal powder, and the pore size of the wick thus formed can be arranged to suit by selecting powders having a particular size. The powder, which is normally spherical, is placed in containers giving the shape required and then either sintered without being further compacted or, if a temporary binder is used, a small amount of pressure may be applied. Sintering is normally carried out at a temperature of 100°C–200°C below the melting point of the sintering material.

The simplest way of making wicks by this method is to sinter the powder in the tube that will form the final heat pipe. This has the advantage that the wick is also sintered to the tube wall and thus makes a stronger structure. In order to leave the central vapour channel open, a temporary mandrel has to be inserted in the tube and the powder is then placed in the annulus between mandrel and tube. In the case of copper powder, a stainless steel mandrel is satisfactory as the copper will not bond to stainless steel and thus the bar can easily be removed after sintering. The bar is held in a central position at each end of the tube by a stainless steel collar.

A typical sintering process is described below. Copper was selected as both the powder material and the heat pipe wall. The particle size chosen was $-150 + 300$ grade, giving particles of 0.05–0.11 mm diameter. The tube was fitted with a mandrel and a collar at one end. The powder was then poured in from the other end.

No attempt was made to compact the powder apart from tapping the tube to make sure there were no gross cavities left. When the tube was full the other collar was put in place and pushed up against the powder. The complete assembly was then sintered by heating in hydrogen at 850°C for 0.5 h. After the tube was cooled and removed, the tube, without the mandrel, was then resintered. (The reason for this was that when the mandrel was in place the hydrogen could not flow easily through the powder and as a result, sintering may not have been completely successful since hydrogen is necessary to reduce the oxide film that hinders the process.) After this operation, the tube was ready for use. Fig. 3.10 shows a cross-section of a completed tube, and Fig. 3.11 shows a magnified view of the structure of the copper wick. The porosity of the finished wick is of the order of 40–50%. A second type of sintering may be carried out to increase the porosity. This necessitates the incorporation of inert filler material to act as pore formers. This is subsequently removed during the sintering process, thus leaving a very porous structure. The filler used was a Perspex powder (that is available as small spheres) which was sieved to remove the $-150 + 300$ (0.050–0.100 mm) fraction, and this was mixed with an equal volume of very fine copper powder (−200/μm).

On mixing, the copper uniformly coats the plastic spheres—this composite powder then showing no tendency to separate into its components.

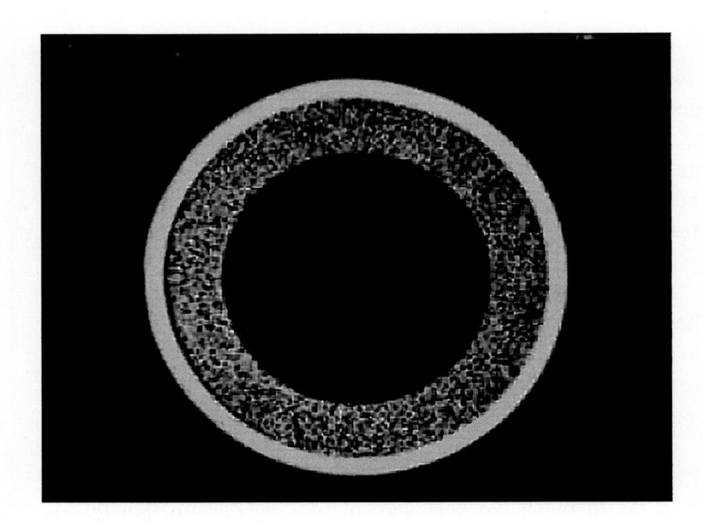

**FIGURE 3.10** Sintered wick cross-section (copper).

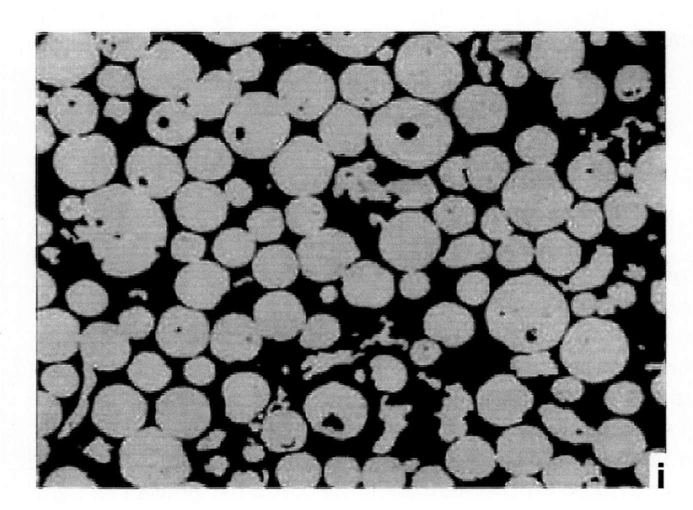

**FIGURE 3.11** Magnified view of sinter structure.

The wick is now made up exactly as the previous tube with the exception that more compaction is required in order to combat the very high shrinkage that takes place during sintering. During the initial stages of the sintering, the plastic is vapourised and diffused out of the copper compact, thus leaving a skeletal structure of fine copper powder with large interconnected pores. The final porosity is probably of the order of 75%–85%.

It is obvious that there are many possible variations of the wicks made by sintering methods. Porosity, capillary rise and volume flow can all be optimised by the correct choice of metal powder size, filler size, filler proportion and by incorporation of channel-forming fillers.

Flat plate heat pipes, sometimes called vapour chambers (where the role of the wick may be solely for liquid distribution across the evaporator), most commonly use sintered wick structures. The Thermacore 'Therma-Base' unit [102] uses a copper sintered wick – the unit being illustrated in Fig. 3.12. An alternative method for achieving something approaching a uniform heat distribution across a plate is to embed heat pipes of tubular form within it, as shown in the Thermacore unit in Fig. 3.13.

Not all flat plate heat pipes are designed to operate against gravity – in which case the equipment can operate with relatively simple wick structures – and one heat pipe that was mounted vertically was used in the thermal control of thermoelectric refrigeration units [103]. Although subject to further development, the use of a sintered copper (40 μm particles) wick with longitudinal channels – a graded wick structure or composite structure – was interesting. As highlighted elsewhere, the impact of the fluid fill on the heat pipe performance was shown to be significant. Results showed that as the fluid loads varied between 5 and 40 g, the thermal resistance initially decreased from 0.25 K/W at

**FIGURE 3.12** The Thermacore Therma-Base flat plate heat pipe, employing a sintered wick. *Courtesy: Thermacore International Inc.*

**FIGURE 3.13** An alternative approach to 'spreading the heat' over a flat surface – the Therma-core heat spreader. *Courtesy: Thermacore International Inc.*

the 5 g inventory to a minimum of 0.15 K/W at the 20 g, before rising to over 0.3 K/W at 40 g. Careful assessment for fluid inventory in wicked heat pipes of all types is therefore essential.

A French heat pipe manufacturer, Atherm, (see Appendix 3) has recently reported on 'industrial manufacturing' of the wicks used in LHPs. These are sintered structures but the geometry/dimensions differ from those in conventional tubular heat pipes. In the Atherm case, the LHP evaporator wick was 30 mm long, with an outside diameter of 10 mm and a 2 mm diameter cylindrical hole in the centre. In order to achieve the close tolerances needed for the LHP unit (which could not tolerate shrinkage of the sintered wick), nickel and bronze preformed sintered parts were machined, and the machining methods such as drilling, turning and milling were found not to block the surface pores and satisfactory capillary action was achieved [104]. The sintered structure had a porosity of 65%–70% and a pore diameter of 8–9 μm.

Work on biporous wicks (a new term for what are called composite wicks in other chapters) has been reported by Byon et al. in Korea [105]. The wicks were formed from spherical copper particles (supplied by ACupowder, Inc.) sintered into a porous medium in a high-temperature furnace with a reducing atmosphere of nitrogen and hydrogen. The sintered porous medium was ground into clusters, sieved into specific sizes and again sintered onto a copper disc (10-mm diameter, 5-mm thick). In order to carry out experiments to compare the performances of the two wick structures, the wick thickness was defined to be 1 mm and set to be 1 mm. The diameter of the spheres in the monoporous wick was 45–200 μm and the same diameter particles were used in the biporous wick, but before final sintering, these were sintered into clusters of 250–675 μm diameter.

### *3.8.2.3 Vapour deposition*

Sintering is not the only technique whereby a porous layer can be formed which is in intimate contact with the inner wall of the heat pipe. Other processes include vapour coating, cathode sputtering and flame spraying. Brown Boveri, in UK Patent 1313525, describes a process known as 'vapour plating' that has been successfully used in heat pipe wick construction. This involves plating the internal surface of the heat pipe structure with a tungsten layer by reacting tungsten hexafluoride vapour with hydrogen, the porosity of the layer being governed by the surface temperature, nozzle movement and distance of the nozzle from the surface to be coated.

### 3.8.2.4 Microlithography and other techniques

The trend towards microheat pipes (see also Chapter 2) has led to the use of manufacturing techniques for such units that copies some of the methods used in the microelectronics area that these devices are targeting. Sandia National Laboratory is one of a number of laboratories using photolithographic methods (or similar techniques) for making heat pipes, or more specifically, the wick. Workers at Sandia [106] wanted to make a microheat pipe that could cool multiple heat sources, keep the favourable permeability characteristics of longitudinally grooved wicks and allow fabrication using photolithography.

Full arguments behind the selection of the anisotropic wick concept (based upon longitudinal liquid flow whilst minimising transverse movement) are given in the referenced paper, but the wick they made was formed with rectangular channels of width 40 μm and height 60 μm and used an electroplating process that plated the copper wick material onto the flat 150 mm substrate wafers.

Final assembly comprised cutting the wick parts, putting in spacers and the fill tube, and then electron beam welding around the periphery (support posts inside were resistance welded).

In the period since this work was reported, progress is such that microfabrication technology, including direct writing (3D printing) methods such as selective laser melting (SLM), has developed to the point where a three-dimensional wick structure of almost any form could be fabricated by a rapid prototyping method in several laboratories around the world.

### 3.8.2.5 Grooves

A type of wick – widely used in spacecraft applications but which is unable to support significant capillary heads in earth gravity – is a grooved system, and the simplest way of producing longitudinal grooves in the wall of a heat pipe is by extrusion or by broaching. Aluminium is the most satisfactory material for extruding because grooves may be comparatively narrow in width but still possess a greater depth. An example of a copper-grooved heat pipe wick is shown in Fig. 3.14. The external cross-section of the heat pipe can also be adapted for a particular application and if the heat pipe is to be mounted on a plate, a flat surface may be incorporated on the wall of the heat pipe to give better thermal contact with the plate.

An alternative groove arrangement involves 'threading' the inside wall of the heat pipe using taps or a single-point cutting tool to give a thread pitch of up to 40 threads per cm. Threaded arteries are attractive for circumferential liquid distribution and may be used in conjunction with a different artery system for axial liquid transport.

Longitudinal grooves may be used as arteries (see Chapter 4 for example) and Fig. 3.15a shows such an artery set machined in the form of six grooves in a former for insertion down the centre of a heat pipe. Prior to insertion, the

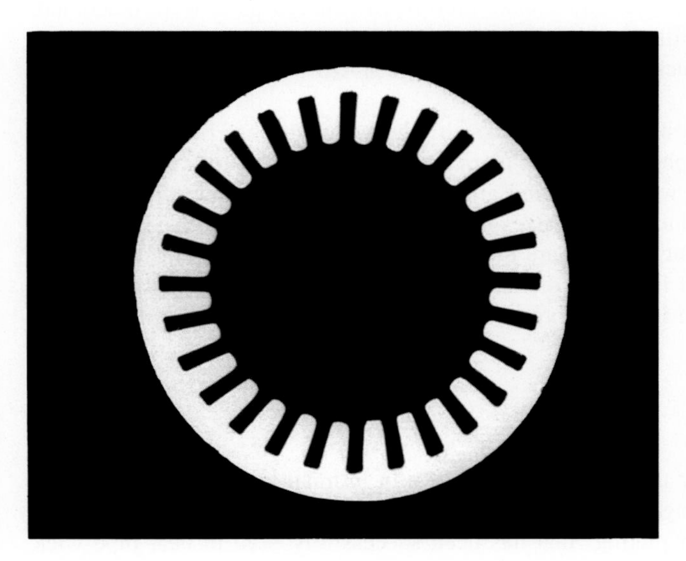

**FIGURE 3.14** Grooved wick in a copper tube, obtained by mandrel drawing.

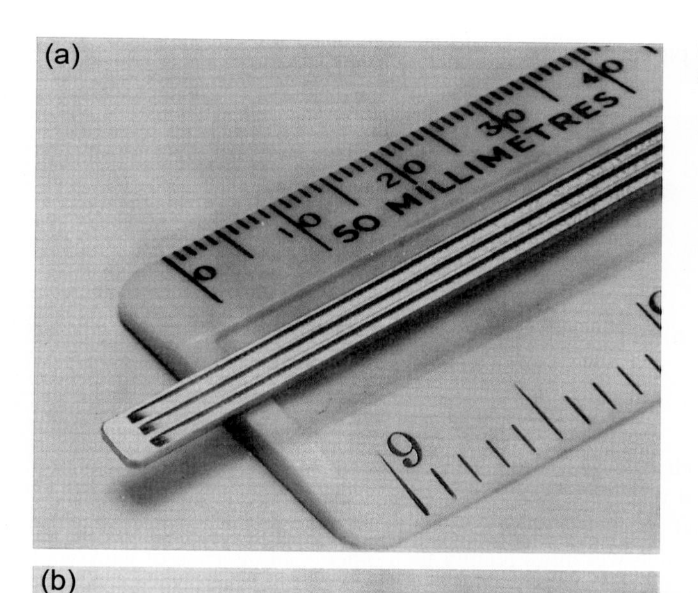


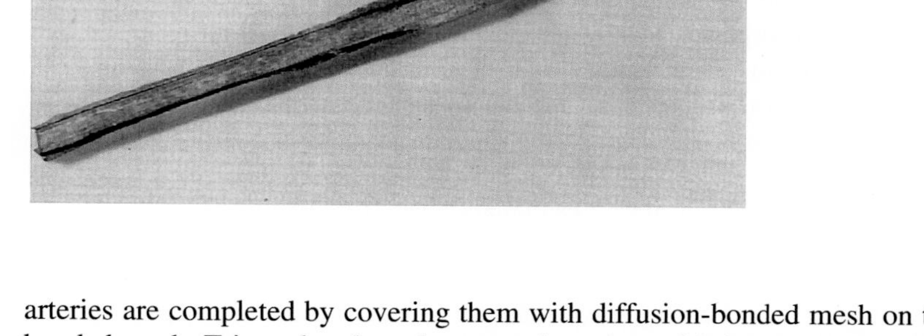


**FIGURE 3.15** (a) Artery set prior to covering with mesh. (b) Artery set prior to covering with mesh.

arteries are completed by covering them with diffusion-bonded mesh on their outer surface. Fig. 3.15b shows diffusion-bonded mesh. Triangular-shaped grooves have been fabricated in silicon for electronics cooling duties. Work at institut national des sciences appliquées (INSA) in France [107,108] used two processes – anisotropic chemical etching followed by direct silicon wafer bonding – for microheat pipe fabrication. In one unit, 55 triangular parallel microheat pipes were constructed (230 μm wide and 170 μm deep) and ethanol was used as the working fluid. A section through the array is shown in Fig. 3.16a, whilst a second unit, shown in Fig. 3.16b, uses arteries, fabricated in an identical manner in a third layer of silicon.

### 3.8.2.6 Felts and foams

Several companies are now producing metal and ceramic felts and metal foam which can be effectively used as heat pipe wicks – particularly where units of noncircular cross-section are required. The properties of some of these materials are given in Table 3.3. Foams are available in nickel, stainless steel and copper, and felt materials include stainless steel and woven ceramic fibres (Refrasil). Foams are available in sheet and rod forms and can be supplied in a variety of pore sizes. Metallic felts are normally produced in sheets and are much more pliable than foams. An advantage of the felt is that by using mandrels and applying a sintering process, longitudinal arteries could be incorporated inside the structure providing low-resistance flow paths. The foam, however, may also double as a structural component.

Knitted ceramic fibres are available with very small pore sizes and are inert to most common working fluids. Because of their lack of rigidity, particularly when saturated with a liquid, it is advisable to use them in conjunction with a wire mesh wick to retain their shape and desired location. The ceramic structure can be obtained in the form of multilayer sleeves, ideal for immediate use as a wick, and a range of sleeve diameters is available. Some stretching of the sleeve can be applied to reduce the diameter, should the exact size not be available.

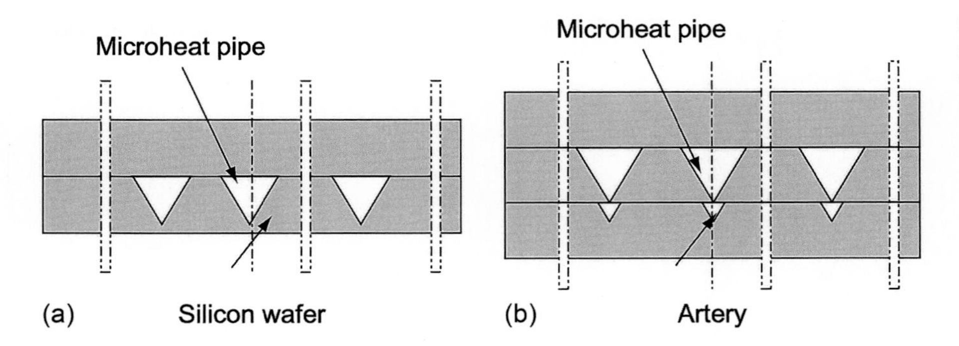


**FIGURE 3.16** Cross-sections of two microheat pipe arrays fabricated in silicon: (a) without artery and (b) with artery [108].

### 3.8.3 Cleaning of container and wick

All the materials used in a heat pipe must be clean because cleanliness achieves two objectives. It ensures that the working fluid will wet the materials and that no foreign matter is present which could hinder capillary action or create incompatibilities.

The cleaning procedure depends upon the material used, the process undergone in manufacturing and locating the wick, and the requirements of the working fluid (some of which wet more readily than others). In the case of wick/wall assembles produced by processes such as sintering or diffusion bonding carried out under an inert gas or vacuum, the components are cleaned during the bonding process and provided that the time between this process and final assembly is short no further cleaning may be necessary.

If the working fluid is a solvent, such as acetone, no extreme precautions are necessary to ensure good wetting and an acid pickle followed by a rinse in the working fluid appears to be satisfactory. However, cleaning procedures become more rigorous as the operating temperature range is increased to incorporate, for example, liquid metals as working fluids.

The pickling process for stainless steel involves immersing the components in a solution of 50% nitric acid and 5% hydrofluoric acid, and this is followed by a rinse in demineralised water. If the units are to be used in conjunction with water, the wick should then be placed in an electric furnace and heated in air to 400°C for 1 h. At this temperature, grease is either volatilised or decomposed and the resulting carbon is burnt off to form carbon dioxide. Since an oxide coating is required on the stainless steel it is not necessary to use an inert gas blanket in the furnace.

Nickel may undergo a similar process to that described above for stainless steel, but pickling should be carried out in a 25% nitric acid solution. Pickling of copper demands a 50% phosphoric acid and 50% nitric acid mixture.

Cleanliness is difficult to quantify and the best test is to add a drop of demineralised water to the cleaned surface. If the drop immediately spreads across the surface or is completely absorbed into the wick, good wetting has occurred and satisfactory cleanliness has been achieved.

Stainless steel wicks in long heat pipes sometimes create problems in that furnaces of sufficient size to contain the complete wick may not be readily available. In this case, a flame cleaning procedure may be used whereby the wick is passed through a Bunsen flame as it is fed into the container.

An ultrasonic cleaning bath is a useful addition for speeding up the cleaning process but is by no means essential for low-temperature heat pipes. With this process (or any other associated with the immersion of components in a liquid to remove contaminants), the debris will float to the top of the bath and must be skimmed off before removing the parts being treated. If this is not done, the parts could be recontaminated as they are removed through this layer. Electropolishing may also be used to aid in the cleaning of metallic components.

Ceramic wick materials are generally exceptionally clean when received from the manufacturer owing to the production process used to form them and therefore need no treatment provided that their handling during assembly of the heat pipe is under clean conditions.

In this respect, it is important, particularly when water is used as the working fluid, to avoid skin contact with the heat pipe components. Slight grease contamination can prevent wetting and the use of surgical gloves for handling is advisable. Wetting can be aided by additives (wetting agents) [109] applied to the working fluid, but this can introduce compatibility problems and may also affect surface tension.

### 3.8.4 Material outgassing

When the wick or wall material is under vacuum, gases will be drawn out (particularly if the components are metallic), and if not removed prior to sealing of the heat pipe these gases could collect in the heat pipe vapour space. The process is known as outgassing.

Whilst outgassing does not appear to be a problem in low-temperature heat pipes applications that are not too arduous, high-temperature units ( $>$400°C) and pipes for space use should be outgassed in the laboratory prior to filling with working fluid and sealing.

The outgassing rate is strongly dependent on temperature, increasing rapidly as the component temperature is raised. Following cleaning, it is advisable to outgas components under vacuum at a baking temperature of about 400°C; then following this baking, the system should be vented with dry nitrogen. The rate of outgassing depends on the heat pipe operating vapour pressure, and if this is high the outgassing rate will be restricted.

If the heat pipe has been partially assembled prior to outgassing and the end caps fitted, it is necessary to make sure that are no leaks with the welds etc., as these could produce misleading results to the outgassing rate. Generally though, analysis of gases escaping through a leak will show a very large air content, whereas those brought out by outgassing will contain a substantial water vapour content. A mass spectrometer can be used to analyse these gases. Leak detection is covered in Section 3.8.6.

The outgassing characteristics of metals can differ considerably. The removal of hydrogen from stainless steel, for example, is much easier to effect than its removal from aluminium. Aluminium is particularly difficult to outgas and can hold comparatively large quantities of noncondensables. In one test, it was found [110] that gas was suddenly released from the aluminium when it approached red heat under vacuum. Two hundred grams of metal gave 89.5 cc of gas at NTP, 88 cc being hydrogen and the remainder carbon dioxide. It is also believed that aluminium surfaces can retain water vapour even when heated to 500°C or dried over phosphorous pentoxide. This could be particularly significant because of the known incompatibility of water with aluminium (see Section 3.8.12 for high-temperature heat pipes).

### 3.8.5 Fitting of wick and end caps

Cleaning of the heat pipe components is best carried out before insertion of the wick as it is then easier to test the wick for wettability. Outgassing may be implemented before assembly or whilst the heat pipe is on the filling rig (see Section 3.8.8).

In cases where the wick is an integral part of the heat pipe wall, as in the case of grooves, sintered powders or diffusion-bonded meshes, cleaning of the heat pipe by flushing through with the appropriate liquid is convenient prior to the welding of the end caps.

If a mesh wick is used and the mesh layers are not bonded to one another or to the heat pipe wall, and particularly when only a fine mesh is used, a coiled spring must be inserted to retain the wick against the wall. This is readily done by coiling the spring tightly around a mandrel giving a good internal clearance in the heat pipe. The mandrel is inserted into the pipe, the spring tension is released, and the mandrel is then removed. The spring will now be holding the wick against the wall. Typically, the spring pitch is about 1 cm. (In instances where two mesh sizes may be used in the heat pipe, say two layers of 200 mesh and one layer of 100 mesh, the liquid–vapour interface must always be in the 200 mesh to achieve maximum capillary rise. It is therefore advisable to wrap the 200 mesh over the end of the 100 mesh, as shown in Fig. 3.17. It is possible to locate the fine mesh against the wall, where it will suppress boiling.)

The fitting of end caps is normally carried out by argon-arc welding. This need not be done in a glove box and is applicable to copper, stainless steel and aluminium heat pipes. The advantage of welding over brazing or soldering is that no flux is required so the inside of cleaned pipes does not suffer from possible contamination. However, possible inadequacies of the argon shield, in conjunction with the high temperatures involved can lead to local material

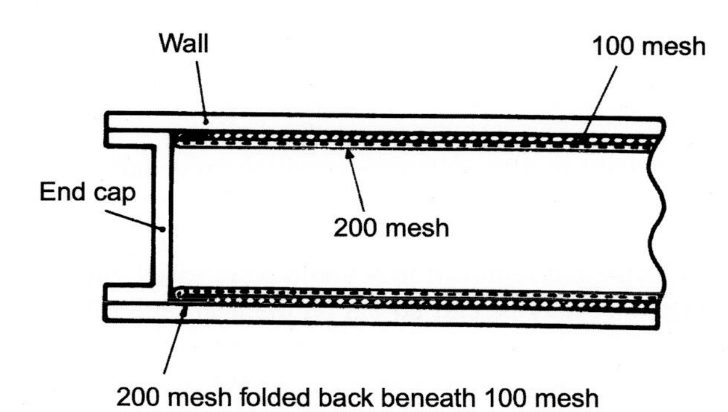


**FIGURE 3.17** Sealing of mesh at the end of heat pipe.

oxidation which may be difficult to remove from the heat pipe interior. Assembly in a glove box filled with argon would overcome this but would be expensive. The use of a thermal absorbent paste (such as Rocol HS) to surround the area of the heat pipe local to the weld can considerably reduce the amount of oxide formed.

Electron beam welding may also be used for heat pipe assembly, but this added expense cannot be justified in most applications.

### 3.8.6 Leak detection

All welds on heat pipes should be checked for leaks. If quality control is to be maintained a rigorous leak check procedure is necessary because a small leak that may not initially affect heat pipe performance could make itself felt over a period of months.

The best way to test a heat pipe for leaks is to use a mass spectrometer that can be used to evacuate the heat pipe to a very high vacuum (better than $10^{-5}$ torr) using a diffusion pump. The weld area is then tested by directing a small jet of helium gas onto it. If a leak is present, the gauge head on the mass spectrometer will sense the presence of helium once it enters the heat pipe. After an investigation of the weld areas and location of the general leak area(s), if a leak is present, a hypodermic needle can be attached to the helium line and careful traversing of the suspected region can lead to very accurate identification of the leak position, possibly necessitating only a very local rewelding procedure to seal it.

Obviously, if a very large leak is present, the pump on the mass spectrometer may not even manage to obtain a vacuum better than $10^{-2}$ or $10^{-3}$ torr. Porosity in weld regions can create conditions leading to this and may point to impure argon or an unsuitable welding filler rod.

It is possible, if the leak is very small, for water vapour from the breath to condense and block, albeit temporarily, the leak. It is therefore important to keep the pipe dry during leak detection.

### 3.8.7 Preparation of the working fluid

It is necessary to treat the working fluid used in a heat pipe with the same care as that given to the wick and container.

The working fluid should be the most highly pure available, but further purification may still be necessary following purchase and this may be carried out by distillation. In the case of low-temperature working fluids such as acetone, methanol and ammonia, the presence of water can lead to incompatibilities, and the minimum possible water content should be achieved.

Some brief quotations from a treatise on organic solvents [111] highlight the problems associated with acetone and its water content:

> *Acetone is much more reactive than is generally supposed. Such mildly basic materials as alumina gel cause aldol condensation to 4-hydroxy-4-methyl-2-pentanone, (diacetone alcohol), and an appreciable quantity is formed in a time if the acetone is warm. Small amounts of acidic material, even as mild as anhydrous magnesium sulphate, cause acetone to condense.*
>
> *Silica gel and alumina increased the water content of the acetone, presumably through the aldol condensation and subsequent dehydration. The water content of acetone was increased from 0.24 to 0.46% by one pass over alumina. All other drying agents tried caused some condensation.*
>
> *Ammonia has a very great affinity for water, and it has been found that a water content of the order of 10 ppm is necessary to obtain satisfactory performance. Several chemical companies are able to supply high-purity ammonia but exposure to air during heat pipe filling must be avoided.*

The above examples are extreme but serve to illustrate the problems that can arise when the handling procedures are relaxed.

A procedure that is recommended for all heat pipe working fluids used up to 200°C is freeze degassing. This process removes all dissolved gases from the working fluid, because if the gases are not removed they could be released during heat pipe operation and collected in the condenser section. Freeze degassing may be carried out on the heat pipe filling rig described in Section 3.8.8 and is a simple process. The fluid is placed in a container in the rig directly connected to the vacuum system and is frozen by surrounding the container with a flask containing liquid nitrogen. When the working fluid is completely frozen, the container is evacuated and resealed and the liquid nitrogen flask is removed. The working fluid is then allowed to thaw and any dissolved gases will be seen to bubble out of the liquid. The working fluid is then refrozen and the process is repeated. All gases will be removed after three or four freezing cycles and the liquid will now be in a sufficiently pure state for insertion into the heat pipe.

### 3.8.8 Heat pipe filling

A flow diagram for a rig that may be used for heat pipe filling is shown in Fig. 3.18, but the rig may also be used to carry out the following processes:

- Working fluid degassing
- Working fluid metering
- Heat pipe degassing
- Heat pipe filling with inert gas.

Before describing the rig and its operation it is worth mentioning the general requirements when designing vacuum rigs. Pipework is generally either glass or stainless steel, and glass has advantages when handling liquids in that the presence of liquid droplets in the ductwork can be observed and their vaporisation under vacuum noted. On the other hand, stainless steel has obvious strength benefits and must be used for all high-temperature work, together with high-temperature packless valves such as Hoke bellows valves. The rig described below is for low-temperature heat pipe manufacture.

Valves used in vacuum rigs should preferably have 'O' ring seals, and it is important to ensure that the ductwork is not too long nor has a small diameter as these can greatly increase evacuation times.

The vacuum pump may be the diffusion type or a sorption pump containing a molecular sieve that produces vacuums as low as $10^{-4}$ torr. It is, of course, advisable to refer to experts in the field of high-vacuum technology when considering designing a filling rig.

#### *3.8.8.1 Description of rig*

The heat pipe filling rig described below is made using glass for most of the pipework. Commencing from the right-hand side, the pump is of the sorption type, which is surrounded by a polystyrene container of liquid nitrogen when a vacuum is desired. Two valves are fitted above the pump, the lower one being used to disconnect the pump when it becomes saturated. (The pump may be cleaned by baking out in a furnace for a few hours.) Above the valve V2, a glass-to-metal seal is located and the rest of the pipework is glass. Two limbs lead from this point, both interrupted by cold traps in the form of small glass flasks which are used to trap stray liquid and any impurities which could affect other parts of the rig or contaminate the pump. The cold traps are formed by surrounding each flask with a container of liquid nitrogen.

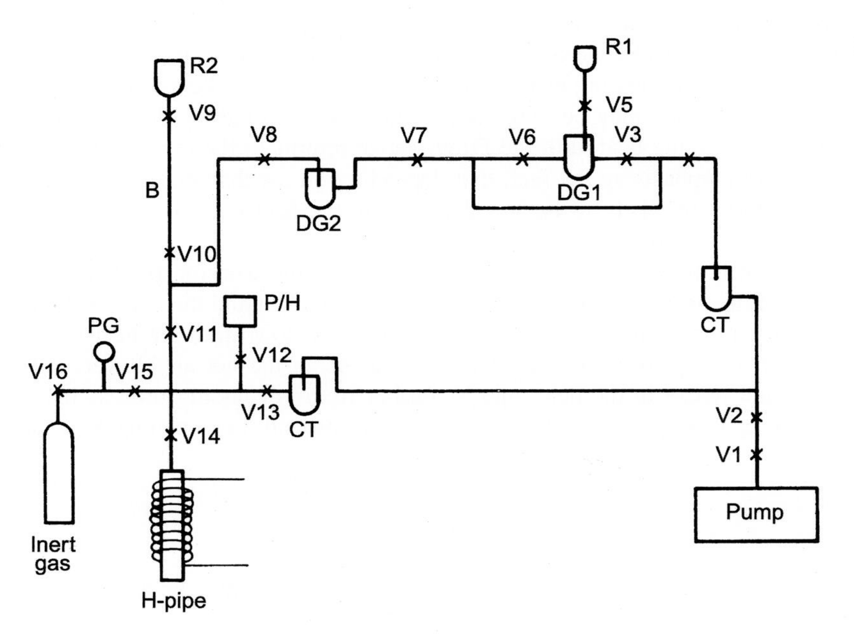


**FIGURE 3.18** A heat pipe filling rig layout.

The upper limb includes provision for adding working fluid to the rig and two flasks are included (DG1 and DG2) for degassing the fluid. The section of the rig used for adding fluid can be isolated once a sufficient quantity of fluid has been passed to flask DG2 and thence to the burette between valves V9 and V10.

The lower limb incorporates a Pirani head that is used to measure the degree of vacuum in the rig. The heat pipe to be filled is fitted below the burette, and provision is also made to electrically heat the pipe to enable outgassing of the unit to be carried out on the rig (see also Section 3.8.4). An optional connection can be made via valve V15 to permit the loading of inert gas into the heat pipe for variable conductance types.

#### *3.8.8.2 Procedure for filling a heat pipe*

The following procedure may be followed using this rig for filling, for example, a copper/ethanol heat pipe.

1. Close all valves linking rig to atmosphere (V5, V9, V14, V15).
2. Attach sorption pump to rig via valves V1 and V2, both of which should be closed.
3. Surround the pump with liquid nitrogen and also top up the liquid nitrogen containers around the cold traps. (It will be found that the liquid $N_2$ initially evaporates quickly, and regular topping up of the pump and traps will be necessary.)
4. After approximately 30 min, open valves V1 and V2, commencing rig evacuation. Evacuate to about 0.010 mmHg, the time to achieve this depending on the pump capacity, rig cleanliness and rig volume.
5. Close valves V4 and V6 and top up reservoir R1 with ethanol.
6. Slowly crack valve V5 to allow ethanol into flask DG1. Reclose V5 and freeze the ethanol using a flask of liquid $N_2$ around DG1.
7. When all the ethanol is frozen, open V4 and evacuate. Close V4 and allow ethanol to melt. All gas will bubble out of the ethanol as it melts. The ethanol is then refrozen.
8. Open V4 to remove gas.
9. Close V4, V3 and V8; open V6 and V7. Place liquid $N_2$ container around flask DG2.
10. Melt the ethanol in DG1 and drive it into DG2. (This is best carried out by carefully heating the frozen mass using a hair dryer. Warming of the ductwork between DG1 and DG2 and up to V4 will assist.)
11. The degassing process may be repeated in DG2 until no more bubbles are released. V4 and V6 are now closed, isolating DG1.
12. Close V7 and V11; open V8 and V10 and drive the ethanol into the burette as in (x). Close V10 and V8 and open V11. The lower limb and upper limb back to V8 are now brought to a high vacuum ($\approx$0.005 mm/Hg).

The heat pipe to be filled should now be attached to the rig. In cases where the heat pipe does not have its own valve, the filling tube may be connected to the rig below V14 using thick-walled rubber tubing, or in cases where this may be attacked by the working fluid by another flexible tube material or a metal compression or 'O' ring coupling. If a soft tube material is used, the joints should be covered with silicone-based vacuum grease to ensure no leaks.

The heat pipe may be evacuated by opening valve V14. Following evacuation – that should only take a few minutes depending on the diameter of the filling tube – the heat pipe may be outgassed by heating. This can be done by surrounding the heat pipe with electric heating tape and applying heat until the Pirani gauge returns to the maximum vacuum obtained before heating commenced. (It is worth emphasising the fact, that depending on the diameter of the heat pipe and filling tube, the pressure recorded by the Pirani is likely to be less than that in the heat pipe. It is therefore preferable from this point of view to have a large diameter filling tube.)

To prepare the heat pipe for filling, the lower end is immersed in liquid nitrogen so that the working fluid, which flows towards the coldest region, will readily flow to the heat pipe base. Valve V10 is then cracked and the correct fluid inventory (in most cases enough to saturate the wick plus a small excess) is allowed to flow down into the heat pipe. Should fluid stray into valve seats or other parts of the rig, local heating of these areas using the hot air blower will evaporate any liquid which should then condense and freeze in the heat pipe. A further freeze degassing process may be carried out with the fluid in the heat pipe, allowing it to thaw with V14 closed, refreezing and then opening V14 to evacuate any gas. The heat pipe may then be sealed.

### 3.8.9 Heat pipe sealing

Unless the heat pipe is to be used as a demonstration unit, or for life-testing (in which case, a valve may be retained on one end), the filling tube must be permanently sealed.

With copper, this is conveniently carried out using a tool that will crimp and cold weld the filling tube. A typical crimp obtained with this type of tool is shown in Fig. 3.19 and the force to operate this is applied manually. The tool is illustrated in Fig. 3.20.

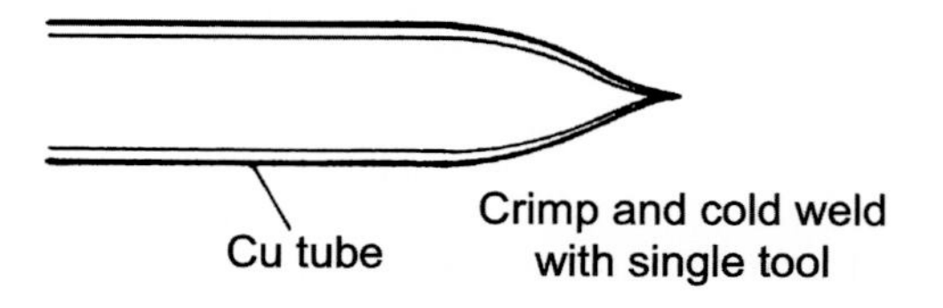


FIGURE 3.19 Crimped and cold welded seal.

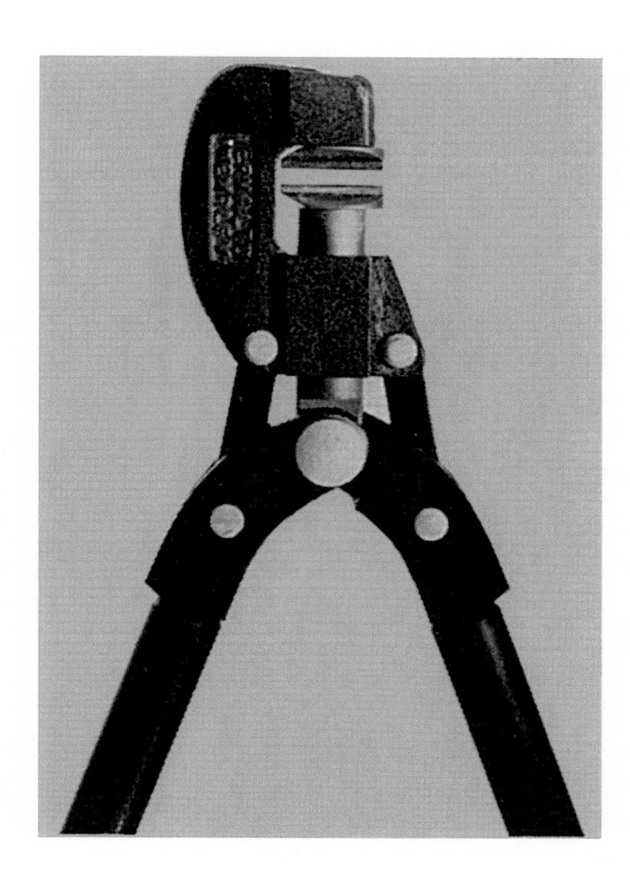

FIGURE 3.20 Crimp and cold welding tool.

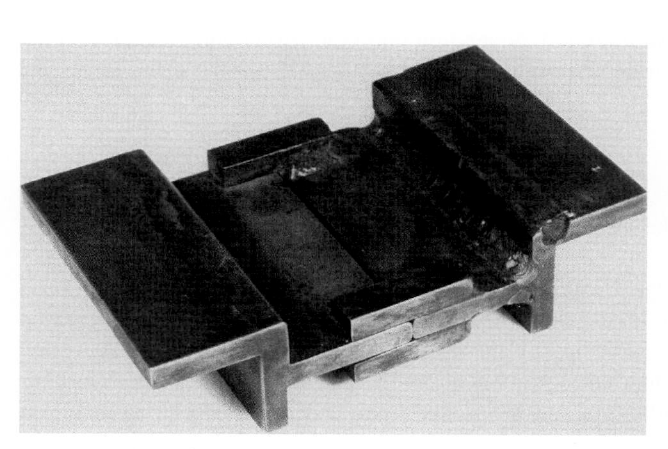

FIGURE 3.21 Jaws for crimping prior to welding.

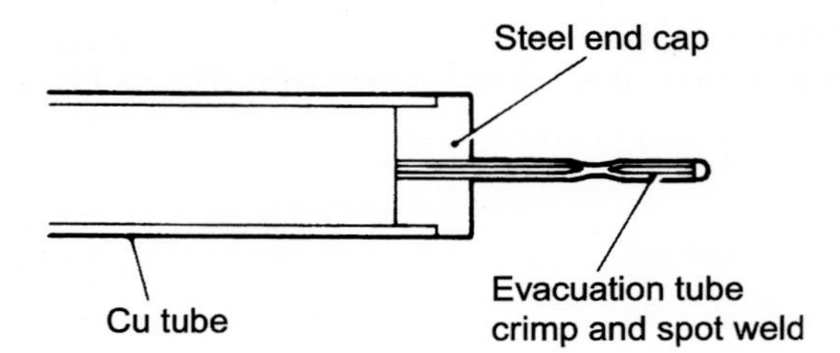


FIGURE 3.22 Crimped and argon-arc welded end.

If stainless steel or aluminium is used as the heat pipe filling tube material, crimping followed by argon-arc welding is a more satisfactory technique. Once the desired vacuum has been attained and the fluid injected, two 0.5 in. (12.7 mm) jaws are brought into contact with the evacuating tube and the latter is flattened.

The heat pipe is then placed between two 0.25 in. thick jaws located at the lower half of the 0.5 in. flattened section. Sufficient load is placed on the evacuating tube to temporarily form a vacuum-tight seal and the remaining 0.25 in. flattened section is simultaneously cut through and welded using an argon-arc torch. The 0.25 in. (6.3 mm) crimping tool, which fits between the jaws of a standard vice, is shown in Fig. 3.21. Results obtained are shown in Fig. 3.22.

Following sealing, the filling tube may be protected by a cap having an outer diameter the same as that of the heat pipe wall. The cap may be filled with solder, a metal-loaded resin or any other suitable material.

### 3.8.10 Summary of assembly procedures

The following is a list of the procedures described above, which should be followed during heat pipe assembly:

1. Select container material.
2. Select wick material and form.
3. Fabricate wick and end caps etc.
4. Clean wick, container and end caps.
5. Outgas metal components.
6. Insert wick and locate.
7. Weld end caps.
8. Check welds for leaks.
9. Select working fluid.
10. Purify working fluid (if necessary).
11. Degas working fluid.
12. Evacuate and fill heat pipe.
13. Seal heat pipe.

It may be convenient to weld the blank end cap before wick insertion, and in cases of sintered and diffusion-bonded wicks, the outgassing may be done with the wick in place within the container.

For the manufacturer considering the production of a considerable number of identical heat pipes – for example, 50 or more units following prototype trials – a number of the manufacturing stages may be omitted. For example, outgassing of metal components may be unnecessary and it may also be found, depending upon the filling and evacuation procedure used, that fluid degassing may be eliminated as a separate activity.

### 3.8.11 Heat pipes containing inert gas

Heat pipes of the variable conductance type (see Section 3.8.11) contain an inert gas in addition to the normal working fluid and so an additional step in the filling process must be carried out. The additional features on the filling rig to cater for inert gas metering are shown in Fig. 3.18.

The working fluid is inserted into the heat pipe in the normal way, and then the system is isolated and the line connecting the heat pipe to the inert gas bottle is opened and the inert gas is bled into the heat pipe. The pressure increases as the inert gas quantity in the heat pipe is raised and this is indicated by the pressure gauge in the gas line. The pressure appropriate to the correct gas inventory may be calculated, taking into account the partial pressure of the working fluid vapour in the heat pipe (see Appendices for mass calculation), and when this is reached, the heat pipe is sealed in the normal manner.

Two aspects of variable conductance (gas-loaded) heat pipes need to be accounted for in manufacture and test procedures. Firstly, most theories for variable conductance heat pipes are based on the assumption that the gas/vapour interface is sharp and diffusion between the two regions is not present. In practice, this is not the case, and in some designs, it is necessary to take diffusion into account.

A second more serious phenomenon, resulting from the introduction of inert gas control into a heat pipe, occurs when gas bubbles enter the wick structure via the working fluid.

These two features are briefly discussed below.

#### 3.8.11.1 Diffusion at the vapour/gas interface

It has been demonstrated that, in some gas-buffered heat pipes, the energy and mass diffusion between the vapour and the noncondensable gas could have an appreciable effect on heat transfer in the interface region and the temperature distribution along the heat pipe.

The diffusion coefficient of the inert gas has an effect on the extent of the diffuse region, with gases having higher diffusion coefficients being less desirable, reducing the maximum heat transport capability of the heat pipe by reducing local condenser temperature. It must be noted that the diffusion coefficient is inversely proportional to density and therefore at lower operating temperatures, particularly during start-up of the heat pipe, the diffuse region may be extensive and of even greater significance. It is therefore important to cater for this during any transient performance analysis and this should be noted during inert gas selection.

#### 3.8.11.2 Gas bubbles in arterial wick structures

Even though in simple heat pipes containing only, the working fluid freeze degassing of the liquid can remove any dissolved gases – in a variable conductance heat pipe, inert gas is always present. Consequently, if the gas dissolves in the working fluid, or finds its way in bubble form into arteries carrying liquid, the performance of the heat pipe can be adversely affected. Saaski [112] carried out theoretical and experimental work on the isothermal dissolution of gas in arterial heat pipes, examining the effects of solubility and diffusivity for helium and argon in ammonia, and methanol.

One of the significant factors determined by Saaski was the venting time $t_v$ of bubbles in working fluids (the time for a bubble to disappear), and this may be calculated from the equation:

$$t_v = \frac{R_0^2}{3\alpha D} \tag{3.12}$$

where $R_0$ is the bubble radius (initial); $\alpha$ the Ostwald coefficient (given by the ratio of the solute concentration in the liquid phase to the concentration in the gaseous phase [113]) and $D$ the diffusion coefficient. The predicted values of $t_v$ are given in Table 3.9.

Table 3.9 shows that the venting times can be considerable when the working fluid is at a low temperature, but in general argon is more easily vented than helium.

**TABLE 3.9 Venting time of gas bubbles in working fluids ($R_0 = 0.05$ cm).**

| Fluid | Temperature | $t_v$ (s) | |
|---|---|---|---|
| | | Helium | Argon |
| Ammonia | −40 | 1200 | 107 |
| | 20 | 63 | 6.7 |
| | 60 | 7 | 1.6 |
| Methanol | −40 | 1030 | 154 |
| | 20 | 133 | 55 |
| | 60 | 50 | 26 |
| Water | 22 | 1481 | 1215 |

**TABLE 3.10 Half-lives of arterial bubbles in various working fluids.**

| Fluid | $t_{1/2}$ (Helium) | $t_{1/2}$ (Argon) |
|---|---|---|
| Ammonia | 7 days | 17 h |
| Methanol | 4.8 h | 1.7 h |
| Water | 3 h | 2.5 h |

The equation above is not valid when noncondensable gas pressure is significant compared to the value of $2a_x/R_0$, where $a_x$ is the surface tension of the working fluid. Saaski stated that the venting time increases linearly with noncondensable gas pressure (other factors being equal) and showed that if, as in a typical gas-controlled heat pipe, the helium pressure is about equal to the ammonia working fluid vapour pressure, the vent time can be 9 days. This is a very long time when compared with the transients to be expected in a VCHP; by changing the working fluid and/or control gas, relatively long venting times may still occur.

Having established venting times for spherical bubbles, Saaski developed a theory to cater for elongated bubbles – the type most likely to form in arteries. They obtained the results in Table 3.10 for the half-lives of elongated arterial bubbles in a VCHP at 20°C (artery radius 0.05 cm, noncondensable gas partial pressure equal to vapour pressure).

The models used to calculate these values were also confirmed experimentally and it was concluded that the venting times are of sufficient length that repriming of an arterial heat pipe containing gas may only be possible if some assistance in releasing gas occlusions can be given during start-up or steady-state operation, either by internal phenomena or by external interference.

Kosson et al. [114] introduced another factor affecting VCHPs, namely variations in pressure within the pipe resulting from oscillations in the diffusion zone. These pressure variations are of the same order as the capillary pressure and can cause vapour flashing within the artery with accompanying displacement of liquid from the artery.

In order to overcome this and other occlusion problems, subcooling of the liquid in the artery was carried out by routing the fluid to the condenser wall so that it experienced sink conditions before returning to the evaporator. As shown in Saaski's results, lowering the liquid temperature improved the venting time. It was also found to reduce the sensitivity of the artery to vapour formation caused by the pressure oscillations described above.

Work recently reported by Lockheed Martin Space Systems [115] suggests that it is possible to model the effect of accelerations on heat pipes. This is particularly important in spacecraft applications – and these may occur in orbit or during the launch phase. This can involve depriming of the wick, and repriming of the form discussed above is relevant and of course necessary. The work by Ambrose studied the acceleration forces that may be encountered whilst the units containing heat pipes are in orbit, and this relates more specifically to the effect of thrusts that may be used to reorientate the spacecraft or change orbit. Predictions were made of heat pipe thermal responses to acceleration pulses and of the wick rewetting behaviour. In the case of rewetting of the evaporators (in a Lunar Reconnaissance Orbiter) located on the heat pipe avionics radiator (involving over 20 heat pipes), the predicted adiabatic rewetting was found to be in good agreement with the 2–4 min recovery time measured in practice.

### 3.8.12 Liquid-metal heat pipes

Early work on liquid-metal heat pipes was concerned with the application to thermionic generators, and for this application, there are two temperature ranges of interest – the emitter range of 1400°C–2000°C and the collector range of 500°C–900°C. In both temperature ranges, liquid-metal working fluids are required and there is a considerable body of information on the fabrication and performance of such heat pipes. More recently, heat pipes operating in the lower temperature range have been used to transport heat from the heater to the multiple cylinders of a Stirling engine and for industrial ovens. A large range of material combinations have been found suitable in this temperature range and compatibility and other problems are well understood. The alkali metals are used with containment materials such as stainless steel, nickel, niobium–zirconium alloys and other refractory metals. Lifetimes of greater than 20,000 h are reported [116]. Grover [117] reports on the use of a light-weight pipe made from beryllium using potassium as the working fluid, and where the beryllium was inserted between the wick and wall of a pipe both made from niobium 1% zirconium. The pipe operated at 750°C for 1200 h with no signs of attack, alloying or mass transport.

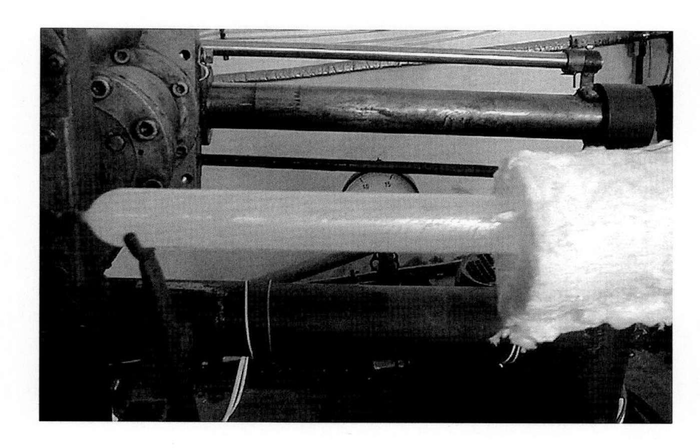

**FIGURE 3.23** Sodium heat pipe during manufacturing process. Dimensions: length, 2800 mm; outside diameter, 38 mm; inside diameter, 32 mm; destination: catalytic reactors.

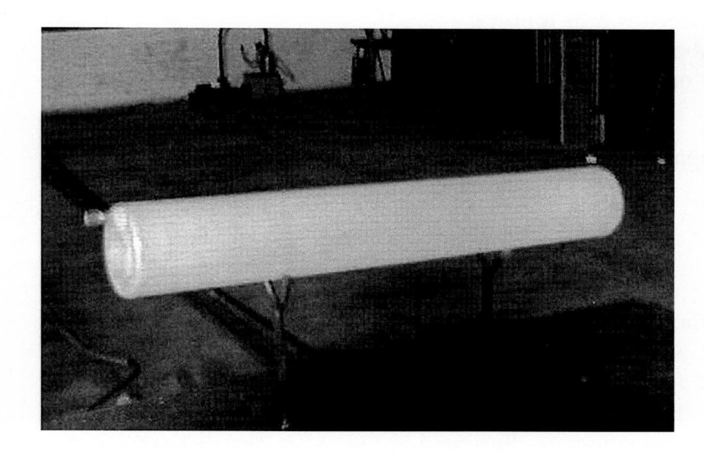

**FIGURE 3.24** Isothermal oven for growing crystals experiments. Working fluid: sodium; dimensions: length, 500 mm; outside diameter, 51 mm; inner working space diameter 25 mm.

The use of Inconel 600 as a container material with sodium as the working fluid was investigated by Japanese researchers with a view to show its long-term durability compared to stainless steel type 316. Standard assembly procedures were used and results of life tests showed some superiority of Inconel 600 over 316 stainless steel, but grain boundary examination suggested after 60,000 h of operation that pitting would lead to pin holes of corrosion [118].

High-performance, long-life, liquid-metal heat pipes can be constructed with some confidence; they are, however, expensive. Hence, before commencing the design of a liquid-metal heat pipe it is important to decide what is to be required from it. Often an application is not required to pump against a gravity head, in which case a thermal syphon is adequate and this greatly reduces the importance of working fluid purity. Similarly, a short operational life at a low rating enables cheaper and less time-consuming fabrication methods to be adopted, and if gas buffering is possible a simpler crimped seal arrangement can be used.

Two examples of heat pipes using liquid sodium as the working fluid are shown in Figs. 3.23 and 3.24. These heat pipes (manufactured by Transterm in Romania) are tubular and annular units, respectively. The unit in Fig. 3.23 is destined for a chemical reactor whilst the second unit, an isothermal oven, may be used for crystal growing.

### 3.8.13 Liquid-metal heat pipes for the temperature range 500°C–1100°C

In this temperature range, potassium and sodium are the most suitable working fluids and stainless steel is selected for the container. The construction and fabrication of a sodium heat pipe [119] will be described to indicate the processes involved. The heat pipe container was made from type 321 (EN58B) stainless steel tube 2.5 cm diameter and 0.9 mm wall thickness. The capillary structure was two layers of 100-mesh stainless steel having a wire diameter of 0.1016 mm and an aperture size of 0.152 mm. The pipe was 0.9 m in length, and the wick was welded by spot welds using a special tool built for the purpose.

### 3.8.13.1 Cleaning and filling

The following cleaning process was followed:

1. Wash with water and detergent.
2. Rinse with demineralised water.
3. Soak for 30 min in a 1:1 mixture of hydrochloric acid and water.
4. Rinse with demineralised water.
5. Soak for 20 min in an ultrasonic bath filled with acetone and repeat with a clean fluid.

After completion of the welds and brazes this procedure was repeated. Argon-arc welding was used throughout and after leak testing the pipe was outgassed at a temperature of 900°C and a pressure of $10^{-5}$ torr for several hours in order to remove gases and vapours.

Various methods may be used to fill the pipe with liquid metal including

1. Distillation, sometimes from a getter sponge to remove oxygen.
2. Breaking an ampoule contained in the filler pipe by distortion of the filler pipe. Distillation is essential if a long life is required. The method adopted for the pipe being described was as follows:
3. 99.9% industrial sodium was placed in a glass filter tube attached to the filling tube of the heat pipe. A bypass to the filter allowed the pipe to be initially evacuated and outgassed. The filling pipe and heat pipe were immersed in the heated liquid paraffin bath to raise the sodium above its melting point. The arrangement is shown in Fig. 3.25.

Finally, the bypass valve is closed and a pressure is applied by means of helium gas to force the molten sodium through the filter and into the heat pipe.

### 3.8.13.2 Sealing

For liquid-metal heat pipes at Reading University, the technique of plug sealing was adopted, as shown in Fig. 3.26.

A special rig has been constructed which allows for the outgassing of an open-ended tube and sodium filling by the filtering method described above. On completion of the filling process, the end sealing plug, supported by a swivel arm within the filling chamber, is swung into position and placed within the heat pipe. The plug is then induction heated to create a brazed vacuum seal. The apparatus and sequence of operation are illustrated in Fig. 3.25. The end sealing plug is finally argon-arc welded after the removal of the heat pipe from the filling apparatus

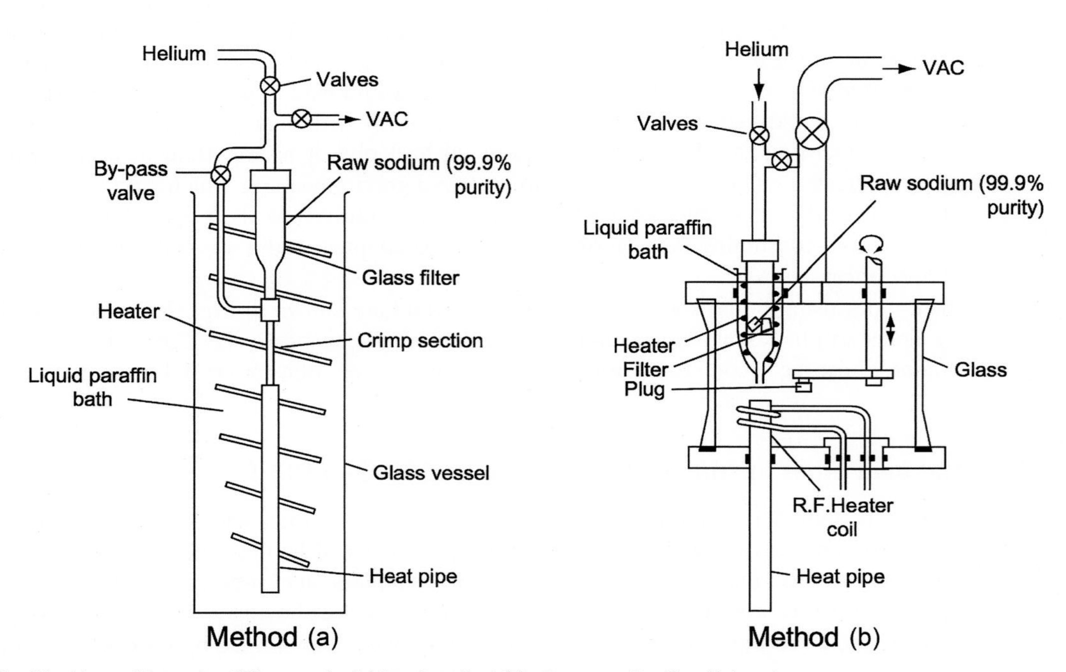


**FIGURE 3.25** Liquid-metal heat pipe filling: method (a) and method (b). *Courtesy: Reading University.*

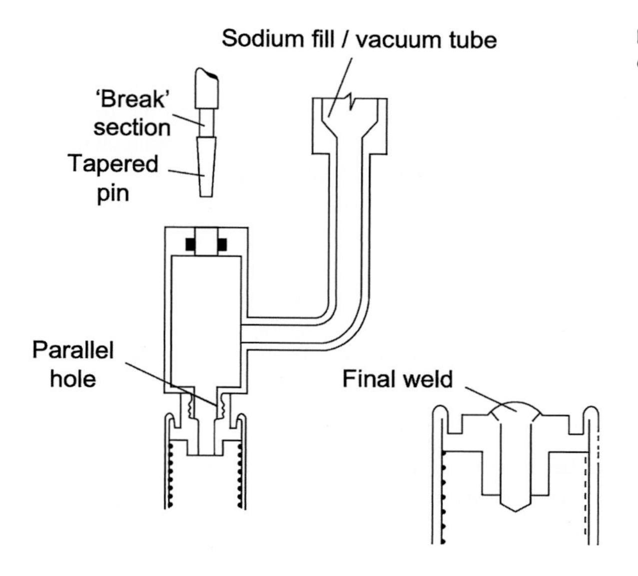


**FIGURE 3.26** Plug sealing technique for sealing liquid-metal heat pipes. *Courtesy: Reading University.*

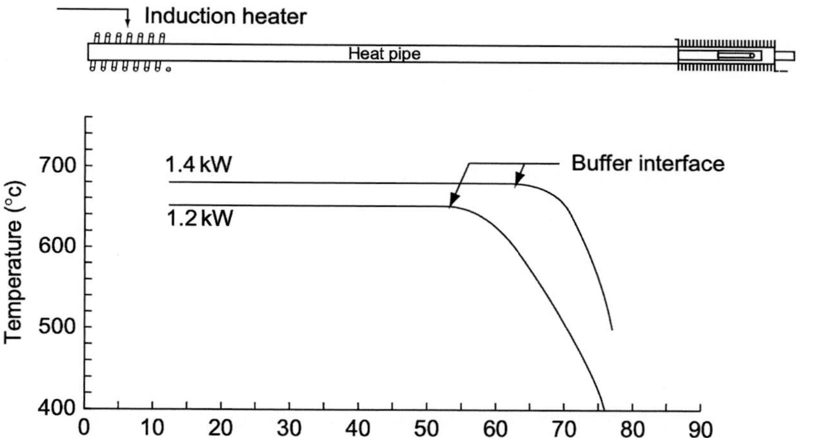


**FIGURE 3.27** Temperature profiles along a sodium heat pipe. *Courtesy: Reading University.*

### *3.8.13.3 Operation*

It has been found that wetting of the wick structure does not occur immediately and that it was necessary to heat the pipe as a thermal syphon for several hours at 650°C. Heating was by an R.F. induction heater over a length of 10 cm. Temperature profiles are given in Fig. 3.27 for heat inputs of 1.2 and 1.4 kW. Before sealing, the heat pipe was filled with helium at a pressure of 20 torr to protect the copper crimp by the resulting gas buffer. It is seen that the gas buffer length is approximately proportional to the power input as might be expected.

The start-up of the heat pipe after conditioning was interesting. In thermal syphon mode, that is with the pipe vertical and heated at the bottom, there were violent temperature variations associated with boiling in the evaporator zone, but this was not experienced when the heater was at the top of the heat pipe. Further sodium work at Reading is described in [120].

Similar work has been reported by other authors and an interesting method for making rigid thin-walled wicked pipes is described by Vinz et al. [121]. Previous work on mesh wicks has included methods such as spot welding, drawing and sintering. The first method does not give uniform adhesion and drawing methods cannot be used for very fine wicks (<200–400 mesh) because of damage. Vinz's method consists of winding a screen strip spirally onto a mandrel and sintering it whilst simultaneous axial pulling and twisting. A gauze of 508 × 3600 mesh has been used successfully to give pore diameters of 10% reproducible to ± 10% and with a free surface for evaporation of 15%–20%.

Broached grooves can be used either alone or with gauze wicks.

### 3.8.13.4 High-temperature liquid-metal heat pipes >1200°C

At the lower end of the range, lithium is preferred as the working fluid with niobium–zirconium or tantalum as the container material. At higher temperatures, silver may be used as the working fluid with tungsten or rhenium as the container material. Data on the compatibility and lifetime of heat pipes made from these materials are given in this chapter. Such refractory materials have a high affinity for oxygen and must be operated in a vacuum or inert gas.

Busse and his collaborators have carried out a considerable programme on fluid heat pipes using lithium and silver as the working fluid, and the techniques used for cleaning, filling, fabrication and sealing are described in Refs [122,123].

More recently [124], a lithium heat pipe system has been studied by Advanced Cooling Technologies in the United States on behalf of Lockheed Martin and the US Air Force Research Laboratory. The system, operating at slightly lower temperatures (to 1100°C), is directed at cooling the wings of spacecraft upon re-entry into the Earth's atmosphere.

### 3.8.13.5 Gettering

Oxides can be troublesome in liquid-metal heat pipes since they will be deposited in the evaporator area, and dissolved oxygen is a particular problem in lithium heat pipes since it causes corrosion of the container material. Oxygen can arise both as an impurity in the heat pipe fluid and also from the container and wick material and a number of authors report the use of getters to help overcome this. For example, Busse et al. [123] used a zirconium sponge from which they distilled lithium into the pipe. Calcium can also be used for gettering.

## 3.8.14 Safety aspects

Whilst there are no special hazards associated with heat pipe construction and operation there are a number of aspects that should be borne in mind. Where liquid metals are employed, standard handling procedures should be adopted. The affinity of alkali metals for water can give rise to problems; a fire was started in one laboratory when a stainless steel pipe distorted releasing sodium and at the same time fractured a water pipe.

Mercury is a highly toxic material and its saturated vapour density at atmospheric pressure is many times the recommended maximum tolerance.

One danger that is sometimes overlooked is the high pressure which may occur in a heat pipe when it is accidentally raised to a higher temperature above its design value and water is particularly dangerous in this respect. The critical pressure of water is 220 bar and occurs at a temperature of 374°C, and when a water in copper heat pipe sealed by a soldered plug was inadvertently overheated, both the 30 cm long heat pipe and the plug were ejected from the clamps at very high velocity. This could well have had fatal results. So it is imperative that a release mechanism (such as a crimp seal) be incorporated in such heat pipes.

Cryogenic heat pipes employing fluids such as liquid air should have special provision for pressure release (or be of sufficient strength) since they are frequently allowed to rise to room temperature when not in use. The critical pressure of nitrogen is 34 bar.

Organisations using specific 'Health and Safety at Work' documentation and procedures will find that many aspects of heat pipe manufacture and use may need bringing to the attention of personnel – such as the toxicity/flammability of some working fluids, the high temperature of some surfaces, and the need to keep within internal pressure guidelines. (Note: in most cases, pressure will not be monitored and can only be assessed from knowledge of the heat pipe temperature and the fluid used within it.)

## 3.8.15 3D-printed heat pipes

3D-printed heat pipes are being developed as part of an ongoing research project at the Faculty of Engineering and Environment, Northumbria University, Newcastle upon Tyne, in collaboration with the University of Liverpool and Thermacore's UK base in Northumberland. The main objective of the project is to develop an aluminium ammonia heat pipe with a sintered wick structure. Currently, available ammonia heat pipes mainly use extruded grooved aluminium tubes, and although there have been a small number of attempts at employing sintered steel or nickel wicks in steel tubes with ammonia as the working fluid [125], this is believed to be the first published data on aluminium heat pipes with ammonia and a sintered wick structure. The main barrier is the difficulty of sintering aluminium powders to make sintered wicks. So far, promising sintered aluminium heat pipe samples have been manufactured using the SLM technique with various wick characteristics. This new method proved to be capable of producing very complicated wick structures with different thickness, porosity, permeability and pore sizes in different regions of a heat pipe in addition to the solid (nonporous) walls. Moreover, the entire heat pipe including the end cap, wall, wick and fill tube can be produced in a single process [126].

The selective laser sintering/melting (SLS/SLM) manufacturing process utilises a laser beam to locally melt a thin layer of metal powder. Then by applying additional powder layers and using 3D CAD and custom beam control software to melt a pattern across each layer, complex 3D components that are not able to be manufactured using conventional machining can be produced. There are many companies offering this facility, including the manufacture of heat exchangers – see, for example, the website of Within [127].

After analysing the solid and porous structures, various designs were proposed for manufacturing heat pipes by SLM. In order to check the feasibility of manufacturing the proposed designs – especially the axial grooved heat pipe (AGHP) – a preliminary 20 mm section of heat pipe was built in titanium. Then, a 60 mm long sample of the same geometry was made from Al6061. These samples were to prove the ability of SLM to manufacture complicated wick structures, and a total of four different wicks structures were manufactured: (1) an axial grooved wick (porous fins), (2) an annular wick, (3) a graded wick (different thickness and porosity in the evaporator and condenser sections of the pipe), and (4) an arterial wick (small ducts fabricated into an annular porous wick to facilitate the return of the working fluid condensate from condenser). The heat pipes were built up from the end cap upwards, with the end cap, wick and wall being made together.

In Fig. 3.28 (from left to right), first a heat pipe is shown as it is modelled in CAD, which is then converted to a machine-readable format using the SLM special CAD software. Next, the build process is shown where layers of powder are laid on the substrate. First, layers are fully melted/fused together in a circular area to form the heat pipe end cap (to the specified thickness) and then, in the following layers, powders in an annular section are fully melted and fused together to form the heat pipe wall. Adjacent to the wall the above-mentioned octahedral structure is formed by fusing the powders together along special passes up to the specified distance from the wall (wick thickness). This forms the capillary structure. Elsewhere the powder remains loose and is removed at the end by shaking the sample. In Fig. 3.29, a finished SLM arterial HP and its cross-section are shown.

## 3.9 Heat pipe life-test procedures

Life-testing and performance measurements on heat pipes, in particular when accelerated testing is required, are the most important factors in their selection.

In spacecraft, for example, the ESA stipulate [128] that heat pipes should still be suitable for operation for 7–10 years in space after 5 years of ground storage and testing during spacecraft development. It is specifically stated that evidence of long-term compatibility of materials and working fluids must be available before spacecraft 'qualification' can be granted. (Requirements are presented in more detail later.)

Life tests on heat pipes are commonly regarded as being primarily concerned with the identification of any incompatibilities that may occur between the working fluid and wick and wall materials. However, the ultimate life test would be in the form of a long-term performance test under conditions appropriate to those of the particular application. Even if this is done, it is still difficult in cases where the wick is pumping against gravity to accelerate the life test by

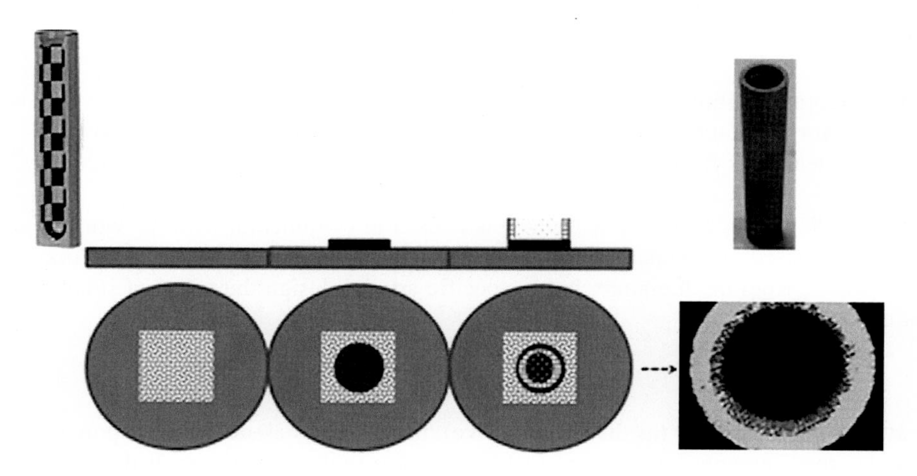

FIGURE 3.28 Heat pipe build process in SLM. The heat pipe is built up from the end cap upwards with the end cap, wall and wick made together on top of a substrate disc.

**FIGURE 3.29** Arterial wick heat pipe made by SLM (14 mm OD and 70 mm length). *OD*, Outside diameter.

increasing, say, the evaporator heat flux as this could cause heat pipe failure because it is then likely to be operating well in excess of its design capabilities. This, therefore, necessitates operation in the reflux mode. (Compatibility data are presented in Tables 3.5 and 3.6.)

There are many factors to be taken into account when setting up a full life test programme, and the relative merits of the alternative techniques are discussed below.

### 3.9.1 Variables to be taken into account during life tests

The number of variables to be considered when examining the procedure for life tests on a particular working fluid/wick/wall combination is very extensive and would require a large and comprehensive number of heat pipes to be fully tested. However, several of these may be discounted because of existing available data on particular aspects, but one important point which must be emphasised is the fact that quality control and assembly techniques inevitably vary from one laboratory to another, and these differences can be manifested in differing compatibility data and performance figures.

#### *3.9.1.1 The working fluid*

The selection of the working fluid must take into account the following factors which can all be investigated by experiments:

1. Purity – the working fluid must be free of dissolved gases and other liquids, for example, water. Such techniques as freeze degassing and distillation are available to purify the working fluid, but it is also important to ensure that the subsequent handling of the working fluid following purification does not expose it to contaminants.
2. Temperature – some working fluids are sensitive to operating temperature. If such behaviour is suspected the safe temperature band must be identified.
3. Heat flux – high heat fluxes can create vigorous boiling action in the wick which can lead to erosion.
4. Compatibility with wall and wick – the working fluid must not react with the wall and wick. This can also be a function of temperature and heat flux; the tendency for reactions to occur generally increases with increasing temperature or flux.
5. Noncondensable gas – in the case of VCHPs, where a noncondensable gas is used in conjunction with the working fluid, the selection of the two fluids must be based on compatibility and also on the solubility of the gas in the working fluid. (In general, these data are available from the literature, but in specific arterial design the effect of solubility may only be apparent after experimentation.)

#### *3.9.1.2 The heat pipe wall*

In addition to the interface with the heat pipe working fluid (as discussed above), the wall and associated components such as end caps have their own particular requirements with regard to life and also their interface with the wick.

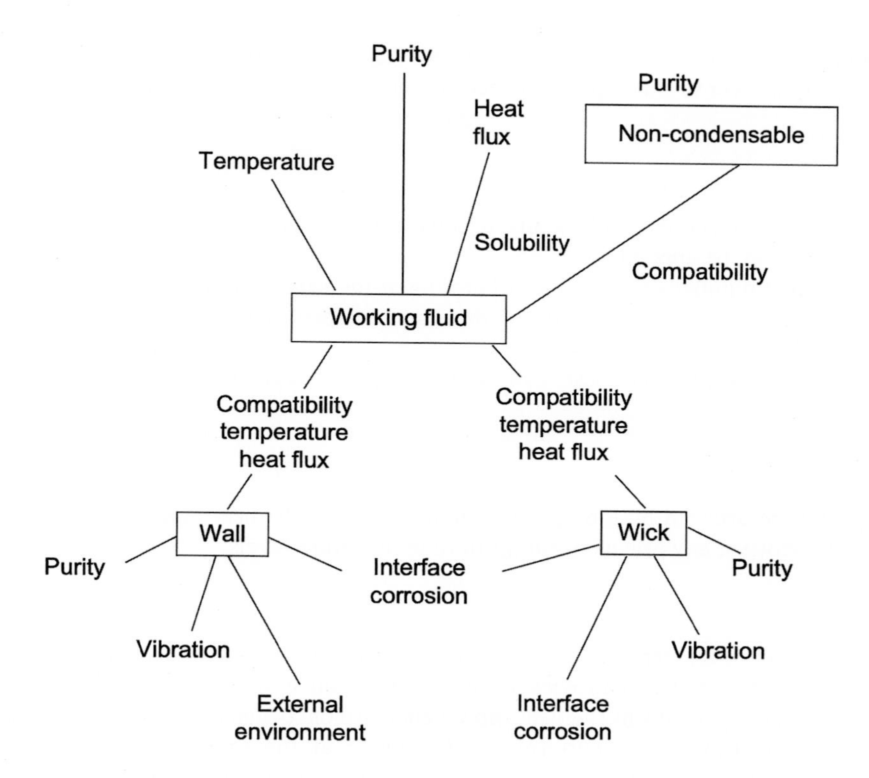


**FIGURE 3.30** Heat pipe life test factors.

The successful operation of the heat pipe must therefore take the following into account:

1. Vibration and acceleration – the structure must be able to withstand any likely vibrations and accelerations, and any qualification procedures designed to ensure that the units meet these specifications should be regarded as an integral part of any life test programme.
2. Quality assurance – the selection of the outer case material should be based on the purity or at least the known alloy specification of the metal used.
3. External environment – the external environment could affect the case material properties or cause degradation of the outer surface. This should also be the subject of life test investigation if any deterioration is suspected.
4. Interface corrosion – it is possible that some corrosion could occur at metallic interfaces, particularly where dissimilar metals are used, in the presence of the working fluid.

#### *3.9.1.3 The wick*

The heat pipe wick is subjected to the same potential hazards as the heat pipe wall with the exception of external attack. Vibration is much more critical, however, and the wick itself contains, in most cases, many interfaces where corrosion could occur (Fig. 3.30).

### 3.9.2 Life test procedures

There are many ways of carrying out life tests, all having the same aim – namely, to demonstrate that the heat pipe can be expected to last for its design life with an excellent degree of certainty.

The most difficult part of any life test programme is the interpretation of the results and the extrapolation of these results to predict long-term performance. (One technique used for extrapolating results obtained from gas generation measurements is described in Section 3.9.3.)

The main disadvantage of carrying out life tests of one specific combination of materials (be the test accelerated or at design load), is the fact that if any reaction does occur insufficient data are probably available to explain the main causes of the degradation. For example, in some life tests carried out at IRD, diacetone alcohol was formed as a result of acetone degradation, However, it was not possible (without further testing over a considerable period) to state whether this phenomenon was a function of operating temperature, because to do so, life tests on identical units operating at several different vapour temperatures will have to be carried out. It is even possible that comprehensive life test programmes may never provide a complete answer to some questions because new aspects may be found during each study.

#### 3.9.2.1 Effect of heat flux

The effect of heat flux on heat pipe lives and performance can only really be investigated using units in the reflux mode, where fluxes well in excess of design values may be applied.

For example, setting up experiments involving a number of heat pipes operating at the same vapour temperature but with differing evaporator heat fluxes enables the later examination of the inner surface of the evaporator for corrosion etc.

Also, if carried out on a representative heat pipe, performance tests can be carried out at regular intervals during these life tests.

#### 3.9.2.2 Effect of temperature

Compatibility and working fluid makeup can both be affected by the operating temperature of the heat pipe, and so it is, therefore, important to be able to discriminate between any effects resulting from temperature levels.

#### 3.9.2.3 Compatibility

In conjunction with considering the effect of heat flux or temperature on the working fluid alone, it is also necessary to investigate the compatibility of the working fluid with the wall and wick materials. The aim is to look for reactions between the materials which could change the surface structure in the heat pipe, generate noncondensable gas or produce impurities in the form of deposits that could affect evaporator performance. Of course, all three phenomena could occur at the same time by differing degrees and this can make the analysis of the degradation much more complex.

Compatibility tests can be carried out at design conditions on a heat pipe operating horizontally or under tilt against gravity, but to be meaningful such tests should continue for years. Then if compatibility is shown to be satisfactory over, say, a 3-year period some conclusions can be made concerning the likely behaviour over a much longer life. Accelerated compatibility tests could also be performed with occasional tests in the heat pipe mode to check on the design performance.

#### 3.9.2.4 Other factors

The life of a heat pipe can be affected by assembly and cleaning procedures, and it is important to ensure that life test pipes are fully representative as far as assembly techniques are concerned. The working fluid used must, of course, be of the highest purity.

Another feature of life testing is the desirability of incorporating valves on the pipes so that samples of gas etc., can be taken out without necessarily causing the unit to cease functioning. However, one disadvantage of using valves is the possibility of introducing a new incompatibility – that of the working fluid and valve material, although this can be ruled out with modern stainless steel valves.

Also when testing in the heat pipe mode, a valve body can be filled with working fluid that may be difficult to remove, and this should be taken into account when carrying out such tests in case depletion of the wick or artery system occurs.

### 3.9.3 Prediction of long-term performance from accelerated life tests

One of the major drawbacks of accelerated life tests has been the uncertainty associated with the extrapolation of the results to estimate performance over a considerably longer periods of time. Baker [129] has correlated data on the generation of hydrogen in stainless steel heat pipes using an Arrhenius plot, with some success, and this has been used to predict noncondensable gas generation over a 20-year period.

The data were based on life tests carried out at different vapour temperatures over a period of 2 years, and the mass of generated hydrogen was measured periodically. Vapour temperatures of 100, 200 and 300°F were used, five heat pipes being tested at each temperature.

Baker applied Arrhenius plots to these results, which were obtained at the Jet Propulsion Laboratory, in the following way.

The Arrhenius model is applicable to activation processes, including corrosion, oxidation, creep and diffusion. Where the Arrhenius plot is valid, the plot of the log of the response parameter ($F$) against the reciprocal of absolute temperature is a straight line.

The response parameter is defined by the equation:

$$F = \text{Constant} \times e^{(-A/kT)} \tag{3.13}$$

where $A$ is the reaction activation energy; $k$ the Boltzmann constant ($1.38 \times 10^{-23}$ J/K) and $T$ the absolute temperature. For the case of the heat pipe, Baker described the gas generation process as:

$$\dot{m}(t, T) = f(t)F(T) \tag{3.14}$$

where $m$ is the mass generation rate; $t$ denotes time and $F(t)$ is given in Eq. (3.13).

Plotting the mass of hydrogen generation in each heat pipe against time (with results at different temperatures) provides data to obtain a universal curve – presenting the mass of hydrogen generated as a function of time × shift factor – which will be a straight line on logarithmic paper. Finally, the shift factors are plotted against the reciprocal of absolute temperature for each temperature examined, and the slope of this curve gives the activation energy $A$ in Eq. (3.13)

The mass of hydrogen generated at any particular operating temperature can then be determined using the appropriate value of the shift factor. Baker concluded that stainless steel/water heat pipes could operate for many years at temperatures of the order of 60°F, but at 200°F the gas generation would be excessive.

It is probable that this model could be applied to other wall/wick/working fluid combinations, the only drawback being the large number of test units needed for accurate predictions. The minimum is of the order of 12 – results being obtained at three vapour temperatures, four heat pipes being tested at each temperature.

Another study was concerned with the evolution of hydrogen in nickel/water heat pipes. Anderson used a corrosion model to enable him to predict the behaviour of heat pipes over extended periods based on accelerated life tests following Baker's method [130].

He argued that oxidation theory predicts, that passivating film growth occurs with a parabolic time dependence and an exponential temperature dependence, and his work offered the following values for $A$, the reaction activation energy:

| | |
|---|---|
| Stainless steel (304)/water | $8.29 \times 10^{-20}$ J |
| Nickel/water | $10.3 \times 10^{-20}$ J |

confirming Baker's model.

Later work in Japan [131,132] concentrated on a statistical treatment of life test data from accelerated tests on copper/water heat pipes. This has been directed, in part, at investigating the formation of small quantities of noncondensable gas ($CO_2$) in such pipes where lifetimes of 20 years (or more) are required. The investigations were carried out on axially grooved heat pipes, some of which used commercial phosphorous, deoxidised copper and others oxygen-free copper.

High-temperature ageing was done for periods of 20, 40 and 150 days and analysis was undertaken by X-ray microscopy and infrared spectroscopy. The infrared absorption spectra showed absorption caused by benzene rings, phenyl groups, an O–H link and a C–O–C link, from which it was judged that the products were organic. After the full ageing period, corrosion was observed on the inside of the commercial copper whilst the oxygen-free copper (OFC) copper showed only slight corrosion, and it was therefore concluded that phosphorous used during the refining of the copper had a profound effect on its corrosion properties.

With regards to the generation of $CO_2$, on the basis of the criterion that the active-inactive (i.e. buffered) boundary in the vapour space is where the temperature drop becomes one half of the total temperature drop, it was possible to estimate the gas column length and temperature drop achieved after 20 years of use of the heat pipe.

After 1000 days a temperature difference of about 2.5°C was yielded at 160°C. Ageing for 470 days at 393K corresponds, according to data in [131], to a 20-year use at 333K, and it was thus concluded that commercially available heat pipes using this phosphorous-containing material could be used satisfactorily if a temperature drop of 3°C is acceptable.

### 3.9.4 A life test programme

A life test programme must provide detailed data on the effects of temperature, heat flux, and assembly techniques on the working fluid, and the working fluid/wall and wick material compatibility.

**TABLE 3.11 Heat pipe life test priorities.**

| Priority | Minimum no. units to be tested | Number with valves | Test specification |
|---|---|---|---|
| 1 | – | – | Cleanliness of materials |
| 1 | – | – | Purity of working fluid |
| 1 | – | – | Sealing of case |
| 1 | – | – | Outgassing |
| 2 | 2 | 2 | Refluxing – vapour temperature at maximum design |
| 1 | 4 at each temp. | all | Refluxing – temperature range up to maximum design (to include bonding temperatures) |
| 3 | 2 | 2 | Refluxing – heat flux at maximum design |
| 2 | 2 | 1 | Heat pipe mode – intermittent tests between refluxing |
| 1 | 2 | 1 | Heat pipe mode – long-term continuous performance test |
| 1[a] | 2 | 0 | Heat pipe mode – vibration test with intermittent performance tests |
| 1 | 2 | 2 | Variable conductance heat pipe – solubility of gas in working fluid and effect on artery |

[a]*Where applicable.*

The alternative techniques for testing have been discussed in Section 3.9 and it now remains to formulate a programme that will enable sufficient data to be accumulated to enable the life of a particular design of heat pipe to be predicted accurately.

Each procedure may be given a degree of priority (numbered 1–3, in decreasing order of importance) and these are presented in Table 3.11 together with the *minimum* number of units required for each test. The table is self-explanatory.

This programme should provide sufficient data to enable the confident prediction (based on an Arrhenius plot) of the long-term performance of a heat pipe, and the maximum allowable operating temperature based on fluid stability. The programme involves a considerable amount of testing, but the cost should be weighed against satisfactory heat pipe performance over its life in applications where reliability is of prime importance.

### 3.9.5 Spacecraft qualification plan

The use of heat pipes in spacecrafts, as discussed earlier, makes heavy demands on heat pipe technology. The qualification plan set out by the ESA and illustrated in Fig. 3.31 involves the construction of 15 sample heat pipes, the majority of which are eventually opened for detailed physical analysis. Of the heat pipes included in the qualification programme, it is stipulated that at least 10% (and not less than 3 units) should be subjected to life tests of 8000 h and guidelines set out the temperature and heat load requirements of these ‘ageing’ tests. It is also pointed out that these tests, by themselves, do not necessarily prove the longevity of the heat pipes for 7–10 years of operation in space.

An example of the qualification procedures for a specific ESA heat pipe type is given in [133]. In addition, Alcatel Space [134] present data of value to those developing heat pipes for spacecraft in the referenced paper summarising a ‘roadmap’. This sets out the objectives and the technical challenges of a development programme – in particular for aluminium/ammonia heat pipes, but also applicable more broadly.

The Company, interestingly, outlines the facilities that enable it to produce typically 2000 heat pipes per annum, and these include:

- filling stations
- automatic welding machine
- testing benches – proof pressure, ageing and thermal cycling
- performance test benches
- bending machines.

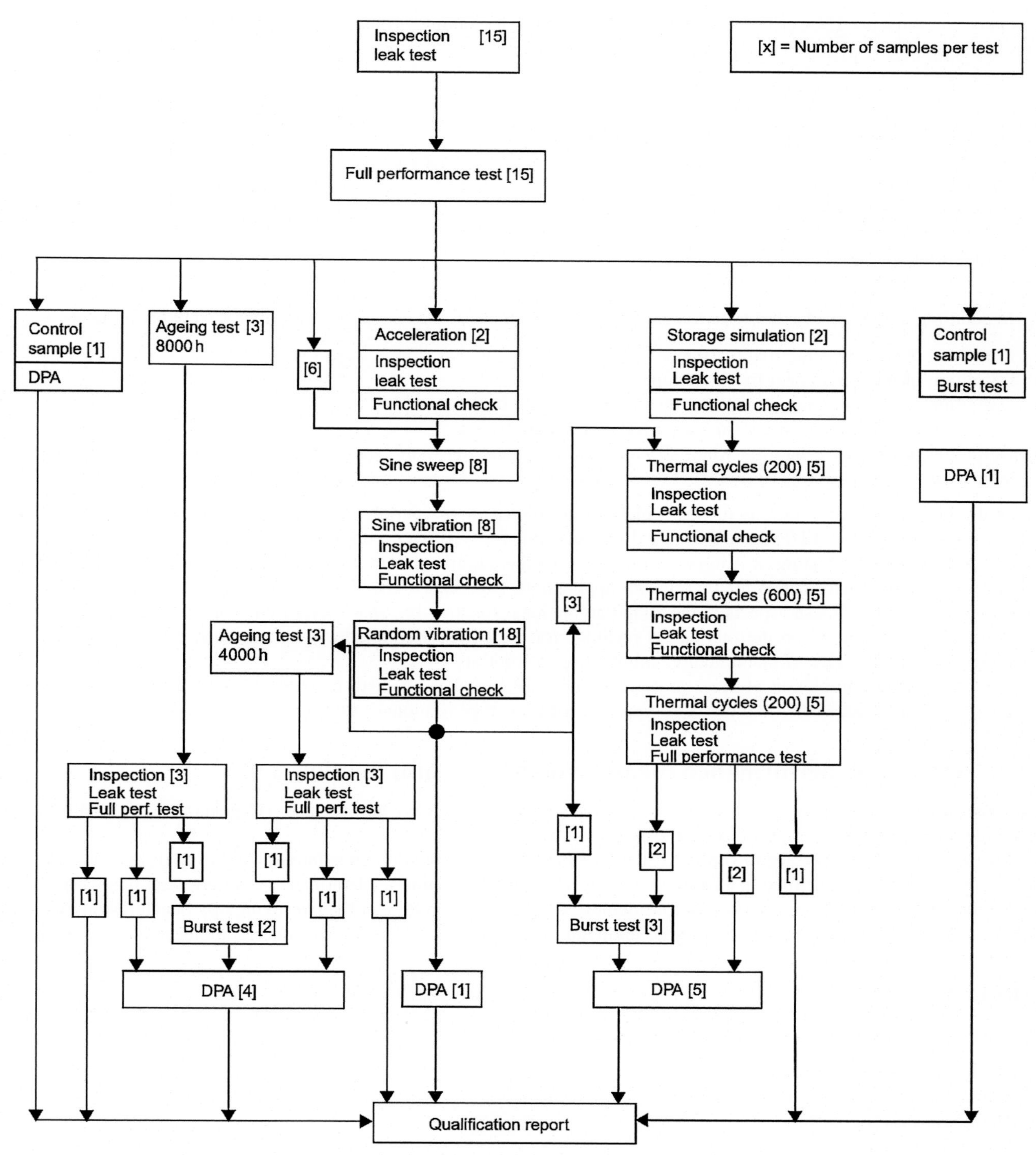


**FIGURE 3.31** European Space Agency heat pipe qualification plan [128].

The thermal tests carried out on the bench include burnout (maximum heat transport capability), heat flux tests and precise measurement of the thermal conductance. The ability to incline the pipe (a necessity for space qualification procedures) is also incorporated in the test bench.

Other countries use ESA criteria for space qualification. For example, in the Ukraine [135], this led to the following tests being proposed:

- Inspection and physical measurement
- Proof pressure testing – leak test
- Performance testing
- Burst test
- Random vibration
- Storage simulation test
- Thermal cycles/shock test
- Ageing test (life test)
- Noncondensable gas definition test.

In addition, a number of other tests to measure various aspects of the heat pipe performance were recommended:

- Definition of heat pipe thermal resistance
- Definition of maximum heat transport capability
- Definition of the temperature distribution along the length of the heat pipe
- Definition of heat pipe priming time after full evaporator dry-out
- Definition of start-up capability when 80% of maximum heat transport capability is applied, over a range of vapour temperatures.

With alternative configurations of heat pipes being applied in spacecraft, different qualification procedures sometimes need to be applied. LHPs have been used in spacecraft for well over a decade, but as has been emphasised in the context of heat pipe life testing, it is important to carry out the qualification of the manufacturing methods and materials as well as the LHP performance for its relevant application. Dos Santos and Riehl [136] reported on the qualification procedures adopted for these aspects at the Brazilian National Institute for Space Research. The authors emphasise that the procedures are focused on the fabrication methods for the LHPs, the material evaluation and certification, and the processes used in all stages of manufacture, culminating in life tests.

After successful completion of these procedures, an LHP can, it is claimed, be built as a certified device for space applications. The target at the time was a 10-year life for uses in geostationary satellites.

## 3.10 Heat pipe performance measurements (see also Section 3.9)

The measurement of the performance of heat pipes is comparatively easy and requires (in general) equipment readily available in any laboratory engaged in heat transfer work.

Measurements are necessary to show that the heat pipe meets the requirements laid down during design. The limitations to heat transport, described in Chapter 4 and presented in the form of a performance envelope, can be investigated, as can the degree of isothermalisation. A considerable number of variables can be investigated by bench testing, including orientation with respect to gravity, vapour temperature, evaporator heat flux, start-up, vibrations and accelerations.

### 3.10.1 The test rig

A typical test rig is shown diagrammatically in Fig. 3.32; which has the following features and facilities:

1. Heater for evaporator section
2. Wattmeter for power input measurement
3. Variac for power control
4. Condenser for heat removal
5. Provision for measuring flow and temperature rise of condenser coolant
6. Provision for tilting heat pipe
7. Thermocouples for temperature measurement and associated readout system
8. Thermal insulation.

The heater may take several forms just as long as heat is uniformly applied and the thermal resistance between the heater and the evaporator section is low. This can be achieved by using rod heaters mounted in a split copper block clamped around the heat pipe or by wrapping insulated heater wire directly on the heat pipe. For many purposes, eddy

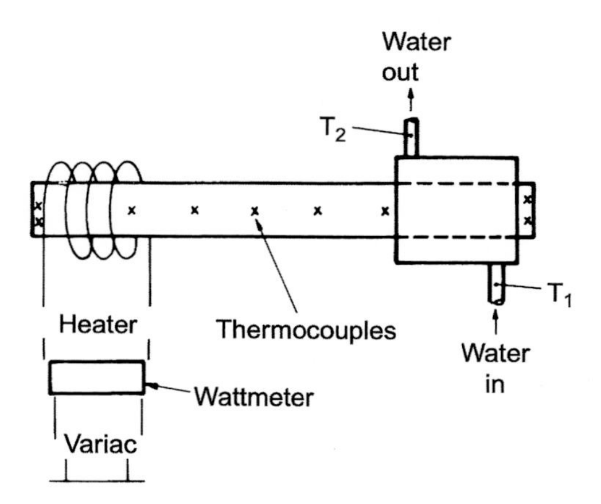


FIGURE 3.32 Heat pipe performance test rig.

current heating is convenient, using the condenser as a calorimeter. Heat losses by radiation and convection to the surroundings should be minimised by applying thermal insulation to the outside of the heater. An accurate wattmeter covering the anticipated power range and a variac for close control of power should be incorporated in the heater circuit. Where orientation may be varied, long leads between the heater and the instruments should be used for convenience.

An effective technique for measuring the power output of heat pipes operating at vapour temperatures appropriate to most organic fluids and water is to use a condenser jacket through which a liquid is passed. For many cases, this can be water. The heat given up to the water can be obtained if the temperature rise between the condenser inlet and outlet is known together with the flow rate. The temperature of the liquid flowing through the jacket may be varied to vary the heat pipe vapour temperature. Where performance measurements are required at vapour temperatures of about 0°C a cryostat may be used.

Cryogenic heat pipes should be tested in a vacuum chamber. This prevents convective heat exchange, and a cold wall may be used to keep the environment at the required temperature. As a protection against radiation heat input, the heat pipe, fluid lines and cold wall should all be covered with superinsulation. If the heat pipe is mounted such that the mounting points are all at the same temperature (cold wall and heat sink) it can be assumed that all heat supplied to the evaporator will be transported by the heat pipe as there will be no heat path to the environment. Further data on cryogenic heat pipe testing can be obtaining from [137,138].

An important factor in many heat pipe applications is the effect of orientation on performance. The heat transport capability of a heat pipe operating with the evaporator below the condenser (thermosyphon or reflux mode) can be up to an order of magnitude higher than that of a heat pipe using the wick to return liquid to the evaporator from a condenser at a lower height. In many cases, the wick may prove incapable of functioning when the heat pipe is tilted so that the evaporator is only a few centimetres above the condenser. Of course, wick selection is based (in part) on the likely orientation of the heat pipe in the particular application.

Provision should be made on the rig to rotate the heat pipe through 180°C whilst keeping the heater and condenser in operation. The angle of the heat pipe should be accurately set and measured. In testing of heat pipes for satellites, a tilt of only 0.5 cm over a length of 1 m may be required to check heat pipe operation, and this requires very accurate rig alignment.

The measurement of temperature profiles along the heat pipe is normally carried out using thermocouples attached to the heat pipe's outer wall. If it is required to investigate transient behaviour, for example during start-up, burnout or on a VCHP, automatic electronic data collection is required. For steady-state operation, a switching box connected to a digital voltmeter or a multichannel chart recorder should suffice, but most laboratories now possess computer-aided data collection and real-time presentation of such data.

### 3.10.2 Test procedures

Once the heat pipe is fully instrumented and set up in the rig, the condenser jacket flow may be started and heat applied to the evaporator section. Preferably heat input should be applied at first in steps, building up to design capability and allowing the temperatures along the heat pipe to achieve a steady state before adding more power. When the steady-state condition is reached, power input, power output (i.e. condenser flow rate and $\Delta T$), and the temperature profile along the heat pipe should be noted.

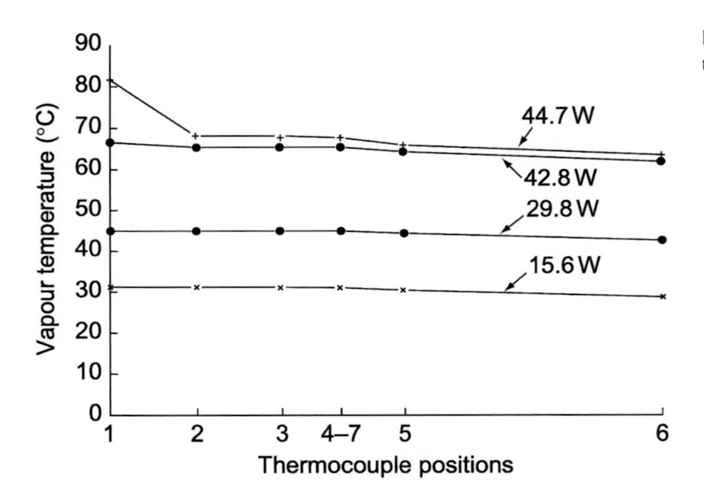


**FIGURE 3.33** Typical temperature profiles along a heat pipe under test.

If temperature profiles as shown in Fig. 3.33 are achieved, the heat pipe is operating satisfactorily. However, several modes of failure can occur, all being recognisable by temperature changes at the evaporator or condenser.

The most common failure is burnout created by excessive power input at the evaporator section. It is brought about by the inability of the wick to feed sufficient liquid to the evaporator and is characterised by a rapid rise in evaporator temperature compared to other regions of the heat pipe. Typically, the early stages of burnout are represented by the upper curve in Fig. 3.33.

Once burnout has occurred the wick has to be reprimed and this is best achieved by cutting off the power input completely. When the temperature difference along the pipe drops to 1°C–2°C the power may be reapplied. The wick must reprime, that is, be rewetted and saturated with working fluid along its complete length if operation against gravity or zero gravity is envisaged. If this is the case, then recovery after burnout must be demonstrated in the tilted condition. In other cases, the recovery may be aided by gravity assistance.

A second failure mechanism recognisable by an increased evaporator temperature (and known as overheating) occurs at elevated temperatures. As explained earlier, each working fluid has an operating temperature range characterised by the Merit number which achieves an optimum value at a particular temperature and then decreases as this temperature is exceeded. This means that the fluid is able to transport less heat and so the temperature of the evaporator becomes higher than the rest of the pipe. In general, the evaporator temperature does not increase as quickly as in a burnout condition but these two phenomena are difficult to distinguish.

Temperature changes at the condenser section can also point to failure mechanisms or a decrease in performance. A sudden drop in temperature at the end of the heat pipe downstream of the cooling jacket occurring at high powers can be attributed to the collection of working fluid in that region, insulating the wall and creating a cold spot. This has been called 'coolout' [139]. Complete failure need not necessarily occur when this happens, but the overall $\Delta T$ will be substantially increased and the effective heat pipe length reduced.

A similar drop in temperature downstream of the condenser jacket can occur in pipes of small diameter (<6 mm bore) when the fluid inventory is greater than that needed to completely saturate the wick. The vapour tends to push the excess fluid to the cooler end of the heat pipe, where, because of the small vapour space volume, a small excess of fluid will create a long cold region. This can occur at low powers and adjustments in fluid inventory may be made if a valve is incorporated in the heat pipe. One way is to use an excess fluid reservoir which acts as a sponge but has pores sufficiently large to prevent it from sucking fluid out of the wick. This technique is used in heat pipes for space use and the reservoir may be located at any convenient part of the vapour space.

Failure can be brought about by incompatibilities of materials, generally in the form of the generation of noncondensable gases that collect in the condenser section. Unlike liquid accumulation, the gas volume is a function of vapour temperature and its presence is easily identified.

Unsatisfactory wick cleaning can inhibit wetting and if partial wetting occurs the heat pipe will burnout very quickly after the application of even small amounts of power.

Recently reported from the National Taiwan University are tests to measure the viscous limits of heat pipes [140]. When considering the start-up of heat pipes and thermosyphons the viscous limit can be important as it may inhibit starting of the high vapour densities/low temperatures (appropriate to the particular working fluid).

Two heat pipes were tested – a short (150 mm × 9 mm diameter) and long (240 mm × 12 mm diameter) one. Water was the working fluid; the wicks were grooves and the tests were in the thermosyphon mode. The test procedure was described as the 'dynamic method'. In order to study the performance during start-up, the heating temperature is controlled in a range that is only marginally above ambient temperature, being then raised in 2–3°C increments above this. Heat input to the evaporator is via circulating water in a heated bath and heat removal by air-forced convection. The point where the condenser temperature starts to rise as the evaporator temperature is increased in small increments is noted, and the time taken for the condenser to start to respond is reduced as the operating temperature rises. The researchers point out that the operating temperature thus affects the 'dynamic start-up', and they define the viscous limit in the dynamic test as the '*minimum heating temperature measured to overcome the pressure drop at the corresponding ambient condition and the associated operating temperature for the heat pipe*'.

For those interested in testing procedures for LHPs, the work at the Chinese Academy of Space Technology in Beijing is relevant [141]. The start-up under supercritical conditions is described for a nitrogen LHP.

### 3.10.3 Evaluation of a copper heat pipe and typical performance

#### 3.10.3.1 Capabilities

A copper heat pipe using water as the working fluid was manufactured and tested to determine the temperature profiles and the maximum capability. The design parameters of the pipe were as follows:

| | |
|---|---|
| Length | 320 mm |
| Outside dia. | 12.75 mm |
| Inside dia. | 10.75 mm |
| Material of case | Copper |
| Wick form | 4 layers 400 mesh |
| Wick wire dia. | 0.025 mm |
| Effective pore radius | 0.031 mm |
| Calculated porosity | 0.686 |
| Wick material | Stainless steel |
| Locating spring length | 320 mm |
| Pitch | 7 mm |
| Wire dia. | 1 mm |
| Material | Stainless steel |
| Working fluid | Water ($10^6$ ft resistivity) |
| Quantity | 2 ml |
| End fittings | Copper |
| Instrumentation thermocouples | 7 |

#### 3.10.3.2 Test procedure

The evaporator section was fitted into the 100-mm long heater block in the test rig, and the condenser section was covered by a 150-mm long water jacket. The whole system was then lagged. First tests were carried out with the heat pipe operating vertically with gravity assistance. The power was applied and on achievement of a steady-state condition the thermocouple readings and temperature rise through the water jacket were noted as was the flow rate.

Power to the heaters was increased incrementally, and the steady-state readings were noted until dry-out was seen to occur. (This was characterised by a sudden increase in the potential of the thermocouple at the evaporator section relative to the readings of the other thermocouples.)

The above procedure was performed for various vapour temperatures and heat pipe orientations with respect to gravity.

#### 3.10.3.3 Test results

Typical results obtained are shown earlier in Fig. 3.33 for the vapour temperature profile along the pipe when operating with the evaporator 10 mm above the condenser.

The table below gives power capabilities for a 9.5 mm (outside diameter) copper heat pipe of length 30 cm, with a composite wick of 100 and 400 mesh, and operating at an elevation (evaporator above condenser) of 18 cm.

| Vapour temp (°C) | Power out (W) |
|---|---|
| 84 | 17 |
| 121 | 30.5 |
| 162.5 | 54 |
| 197 | 89 |

The working fluid was again water, and a capability of 165 W was measured with the horizontal operation (290 W with gravity assistance).

### 3.10.3.4 Tests on thermosyphons to compare working fluids

The tests on several thermosyphons are described here. The work was carried out at Heriot-Watt University in Edinburgh, UK as part of a postgraduate short research project directed at looking at fluids that could replace R134a which is a rather strong global warming gas [142].

The thermosyphons in an air–air heat exchanger application currently use R134a as the working fluid for the target temperature ranges of −10°C–50°C on the cold side and 60°C–80°C on the hot side. Storage temperatures can be as low as −30°C. Whilst performance is adequate with R134a, there are compelling reasons to investigate a replacement:

- At the upper end of the operating temperature range, the vapour pressure reaches about 30 bar.
- R134a is one of the replacement refrigerants for the CFCs but it still has a large GWP.
- Fluids such as R1234ze [143] have been proposed as replacements for HFCs. (These fluids and other heavy fuel oils (HFOs) are currently proposed for refrigeration, air conditioning and heat pump systems).
- Theory suggests other fluids could perform better.

The final choice of fluids was:

- Water – expected to give the best performance.
- Methanol – has a high Merit number.
- Water–5% ethylene glycol – could potentially combine the performance of water and avoid problems associated with freezing.
- Two R134a pipes – one as a control for all the experiments and one for calibration of the test rig.

A full selection of fluids that might be considered for thermosyphons at the temperatures of interest are given (with properties and comments) in the Tables in Appendix 1.

Using the facility described in [142], experiments were conducted keeping the cooling water at about 30°C and heating the evaporator from about 35°C to 65°C. As $T_e$ increased (and thus $\Delta T$) the duty increased and Fig. 3.34 shows typical results.

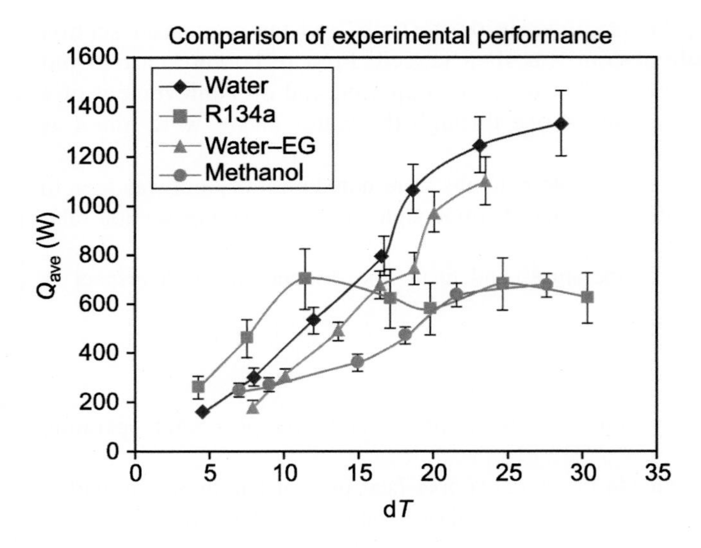


**FIGURE 3.34** Experimental results for heating the evaporator whilst keeping the condenser at a constant 30°C.

All the pipes appeared to reach a limit in performance. For R134a, it occurred at a lower $\Delta T$, about 12°C; the methanol pipe had a limit at close to a $\Delta T$ of 25°C; the water and water − 5% ethylene glycol pipes appeared to reach a limit at the upper end of the $\Delta T$ at which the rig could operate. Like the methanol pipe, there was audible pinging − which is a sign that the flooding limit has been reached. This is also likely to be the limit for the R134a pipe but it was silent. The vapour pressure is much higher in the R134a pipe than in the others, and this may be responsible for this difference.

Water gave the highest duty, reaching over 1300 W. The addition of 5% ethylene glycol in the water −5% ethylene glycol pipe lowered performance by a small amount, the highest duty being over 1200 W. Methanol had a maximum duty of about 750 W compared to 700 W for R134a.

R134a has the best duty at lower values of $\Delta T$, up to approximately a $\Delta T$ of 14°C when water becomes the best choice. Water − 5% ethylene glycol becomes better than R134a above a $\Delta T$ of approximately 16°C. Methanol has a similar duty to R134a above a $\Delta T$ of 20°C, and significantly less at lower $\Delta T$'s. Water − 5% ethylene glycol followed a similar pattern to that of water, but the curve was displaced to a duty 100–150 W lower.

The performance predicted by the engineering sciences data unit (ESDU) equations is shown in Fig. 3.35 for the same conditions as in Fig. 3.34. It shows similar trends to the experimental results for water and water − 5% ethylene glycol, but different trends for methanol, and especially R134a.

In all cases, the first predicted limit encountered was the flooding limit. Water − 5% ethylene glycol has a higher flooding limit than pure water, but this does not agree with the experiment results since it appeared the flooding limit had been reached for both fluids with water giving a higher duty. Methanol has a much higher predicted limit than was found in practice. Fig. 3.36 shows the predicted limits graphically.

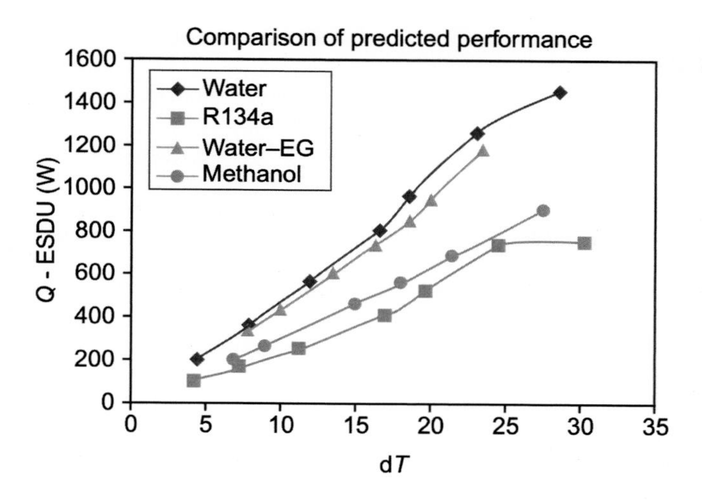


**FIGURE 3.35** Predicted results for the same $\Delta T$s for each fluid as the results in Fig. 3.34.

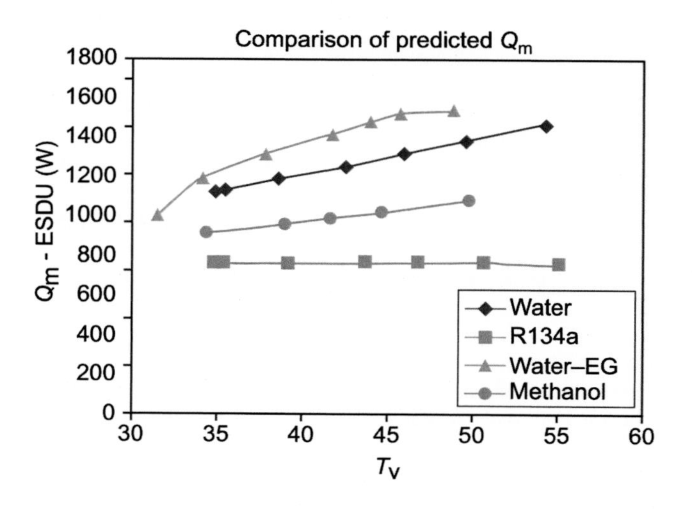


**FIGURE 3.36** Predicted limiting duty for each fluid at each $T_v$ corresponding to the results in Fig. 3.34.

It was concluded that it is possible to recommend using water − 5% ethylene glycol in thermosyphons for use in a range of ambient conditions representative of most regions of the world. Tests to confirm the behaviour at low storage temperatures should be undertaken, and these should be relevant to the application conditions for the intended use of these thermosyphons. Water − 5% ethylene glycol meets the environmental criteria and is suitable for the operating and storage conditions subject to experimental confirmation of the latter aspect. Water − 5% ethylene glycol does not outperform R134a at all operating conditions, so this should be borne in mind when designing the heat exchangers. It can offer significantly better duty, but only with a $\Delta T$ above 16°C between the evaporator and condenser.

The prediction model was found to work best for water for which it gives values mostly within experimental error. For water − 5% ethylene glycol, it gives reasonable agreement but underestimates for most cases when it is outside the experimental range. For methanol the agreement was not so good and there was a limit reached by the test pipe which was not predicted by the model. For R134a the model did not match the results well at all. The magnitude of the maximum heat flux was correct, but it occurred for different conditions than it did in the experiments. It remains unclear as to why the model does not work well for R134a.

## References

[1] N. Zhang, Innovative heat pipe systems using a new working fluid, Int. Commun. Heat. Mass. Transf. 28 (2001) 1025−1033. Available from: https://doi.org/10.1016/S0735-1933(01)00306-2.

[2] Anon, Thermophysical properties of heat pipe working fluids: operating range between 260°C and 300°C, Data Item No. 80017, Engineering Sciences Data Unit, London, 1980.

[3] M. Esen, Thermal performance of a solar cooker integrated vacuum-tube collector with heat pipes containing different refrigerants, Sol. Energy 76 (2004) 751−757. Available from: https://doi.org/10.1016/J.SOLENER.2003.12.009.

[4] H. Jouhara, A.J. Robinson, Experimental investigation of small diameter two-phase closed thermosyphons charged with water, FC-84, FC-77 and FC-3283, Appl. Therm. Eng. 30 (2010) 201−211. Available from: https://doi.org/10.1016/j.applthermaleng.2009.08.007.

[5] R. Dobson, Relative thermal performance of supercritical $Co_2$, $H_2O$, $N_2$, and He charged closed-loop thermosyphon-type heat pipes, Heat. Pipe Sci. Technol. An. Int. J. (2012) 3.

[6] M. Groll, D. Heine, T. Spendel, Heat recovery units employing reflux heat pipes as components (Technical Report) | ETDEWEB, 1984.

[7] M. Groll, Heat pipe research and development in western Europe, Heat. Recov. Syst. CHP 9 (1989) 19−66. Available from: https://doi.org/10.1016/0890-4332(89)90139-7.

[8] S. Roesler, D. Heine, M. Groll, Performance of closed two-phase thermosyphons with high- temperature organic working fluids, in: Proceedings of the Seventh International Heat Pipe Conference, Minsk, Hemisphere, New York, NY, 1990.

[9] J. Zhang, J. Li, J. Yang, T. Xu, Analysis of heat transfer in the condenser of naphthalene ther- mosyphon at small inclination, Paper H3-1, in: Proceedings of Tenth International Heat Pipe Conference, Stuttgart, 21−25 September 1997.

[10] F.H. Milanez, M.B.H. Mantelli, A two-phase loop thermosyphon with naphthalene as working fluid, in: Proceedings of the Sixteenth International Heat Pipe Conference, Lyon, France, 20−24 May 2012.

[11] A. Devarakonda, J.K. Olminsky, An evaluation of halides and other substances as potential heat pipe fluids, in: Collection of Technical Papers - 2nd International Energy Conversion Engineering Conference, vol. 1, American Institute of Aeronautics and Astronautics (AIAA), 2004, pp. 471−477. Available from: https://doi.org/10.2514/6.2004-5575.

[12] L.M. Libby, W.F. Libby, One-parameter equation of state for metals and certain other solids, Proc. Natl Acad. Sci. 69 (1972) 3305−3306. Available from: https://doi.org/10.1073/PNAS.69.11.3305.

[13] W.G. Anderson, S. Tamanna, C. Tarau, J.R. Hartenstine, D. Ellis, Intermediate temperature heat pipe life tests, Screen. (Lond.) (2012) 100. 22.

[14] J.E. Deverall, Mercury as a heat pipe fluid, 1970.

[15] D.A. Reay, Mercury wetting of wicks, in: Proceedings of the 4th C.H.I.S.A. Conference, Prague. International Congress of Chemical and Process Engineering CHISA (Chemical Engineering, Chemical Equipment Design and Automation), 1972.

[16] T. Yamamoto, K. Nagata, M. Katsuta, Y. Ikeda, Experimental study of mercury heat pipe, Exp. Therm. Fluid Sci. 9 (1994) 39−46. Available from: https://doi.org/10.1016/0894-1777(94)90006-X.

[17] P. Marcarino, A. Merlone, Gas-controlled heat-pipes for accurate temperature measurements, Appl. Therm. Eng. 23 (2003) 1145−1152. Available from: https://doi.org/10.1016/S1359-4311(03)00045-0.

[18] W. Bienert, Heat pipes for solar energy collectors, in: Proceedings of the First International Heat Pipe Conference, Stuttgart, Paper 12−1, October 1973.

[19] J.E. Kemme, C.A. Busse, Performance investigations of liquid metal heat pipes for space and terrestrial applications. AIAA Paper 78−431, in: Proceedings of the Third International Heat Pipe Conference, Palo Alto, CA, May 1978.

[20] D. Quataert, C.A. Busse, F. Geiger, Long time behaviour of high temperature tungsten- rhenium heat pipes with lithium or silver as the working fluid, Paper 4-4, in: Proceedings of the Third International Heat Pipe Conference, Palo Alto, CA, May 1978.

[21] J.M. Tournier, M.S. El-Genk, Startup of a horizontal lithium-molybdenum heat pipe from a frozen state, Int. J. Heat. Mass. Transf. 46 (2003) 671−685. Available from: https://doi.org/10.1016/S0017-9310(02)00324-1.

[22] C.Y. Tsai, H.T. Chien, P.P. Ding, B. Chan, T.Y. Luh, P.H. Chen, Effect of structural character of gold nanoparticles in nanofluid on heat pipe thermal performance, Mater. Lett. 58 (2004) 1461–1465. Available from: https://doi.org/10.1016/J.MATLET.2003.10.009.
[23] N. Putra, W.N. Septiadi, H. Rahman, R. Irwansyah, Thermal performance of screen mesh wick heat pipes with nanofluids, Exp. Therm. Fluid Sci. 40 (2012) 10–17. Available from: https://doi.org/10.1016/J.EXPTHERMFLUSCI.2012.01.007.
[24] A.B. Solomon, K. Ramachandran, B.C. Pillai, Thermal performance of a heat pipe with nanoparticles coated wick, Appl. Therm. Eng. 36 (2012) 106–112. Available from: https://doi.org/10.1016/J.APPLTHERMALENG.2011.12.004.
[25] K.Y. Leong, R. Saidur, T.M.I. Mahlia, Y.H. Yau, Performance investigation of nanofluids as working fluid in a thermosyphon air preheater, Int. Commun. Heat. Mass. Transf. 39 (2012) 523–529. Available from: https://doi.org/10.1016/J.ICHEATMASSTRANSFER.2012.01.014.
[26] J.A. Lidbury, A Helium Heat Pipe, NDG-72-11, Rutherford Laboratory, England, 1972.
[27] M. Groll, Wärmerohre als Baudemente in der Wärme-und Kältetechnik, Brennst-Waerme-Kraft (1973) 25.
[28] P.J. Marto, W.L. Mosteller, Effect of nucleate boiling on the operation of low temperature heat pipes, ASME (American Society of Mechnical Engineers) Paper 69-HT-24.
[29] E.C. Phillips, Low-Temperature Heat Pipe Research Pogram - NASA Technical Reports Server (NTRS), 1969.
[30] D. Keser, Experimental determination of properties of saturated sintered wicks, in: 1st International Heat Pipe Conference, International Atomic Energy Agency, Stuttgart, 1973.
[31] K. Moritz, R. Pruschek, Limits of energy transport in heat pipes, Chem. Ing. Technik (1969) 41.
[32] P. Vinz, C.A. Busse, Axial heat transfer limits of cylindrical sodium heat pipes between 25 W/cm$^2$ and 15.5 kW/cm$^2$ in: 1st International Heat Pipe Conference, International Atomic Energy Agency, Stuttgart, 1973.
[33] V.A. Busse, Heat pipe research in Europe, Euratom Report, EUR 4210 f, 1969.
[34] D. Quataert, C.A. Busse, F. Geiger, Long term behaviour of high temperature tungsten-rhenium heat pipes with lithium or silver as working fluid in: 1st International Heat Pipe Conference, International Atomic Energy Agency, Stuttgart, 1973.
[35] A.M. Schroff, M. Armand, LeCaloduc, Rev. Tech. Thomson-CSF 1 (4) (1969), (in French).
[36] R.A. Farran, K.E. Starner, Determining wicking properties of compressible materials for heat pipe applications, in: Proceedings of Aviation and Space Conference, Beverley Hills, California, International Atomic Energy Agency, 1968.
[37] J.K. Ferrell, J. Alleavitch, Vaporisation heat transfer in capillary wick structures, Raleigh, USA, 1969.
[38] R.A. Freggens, Experimental determination of wick properties for heat pipe applications. in: 4th Intersociety Energy Conference Engineering Conference, International Atomic Energy Agency, Washington, DC, 1969, pp. 888–897.
[39] J.D. Hinderman, E.C. Phillips, Determination of properties of capillary media useful in heat pipe design, NASA Technical Reports Server (NTRS), in: American Society of Mechanical Engineers and American Institution of Chemical Engineers, Heat Transfer Conference, 1969.
[40] H. Birnbreier, G. Gammel, Measurement of the effective capillary radius and the permeability of different capillary structures. in: 1st International Heat Pipe Conference, International Atomic Energy Agency, Stuttgart, 1973.
[41] L.S. Langston, H.R. Kunz, Liquid transport properties of some heat pipe wicking materials, ASME (American Society of Mechanical Engineers) Paper 69-HT-17, 1969.
[42] B.G. Mc Kinney, An experimental and analytical study of water heat pipes for moderate temperature ranges - NASA Technical Reports Server (NTRS), 1969.
[43] A.T. Calimbas, R.H. Hulett, An avionic heat pipe, ASME (American Sociery of Mechanical Engineers) Paper 69-HT-16, New York, NY, 1969.
[44] S. Katzoff, Heat pipes and vapour chambers for thermal control of spacecraft, AIAA Paper, 1967, pp. 67–310.
[45] C.J. Hoogendoorn, S.G. Nio, Permeability studies on wire screens and grooves, in: International Heat Pipe Conference, International Atomic Energy, Stuttgart, 1973.
[46] K.R. Chun Some experiments on screen wick dry-out limits, ASME Paper, 1971.
[47] M.N. Ivanovskii, et al., Investigation of heat and mass transfer in a heat pipe with a sodium coolant, High. Temp. Sci. 8 (1970) 299–304.
[48] R.R. Williams, D.K. Harris, The heat transfer limit of step-graded metal felt heat pipe wicks, Int. J. Heat. Mass. Transf. 48 (2005) 293–305. Available from: https://doi.org/10.1016/J.IJHEATMASSTRANSFER.2004.08.024.
[49] M.G. Semena, A.P. Nishchik, Structure parameters of metal-fiber heat pipe wicks, J. Eng. Phys. 35 (1978) 1268–1272. Available from: https://doi.org/10.1007/BF00859673.
[50] A.G. Kostornov, A.A. Shapoval, M.J. Lalor, O. Mgaloblishvili, J.C. Legros, A study of heat transfer in heat pipe evaporators with metal fiber capillary structures, J. Enhanced Heat. Transf. (2012) 19.
[51] T. Takaoka, et al., Development of long heat pipes and heat pipe applied products, Fujikura Technical Rev. (1985) 77–93.
[52] A. Schimizu, et al., Characteristics of a heat pipe with carbon fibre wick, in: Proceedings of the Seventh International Heat Pipe Conference, Minsk, 1990, Hemisphere, New York, NY, 1991.
[53] J.V. Kaudinga, et al., Experimental investigation of a heat pipe with carbon fibre wick, in: Proceedings of the Seventh International Heat Pipe Conference, Minsk, 1990, Hemisphere, New York, NY, 1991.
[54] C. Byon, S.J. Kim, Capillary performance of bi-porous sintered metal wicks, Int. J. Heat. Mass. Transf. 55 (2012) 4096–4103. Available from: https://doi.org/10.1016/J.IJHEATMASSTRANSFER.2012.03.051.
[55] C. Figus, L. Ounougha, P. Bonzom, W. Supper, C. Puillet, Capillary fluid loop developments in Astrium, Appl. Therm. Eng. 23 (2003) 1085–1098. Available from: https://doi.org/10.1016/S1359-4311(03)00038-3.
[56] L. Ottenstein, D. Butler, J. Ku, Flight testing of the capillary pumped loop 3 experiment in: Proceedings of the 2002 International Two-phase Thermal Control Technology Workshop, Mitchellville, MA, 24–26 September 2002.

[57] T.D. Swanson, G.C. Birur, NASA thermal control technologies for robotic spacecraft, Appl. Therm. Eng. 23 (2003) 1055–1065. Available from: https://doi.org/10.1016/S1359-4311(03)00036-X.

[58] F. Dobran, Heat pipe research and development in the Americas, Heat. Recovery Syst. CHP 9 (1989) 67–100.

[59] D.A. Reay, Heat transfer enhancement—a review of techniques and their possible impact on energy efficiency in the U.K, Heat. Recovery Syst. CHP 11 (1991) 1–40. Available from: https://doi.org/10.1016/0890-4332(91)90185-7.

[60] T. Zhang, P. Debock, E.W. Stautner, T. Deng, C. Immer, Demonstration of liquid nitrogen wicking using a multi-layer metallic wire cloth laminate, Cryogenics (Guildf.) 52 (2012) 301–305. Available from: https://doi.org/10.1016/J.CRYOGENICS.2012.01.015.

[61] R.L. Gorring, S.W. Churchill, Thermal conductivity of heterogeneous materials, Chem. Eng. Prog. (1961) 57.

[62] R. Kempers, D. Ewing, C.Y. Ching, Effect of number of mesh layers and fluid loading on the performance of screen mesh wicked heat pipes, Appl. Therm. Eng. 26 (2006) 589–595. Available from: https://doi.org/10.1016/J.APPLTHERMALENG.2005.07.004.

[63] K. Kar, A. Dybbs, Effective thermal conductivity of fully and partially saturated metal wicks, 1978. https://doi.org/10.1615/IHTC6.2750.

[64] J.C. Maxwell, A treatise on electricity and magnetism, 3rd, vol. 1, OUP, 1954 Reprinted by Dover, New York, NY, 1981.

[65] H. Kozai, The effective thermal conductivity of screen wick, in: 3rd International Heat Pipe Symposium, Tsukuba, Japan, Japan Association for Heat Pipes, 1988.

[66] J. Xu, Y. Zou, M. Fan, L. Cheng, Effect of pore parameters on thermal conductivity of sintered LHP wicks, Int. J. Heat. Mass. Transf. 55 (2012) 2702–2706. Available from: https://doi.org/10.1016/J.IJHEATMASSTRANSFER.2012.01.028.

[67] P. Joy, Optimum cryogenic heat-pipe design, ASME Pap, 1970. https://doi.org/10.1007/978-1-4684-7826-6_45.

[68] P.E. Eggers, A.W. Serkiz, Development of cryogenic heat pipes, J. Eng. Gas. Turbine Power 93 (1971) 279–286. Available from: https://doi.org/10.1115/1.3445564.

[69] N. Kosowski, R. Kosson, Experimental performance of grooved heat pipes at moderate temperatures, in: 6th Thermophysics Conference, 1971. https://doi.org/10.2514/6.1971-409.

[70] Heat pipes - properties of common small-pore wicks, Data items no. 79013, 1979.

[71] M.G. Semena, V.K. Zaripov, Influence of the diameter and length of fibres on material heat transfer of metal fibre wicks of heat pipes, Therm. Eng. 24 (1977) 69–72T.

[72] A. Acton, Correlating equations for the properties of metal felt wicks, in: Advances in Heat Pipe Technology, in: Proceedings of the IV International Heat Pipe Conference, Pergamon Press, Oxford, 1981.

[73] P.H.D. Santos, E. Bazzo, A.A.M. Oliveira, Thermal performance and capillary limit of a ceramic wick applied to LHP and CPL, Appl. Therm. Eng. 41 (2012) 92–103. Available from: https://doi.org/10.1016/J.APPLTHERMALENG.2012.02.042.

[74] X. Yang, Y.Y. Yan, D. Mullen, Recent developments of lightweight, high performance heat pipes, Appl. Therm. Eng. 33–34 (2012) 1–14. Available from: https://doi.org/10.1016/J.APPLTHERMALENG.2011.09.006.

[75] S. van Oost, B. Aalders, Cryogenic heat pipe ageing, in: Proceedings of the 10th International Heat Pipe Conference, International Atomic Energy Agency, Stuttgart, 1997, pp. 21–25.

[76] A. Basiulis, M. Filler, Operating characteristics and long life capabilities of organic fluid heat pipes, in: 6th Thermophysics Conference, Reston, Virigina, American Institute of Aeronautics and Astronautics, 1971. Available from https://doi.org/10.2514/6.1971-408.

[77] H. Kreeb, M. Groll, P. Zimmermann, Life test investigations with low temperature heat pipes. in: 1st International Heat Pipe Conference, International Atomic Energy Agency, Stuttgart, 1973.

[78] C.A. Busse, A. Campanile, J. Loens, Hydrogen generation in water heat pipes at 250°C, in: 1st International Heat Pipe Conference, Stuttgart, 1973.

[79] E.E. Gerrels, J.W. Larson, Brayton cycle vapour chamber (heat pipe) radiator study, NASA CR- 1677, General Electric Company, Philadelphia, PA, NASA, February 1971.

[80] T.F. Freon, E.I. Solvent, Dupont de Nemours and Company Inc., Technical Bulletin FSR-1, 1965.

[81] Anon, Heat pipes-general information on their use, operation and design, Data Item No. 80013, Engineering Sciences Data Unit, London, 1980.

[82] J. Rosenfeld, K. Minnerly, C. Dyson, Ten year Operating Test Results and Post-Test Analysis of a 1/10 Segment Stirling Sodium Heat Pipe, Phase III, NASA/CR–2012-217430, 2012.

[83] C. Tarau, W.G. Anderson, Sodium variable conductance heat pipe for radioisotope. Stirling systems — design and experimental results, in: Proceedings of the Eighth Annual International Energy Conversion Engineering Conference, AIAA Paper 2010-6758, Nashville, TN, July 2010.

[84] A. Bricard, et al., High temperature liquid metal heat pipes, in: Proceedings of the Seventh International Heat Pipe Conference, Minsk, 1990, Hemisphere, New York, NY, 1991.

[85] M.A. Merrigan, An investigation of lead heat pipes, in: Proceedings of the Seventh International Heat Pipe Conference, Minsk, 1990, Hemisphere, New York, NY, 1991.

[86] A. Basiulis, R.C. Prager, T.R. Lamp, Compatability and reliability of heat pipe materials, in: Proceedings of the 2nd International Heat Pipe Conference, Bologna, ESA Report SP 112, 1976.

[87] W. Anderson, Hydrogen evolution in nickel-water heat pipes, 1973. https://doi.org/10.2514/6.1973-726.

[88] I. Novotna et al., Compatibility of steel-water heat pipes, in: Proceedings of the 3rd International Heat Pipe Symposium, Tsukuba, Japan, Japan Association for Heat Pipes, 1988.

[89] I. Novotna et al., A contribution to the service life of heat pipes, in: Proceedings of the 7th International Heat Pipe Conference, Minsk, Begel Corporation, 1990.

[90] B.M. Rassamakin, N.D. Gomelya, N.D. Khairnasov, N.V. Rassamakina, Choice of the effective inhibitors of corrosion and the results of the resources tests of steel and aluminium thermosyphon with water, in: Proceedings of the 10th International Heat Pipe Conference, American Instutute of Aeronautics and Astronautics, Stuttgart, 1997, pp. 21–25.
[91] Z. Rong Di, et al., Experimental investigation of the compatibility of mild carbon steel and water heat pipes, in: Proceedings of the 6th International Heat Pipe Conference, American Instutute of Aeronautics and Astronautics, Grenoble, 1988.
[92] Y.M. Liang, et al., Applications of heat pipe heat exchangers in energy saving and environmental protection. in: Proceedings of the International Conference on Energy and Environment, American Instutute of Aeronautics and Astronautics, Shanghai, 1995, pp. 8–10.
[93] H. Zhang, J. Zhuang, Research, development and industrial application of heat pipe technology in China, Appl. Therm. Eng. 23 (2003) 1067–1083. Available from: https://doi.org/10.1016/S1359-4311(03)00037-1.
[94] Anon., Technology specification for carbon steel-water gravity heat pipe, Su Q/B- 25–86, 1986.
[95] A. Bricard, Recent advances in heat pipes for hybrid heat pipe heat exchangers, in: 7th International Heat Pipe Conference, Minsk, Begel Corporation, 1990.
[96] Anon, On the influence of the overshoot of oxygen on corrosion of carbon steels in heat supply systems, Teploenergetika 12 (1992) 36–38.
[97] F. Geiger, D. Quataert, Corrosion studies of tungsten heat pipes at temperatures up to 2650°C, in: Proceedings of the 2nd International Heat Pipe Conference, Bologna, ESA Report SP 112, 1976.
[98] J.E. Kemme, Heat Pipe Capability Experiments, Atomic Energy Commission, U.S, 1966. Available from: https://doi.org/10.2172/4473429.
[99] E. Van Andel, Heat pipe design theory, Euratom Center for Information and Documentation, Report EUR No. 4210 e, f, 1969.
[100] C.A. Busse, Theory of the ultimate heat transfer limit of cylindrical heat pipes, Int. J. Heat. Mass. Transf. 16 (1973) 169–186. Available from: https://doi.org/10.1016/0017-9310(73)90260-3.
[101] Guan-Wei Wu., et al., A high thermal dissipation performance poly- ethylene terephthalate heat pipe, in: Proceedings of the Sixteenth International Heat Pipe Conference, Lyon, France, 20–24 May 2012.
[102] D. Mehl, Therma-base vapor chamber heat sinks eliminate semiconductor hot spots. https://www.thermacore.com. Accessed January 2013.
[103] J. Esarte, M. Domíguez, Experimental analysis of a flat heat pipe working against gravity, Appl. Therm. Eng. 23 (2003) 1619–1627. Available from: https://doi.org/10.1016/S1359-4311(03)00111-X.
[104] T. Albertin, et al., Industrial manufacturing of loop heat pipe porous media, in: Proceedings of the Sixteenth International Heat Pipe Conference, Lyon, France, 20–24 May 2012.
[105] C. Byon, S.J. Kim, Effects of geometrical parameters on the boiling limit of bi-porous wicks, Int. J. Heat. Mass. Transf. 55 (2012) 7884–7891. Available from: https://doi.org/10.1016/J.IJHEATMASSTRANSFER.2012.08.016.
[106] M.J. Rightley, C.P. Tigges, R.C. Givler, Cv Robino, J.J. Mulhall, P.M. Smith, Innovative wick design for multi-source, flat plate heat pipes, Microelectron. J. 34 (2003) 187–194. Available from: https://doi.org/10.1016/S0026-2692(02)00187-8.
[107] M. le Berre, S. Launay, V. Sartre, M. Lallemand, Fabrication and experimental investigation of silicon micro heat pipes for cooling electronics, J. Micromech. Microeng. 13 (2003) 436–441.
[108] S. Launay, V. Sartre, M. Lallemand, Experimental study on silicon micro-heat pipe arrays, Appl. Therm. Eng. 24 (2004) 233–243. Available from: https://doi.org/10.1016/J.APPLTHERMALENG.2003.08.003.
[109] Brown-Boveri, Cie Ag, UK Patent No. 1281272, April 1969.
[110] U.R. Evans, The Corrosion And Oxidation of Metals, St. Martin's Press, 1968.
[111] J.A. Riddick, E.E. Toops, A. Weissberger, Organic solvents, Interscience Publishers, New York, 1955.
[112] E.W. Saaski, Gas occlusions in arterial heat pipes, AIAA Paper 73-724, AIAA, New York, NY, 1973.
[113] E.W. Saaski, Investigation of bubbles in arterial heat pipes, NASA CR-114531, 1973.
[114] R. Kosson, et al., Development of a high capacity variable conductance heat pipe, AIAA Paper 73- 728, AIAA, New York, NY, 1973.
[115] J. Ambrose, Modeling transient acceleration effects on heat pipe performance. Heat Pipe Science and Technology, An. Int. J. (2012) 3.
[116] G.G. Birnbreier, Long time tests of Nb 1% Zr heat pipes filled with sodium and caesium, in: International Heat Pipe Conference, Stuttgart, October, 1973.
[117] G.M. Grover, J.E. Kemme, E.S. Keddy, Advances in heat pipe technology, in: International Symposium on Thermionic Electrical Power Generation, Stresa, Italy, May 1968.
[118] S. Matsumoto, T. Yamamoto, M. Katsuta, Heat transfer characteristic change and mass transfer under long-term operation in sodium heat pipe, in: Proceedings of Tenth International Heat Pipe Conference, Paper I-4, Stuttgart, 2125 September 1997.
[119] G.R. Rice, J.D. Jennings, Heat pipe filling, in: International Heat Pipe Conference, Stuttgart, October 1973.
[120] G. Rice, D. Fulford, Capillary pumping in sodium heat pipes, in: Proceedings of Seventh International Heat Pipe Conference, Minsk, 1990, Hemisphere, New York, NY, 1991.
[121] P. Vinz, C. Cappelletti, F. Geiger, Development of capillary structures for high performance sodium heat pipes, in: International Heat Pipe Conference, Stuttgart, October 1973.
[122] D. Quataert, C.A. Busse, F. Geiger, Long time behaviour of high temperature tungsten-rhenium heat pipes with lithium and silver as the working fluid, in: International Heat Pipe Conference, Stuttgart, October 1973.
[123] C.A. Busse, F. Geiger, H. Strub, High temperature lithium heat pipes, in: International Symposium on Thermionic Electrical Power Generation, Stresa, Italy, May 1968.
[124] R. Coppinger, Lithium capillary system to cool wings on re-entry, Flight Int. 168 (2005) 26.

[125] L. Bai, G. Lin, D. Wen, J. Feng, Experimental investigation of startup behaviors of a dual compensation chamber loop heat pipe with insufficient fluid inventory, Appl. Therm. Eng. 29 (2009) 1447–1456. Available from: https://doi.org/10.1016/J.APPLTHERMALENG.2008.06.019.

[126] M. Ameli, B. Agnew, P.S. Leung, B. Ng, C.J. Sutcliffe, J. Singh, et al., A novel method for manufacturing sintered aluminium heat pipes (SAHP, Appl. Therm. Eng. 52 (2013) 498–504. Available from: https://doi.org/10.1016/J.APPLTHERMALENG.2012.12.011.

[127] A company doing additive layer manufacturing, including heat exchangers. <http://www.within-lab.com/>, 2013 (accessed 08.01.13).

[128] Anon, Heat pipe qualification requirements, European Space Agency, Report ESA PSS-49 (TST- OL). Issue No. 1, January 1979.

[129] E. Baker, Prediction of long-term heat-pipe performance from accelerated life tests. 2012;11:1345–7. Available from https://doi.org/10.2514/3.6923.

[130] W.T. Anderson, Hydrogen evolution in nickel–water heat pipes, AIAA (American Institute of Aeronautics and Astronautics) Paper 73-726, 1973.

[131] M. Murakami, et al., Statistical prediction of long term reliability of copperwater heat pipes from accelerated test data, in: Proceedings of Sixth International Heat Pipe Conference, Grenoble, 1987.

[132] Y. Kojima, M. Murakami, A statis- tical treatment of accelerated heat pipe life test data, in: Proceedings of Seventh International Heat Pipe Conference, Minsk, 1990; Hemisphere, New York, NY, 1991.

[133] M. Dubois, B. Mullender, W. Supper, Development and space qualification of high capacity grooved heat pipes, in: Proceedings of Tenth International Heat Pipe Conference, Stuttgart, 21–25 September 1997.

[134] C. Hoa, B. Demolder, A. Alexandre, Roadmap for developing heat pipes for alcatel space's satellites, Appl. Therm. Eng. 23 (2003) 1099–1108. Available from: https://doi.org/10.1016/S1359-4311(03)00039-5.

[135] V. Baturkin, S. Zhuk, D. Olefirenko, A. Rudenko, Thermal qualification tests of longitudinal ammonia heat pipes for using in thermal control systems of small satellites, in: Proceedings of Fourth Minsk International Seminar 'Heat Pipes, Heat Pumps, Refrigerators', Minsk, Belarus, 4–7 September 2000.

[136] N. Dos Santos, R.R. Riehl, Qualification procedures of loop heat pipes for space applications, in: Proceedings of the Fourteenth International Heat Pipe Conference, Florianopolis, Brazil, 22–27 April 2007.

[137] G.L. Kissner, Development of a cryogenic heat pipe, in: First International Heat Pipe Conference, Paper 10-2, Stuttgart, October 1973.

[138] B.E. Nelson, W. Petrie, Experimental evaluation of a cryogenic heat pipe/radiator in a vacuum chamber, in: First International Heat Pipe Conference, Paper 10-2a, Stuttgart, October 1973.

[139] J.P. Marshburn, NASA Technical note NASA TN 0-7219, 1973.

[140] C.C. Shih, et al., The dynamic test for investigating the viscous limitation of heat pipes, in: Proceedings of the Sixteenth International Heat Pipe Conference, Lyon, France, 20–24 May 2012.

[141] M.J. Yin, et al., Research of nitro- gen cryogenic loop heat pipe, in: Proceedings of the Sixteenth International Heat Pipe Conference, Lyon, France, 20–24 May 2012.

[142] R.W. MacGregor, P.A. Kew, D.A. Reay, Investigation of low global warming potential working fluids for a closed two-phase thermosyphon, Appl. Therm. Eng. 51 (2013) 917–925. Available from: https://doi.org/10.1016/J.APPLTHERMALENG.2012.10.049.

[143] J.S. Brown, C. Zilio, A. Cavallini The fluorinated olefin R-1234ze(Z) as a high-temperature heat pumping refrigerant. Int. J. Refrig. 2009;32:1412–1422. Available from https://doi.org/10.1016/J.IJREFRIG.2009.03.002.

Chapter 4

# Heat transfer and fluid flow theory

## 4.1 Introduction

As discussed in Chapter 2, there are many variants of heat pipe, but in all cases the working fluid must circulate when a temperature difference exists between the evaporator and condenser. In this chapter, the operation of the classical wicked heat pipe is discussed. Various analytical techniques are then outlined in greater detail; these techniques are then applied to the classical heat pipe and the gravity-assisted thermosyphon.

## 4.2 Operation of heat pipes

### 4.2.1 Wicked heat pipes

The overall thermal resistance of a heat pipe, defined by Eq. (4.1), should be low; however, it is first necessary to ensure that the device will function correctly.

$$R = \frac{T_{\text{hot}} - T_{\text{cold}}}{\dot{Q}} \tag{4.1}$$

The operating limits for a wicked heat pipe, first described in Ref. [1] are illustrated below:

Each of these limits may be considered in isolation. In order for the heat pipe to operate the maximum capillary pumping pressure, $\Delta P_{c,\max}$, must be greater than the total pressure drop in the pipe. This pressure drop is made up of three components.

1. The pressure drop $\Delta P_l$ required to return the liquid from the condenser to the evaporator
2. The pressure drop $\Delta P_v$ necessary to cause the vapour to flow from the evaporator to the condenser
3. The pressure due to the gravitational head, $\Delta P_g$, which may be zero, positive or negative, depending on the inclination of the heat pipe.

For correct operation,[1]

$$\Delta P_{c,\max} \geq \Delta P_l + \Delta P_v + \Delta P_g \tag{4.2}$$

If this condition is not met, the wick will dry out in the evaporator region and the heat pipe will not operate. The maximum allowable heat flux for which Eq. (4.2) holds is referred to as the capillary limit. Methods of evaluating the four terms in Eq. (4.2) will be discussed in detail in this chapter. Typically, the capillary limit will determine the maximum heat flux over much of the operating range; however, the designer must check that a heat pipe is not required to function outside the envelope defined by the other operating limits, either at design conditions or at start-up.

During start-up and with certain high-temperature liquid-metal heat pipes, the vapour velocity may reach sonic values. The sonic velocity sets a limit on the heat pipe performance. At velocities approaching sonic, compressibility effects must be taken into account in the calculation of the vapour pressure drop.

The viscous or vapour pressure limit is also generally most important at start-up. At low temperature, the vapour pressure of the fluid in the evaporator is very low; as the condenser pressure cannot be less than zero, the difference in vapour pressure is insufficient to overcome viscous and gravitational forces, thus preventing satisfactory operation.

At high heat fluxes, the vapour velocity necessarily increases; if this velocity is sufficient to entrain liquid returning to the evaporator, then performance will decline. Hence the existence of an entrainment limit.

---

1. Note that this implies that the capillary pressure rise is considered as positive, whilst pressure drops due to liquid and vapour flow or gravity are considered positive.

Heat Pipes. DOI: https://doi.org/10.1016/B978-0-12-823464-8.00011-5

The above limits relate to axial flow through the heat pipe. The final operating limit discussed will be the boiling limit. The radial heat flux in the evaporator is accompanied by a temperature difference which is relatively small until a critical value of heat flux is reached above which vapour blankets the evaporator surface resulting in an excessive temperature difference.

The position of the curves and shape of the operating envelope shown in Fig. 4.1 depends upon the wick material, working fluid and geometry of the heat pipe.

### 4.2.2 Thermosyphons

The two-phase thermosyphon is thermodynamically similar to the wicked heat pipe but relies on gravity to ensure liquid return from the condenser to the evaporator. A wick or wicks may be incorporated in at least part of the unit to reduce entrainment and improve liquid distribution within the evaporator.

### 4.2.3 Loop heat pipes and capillary-pumped loops

As will be seen later, the characteristics which produce a wick having the capability to deliver a large capillary pressure also lead to high values of pressure drop through a lengthy wick. This problem can be overcome by incorporating a short wick within the evaporator section and separating the vapour flow to the condenser from the liquid return.

## 4.3 Theoretical background

In this section, the theory underpinning the evaluation of the terms in Eq. (4.1) and Eq. (4.2) and the determination of the operating limits shown in Fig. 4.1 is discussed.

### 4.3.1 Gravitational head

The pressure difference, $\Delta P_g$, due to the hydrostatic head of liquid may be positive, negative or zero, depending on the relative positions of the condenser and evaporator. The pressure difference may be determined from

$$\Delta P_g = \rho_l g l \sin\varphi \tag{4.3}$$

where $\rho_l$ is the liquid density (kg/m$^3$), $g$ the acceleration due to gravity (9.81 m/s$^2$), $l$ length of the heat pipe (m) and $\varphi$ the angle between the heat pipe and the horizontal defined such that $\varphi$ is positive when the condenser is lower than the evaporator.

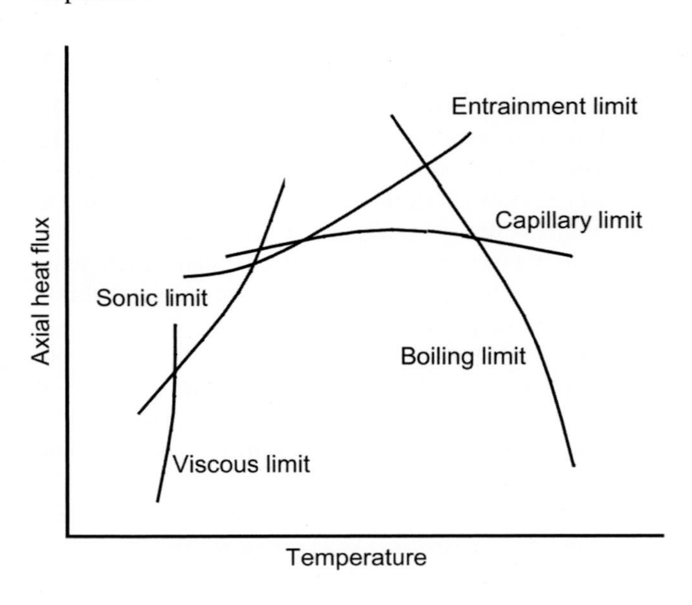


**FIGURE 4.1** Limitations to heat transport in a heat pipe.

### 4.3.2 Surface tension and capillarity

#### 4.3.2.1 Introduction

Molecules in a liquid attract one another. A molecule in a liquid will be attracted by the molecules surrounding it, and, on average, a molecule in the bulk of the fluid will not experience any resultant force. In the case of a molecule at or near the surface of a liquid, the forces of attraction will no longer balance out and the molecule will experience a resultant force inwards. Because of this effect, the liquid will tend to take up a shape having minimum surface area; in the case of a free-falling drop in a vacuum, this would be a sphere. Due to this spontaneous tendency to contract a liquid, the surface behaves rather like a rubber membrane under tension. In order to increase the surface area, work must be done on the liquid. The energy associated with this work is known as the free surface energy, and the corresponding free surface energy/unit surface area is given the symbol $\sigma_l$. For example, if a soap film is set up on a wire support, as in Fig. 4.2a and the area is increased by moving one side a distance $dx$, the work done is equal to $Fdx$; hence, the increase in surface energy is $2\sigma_l l dx$.

Factor 2 arises because the film has two free surfaces. Hence if $T$ is the force/unit length for each of the two surfaces $2Tldx = 2\sigma_l ldx$ or $T = \sigma_l$. This force/unit length is known as the surface tension. It is numerically equal to the surface energy/unit area measured in any consistent set of units, for example, N/m.

Values of surface tension for a number of common working fluids are given in the appendices.

As latent heat of vaporisation, $L$, is a measure of the forces of attraction between the molecules of a liquid, we might expect surface energy or surface tension $\sigma_l$ to be related to $L$. This is found to be the case. Solids also will have a free surface energy, and in magnitude, it is found to be similar to the value for the same material in the molten state.

When a liquid is in contact with a solid surface, molecules in the liquid adjacent to the solid will experience forces from the molecules of the solid in addition to the forces from other molecules in the liquid. Depending on whether these solid/liquid forces are attractive or repulsive, the liquid—solid surface will curve upwards or downwards, as indicated in Fig. 4.2b. The two best-known examples of attractive and repulsive forces, respectively, are water and mercury. Where the forces are attractive, the liquid is said to 'wet' the solid. The angle of contact made by the liquid surface with the solid is known as the contact angle, $\theta$. For wetting, $\theta$ will lie between 0 and $\pi/2$ radians and for nonwetting liquids $\theta > \pi/2$.

The condition for wetting to occur is that the total surface energy is reduced by wetting: $\sigma_{sl} + \sigma_{lv} < \sigma_{sv}$ where the subscripts, s, $l$ and $v$ refer to solid, liquid and vapour phases, respectively, as shown in Fig. 4.3a.

Wetting will not occur if $\sigma_{sl} + \sigma_{lv} > \sigma_{sv}$ as in Fig. 4.3c, whilst the intermediate condition of partial wetting $\sigma_{sl} + \sigma_{lv} = \sigma_{sv}$ is illustrated in Fig. 4.3b.

#### 4.3.2.2 Pressure difference across a curved surface

A consequence of surface tension is that the pressure on the concave surface is greater than that on the convex surface. With reference to Fig. 4.2c and Fig. 4.4, this pressure difference $\Delta P$ is related to the surface energy $\sigma_l$ and radius of curvature $R$ of the surface.

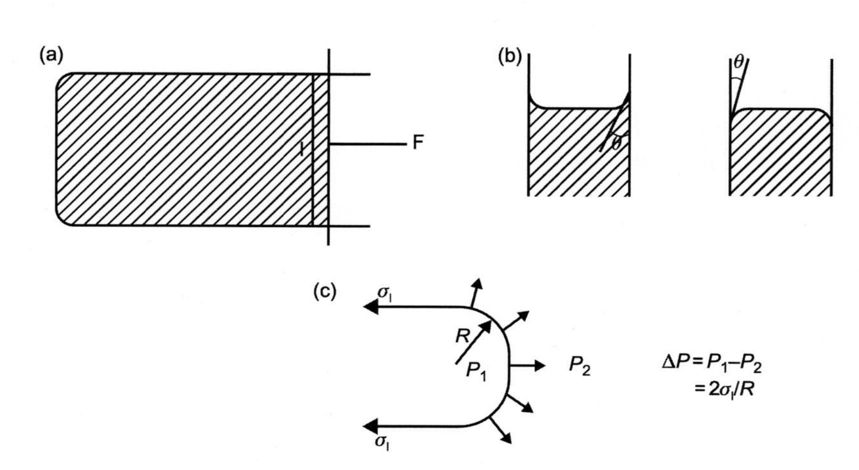


**FIGURE 4.2** Representation of surface tension and pressure difference across a curved surface. (a) Soap film on wire support, (b) liquid solid surface curving upward/downward, (c) nonwetting condition.

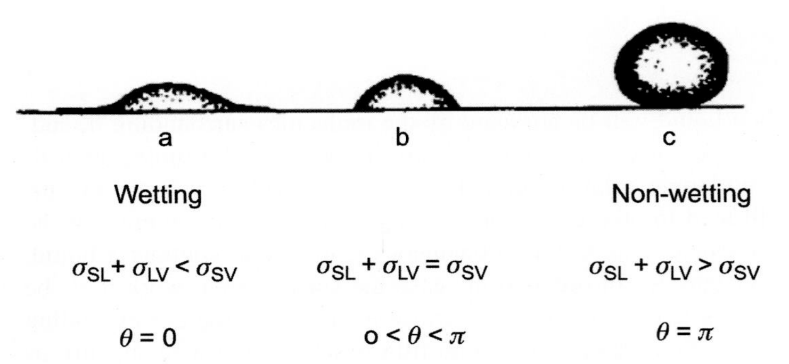


**FIGURE 4.3** Wetting and nonwetting contact.

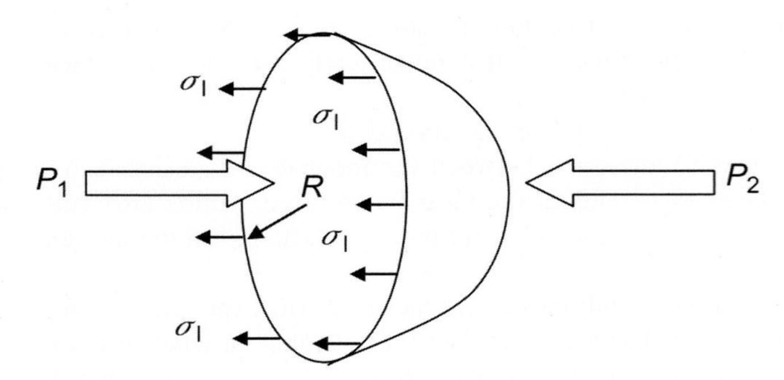


**FIGURE 4.4** Pressure difference across a curved liquid surface.

The hemispherical surface tension force acting around the circumference is given by $2\pi R\sigma_l$ and must be balanced by the net force on the surface due to the pressures which is $(P_1 - P_2)\pi R^2$ or $\Delta P\pi R^2$. Thus

$$\Delta P = \frac{2\sigma_l}{R} \tag{4.4}$$

If the surface is not spherical, but can be described by two radii of curvature, $R_1$ and $R_2$ at right angles, then it can be shown that Eq. (4.4) becomes

$$\Delta P = 2\sigma_l\left(\frac{1}{R_1} + \frac{1}{R_2}\right) \tag{4.5}$$

Due to this pressure difference, if a vertical tube, radius $r$, is placed in a liquid which wets the material of the tube, the liquid will rise in the tube to a height above the plane surface of the liquid as shown in Fig. 4.5.

The pressure balance gives, for a tube radius $r$:

$$(\rho_l - \rho_v)gh = \frac{2\sigma_l}{r}\cos\theta \approx \rho_l gh \tag{4.6}$$

For the case of a noncircular tube,

$$\frac{1}{r} = \left(\frac{1}{R_1} + \frac{1}{R_2}\right) \tag{4.7}$$

where $\rho_l$ is the liquid density, $\rho_v$ is the vapour density and $\theta$ is the contact angle. This effect is known as capillary action or capillarity and is the basic driving force for the wicked heat pipe. For nonwetting liquids, cos $\theta$ is negative and the curved surface is depressed below the plane of the liquid level. In heat pipes, wetting liquids are always used, the capillary lift increasing with liquid surface tension and decreasing contact angle.

#### *4.3.2.3 Change in vapour pressure at a curved liquid surface*

From Fig. 4.5, it can be seen that the vapour pressure at the concave surface is less than that at the plane liquid surface by an amount equal to the weight of a vapour column of unit area, length $h$.

$$P_c - P_o = g\rho_v h$$

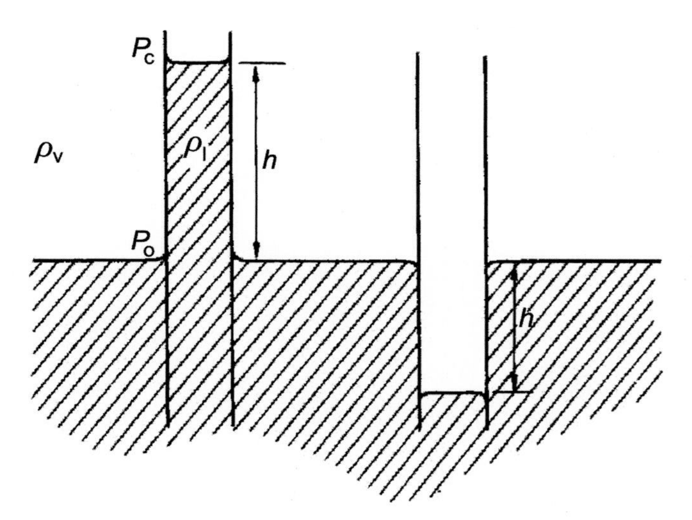


**FIGURE 4.5** Capillary rise in a tube (wetting and nonwetting fluids).

assuming that $\rho_v$ is constant. Combining this with Eq. (4.6):

$$P_c - P_o = \frac{2\sigma_l}{r} \frac{\rho_v}{(\rho_l - \rho_v)} \cos\theta \tag{4.8}$$

This pressure difference $P_c - P_o$ is small compared to the capillary head $(2\sigma_l/r)\cos\theta$ and may normally be neglected when considering the pressures within a heat pipe. It is, however, worth noting that the difference in vapour pressure between the vapour in a bubble and the surrounding liquid is an important phenomenon in boiling heat transfer.

#### 4.3.2.4 Measurement of surface tension

There are a large number of methods for the measurement of surface tension of a liquid, and these are described in the standard texts [2,3]. For our present purpose, we are interested in the capillary force, $\sigma_l\cos\theta$, which is dependent on both liquid and surface properties. The simplest measurement is that of capillary rise $h$ in a tube, which gives

$$\sigma_l \cos\theta \approx \frac{\rho_l g h r}{2} \tag{4.9}$$

In practical heat pipe design, it is also necessary to know $r$, the effective pore radius. This is by no means easy to estimate for a wick made up of a sintered porous structure or from several layers of gauze. By measuring the maximum height the working fluid will attain, it is possible to obtain information on the capillary head for fluid wick combinations. Several workers have reported measurements on maximum height for different structures, and some results are given in Chapter 3. The results may differ for the same structure depending on whether the film was rising or falling; the reason for this is brought out in Fig. 4.6.

Another simple method, for the measurement of $\sigma_l$, due to Jäger and shown schematically in Fig. 4.7, is sometimes employed. This involves the measurement of maximum bubble pressure. The pressure is progressively increased until the bubble breaks away and the pressure falls. When the bubble radius reaches that of the tube, pressure is a maximum $P_{max}$ and at this point,

$$P_{max} = \rho_l h g + \frac{2\sigma_l}{r} \tag{4.10}$$

This method has proved appropriate [4] for the measurement of the surface tension of liquid metals.

The surface tension of two liquids can be compared by comparing the mass of the droplets falling from a narrow vertical tube. If the mass of the droplets for liquids 1 and 2 are $m_1$ and $m_2$, respectively, then

$$\frac{m_1\rho_{l1}}{m_2\rho_{l2}} = \frac{\sigma_{f1}}{\sigma_{f2}} \tag{4.11}$$

**FIGURE 4.6** Rising and falling column interface.

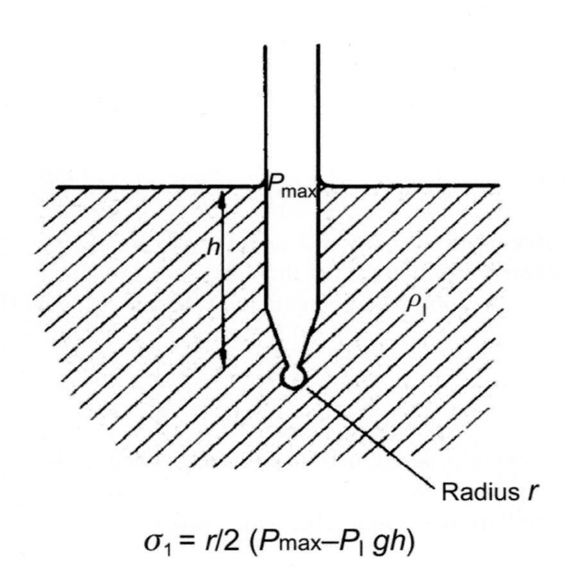


**FIGURE 4.7** Jäger's method for surface tension measurement.

### *4.3.2.5 Temperature dependence of surface tension*

Surface tension decreases with increasing temperature; it is therefore important to take temperature effects into account if using the results of measurements at typical ambient temperatures.

Values of surface tension for over 2000 pure fluids have been tabulated by Jasper [5], and temperature corrections of the form $\sigma = a + bT$ are suggested.

For water, the following interpolating equation [6] gives good values of surface tension:

$$\sigma_l = B(1 - T_r)^{\mu}(1 - b(1 - T_r)) \tag{4.12}$$

where $T_r = T/T_c$ is the reduced temperature, $Tc = 647.096$ K, $B = 235.8$ mN/m, $b = 0.625$ and $\mu = 1.256$.

Eq. (4.12) is valid between the triple point (0.01°C) and the critical temperature and is in agreement with measured data to within experimental uncertainty.

Eötvös proposed a relationship which was later modified by Ramsay and Shields to give

$$\sigma_l \left(\frac{M}{\rho_l}\right)^{\frac{2}{3}} = H(T_c - 6 - T) \tag{4.13}$$

where $M$ is the molecular weight, $T_c$ is the critical temperature (K), $T$ is the fluid temperature (K) and $H$ is a constant, the value of which depends upon the nature of the liquid.

The Eotvös–Ramsay–Sheilds equation does not give agreement with the experimentally observed behaviour of liquid metals and molten salts. Bohdanski and Schins [7] have derived an equation which applies to alkali metals. Fink and Leibowitz [8] recommended

$$\sigma_l = \sigma_o \left(1 - \frac{T}{T_c}\right)^n (\text{mN/m}) \tag{4.14}$$

for sodium, where $\sigma_o = 240.5$ mN/m, $n = 1.126$ and $T_c = 2503.7$ K.

Alternatively, the surface tension of liquid metals may be estimated from the data provided by Iida and Guthrie [9] and summarised in Table 4.1. The value of surface tension may then be calculated:

$$\sigma_l = \sigma_{lm} + (T - T_m)d\sigma/dT$$

A method of evaluating surface tension based on the number and nature of chemical bonds was first suggested by Walden [10] and developed by Sugden [11] and Quale [12].

$$\sigma^{0.25} = \frac{\text{P}}{M}(\rho_l - \rho_v) \tag{4.15}$$

where $P$ was defined by Walden as a parachor. Values of the increments to be used in evaluating the parachor, adapted to give values of surface tension in N/m, are given in Ref. [13].

### 4.3.2.6 Capillary pressure $\Delta P_c$

Eq. (4.4) shows that the pressure drop across a curved liquid interface is

$$\Delta P = \frac{2\sigma_l}{R}$$

From Fig. 4.8 we can see that $R\cos\theta = r$ where $r$ is the effective radius of the wick pores and $\theta$ the contact angle. Hence the capillary head at the evaporator, $\Delta P_e'$ is

$$\Delta P_e' = 2\sigma_l \frac{\cos\theta_e}{r_e} \tag{4.16a}$$

Similarly, for the condenser,

$$\Delta P_c' = 2\sigma_l \frac{\cos\theta_c}{r_c} \tag{4.16b}$$

and the capillary driving pressure, $\Delta P_c$, is given by $\Delta P_e' - \Delta P_c'$.

It is worth noting that $\Delta P_c$ is a function only of the conditions where a meniscus exists. It does not depend on the length of the adiabatic section of the wick. This is particularly important in the design of loop heat pipes.

**TABLE 4.1** Surface tension of selected liquid metals [9].

| | | M | Melting point $T_m$ | Boiling point (1 bar) | $\sigma_{lm}$ at melting point | $d\sigma/dT$ |
|---|---|---|---|---|---|---|
| | | kg/kmol | K | K | mN/m | mN/mK |
| Lithium | Li | 7 | 452.2 | 1590 | 398 | −0.14 |
| Sodium | Na | 23 | 371.1 | 1151 | 191 | −0.1 |
| Potassium | K | 39 | 336.8 | 1035 | 115 | −0.08 |
| Caesium | Cs | 133 | 301.65 | 1033 | 70 | −0.06 |
| Mercury | Hg | 200 | 234.3 | 630 | 498 | −0.2 |

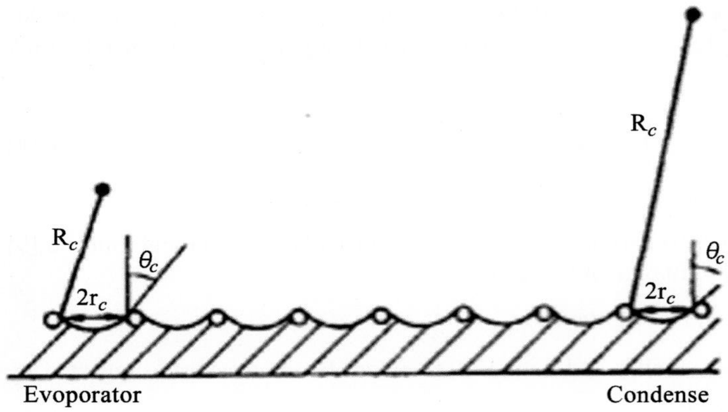

**FIGURE 4.8** Wick and pore parameters in evaporator and conden.

### 4.3.3 Pressure difference due to friction forces

In this section, we will consider the pressure differences caused by frictional forces in liquids and vapours flowing in a heat pipe. Firstly, it is convenient to define some of the terms which will be used later in the chapter.

#### *4.3.3.1 Laminar and turbulent flow*

If one imagines a deck of playing cards or a sheaf of paper, initially stacked to produce a rectangle, to be sheared as shown in Fig. 4.9, it can be seen that the individual cards, or lamina, slide over each other. There is no movement of material perpendicular to the shear direction.

Similarly, in laminar fluid flow, there is no mixing of the fluid and the fluid can be regarded as a series of layers sliding past each other.

Consideration of a simple laminar flow allows us to define viscosity. Fig. 4.10 illustrates the velocity profile for a laminar flow of a fluid over a flat plate.

The absolute or dynamic viscosity of a fluid, $\mu$, is defined by

$$\tau = \mu \frac{dv}{dy}$$

where $\tau$ is the shear stress. At the wall, the velocity of the fluid must be zero, and the wall shear stress is given by

$$\tau_w = \mu \left( \frac{dv}{dy} \right)_w$$

In practice, laminar flow is observed at low speeds, in small tubes or channels, with highly viscous fluids and very close to solid walls. It is the flow normally observed when liquid flows through the wick of a heat pipe.

If the fluid layers seen in laminar flow break up and fluid mixes between the layers then the flow is said to be turbulent. The turbulent mixing of fluid perpendicular to the flow direction leads to a more effective transfer of momentum and internal energy between the wall and the bulk of the fluid. This is illustrated in Fig. 4.11.

The heat transfer and pressure drop characteristics of laminar and turbulent flows are very different. In forced convection, the magnitude of the Reynolds number, defined below, provides an indication of whether the flow is likely to be laminar or turbulent.

$$\text{Re} = \frac{\rho v d_{rep}}{\mu}$$

where $d_{rep}$ is a representative linear dimension. If Reynold's number is written as

$$\text{Re} = \frac{\rho v^2}{\mu v / d}$$

it can be seen to be a measure of the relative importance of inertial and viscous forces acting on the fluid.

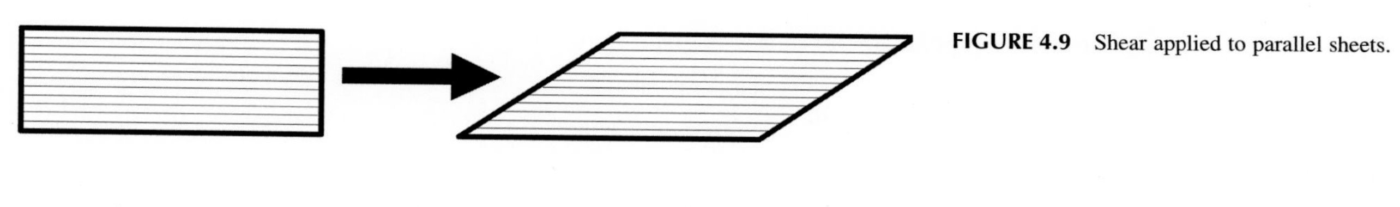

**FIGURE 4.9** Shear applied to parallel sheets.

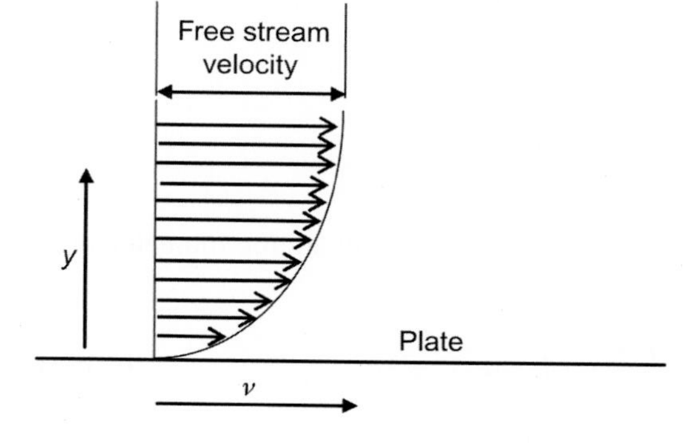


**FIGURE 4.10** Velocity profile in laminar flow over a flat plate.

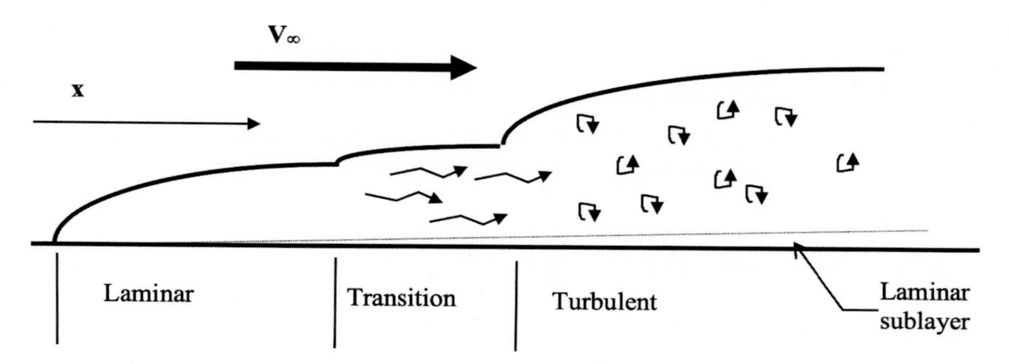


**FIGURE 4.11** Development of the boundary layer over a flat plate.

For flow over a flat plate, as shown in Fig. 4.11, we may determine whether the flow in the boundary layer is likely to be laminar or turbulent by applying the following conditions:

$$Re_x\left(=\frac{\rho V_\infty x}{\mu}\right) < 10^5 \text{ Laminar flow}$$

$$Re_x\left(=\frac{\rho V_\infty x}{\mu}\right) > 10^6 \text{ Turbulent flow}$$

where $x$ is the distance from the leading edge of the plate.

For values of Reynolds number between $10^5$ and $10^6$ the situation is complicated by two factors. Firstly, the transition is not sharp, it occurs over a finite length of plate. In the transition region, the flow may intermittently take on turbulent and laminar characteristics. Secondly, the position of the transition zone depends not only upon the Reynolds number; it is also influenced by the nature of the flow in the free stream and the nature of the surface. Surface roughness or protuberances on the surface tend to trip the boundary layer from laminar to turbulent.

For flow in pipes, channels or ducts, the situation is similar to that for a flat plate in the entry region, but in long channels, the boundary layers from all walls meet, and fully developed temperature and velocity profiles are established.

For fully developed flow in pipes or channels, the transition from laminar to turbulent flow occurs at a Reynolds number, $\text{Re}_d = (\rho v d_e/\mu)$, of approximately 2100. The dimension $d_e$ is on the channel equivalent or hydraulic diameter:

$$d_e = \frac{4 \times \text{Cross-sectional area}}{\text{Wetted perimeter}} \tag{4.17a}$$

As expected, for a circular duct or pipe, diameter $d$, this is given by

$$d_e = \frac{4\pi d^2/4}{\pi d} = d \tag{4.17b}$$

For a square duct, side length $x$,

$$d_e = \frac{4x^2}{4x} = x \tag{4.17c}$$

and for a rectangular duct, width $a$ and depth $b$:

$$d_e = \frac{4ab}{2(a+b)} \tag{4.17d}$$

For flow through an annulus having inner and outer diameters $d_1$ and $d_2$, respectively, the hydraulic diameter may be calculated:

$$d_e = \frac{4\pi\left(d_2^2 - d_1^2\right)/4}{\pi(d_2 + d_1)} = \frac{4\pi(d_2 - d_1)(d_2 + d_1)/4}{\pi(d_2 + d_1)} = (d_2 - d_1) \tag{4.17e}$$

which is equal to twice the thickness of the annular gap.

The velocity profile in laminar flow in a tube is parabolic, whilst in turbulent flow the velocity gradient close to the wall is much steeper, as shown in Fig. 4.12.

### 4.3.3.2 *Laminar flow – the Hagen–Poiseuille equation*

The steady-state laminar flow of an incompressible fluid of constant viscosity $\mu$, through a tube of circular cross-section, radius $a$, is described by the Hagen–Poiseuille equation. This equation relates the velocity, $v_r$, of the fluid at radius $r$ to the pressure gradient $(dp/dl)$ along the tube.

$$v_r = \frac{a^2}{4\mu}\left[1 - \left(\frac{r}{a}\right)^2\right]\left(-\frac{dp}{dl}\right) \tag{4.18}$$

The velocity profile is parabolic, varying from the maximum value, $v_m$, given by

$$v_m = \frac{a^2}{4\mu}\left(-\frac{dp}{dl}\right) \tag{4.19}$$

on the axis of the tube to zero adjacent to the wall. The average velocity, $v$, is given by

$$v = \frac{a^2}{8\mu}\left(-\frac{dp}{dl}\right) \tag{4.20}$$

or rearranging

$$\frac{dp}{dl} = -\frac{8\mu v}{a^2}$$

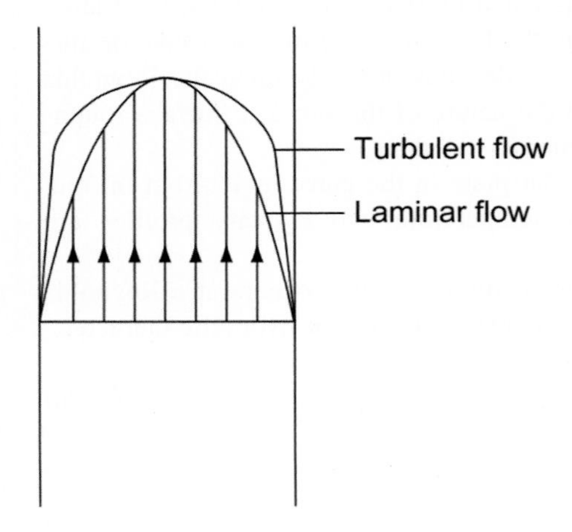


**FIGURE 4.12** Velocity distribution in a circular tube for laminar and turbulent flow.

In a one-dimensional treatment, the average velocity may be used throughout. The volume flowing per second, $S$, is

$$S = \pi a^2 v = -\frac{\pi a^4}{8\mu}\left(-\frac{dp}{dl}\right) \tag{4.21}$$

and if $\rho$ is the fluid density, the mass flow $\dot{m}$ is given by

$$\dot{m} = \rho \pi a^2 v = -\rho \frac{\pi a^4}{8\mu}\left(-\frac{dp}{dl}\right) \tag{4.22}$$

For incompressible, fully developed flow, the pressure gradient is constant, so the term $(-dp/dl)$ in Eqs (4.18–4.22) may be replaced by $\frac{P_1 - P_2}{l}$ where $P_1 - P_2$ is the pressure drop $\Delta P_f$ over a length $l$ of the channel.

The kinetic head, or flow energy, may be compared to the energy lost due to viscous friction over the channel length $l$. Both may be expressed in terms of the effective pressure difference, $\Delta P$.

The kinetic energy term and the viscous term are given by

$$\Delta P_{KE} = \frac{1}{2}\rho v^2$$

$$\Delta P_F = \frac{8\mu v}{a^2}$$

respectively.

$$\frac{\Delta P_{KE}}{\Delta P_F} = \frac{\rho v a^2}{16\mu l} = \mathrm{Re}\frac{a}{32l} = \frac{\mathrm{Re}}{64}\frac{d}{l}$$

Thus assuming the flow is laminar, the kinetic and viscous terms are equal when

$$l = \frac{\mathrm{Re}}{64}d \tag{4.23}$$

For high $l/d$ ratios, viscous pressure drop dominates.

#### *4.3.3.3 Turbulent flow – the Fanning equation*

The frictional pressure drop for turbulent flow is usually related to the average fluid velocity by the Fanning equation

$$\left(-\frac{dp}{dl}\right) = \frac{4}{d}f\frac{1}{2}\rho v^2 \tag{4.24a}$$

$$\frac{P_1 - P_2}{l} = \frac{4}{d}f\frac{1}{2}\rho v^2 \tag{4.24b}$$

where $f$ is the Fanning friction factor. $f$ is related to the Reynolds number in the turbulent region and a commonly used relationship is the Blasius equation.

$$f = \frac{0.0791}{Re^{0.25}},\ 2100 < \mathrm{Re} < 10^5 \tag{4.25}$$

The Fanning equation may be applied to laminar flow if

$$f = \frac{16}{Re},\ \mathrm{Re} < 2100 \tag{4.26}$$

### 4.3.4 Flow in Wicks

#### *4.3.4.1 Pressure difference in the liquid phase*

The flow regime in the liquid phase is almost always laminar. As the liquid channels will not in general be straight nor of circular cross-section and will often be interconnected, the Hagen–Poiseuille equation must be modified to take account of these differences.

Since mass flow will vary in both the evaporator and the condenser region, an effective length rather than the geometrical length must be used for these regions. If the mass change per unit length is constant, the total mass flow will increase, or decrease, linearly along the regions, being zero at the end. We can therefore replace the lengths of the evaporator $l_e$ and the condenser $l_c$ by $\frac{l_e}{2}$ and $\frac{l_c}{2}$. The total effective length for fluid flow will then be $l_{\text{eff}}$ where

$$l_{\text{eff}} = l_a + \frac{l_e + l_c}{2} \tag{4.27}$$

Tortuosity within the capillary structure must be taken into account separately and is discussed below.

There are three principal capillary geometries

**(1)** Wick structures consisting of a porous structure made up of interconnecting pores. Gauzes, felts and sintered wicks come under this heading; these are frequently referred to as homogeneous wicks.
**(2)** Open grooves.
**(3)** Covered channels consisting of an area for liquid flow closed by a finer mesh capillary structure. Grooved heat pipes with gauze covering the groove and arterial wicks are included in this category; these wicks are sometimes described as composite wicks.

Some typical wick sections are shown in Fig. 4.13 and expressions for pressure drop within particular structures are discussed.

Recent innovations in wick design may involve the use of nanoparticles to modify the surface characteristics of the wick, and it is reported [14] that this can result in a reduction in heat transfer overall resistance. It has been demonstrated that nanostructured microposts can be fabricated [15], which give excellent capillary performance and improvements in both heat transfer coefficient and critical heat flux [16].

### 4.3.4.2 Homogeneous wicks

If $\varepsilon$ is the fractional void of the wick, that is the fraction of the cross-section available for the fluid flow, then the total flow cross-sectional area $A_f$ is given by

$$A_f = A\varepsilon = \pi\left(r_w^2 - r_v^2\right)\varepsilon \tag{4.28}$$

where $r_w$ and $r_v$ are the outer and inner radius of the wick, respectively.

If $r_c$ is the effective pore radius, then the Hagen–Poiseuille equation (Eq. (4.22)) may be written as

$$\dot{m} = \frac{\pi\left(r_w^2 - r_v^2\right)\varepsilon r_c^2 \rho_l}{8\mu_l} \frac{\Delta P_l}{l_{eff}} \tag{4.29}$$

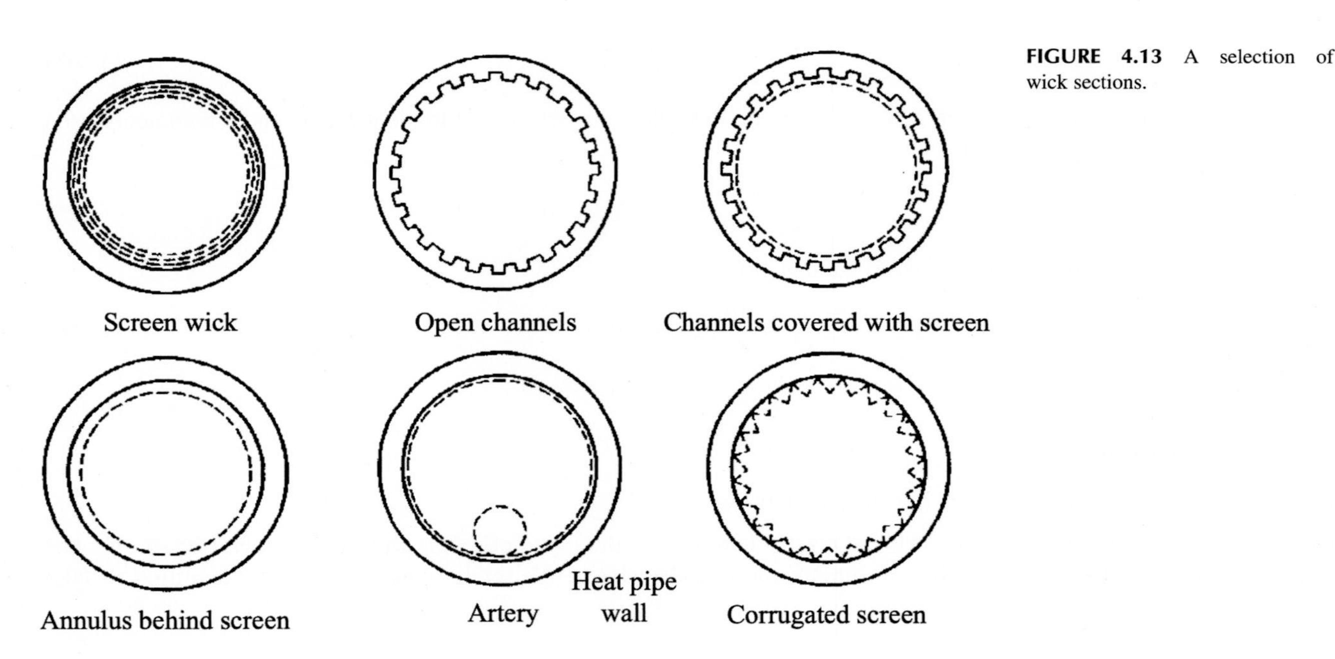


**FIGURE 4.13** A selection of wick sections.

or, relating the heat and mass flows, $\dot{Q} = \dot{m}L$, where $L$ is the latent heat of vaporisation and rearranging:

$$\Delta P_l = \frac{8\mu_l \dot{Q} l_{eff}}{\pi\left(r_w^2 - r_v^2\right)\varepsilon r_c^2 \rho_l L} \tag{4.30}$$

For porous media, Eq. (4.30) is usually written as

$$\Delta P_l = \frac{b\mu_l \dot{Q} l_{eff}}{\pi\left(r_w^2 - r_v^2\right)\varepsilon r_c^2 \rho_l L} \tag{4.31}$$

The number 8, derived for round tubes, is replaced by the dimensionless constant, $b$, typically $10 < b < 20$ to include a correction for tortuosity.

Whilst this relation can be useful for a theoretical treatment, it contains three constants, $b$, $\varepsilon$ and $r_c$ which are all difficult to measure in practice. It is therefore useful to relate the pressure drop and flow rate for a wick structure by using a form of Darcy's 'Law':

$$\Delta P_l = \frac{\mu_l l_{eff} \dot{m}}{\rho_l K A} \tag{4.32}$$

where $K$ is the wick permeability.

Comparison of Eq. (4.32) with Eq. (4.30), shows that Darcy's 'Law' is the Hagen–Poiseuille equation with correction terms included in the constant $K$ to take into account pore size, pore distribution and tortuosity. It serves to provide a definition of permeability, $K$, a quantity which can be easily measured.

The Blake–Kozeny equation is sometimes used in the literature. This equation relates the pressure gradient across a porous body, made up from spheres diameter $D$, to the flow of liquid. Like Darcy's 'Law' it is merely the Hagen–Poiseuille equation with correction factors. The Blake–Kozeny equation may be written:

$$\Delta P_l = \frac{150\mu_l(1-\varepsilon')^2 l_{eff} v}{D^2 \varepsilon'^3} \tag{4.33}$$

and is applicable only to laminar flow, which requires that:

$$\text{Re}' = \frac{\rho v D}{\mu(1-\varepsilon')} < 10$$

where $v$ is the superficial velocity $\dot{m}/\rho_l A$ and $\varepsilon' =$ (volume of voids/volume of body).

### 4.3.4.3 Nonhomogeneous wicks

*Longtitudinal groove wick.* For grooved wicks, the pressure drop in the liquid is given by

$$\Delta P_l = \frac{8\mu_l \dot{Q} l}{\pi\left(\frac{1}{2}d_e\right)^4 N \rho_l L} \tag{4.34}$$

where $N$ is the number of grooves and $d_e$ is the effective diameter defined by Eq. (4.17). At high vapour velocities, shear forces will tend to impede the liquid flow in open grooves. This may be avoided by using a fine pore screen to form a composite wick structure.

*Composite wicks.* Such a system as arterial or composite wicks requires an auxiliary capillary structure to distribute the liquid over the evaporator and condenser surfaces. The pressure drop in wicks constructed by an inner porous screen separated from the heat pipe wall to give an annular gap for the liquid flow may be obtained from the Hagen–Poiseuille equation applied to parallel surfaces, provided that the annular width $w$ is small compared to the radius of the pipe vapour space $r_v$.

In this case,

$$\Delta P_l = \frac{6\mu_l \dot{Q} l}{\pi r_v w^3 \rho_l L} \tag{4.35}$$

This wick structure is particularly suitable for liquid-metal heat pipes. Variants are also used in lower temperature high-performance heat pipes for spacecraft. Crescent annuli may be used, in which it is assumed that the screen is moved down to touch the bottom of the heat pipe wall leaving a gap of $2w$ at the top. In this case:

$$\Delta P_l = \frac{6\mu_l \dot{Q} l}{8\pi r_v w^3 \rho_l L} \tag{4.36}$$

In Eq. (4.35) and Eq. (4.36), the length should be taken as the effective length defined in Eq. (4.27).

## 4.3.5 Vapour phase pressure difference, $\Delta P_v$

### 4.3.5.1 Introduction

The total vapour phase difference in pressure will be the sum of the pressure drops in the three regions of a heat pipe, namely the evaporator drop $\Delta P_{ve}$, the adiabatic section drop $\Delta P_{va}$ and the pressure drop in the condensing region $\Delta P_{vc}$. The problem of calculating the vapour pressure drop is complicated in the evaporating and condensing regions by radial flow due to evaporation or condensation. It is convenient to define a further Reynolds number, the Radial Reynolds number:

$$\mathrm{Re}_r = \frac{\rho_v v_r r_v}{\mu_v} \tag{4.37}$$

to take account of the radial velocity component $v_r$ at the wick where $r = r_v$.

By convention, the vapour space radius $r_v$ is used rather than the vapour space diameter, which is customary in the definition of axial Reynolds number. $\mathrm{Re}_r$ is positive in the evaporator section and negative in the condensing section. In most practical heat pipes, $\mathrm{Re}_r$ lies in the range of 0.1–100.

$\mathrm{Re}_r$ is related to the radial rate of mass injection or removal per unit length $(d\dot{m}/dz)$ as follows:

$$\mathrm{Re}_r = \frac{1}{2\pi\mu_v}\frac{d\dot{m}}{dz} \tag{4.38}$$

The radial and axial Reynolds numbers are related for uniform evaporation or condensation rates by the equation,

$$\mathrm{Re}_r = \frac{\mathrm{Re}}{4}\frac{r_v}{z} \tag{4.39}$$

where $z$ is the distance from either the end of the evaporator section or the end of the condenser section.

In Section 4.3.3.2, we showed in Eq. (4.23) that, provided the flow is laminar, the pressure drop due to viscous forces in a length $l$ is equal to the kinetic head when

$$l = \frac{\mathrm{Re}\, d}{64} = \frac{\mathrm{Re}\, r_v}{32} \tag{4.40}$$

If we substitute

$$\mathrm{Re} = \frac{4\mathrm{Re}_r l}{r_v}$$

for the evaporator or condenser region, we find that the condition reduces to

$$\mathrm{Re}_r = 8 \tag{4.41}$$

Fig. 4.14, taken from Busse [17], shows $\mathrm{Re}_r$ as a function of power/unit length for various liquid-metal working fluids. Clearly, the kinetic head may be a significant component of the vapour pressure drop in the evaporator and result in an appreciable pressure rise in the condenser when working with liquid metals.

### 4.3.5.2 Incompressible flow: (simple one-dimensional theory)

In the following treatment, we will regard the vapour as an incompressible fluid. This assumption implies that the flow velocity v is small compared to the velocity of sound $c$ in the vapour, that is, the Mach number:

$$\frac{v}{c} < 0.3$$

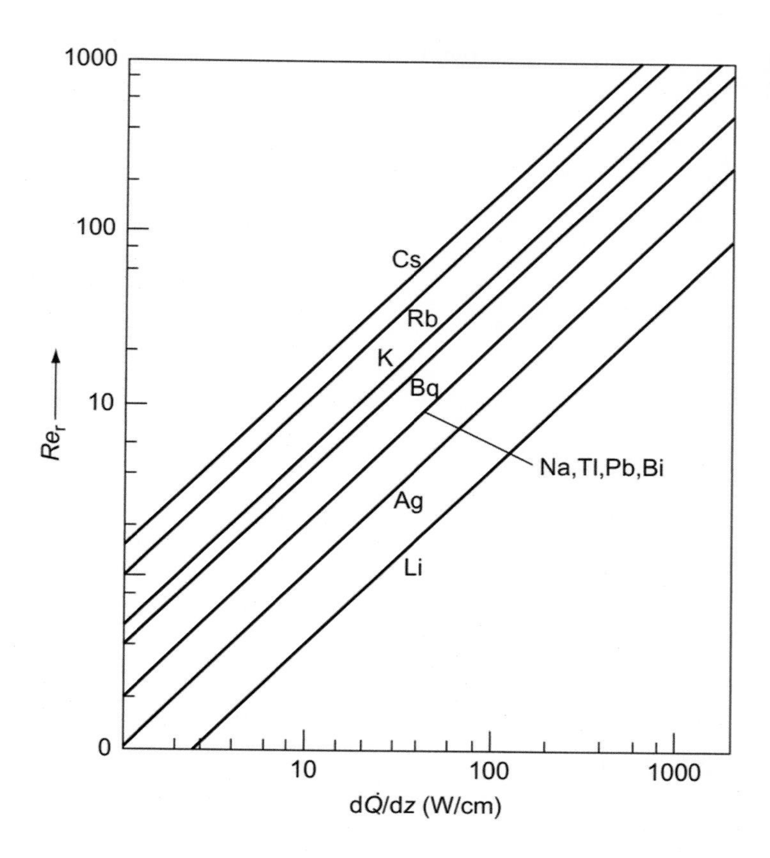


**FIGURE 4.14** Radial Reynolds number versus heat input per unit length of heat pipe (liquid-metal working fluids).

Alternatively, this condition implies that the treatment is valid for heat pipes in which $\Delta P_v$ is small compared with $P_v$, the average vapour pressure in the pipe. This assumption is not necessarily valid during start-up, nor is it always true in the case of high-temperature liquid-metal heat pipes. The effect of compressibility of the vapour will be considered in Section 4.3.5.6.

In the evaporator region, the vapour pressure gradient will be necessary to carry out two functions.

**i.** To accelerate the vapour entering the evaporator section up to the axial velocity, $v$, since, on entering the evaporator, this vapour will have radial velocity but no axial velocity. The necessary pressure gradient we will call the inertial term $\Delta P_v{}'$

**ii.** To overcome frictional drag forces at the surface $r = r_v$ at the wick. This is the viscous term $\Delta P_v{}''$

We can estimate the magnitude of the inertial term as follows. If the mass flow/unit area of cross-section at the evaporator is $\rho v$, then the corresponding momentum flux/unit area will be given by $\rho v \times v$ or $\rho v^2$. This momentum flux in the axial direction must be provided by the inertial term of the pressure gradient.

Hence,

$$\Delta P_v{}' = \rho v^2 \tag{4.42}$$

Note that $\Delta P_v{}'$ is independent of the length of the evaporator section. The way in which $\Delta P_v{}'$ varies along the length of the evaporator is shown in Fig. 4.15a.

If we assume laminar flow, we can estimate the viscous contribution to the total evaporator pressure loss by integrating the Hagen–Poiseuille equation. If the rate of mass entering the evaporator per unit length $d\dot{m}/dz$ is constant, we find by integrating Eq. (4.22) along the length of the evaporator section,

$$\Delta P_v{}'' = \frac{8\mu_v \dot{m}}{\rho_v \pi r_v^4} \frac{l_e}{2} \tag{4.43}$$

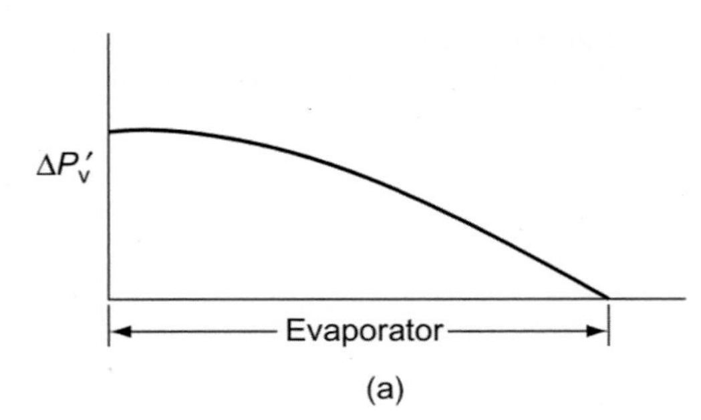


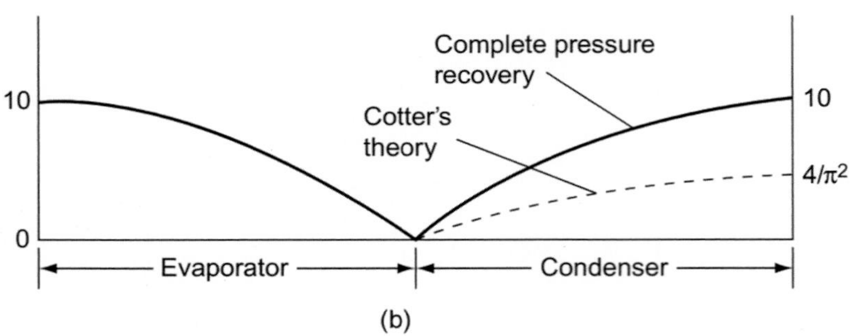


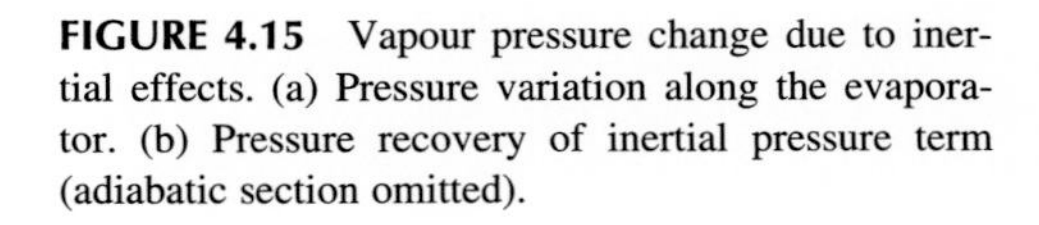
**FIGURE 4.15** Vapour pressure change due to inertial effects. (a) Pressure variation along the evaporator. (b) Pressure recovery of inertial pressure term (adiabatic section omitted).

Thus the total pressure drop in the evaporator region $\Delta P_{ve}$ will be given by the sum of the two terms:

$$\Delta P_{ve} = \Delta P_v' + \Delta P_v''$$

$$\Delta P_{ve} = \rho v^2 + \frac{8\mu_v \dot{m}\, l_e}{\rho \pi r_v^4\; 2} \tag{4.44}$$

The condenser region may be treated in a similar manner, but in this case axial momentum will be lost as the vapour stream is brought to rest so the inertial term will be negative, that is there will be pressure recovery. For the simple theory, the two inertial terms will cancel and the total pressure drop in the vapour phase will be due entirely to the viscous terms. It is shown later that it is not always possible to recover the initial pressure term in the condensing region.

In the adiabatic section, the pressure difference will contain only the viscous term which will be given either by the Hagen–Poiseuille equation or the Fanning equation depending on whether the flow is laminar or turbulent.

For laminar flow:

$$\Delta P_{va} = \frac{8\mu_v \dot{m}}{\rho \pi r_v^4} l_a \mathrm{Re} < 2100 \tag{4.45}$$

For turbulent flow:

$$\Delta P_{va} = \frac{2}{r_v} f \frac{1}{2} \rho_v v^2 l_a \; Re > 2100 \tag{4.46}$$

where $f = 0.0791/Re^{0.25}$, $2100 < \mathrm{Re} < 10^5$ (Eq. (4.25)).

Hence the total vapour pressure drop, $\Delta P_v$ is given by

$$\Delta P_v = \Delta P_{ve} + \Delta P_{vc} + \Delta P_{va} = \rho v^2 + \frac{8\mu_v \dot{m}}{\rho \pi r_v^4}\left[\frac{l_e + l_c}{2} + l_a\right] \tag{4.47}$$

for laminar flow with no pressure recovery, and

$$\Delta P_v = \Delta P_{ve} + \Delta P_{vc} + \Delta P_{va} = \frac{8\mu_v \dot{m}}{\rho \pi r_v^4}\left[\frac{l_e + l_c}{2} + l_a\right] \tag{4.48}$$

for laminar flow with full pressure recovery.

Eq. (4.47) and Eq. (4.48) enable the calculation of vapour pressure drops in simple heat pipe design and are used extensively.

#### 4.3.5.3 Incompressible flow: one-dimensional theories of Cotter and Busse

In addition to the assumption of incompressibility, the above treatment assumes a fully developed flow velocity profile and complete pressure recovery. It does, however, give broadly correct results. A considerable number of papers have been published giving a more precise treatment of the problem. Some of these will now be summarised in this and the following section.

The earliest theoretical treatment of the heat pipe was by Cotter [1]. For $\mathrm{Re}_r \ll 1$, Cotter used the following result obtained by Yuan and Finkelstein for laminar incompressible flow in a cylindrical duct with uniform injection or suction through a porous wall.

$$\frac{dP_v}{dz} = \frac{8\mu_v \dot{m}}{\pi \rho_v r_v^4}\left[1 + \frac{3}{4}\mathrm{Re}_r - \frac{11}{270}\mathrm{Re}_r^2\right]$$

He obtained the following expression.

$$\Delta P_{ve} = \frac{4\mu_v l_e \dot{Q}}{\pi \rho_v r_v^4 L} \tag{4.49}$$

which is equivalent to equation Eq. (4.43).

For $Re_r >> 1$, Cotter used the pressure gradient obtained by Knight and McInteer for flow with injection or suction through porous parallel plane walls. The resulting expression for pressure gradient is

$$\Delta P_{ve} = \frac{\dot{m}^2}{8\rho_v r_v^4}$$

which may be rewritten as

$$\Delta P_{ve} = \frac{\left(\rho_v \pi r_v^2 v\right)^2}{8\rho_v r_v^4} = \frac{\pi^2}{8}\rho_v v^2 \approx \rho_v v^2 \tag{4.50}$$

This is equivalent to Eq. (4.42), derived previously, suggesting that inertia effects dominate the pressure drop in the evaporator.

A different velocity profile was used by Cotter in the condenser region, this gave pressure recovery in the condenser of:

$$\Delta P_{vc} = -\frac{4}{\pi^2}\frac{\dot{m}^2}{8\rho_v r_v^4} \tag{4.51}$$

which is $4/\pi^2$, or 0.405, of $\Delta P_{ve}$, giving only partial pressure recovery.

In the adiabatic region, Cotter assumed fully developed laminar flow; hence, Eq. (4.46) was used:

$$\Delta P_{ve} = \frac{\dot{m}^2}{8\rho_v r_v^4}$$

The full expression for the vapour pressure drop combined Eqs (4.46), (4.50) and (4.51) to give

$$\Delta P_v = \left(1 - \frac{4}{\pi^2}\right)\frac{\dot{m}}{8\rho_v r_v^4} + \frac{8\mu_v \dot{m} l_a}{\pi \rho_v r_v^4} \tag{4.52}$$

Busse also considered the one-dimensional case, assuming a modified Hagen–Poiseuille velocity profile and using this to obtain a solution of the Navier–Stokes equation[2] [18] for a long heat pipe and obtained similar results.

#### 4.3.5.4 Pressure recovery

We have seen that the pressure drop in the evaporator and condenser regions consists of two terms, an inertial term and a term due to viscous forces. Simple theory suggests that the inertial term will have the opposite sign in the condenser region and should cancel out that of the evaporator, Fig. 4.15b. There is experimental evidence for this pressure

2. The Navier–Stokes equation, together with the continuity equation, is a relationship which relates the pressure, viscous and inertial forces in a three-dimensional, time-varying flow. A derivation and statement of the Navier–Stokes equation is given in, for example, [18].

recovery. Grover et al. [19] provided an impressive demonstration with a sodium heat pipe. In these experiments, they achieved 60% pressure recovery. The radial Reynolds number was greater than 10. For simplicity, in Fig. 4.15b, the viscous component of the pressure drop has been omitted. The liquid pressure drop is also shown in Fig. 4.16a–c. Ernst [20] has pointed out that if the pressure recovery in the condenser region is greater than the liquid pressure drop, Fig. 4.16b, then the meniscus in the wick will be convex. Whilst this is possible in principle, under normal heat pipe operation there is excess in the condenser region so that this condition cannot occur. For this reason, if $|\Delta P_{vc}| > |\Delta P_{lc}|$, it is usual to neglect pressure recovery and assume that there is no resultant pressure drop or gain in the condenser region, this is indicated in Fig. 4.16c.

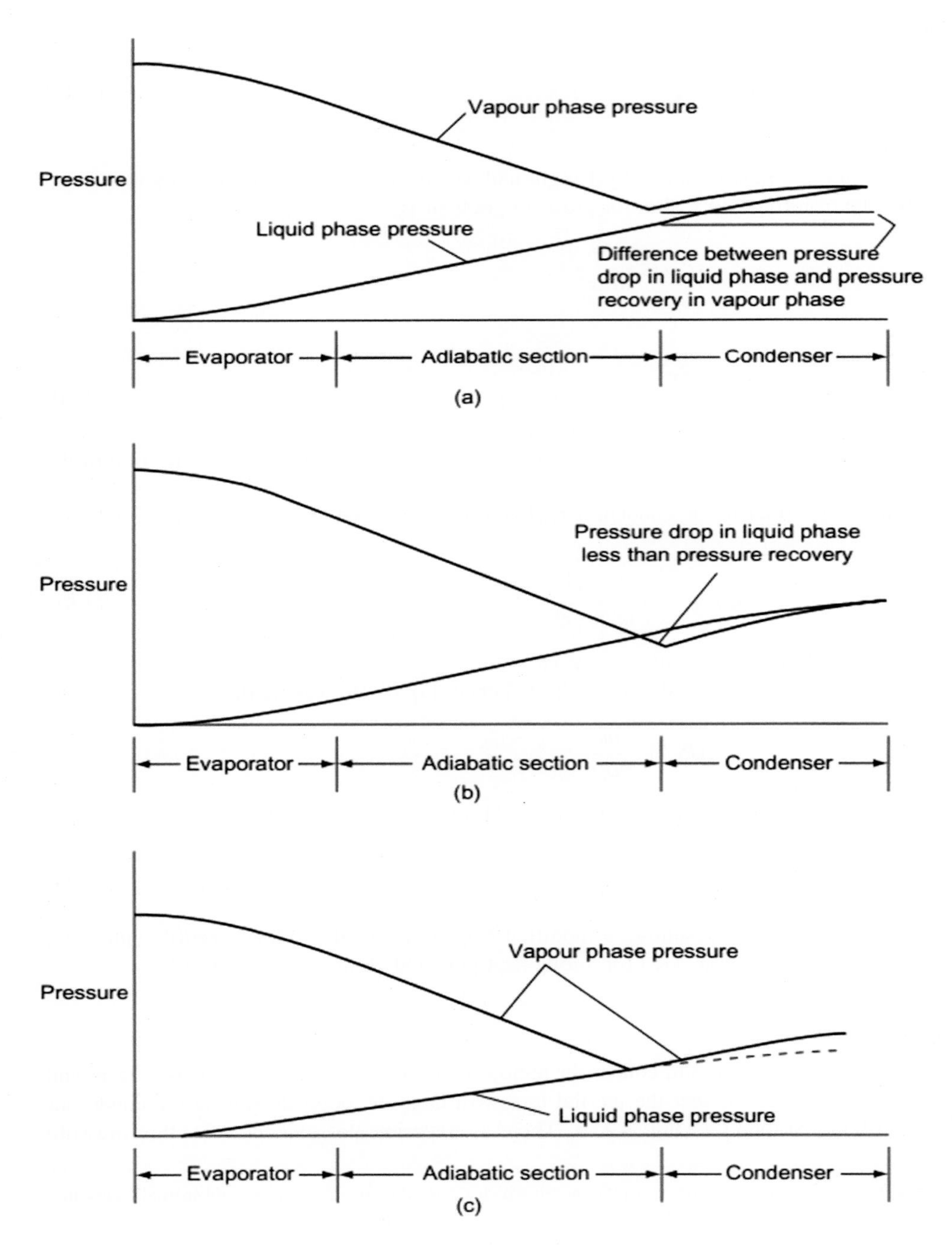


**FIGURE 4.16** Pressure profiles showing pressure recovery in a heat pipe [20]

### 4.3.5.5 Two-dimensional incompressible flow

The previous discussion has been restricted to one-dimensional flow. In practical heat pipes, the temperature and pressure are not constant across the cross-section and this variation is particularly important in the condenser region. A number of authors have considered this two-dimensional problem. Bankston and Smith [21] and Rohani and Tien [22] have solved the Navier–Stokes equation by numerical methods. Bankston and Smith showed that axial velocity reversal occurred at the end of the condenser section under conditions of high evaporation and condensation rates. Reverse flow occurs for $|Re_r| > 2.3$. In spite of this extreme divergence from the assumption of uniform flow, one-dimensional analysis yield good results for $|Re_r| < 10$. This is illustrated in Fig. 4.17.

### 4.3.5.6 Compressible flow

So far, we have neglected the effect of the compressibility of the vapour on the operation of the heat pipe. Compressibility can be important during start-up and also in high-temperature liquid-metal heat pipes; it is discussed in this section.

In a cylindrical heat pipe, the axial mass flow increases along the length of the evaporator region to a maximum value at the end of the evaporator; it will then decrease along the condenser region. The flow velocity will rise to a maximum value at the end of the evaporator region where the pressure will have fallen to a minimum. Deverall et al. [23] have drawn attention to the similarity in flow behaviour between such a heat pipe and that of a gas flowing through a converging–diverging nozzle. In the former, the area remains constant but the mass flow varies; in the latter, the mass flow is constant but the cross-sectional area is changed. It is helpful to examine the behaviour of the convergent-divergent nozzle in more detail before returning to the heat pipe. Let the pressure of the gas at the entry to the nozzle be kept constant and consider the effect of reducing the pressure at the outlet. With reference to the curves in Fig. 4.18, we can see the effect of increasing the flow through a nozzle. For the flow of Curve A, the pressure difference between inlet and outlet is small. The gas velocity will increase to a maximum value in the position of minimum cross-section or throat, falling again in the divergent region. The velocity will not reach the sonic value. The pressure passes through a minimum in the throat. If the outlet pressure is now reduced, the flow will increase and the situation shown in Curve B can be attained. Here the velocity will increase through the convergent region, rising to the sonic velocity in the throat. As before, the velocity will reduce during travel through the diverging section and there will be some pressure recovery. If the outlet pressure is further reduced, the flow rate will remain constant and the pressure profile will follow Curve C. The gas will continue to accelerate after entering the divergent region and will become supersonic. Pressure recovery will occur after a shock front. Curve D shows that for a certain exit pressure the gas can be caused to accelerate throughout the diverging region. Further pressure reduction will not affect the flow pattern in the nozzle region. It should be noted that after Curve C, pressure reduction does not affect the flow pattern in the converging section; hence, the mass flow does not increase after the throat velocity has attained the sonic value. This condition is referred to as choked flow.

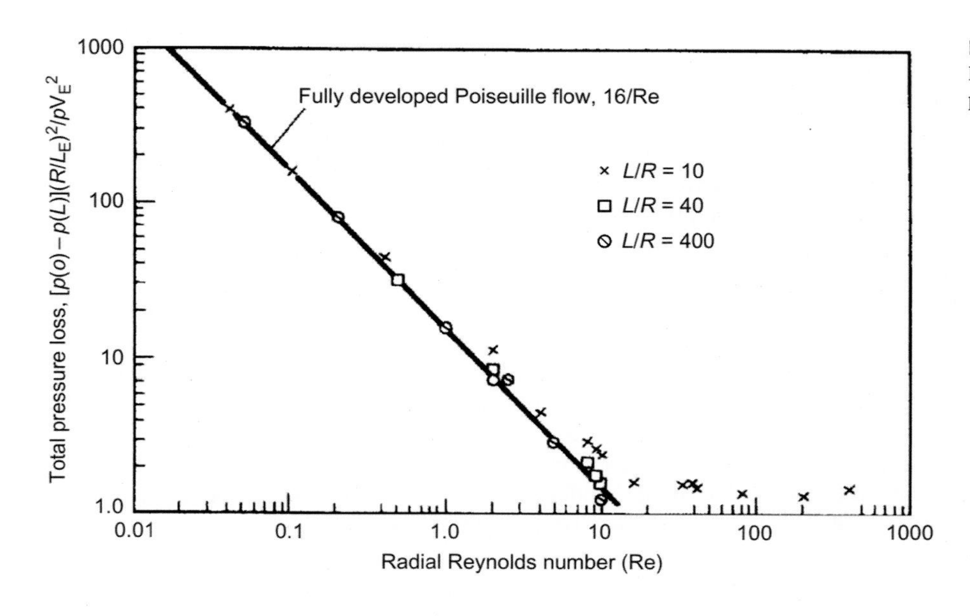


**FIGURE 4.17** Comparison of pressure loss in symmetrical heat pipes with that predicted for Poiseuille [21].

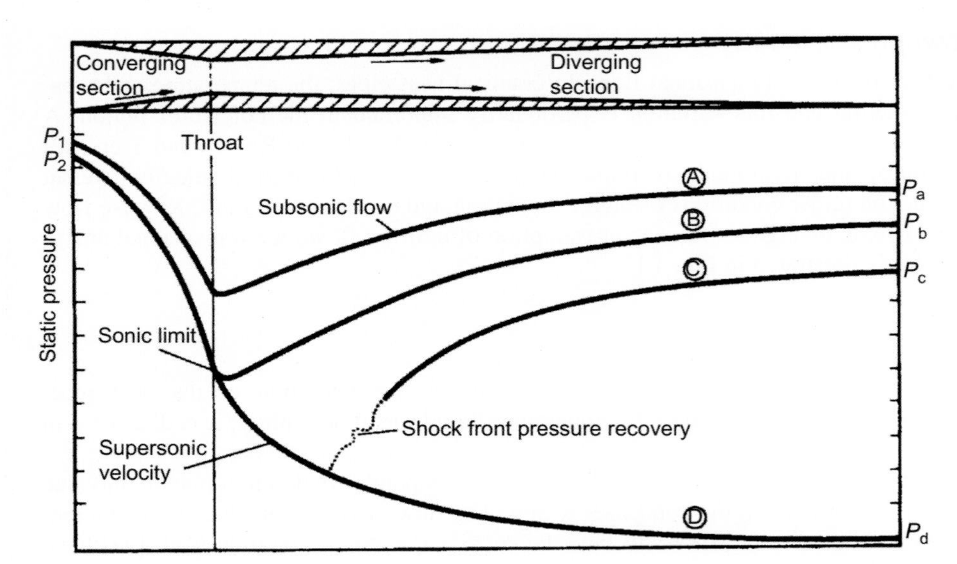


**FIGURE 4.18** Pressure profiles in a converging–diverging nozzle.

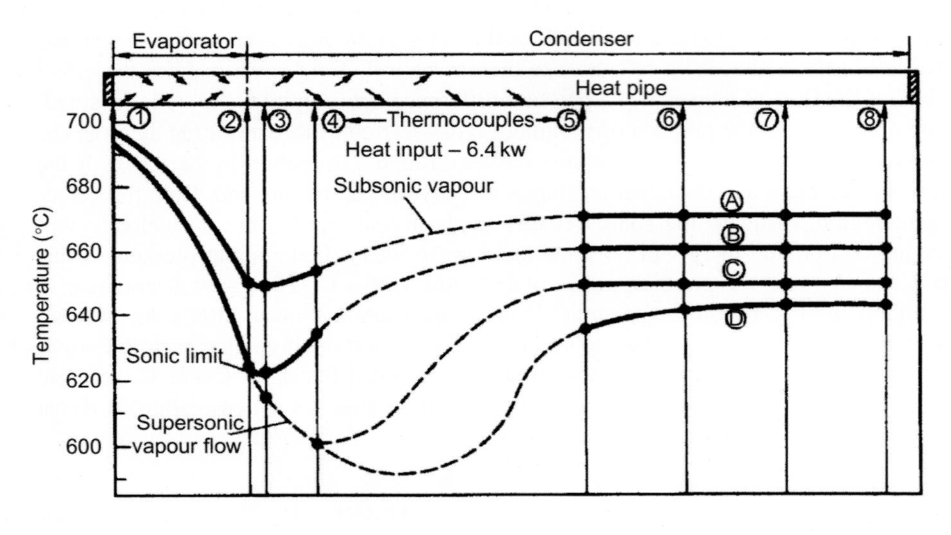


**FIGURE 4.19** Temperature profiles in a sodium heat pipe [24].

Kemme [24] has shown very clearly that a heat pipe can operate in a very similar manner to the diverging nozzle. His experimental arrangement is shown in Fig. 4.19. Kemme used sodium as the working fluid and maintained a constant heat input of 6.4 kW. He measured the axial temperature variation, but since this is related directly to pressure, his temperature profile can be considered to be the same as the pressure profile.

Kemme arranged a variation of the heat rejection at the condenser by means of a gas gap, the thermal resistance of which could be altered by varying the Argon–Helium ratio of the gas. Kemme's results are shown in Fig. 4.19. Curve A demonstrates subsonic flow with pressure recovery; Curve B, obtained by lowering the condenser temperature, achieved sonic velocity at the end of the evaporator and hence operated under choked flow conditions. Further decrease in the thermal resistance between the condenser and heat sink simply reduced the condenser region temperature but did not increase the heat flow which was limited by the choked flow condition and fixed axial temperature drop in the evaporator [25]. It should be noted that under these conditions of sonic limitation, considerable axial temperature and pressure changes will exist and the heat pipe operation will be far from isothermal.

Deverall et al. [23] have shown that a simple one-dimensional model provides a good description of the compressible flow behaviour. Consider the evaporator section using the nomenclature of Fig. 4.19.

The pressure $P_1 = P_o$, where $P_o$ is the stagnation pressure of the fluid. The pressure dropalong the evaporator is given by Eq. (4.42)

$$P_2 - P_o = \rho v^2 \tag{4.53}$$

The equation state for a gas at low pressure may be written:

$$P = \rho_v RT \tag{4.54}$$

and the sonic velocity, $C$, may be expressed:

$$C = \sqrt{\gamma RT} \tag{4.55}$$

The Mach number, $M$, is defined:

$$M = \frac{v}{C}$$

Hence, substituting into Eq. (4.53)

$$\frac{P_o}{P_2} - 1 = \frac{\rho_v v^2}{P_2} = \frac{M^2 \gamma R T_2}{R T_2} = \gamma M^2 \tag{4.56}$$

Defining the total temperature $T_o$ such that:

$$\dot{m} c_p T_o = \dot{m}\left(c_p T_2 + \frac{v^2}{2}\right) \tag{4.57}$$

and remembering that $c_p = \left(\frac{\gamma}{\gamma - 1}\right)$

$$\frac{T_o}{T_2} = 1 + M^2 \frac{\gamma - 1}{2} \tag{4.58}$$

The density ratio may be expressed as follows:

$$\frac{\rho_o}{\rho_2} = \frac{P_o T_2}{P_2 T_o} = \frac{1 + \gamma M^2}{1 + \left(\frac{\gamma - 1}{\gamma}\right) M^2} \tag{4.59}$$

Finally, the energy balance for the evaporator section gives

$$\dot{Q} = \rho_v A v L = \rho_v A M C L \tag{4.60}$$

where $C$ is the sonic velocity at $T_2$. The heat flow may be expressed in terms of the sonic velocity $C_o$ corresponding to the stagnation temperature $T_o$. In this case,

$$\dot{Q} = \frac{\rho_v A M C_o L}{\sqrt{2(\gamma + 1)}} \tag{4.61}$$

The pressure, temperature and density ratio for choked flow may be obtained by substituting $M = 1$ into Eqs (4.59–4.61), respectively, and the values are presented in Table 4.2.

**TABLE 4.2** Effect of $\gamma$ on compressibility parameters.

| | Monatomic gas | Diatomic gas | Triatomic gas |
|---|---|---|---|
| $\gamma$ | 1.66 | 1.4 | 1.3 |
| $\frac{P_o}{P_{2,c}} = 1 + \gamma$ | 2.66 | 2.4 | 2.3 |
| $\frac{T_o}{T_{2,c}} = \frac{1+\gamma}{2}$ | 1.33 | 1.2 | 1.15 |
| $\frac{\rho_o}{\rho_{2,c}} = 2$ | 2 | 2 | 2 |

The sonic limit is discussed more fully in [26]. It is noted that the analysis of compressible flow is generally assumed to be either isothermal or isentropic. The authors also propose a model assuming that the conditions remain saturated throughout the adiabatic region. It is concluded that the results obtained are generally accurate, irrespective of the underlying assumption. However, based on the experimental work of Gagneux [27], it is important that the appropriate reference temperature is chosen and this should be the temperature at the base of the evaporator.

#### 4.3.5.7 Summary of vapour flow

The equations presented in this section permit the designer to predict the pressure drop due to the vapour flow in a heat pipe. These relatively simple equations give acceptable results for most situations encountered within heat pipes. An excellent summary of the analysis of vapour side pressure drop is presented in [28]. For applications which present exceptional problems, the designer should consider using one of the many computational fluid dynamics (CFD) computer packages on the market which can deal with three dimensional and compressible flows.

### 4.3.6 Entrainment

In a heat pipe, the vapour flows from the evaporator to the condenser and the liquid is returned by the wick structure. At the interface between the wick surface and the vapour the latter will exert a shear force on the liquid in the wick. The magnitude of the shear force will depend on the vapour properties and velocity, and its action will be to entrain droplets of liquid and transport them to the condenser end. This tendency to entrain is resisted by the surface tension in the liquid. Entrainment will prevent the heat pipe operating and represents one limit to performance. Kemme observed entrainment in a sodium heat pipe and reports that the noise of droplets impinging on the condenser end could be heard.

The Weber number, We, which is representative of the ratio between inertial vapour forces and liquid surface tension forces, provides a convenient measure of the likelihood of entrainment.

The Weber number is defined as

$$\mathrm{We} = \frac{\rho_v v^2 z}{2\pi\sigma_l} \tag{4.62}$$

where $\rho_v$ is the vapour density, $v$ is the vapour velocity, $\sigma_l$ is surface tension and $z$ is a dimension characterising the vapour–liquid surface. In a wicked heat pipe, $z$ is related to the wick spacing. It should be noted that some authors, including Kim and Peterson [29], define the Weber number as

$$\mathrm{We}' = \frac{\rho_v v^2 z}{\sigma_l} = 2\pi We \tag{4.63}$$

A review of 11 models of entrainment by Kim and Peterson reported critical Weber numbers (as defined by Eq. (4.63)) principally in the range of 0.2–10.

It may be assumed that entrainment may occur when We is of the order 1. The limiting vapour velocity, $v_c$, is thus given by

$$v_c = \sqrt{\frac{2\pi\sigma_l}{\sigma_v z}} \tag{4.64}$$

and, relating the axial heat flux to the vapour velocity using

$$\dot{q} = \rho_v L v$$

the entrainment limited axial flux is given by

$$\dot{q} = \sqrt{\frac{2\pi\rho_v L^2\sigma_l}{z}} \tag{4.65}$$

Extracting the fluid properties from Eq. (4.65) suggests $\rho_v L^2\sigma_l$ as a suitable figure of merit for working fluids from the point of view of entrainment.

Cheung [30] has plotted this figure of merit against temperature for a number of liquid metals, and this is reproduced as Fig. 4.20.

A number of authors report experimental results on the entrainment limit. Typically, they select a value of $z$ which fits their results and show that the temperature dependence of the limit is as predicted by Eq. (4.62).

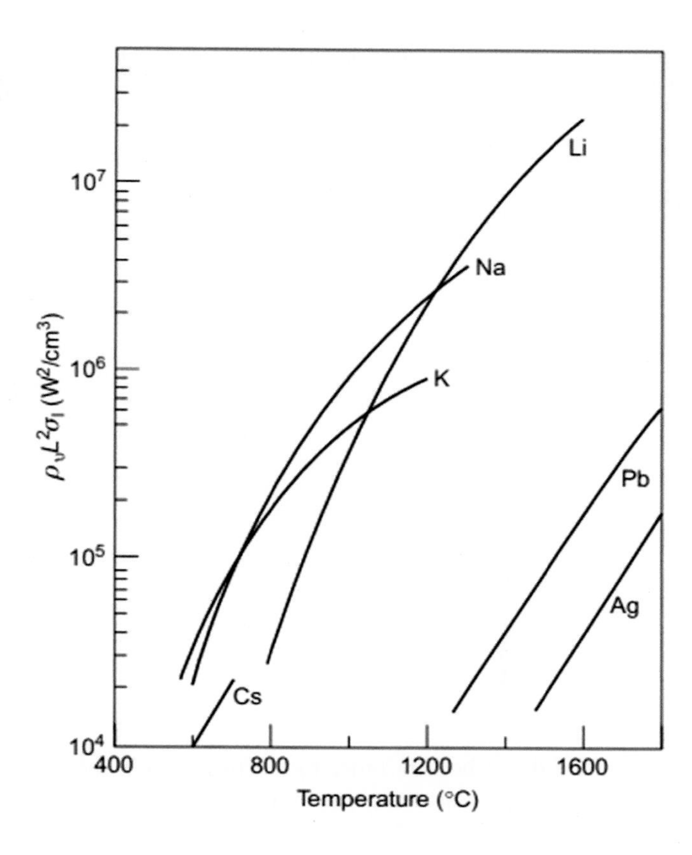


**FIGURE 4.20** Entrainment merit number for liquid metals [30].

Kim and Peterson [29], used the Weber number,

$$\mathrm{We}' = \frac{\rho_v v^2 z}{\sigma_l}, \tag{4.66}$$

defined in Eq. (4.63), and the viscosity number

$$N_{vl} = \frac{\mu_l}{\sqrt{\rho_l \sigma_l \lambda_c / 2\pi}} \tag{4.67}$$

where $\lambda_c$ is the critical wavelength derived from the Rayleigh–Taylor instability [31]. $N_{vi}$ represents the stability of a liquid interface. They derived a correlation, Eq. (4.68), from their experimental measurements with water on mesh wicks for the critical Weber number:

$$\mathrm{We}_{vc} = 10^{-1.163} N_{\mathrm{vl}}^{-0.744} \left(\frac{\lambda_c}{d_1}\right)^{-0.509} \left(\frac{D_h}{d_2}\right)^{0.276} \tag{4.68}$$

where $d_1$ and $d_2$ are the mesh wire spacing and thickness, respectively, and $D_h$ is the equivalent diameter of the vapour space.

## 4.3.7 Heat transfer and temperature difference

### 4.3.7.1 Introduction

In this section, we consider the transfer of heat and the associated temperature drops in a heat pipe. The latter can be represented by thermal resistances, and an equivalent circuit is shown in Fig. 4.21.

Heat can both enter and leave the heat pipe by conduction from a heat source/sink, by convection, or by thermal radiation. The pipe may also be heated by eddy currents or by electron bombardment and cooled by electron emission. Further temperature drops will occur by thermal conduction through the heat pipe walls at both the evaporator and condenser regions. The temperature drops through the wicks arise in several ways and are discussed in detail later in this section. It is found that a thermal resistance exists at the two vapour–liquid surfaces and also in the vapour column.

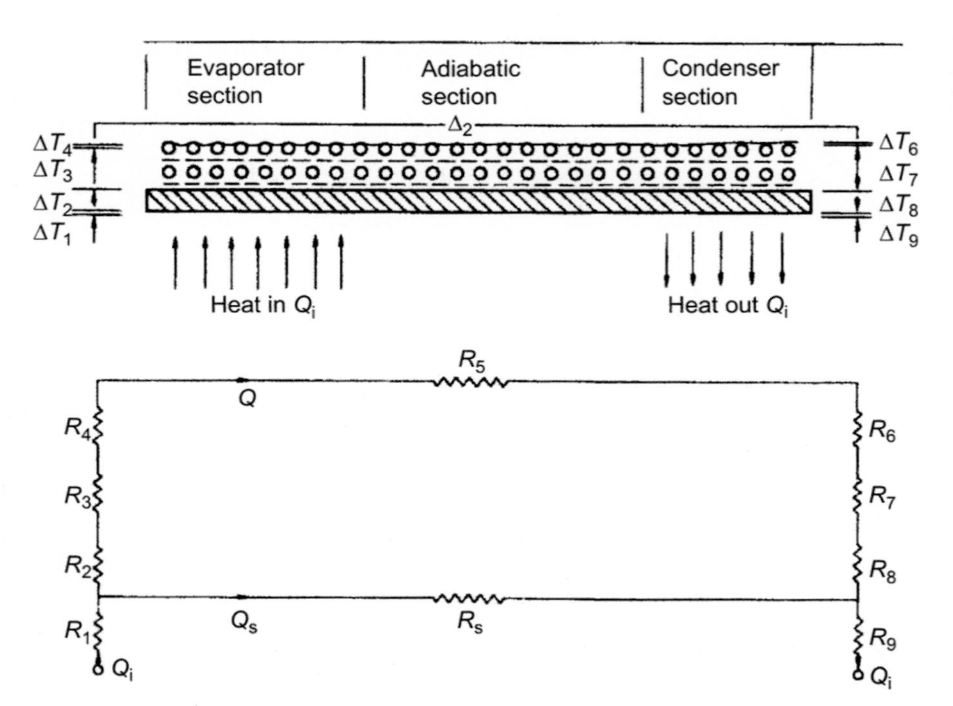


**FIGURE 4.21** Temperature drops and equivalent thermal resistances in a heat pipe.

The processes of evaporation and condensation are examined in some detail both in order to determine the effective thermal resistances and also to identify the maximum heat transfer limits in the evaporator and condenser regions. Finally, the results for thermal resistance and heat transfer limits are summarised for the designer.

### *4.3.7.2 Heat transfer in the evaporator region*

For low values of heat flux the heat will be transported to the liquid surface partly by conduction through the wick and liquid and partly by natural convection. Evaporation will be from the liquid surface. As the heat flux is increased, the liquid in contact with the wall will become progressively superheated and bubbles will form at nucleation sites. These bubbles will transport some energy to the surface by the latent heat of vaporisation and will also greatly increase convective heat transfer. With further increase of flux, a critical value will be reached, burnout, at which the wick will dry out and the heat pipe will cease to operate. Before discussing the case of wicked surfaces, the data on heat transfer from plane, unwicked surfaces will be summarised. Experiments on wicked surfaces are then described and correlations given to enable the temperature drop through the wick and the burnout flux to be estimated. The subject is complex and further work is necessary to provide an understanding of the processes in detail.

### *4.3.7.3 Boiling heat transfer from plane surfaces*

#### 4.3.7.3.1 Bubble dynamic

In boiling heat transfer, also designated as pool boiling, large heat transfer coefficients can be obtained due to the influence of bubble dynamics on the heat transfer. Indeed, from both micro and macroscales, bubbles locally improve the heat transfer and can be used to enhance the heat transfer performance of a heat pipe evaporator. In pool boiling, there are two processes that can lead to the formation of a vapour bubble. In the case where the bubbles appear near the wall, the mass transfer process is designated as *heterogeneous nucleation*. In contrast, a bubble appearing in the bulk of liquid is called a *homogeneous nucleation* process. In practice, homogeneous nucleation only appears at high heat flux as the amount of energy required is higher than during heterogeneous nucleation. In pool boiling, heterogeneous nucleation largely predominates, and bubbles form near the wall. According to the model from Hewitt [32], the evaporator surface contains multiple microcavities inside which a portion of vapour is trapped. When a heat flux is received, this gas nucleus will grow from the cavity and form a bubble that will eventually detach from the wall once it reaches a large diameter. When the bubble leaves the site, a small portion of vapour remains in the cavity which acts as the start for the next bubble. An idealised conical cavity acting as nucleation site is shown in Fig. 4.22.

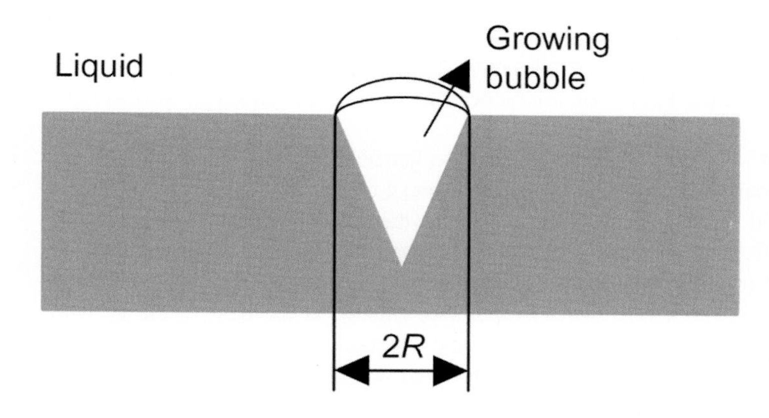


**FIGURE 4.22** Idealised cavity acting as a nucleation site.

In the case of a bubble forming inside a bulk of liquid, the bubble owes its existence to its surface tension force which is applied on the surrounding liquid. To define a criterion for the existence of a bubble, a force balance between the liquid pressure, vapour pressure and bubble surface tension is made [33]:

$$\sigma dA - P_B dV_B - P dV_L = 0 \tag{4.69}$$

where $\sigma$ is the surface tension (J/m$^2$), $A$ is the sphere surface of the bubble (m$^2$), $P_B$ and $P$ are the vapour and liquid pressure (J/m$^3$) and $V_B$ and $V_L$ are the vapour and liquid volumes (m$^3$). For a system with constant temperature and volume, the Young–Laplace equation leads to the pressure inside the bubble:

$$P_B = P + \frac{2\sigma_l}{R} \tag{4.70}$$

where $P$ is the liquid pressure, $r$ is the radius of curvature of the bubble and $\sigma_l$ is the surface tension of the liquid. The radius of curvature is a maximum when the bubble forms a hemispherical cap over the cavity, that is, $r = R$, the radius of the mouth of the cavity. This is the condition for $P_B$ to be a maximum. From the force balance under equilibrium, the condition for the growth of a bubble can be deduced. If the bubble is to grow, then the wall temperature must be sufficiently high to vaporise the liquid at a pressure $P_B$. In order for the bubble to grow, it becomes

$$T_W > T_{sat} + \frac{dT}{dP}(P_B - P) \tag{4.71}$$

The Clapeyron equation states that the slope of the vapour pressure curve is given by

$$\frac{dP}{dT} = \frac{L}{(v_v - v_l)T_{sat}} \tag{4.72}$$

if $v_v$ is very much greater than $v_l$, we can simplify this to

$$\frac{dT}{dP} = \frac{v_v T_{sat}}{L} \tag{4.73}$$

The critical radius of a bubble can then be obtained as follows:

$$R = \frac{2\sigma_l}{\Delta T_{sat}} \frac{T_{sat}}{\rho_v L} \tag{4.74}$$

Based on the force balance for a bubble at equilibrium, for the bubble to grow, the following criterion is deduced:

$$T_W > T_{sat} + \frac{2\sigma_l}{R}\frac{v_v T_{sat}}{L} = \frac{2\sigma_l}{R}\frac{\Delta T_{sat}}{\rho_v L} \tag{4.75}$$

For water boiling at 1 bar, for a bubble to grow, $\Delta T_{sat}$ is commonly of the order of 5 K.

Hsu [34] has conducted a more rigorous analysis and showed that

$$\Delta T = \frac{3.06\sigma_l T_{sat}}{\rho_v L \delta} \tag{4.76}$$

where $\delta$ is the thermal boundary layer thickness; as a first approximation, this may be taken as the average diameter of cavities on the surface. For typical heat pipe evaporators, this is approximately 25 μm. Taking $\delta = 25$ μm allows the superheat for nucleation to be calculated for various fluids, values of this superheat for various fluids at atmospheric temperature are shown in Table 4.3.

In the case of a bubble growing from the evaporator's wall, the superheat needed to generate a bubble can be reduced as the near-wall phenomena are improving the local heat transfer. Carey [35] proposed a model of the bubble growth from a hot surface. According to *Carey*'s model, a layer of hot liquid forms near the wall and favours the development of vapour bubbles. After the departure of the previous bubble, cold liquid from the bulk volume is able to reach the wall and rapidly warms up. As the thermal layer forms again, the hot liquid layer near the wall reduces the amount of energy needed for the bubble growth to start. The growth of the bubble is also eased due to a microlayer evaporation situated at the base of the growing bubble ($t_2$). This very thin layer of liquid situated between the hot wall and the bubble evaporates quickly and enhances the phase change process. Once the bubble reaches its maximum diameter and departs from the wall ($t_4$), a portion of the hot thermal layer is removed from the wall and carried away by the bubble ($t_5$). This hot liquid from the thermal layer then mixes with the colder bulk liquid and significantly increases the heat transfer coefficient. Finally, the rising bubbles inside the liquid pool generate a stirring action inside the liquid pool which also favours the heat transfer. Bubbles also carry energy under the form of latent heat. In the case where the bulk liquid is at temperatures lower than the saturation temperature, the vapour bubble will be consumed by the liquid bulk and collapse in the volume. By doing so, the latent energy carried by the bubble is transmitted to the liquid pool. These mechanisms show that the bubble activity significantly increases the local heat transfer coefficient which explains that the highest heat transfer coefficients are obtained at high bubble activity.

As an increased bubble activity leads to higher heat transfer coefficients at the evaporator, the need of having as many active nucleation sites as possible is obvious. This highlights the importance of the surface aspect of the evaporator solid wall on the heat pipe performance. Indeed, a rough surface with many active nucleation sites will lead to better heat transfer performance than a smooth surface with a limited number of nucleation sites. However, it must be known that all microcavities on a surface do not generate bubbles and may be inactive. To become an active nucleation site, a portion of vapour needs to be trapped inside the cavity. According to the model by Wang and Dhir [36], a condition for the entrapment of vapour in the cavity is that the contact angle of an advancing liquid wave $\theta_a$ is larger than the mouth angle of the cavity $\psi_m$.

$$\theta_a > \psi_m \tag{4.77}$$

This is illustrated in Fig. 4.23.

For water, the smallest active cavity is estimated to be approximately $6.5 \times 10^{-6}$ m radius. This demonstrates that typical active cavities are of the order of 1–10 μm.

In the objective of predicting the boiling activity from a given surface, researchers have proposed correlations to estimate the number of active nucleation sites $N_a$. Zhokhov [37] proposed a correlation relating the number of active nucleation sites to the fluid properties and saturation temperature only:

$$\sqrt{N_a} = 25 \times 10^{-8} \left( \frac{i_{lv} \rho_v \Delta T_{sat}}{T_{sat} \sigma} \right)^{1.5} \tag{4.78}$$

**TABLE 4.3** Superheat for the initiation of nucleate boiling at atmospheric pressure, calculated from Eq. (4.76).

| Fluid | Boiling point (K) | Vapour density (kg/m$^3$) | Latent heat (kJ/kg) | Surface tension (N/m) | $\Delta T$ (°C) |
|---|---|---|---|---|---|
| Ammonia ($NH_3$) | 293.7 | 0.3 | 1350 | 0.028 | 2.0 |
| Ethyl alcohol ($C_2H_5OH$) | 338 | 2.0 | 840 | 0.021 | 0.51 |
| Water | 373 | 0.60 | 2258 | 0.059 | 1.9 |
| Potassium | 1047 | 0.486 | 1938 | 0.067 | 8.9 |
| Sodium | 1156 | 0.306 | 3913 | 0.113 | 24.4 |
| Lithium | 1613 | 0.057 | 19700 | 0.26 | 44.6 |

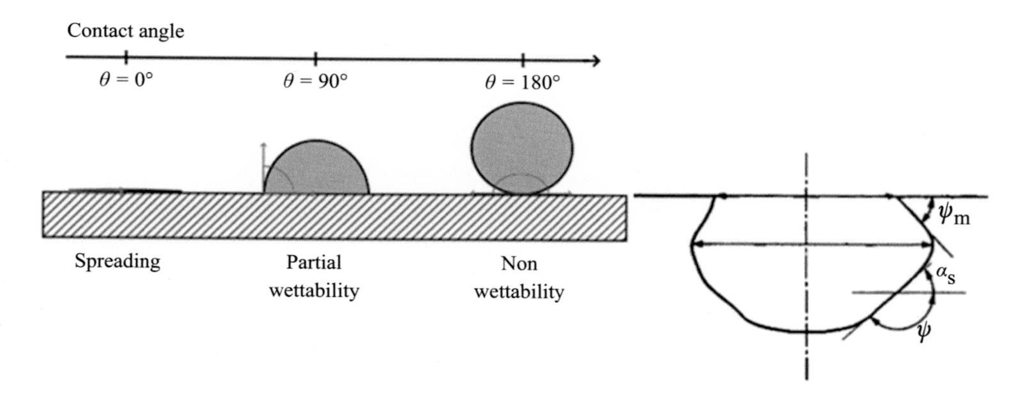


**FIGURE 4.23** Contact angle and mouth angle of a cavity.

Yet, Wang and Dhir [36] observed that the number of active nucleation sites depends on the diameter of the cavities $r_{cav}$ of the evaporator's wall. Indeed, larger cavity diameters may favour the penetration of liquid inside the cavity and can prevent the entrapment of gas inside the cavity. According to the experiments from Wang and Dhir [36] for water on a copper surface with a contact angle $\theta = 90$ degrees, the density of active nucleation sites with a diameter greater than 2 μm was about 104 sites/cm$^2$ whilst it decreases to about 1 site/cm$^2$ for nucleation sites with a diameter greater than 10 μm. In this regard, Mikic and Rohsenow [38] proposed to relate the number of active nucleation sites $N_a$ to the cavity radius $r_{\text{cav}}$:

$$N_a = C\left(\frac{r_{\text{cav,max}}}{r_{\text{cav}}}\right)^m = C{r_{\text{cav,max}}}^m\left(\frac{i_{lv}\rho_v}{2T_{\text{sat}}\sigma}\right)^m {\Delta T_{\text{sat}}}^m \tag{4.79}$$

where $C$ is an experiment-based constant often taken as $C = 1$ unit/m$^2$, $r_{\text{cav,max}}$ is the maximal active cavity radius on the surface (m) and $m$ is a constant (typically, $m = 6.5$).

#### 4.3.7.3.2 Boiling curve

Moving from a microscale to a macroscale, boiling heat transfer was also studied by observing the overall liquid pool. In 1934 Nukiyama [39] performed a pool-boiling experiment, passing an electric current through a platinum wire immersed in water. The heat flux was controlled by the current through and voltage across the wire, and the temperature of the wire was determined from its resistance. Similar results are obtained when boiling from a plane surface or from the surface of a cylinder. An apparatus for measuring pool boiling, heat transfer is shown schematically in Fig. 4.24. Nukiyama then proposed a boiling curve of the form shown in Fig. 4.25.

As we have a liquid and vapour coexisting in the apparatus, both must be at (or during boiling, very close to) the saturation temperature of the fluid at the pressure in the container. If the surface temperature of the heater, $T$, the temperature of the fluid, $T_{sat}$, the rate of energy supply to the heater, $\dot{Q}$ and the heater surface area, $A$, are measured, then a series of tests may be carried out and a graph of $\log\dot{Q}$ or more usually $\log\dot{q} = \log(\dot{Q}/A)$, plotted against $\log\Delta T_{\text{sat}}$, where $\Delta T_{\text{sat}} = (T - T_{\text{sat}})$, often referred to as the wall superheat.

For the case of controlled heat flux (e.g. electric heating), the various regimes may be described:

For increasing heat flux, in the region 'A'–'B' heat transfer from the heater surface is purely by single-phase natural convection. Superheated liquid rises to the surface of the reservoir, and evaporation takes place at this surface. As the heat flux is increased beyond the value at 'B' bubbles begin to form on the surface of the heater, depart from the heater surface and rise through the liquid, this process is referred to as nucleate boiling. At this stage, a reduction of heater surface temperature to 'C' may be observed. Reducing the heat flux would now result in the heat flux temperature difference relationship following the curve 'C'–'B*'. This type of phenomenon, for which the relationship between a dependent and independent variable is different for increasing and decreasing values of the independent variable, is known as hysteresis.

During the nucleate boiling regime 'C'–'D', the heat transfer coefficient increases significantly. This can be explained by the evolution of the vapour structures forming from the evaporator's wall as described by Gaertner [40].

During the early stage of nucleate boiling, also known as partial boiling, isolated bubbles grow and depart from the wall independently. Yet, after the transition to fully developed boiling, the successive bubbles from a nucleation site start to merge and form vapour columns. Eventually, the vapour columns of neighbour's nucleation sites may merge and form small vapour mushrooms. At this stage, the nucleate pool-boiling heat transfer coefficient becomes independent of several variables such as the inclination angle of the wall, the liquid depth and gravitational acceleration. The formation of vapour mushrooms increases further the local heat transfer coefficient as a portion of liquid at the base of

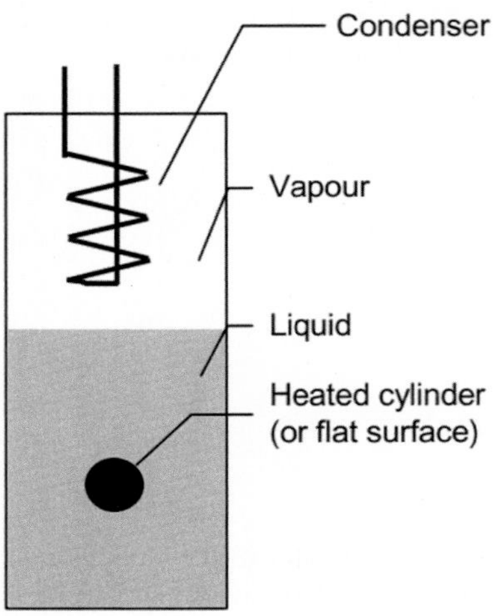

Condenser
Vapour
Liquid
Heated cylinder
(or flat surface)

**FIGURE 4.24** Schematic diagram of pool-boiling experiment.

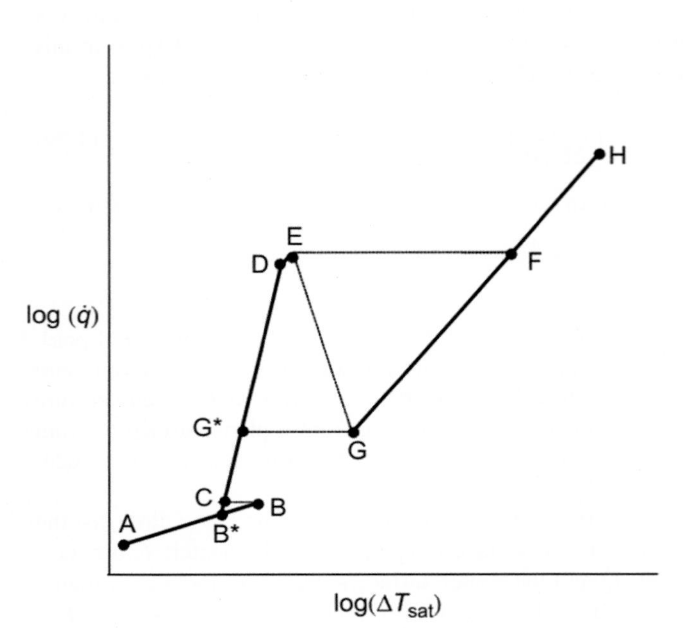


**FIGURE 4.25** Boiling curve for pool boiling from plane surface, cylinder or wire (not to scale).

the mushroom is trapped and evaporates rapidly. With an increase of the heat flux, larger vapour mushrooms develop and merge. The evolution of the pool-boiling heat transfer coefficient in this regime is highly related to the evaporation of the macrolayer of liquid which is trapped at the base of the large vapour mushrooms. Eventually, the increase of the nucleate boiling heat transfer coefficient will slow down as a further increase of the heat flux causes the appearance of dry patches where the hot evaporator wall is directly in contact with vapour.

Further increase in heat flux leads to increased heater surface temperature to point 'D'. Further increase beyond the value at 'D' leads to vapour generation at such a rate that it impedes the flow of liquid back to the surface and transition boiling occurs between 'D' and 'E'. At 'E', a stable vapour film forms over the surface of the heater, and this has the effect of an insulating layer on the heater resulting in a rapid increase in temperature from: 'E' to 'F'. The heat flux at 'E' is known as the critical heat flux. The large temperature increase which occurs if an attempt is made to maintain the heat flux above the level of the critical heat flux is frequently referred to as burnout.

However, if physical burnout does not occur, it is possible to maintain boiling at point 'F' and then adjust the heat flux, the heat flux temperature difference relationship will then follow the line 'G'–'H'. This region on the boiling curve corresponds to the stable film boiling regime. Reduction of the heat flux below the value at 'G' causes a return to the nucleate boiling regime at 'G*'.

If the temperature of the heater, rather than the heat flux, was to be controlled, then increasing temperature above that corresponding to the critical heat flux would result in a decrease in heat flux with increasing temperature from 'E' to 'G', followed by an increase along the line G–H. The point 'G' is sometimes referred to as the Leidenfrost Point. Temperature controlled heating of a surface is found in many heat exchangers and boilers – the temperature of the wall being necessarily below the temperature of the other fluid in the heat exchanger. Experimentally, it is difficult to maintain surface temperatures over a wide range with the corresponding range of heat fluxes. To obtain boiling curves for varying $\Delta T_{\text{sat}}$, it is usual to plunge an ingot of high-conductivity material into a bath of the relevant fluid. The surface temperature is measured directly, and the heat flux can then be calculated from the geometry of the ingot and the rate of change of temperature.

The factors which influence the shape of the boiling curve for a particular fluid include fluid properties, heated surface characteristics and physical dimensions and orientation of the heater. The previous history of the system also influences the behaviour, particularly at low heat flux. Several relationships, defining both the extent of each region and the appropriate shape of the curve for that region, are required to describe the entire curve. It is the nucleate boiling region, 'C'–'D' which is of greatest importance in most engineering applications and heat pipes. However, it is clearly important that the designer ensures that the critical heat flux is not inadvertently exceeded, and there are some systems which operate in the film boiling regime. Many correlations describing each region of the boiling curve have been published.

#### 4.3.7.3.3 Pool-boiling heat transfer correlations

Nucleate boiling is very dependent on the heated surface and factors such as the release of absorbed gas, surface roughness, surface oxidation and wettability greatly affect the surface-to-bulk liquid temperature difference. The nature of the surface may change over a period of time – a process known as conditioning. (The effect of pressure is also important.) For these reasons, reproducibility of results is often difficult. However, a number of authors have proposed correlations, some empirical and some based on physical models. Some of these are listed below. The straight-line C–D on Fig. 4.25 and experimental results for nucleate boiling indicate that correlations may be represented by an equation of the form

$$\dot{q} = a\Delta T_{\text{sat}}{}^{m}$$

This may be rearranged in terms of a heat transfer coefficient, $\alpha$,

$$\alpha = \dot{q}/\Delta T_{sat} \tag{4.80}$$

To date, about 30 correlations aimed at predicting the pool-boiling heat transfer coefficient can be found in the literature. Selecting an appropriate correlation can therefore be tedious. Moreover, significant differences can be found in the predictions of published correlations and, for the same experiment, the predicted nucleate pool-boiling heat transfer coefficient can vary from 500 W/m$^2$ K to values higher than 10,000 W/m$^2$ K. In this regard, based on Refs [41–53], a critical analysis of available correlation was conducted, and the most recommended models are described below:

Rohsenow model [54]:

$$\alpha = \left(\frac{\dot{q}}{L}\right)^{1-r}\left[\mu_l\Big/\sqrt{\frac{\sigma}{g(\rho_l-\rho_v)}}\right]^{r}\frac{c_{p,l}}{C_{sf}}\text{Pr}_l{}^{-s} \tag{4.81}$$

where $r = 1/3$, $\begin{cases} s = n = 1 \text{ for water} \\ s = n = 1.7 \text{ for other fluids} \end{cases}$ where $\dot{q}$ is the heat flux (W/m$^2$), $L$ the latent heat of vaporisation (J/kg), $\mu_l$ the liquid dynamic viscosity (Pa s), $\sigma$ the surface tension (N/m), $g$ the gravitational acceleration (m/s$^2$), $\rho_l$ and $\rho_v$ the liquid and vapour densities (kg/m$^3$), $c_{p,l}$ the liquid specific heat (J/kg K), $C_{sf}$ a liquid/surface coefficient and $\text{Pr}_l$ the liquid Prandtl number.

*Rohsenow* is considered as the reference correlation to predict nucleate pool-boiling heat transfer coefficient. This correlation is the most popular one, and its reliability has been proved throughout its extensive use. The predictive error made by this correlation is usually lower than 30%, and it is valid for a wide range of solid material/fluid combinations. This correlation has also been reported to be accurate for small diameter pools, is insensitive of the heater and surface orientation, and remains fairly accurate at very large heat fluxes. In this correlation, Rohsenow [54] included a coefficient $C_{sf}$ which accounts for the surface aspect. This provides a good flexibility of this correlation which can be easily adapted by changing the value of the coefficient $C_{sf}$ as the surface aspects are usually difficult to anticipate. The value of the constant $C_{sf}$ depends upon the fluid and the surface and typical values range between 0.0025 and 0.015. Since, for a given value of $\Delta T_{\text{sat}}$, the heat flux is proportional to $C_{sf}{}^3$, the correlation is very sensitive to selection of the

**TABLE 4.4 Values of $C_{sf}$ for use in Eq. (4.81) [54].**

| Fluid | Surface | $C_{sf}$ |
|---|---|---|
| Water | Nickel | 0.006 |
| Water | Platinum | 0.013 |
| Water | Copper | 0.013 |
| Water | Brass | 0.006 |
| Carbon tetrachloride | Copper | 0.013 |
| Benzene | Chromium | 0.010 |
| *n*-Penthane | Chromium | 0.015 |
| Ethanol | Chromium | 0.0027 |
| Isopropanol | Copper | 0.0025 |
| *n*-Butanol | Copper | 0.0030 |

correct value. Researchers have therefore carried out experiments to estimate the value of the $C_{sf}$ coefficient for different solid–liquid interactions. Some indicative values of $C_{sf}$ for use in Eq. (4.81) are given in Table 4.4.

One must be aware that the value of the $C_{sf}$ coefficient can vary depending on the roughness of the surface.

Imura et al. model [55]:

$$\alpha = 0.32\left(\frac{\rho_l^{0.65} k_l^{0.3} c_{p,l}^{0.7} g^{0.2}}{\rho_v^{0.25} L^{0.4} \mu_l^{0.1}}\right)\left(\frac{P_v}{P_{\text{atm}}}\right)^{0.3} \dot{q}^{0.4} \tag{4.82}$$

where $\rho_l$ and $\rho_v$ the liquid and vapour densities (kg/m$^3$), $k_l$ the liquid thermal conductivity (W/m K), $c_{p,l}$ the liquid specific heat (J/kg K), $g$ the gravitational acceleration (m/s$^2$), $L$ the latent heat of vaporisation (J/kg), $\mu_l$ the liquid dynamic viscosity (Pa s), $P_v$ the vapour pressure (Pa), $P_{\text{atm}}$ the atmospheric pressure (Pa) and $\dot{q}$ is the heat flux (W/m$^2$). The correlation by Imura et al. [55] has the advantage of being easy to use and only requires fluid-related properties. This correlation was reported to have an excellent accuracy for thermosyphons with high filling ratios. However, this correlation may show some inaccuracies for liquid pools with small diameters. On average, this correlation remains accurate for most of the pool-boiling experiments.

Stephan and Abdelsalam model [56]:

$$\begin{cases} \text{for water:} \\ \alpha = 0.246\dfrac{k_l}{D_d} \times 10^{-7} \times X_1^{0.673} X_3^{1.26} X_4^{-1.58} X_8^{5.22} \\ \text{for hydrocarbons:} \\ \alpha = 0.0546\dfrac{k_l}{D_d} \times X_1^{0.67} X_4^{0.248} X_5^{1.17} X_8^{-4.33} \\ \text{for cryogenic fluids:} \\ \alpha = 4.82\dfrac{k_l}{D_d} \times X_1^{0.624} X_3^{0.374} X_4^{0.329} X_5^{0.257} X_7^{0.117} \\ \text{forrefrigerants:} \\ \alpha = 207\dfrac{k_l}{D_d} \times X_1^{0.745} X_5^{0.581} X_6^{0.533} \end{cases} \tag{4.83}$$

where

$$X_1 = \left(\frac{\dot{q} D_d}{k_l T_v}\right), X_2 = \left(\frac{\alpha_l^2 \rho_l}{\sigma D_d}\right), X_3 = \left(\frac{c_p T_v D_d^2}{\alpha_l^2}\right),$$

$$X_4 = \left(\frac{LD_d{}^2}{\alpha_l{}^2}\right), X_5 = \left(\frac{\rho_v}{\rho_l}\right), X_6 = \left(\frac{c_p\mu_l}{k_l}\right),$$

$$X_7 = \left(\frac{\rho_{l,w}c_{p,l,w}k_{l,w}}{\rho_l c_{p,l}k_l}\right), X_8 = \left(\frac{\rho_l - \rho_v}{\rho_l}\right)$$

More recently, in the objective of developing a correlation that would suit the majority of pool-boiling heat transfers, Stephan and Abdelsalam [56] conducted a regression analysis over 5000 pool-boiling experimental points from various sources. The authors developed four correlations for different types of fluids including water, hydrocarbons, cryogenic fluids and refrigerants which makes it particularly suitable for these types of fluids. The correlation from Stephan and Abdelsalam [56] was reported to have a good flexibility and remain accurate for fluids such as benzene, isopropanol and R-113. The accuracy of this correlation is therefore guaranteed for a majority of cases. However, these correlations are relatively complex and more tedious to use, which explains why the correlations by Rohsenow [54] and Imura et al. [55] are usually preferred.

Cooper model [57]:

$$\alpha = 55P_r^{(0.12-0.2\log\varepsilon)}(-\log P_r)^{-0.55}M^{-0.5}\dot{q}^{0.67} \tag{4.84}$$

Like other researchers, Cooper considered the influence of the surface on the pool-boiling heat transfer coefficient by including the surface roughness $\varepsilon$ in microns. Typically, a value of 1 may be used, thus simplifying the equation. $P_r$ is the reduced pressure, defined as $P/P_{cr}$, and $P_{cr}$ is the critical pressure of the fluid. The Cooper correlation is a dimensional equation; therefore, the units must be consistent with the constants given. For the forms quoted here, pressures are in bar and heat flux in $W/m^2$, giving heat transfer coefficients in $W/m^2$ K. The main advantage of the Cooper correlation is its simplicity but its accuracy and reliability is lower. Comprehensive references on liquid-metal boiling are in Subbotin [58] and Dwyer [59].

#### 4.3.7.3.4 Burnout correlations

As for nucleate boiling, the critical peak flux or boiling crisis flux is also very dependent on surface conditions. For water at atmospheric pressure, the peak flux lies in the range 950–130 $kW/m^2$ and this is between 3 and 8 times the value obtained for organic liquids. The corresponding temperature difference for both water and organics is between 20°C and 50°C. Liquid metals have the advantage of low viscosity and high thermal conductivity, and the alkali metals in the pressure range 0.1–10 bar give peak flux values of 100–300 $kW/m^2$ with a corresponding temperature difference of around 5°C.

A number of authors have provided relationships to enable the critical heat flux $\dot{q}_{cr}$ to be predicted. One of these was developed by Rohsenhow and Griffith [60] who obtained the following expression:

$$\dot{q}_{cr} = 0.012L\rho_v\left[\frac{\rho_l - \rho_v}{\rho_v}\right]^{0.6} \tag{4.85}$$

Another correlation due to Caswell and Balzhieser [61] applies to both metals and nonmetals.

$$\dot{q}_{cr} = 1.02 \times 10^{-6}\frac{L^2\rho_v k_l}{c_p^{\gamma}}\left[\frac{\rho_l - \rho_v}{\rho_v}\right]^{0.65}\mathrm{Pr}^{0.71} \tag{4.86}$$

### *4.3.7.4 Boiling from wicked surfaces*

There is considerable literature on boiling from wicked surfaces. The work reported includes measurements on plane surfaces and on tubes, the heated surfaces can be horizontal or vertical, and either totally immersed in the liquid or evaporating in the heat pipe mode. Water, organics and liquid metals have all been investigated. The effect of the wick is to further complicate the boiling process because, in addition to the factors referred to in the section on boiling from smooth surfaces, the wick provides sites for additional nucleation and significantly modifies the movement of the liquid and vapour phases to and from the heated surface.

At low values of heat flux, the heat transfer is primarily by conduction through the flooded wick. This is demonstrated in the results of Philips and Hinderman [62] who carried out experiments using a wick of 220.5 nickel foam 0.14 cm thick and one layer of stainless-steel screen attached to a horizontal plate. Distilled water was used as the

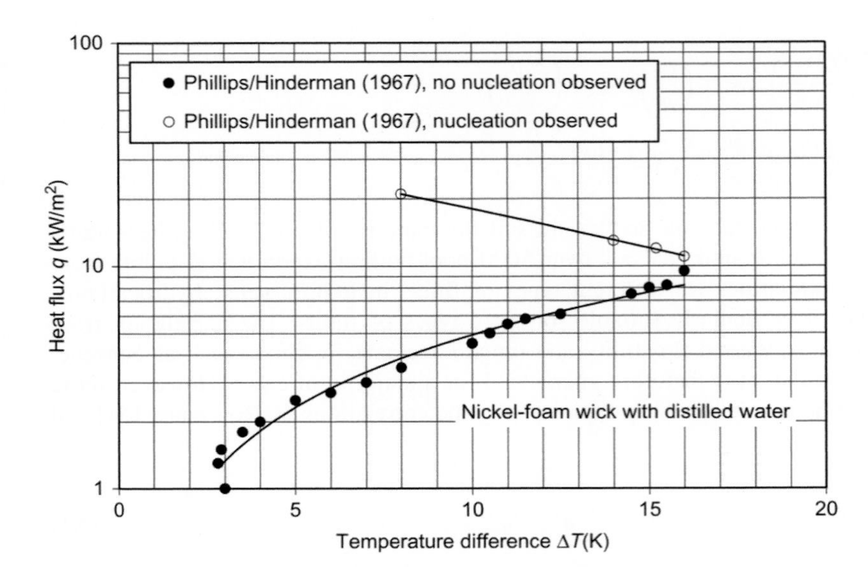


FIGURE 4.26 Heat transfer from a submerged wick data from Ref. [62].

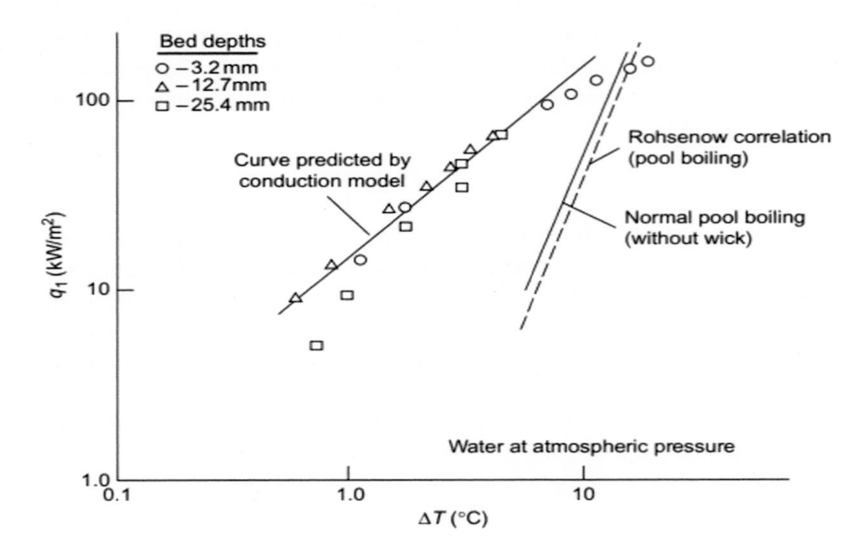


FIGURE 4.27 Evaporation from a submerged wick compared with boiling from a smooth surface [63].

working fluid. Their results are shown in Fig. 4.26. The solid curve is the theoretical curve for conduction through a layer of water of the wick thickness.

At higher values of heat flux nucleation occurs. Ferrell and Alleavitch [63] studied the heat transfer from a horizontal surface covered with beds of Monel beads. Results are reported on 30–40 mesh and 40–50 mesh using water at atmospheric pressure as the working fluid and a total immersion to a depth of 7.5 cm. The bed depth ranged from 3 to 25 mm. They concluded that the heat transfer mechanism was conduction through a saturated wick liquid-matrix to a vapour interface located in the first layer of beads. Agreement was good between the theoretical predictions of this model and the experimental results. No boiling was observed. Fig. 4.27 shows these results together with the experimental values obtained for the smooth horizontal surface. It is seen that the latter agrees well with the Rohsenhow correlation but that for low values of temperature difference the heat flux for the wicked surface is much greater than for the smooth surface. This effect has been observed by other workers, for example, Corman and Welmet [64]; the curves cross over at higher values of heat flux, probably because of increased difficulty experienced by the vapour in leaving the surface.

This concept has been developed further by Brautsch and Kew [65], who examined boiling from a bare surface and the surface covered with mesh wicks. Samples of these results are shown in Fig. 4.28.

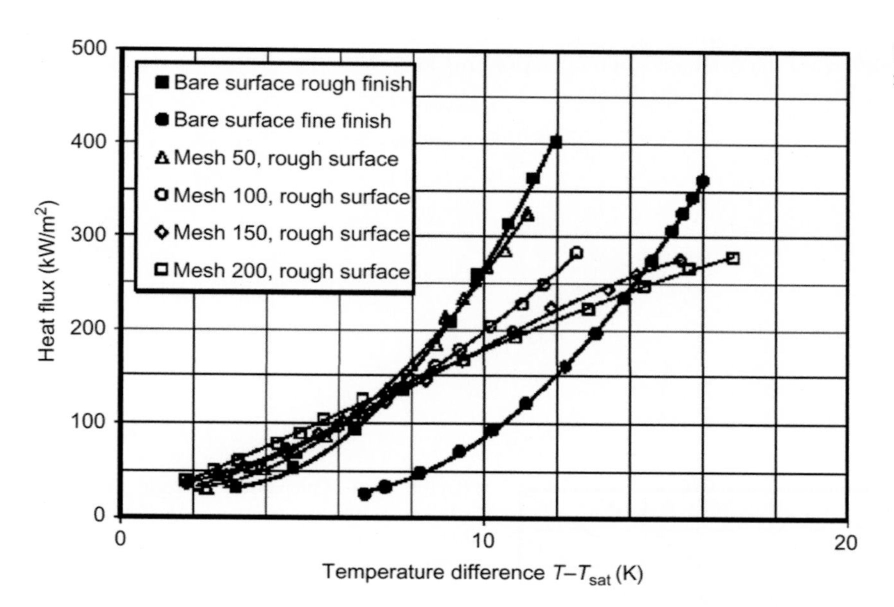


**FIGURE 4.28** Effects of surface finish and mesh on heat transfer [65].

Brautsch [65–67] correlated the heat transfer from the mesh-covered surface by comparing it with a reference heat transfer coefficient obtained for boiling of water and R141b from a smooth plate. Having observed the boiling process using high-speed video, it was noted that the mesh acted to enhance nucleation due to the provision of additional nucleation sites, but the presence of the mesh impeded departure of bubbles from the surface. This yielded an equation of the form:

$$\frac{\alpha}{\alpha_0} = E(1 - B) \tag{4.87}$$

where $E$ and $B$ were defined as enhancement and blocking factors, respectively, and correlated by

$$E = \left[1 + \left(\frac{d_b}{d_{m,i}}\right)^m \left(\frac{R_a}{R_{a_0}}\right)^n \frac{q_0}{q}\right]^r \tag{4.88}$$

and

$$B = \left(\frac{d_b}{d_{m,i}}\right)^u \left(\frac{R_a}{R_{a_0}}\right)^v K^{-1}\left(\left(\frac{q}{q_0}\right)^2 + \frac{q}{q_0}\right) \tag{4.89}$$

with

$$d_b = 0.851\beta_0\sqrt{\frac{2\sigma}{g(\rho_l - \rho_v)}} \tag{4.90}$$

where $\beta_o$ is the contact angle (radians), and the exponents $m$, $n$, $r$, $u$, $v$ and the fitting factor $K$ are summarised in Table 4.5.

The Cooper correlation (equation) was found to predict the reference heat transfer coefficient, $\alpha_0$, well (Fig. 4.29).

This correlation showed good agreement with experimental data from Asakavicius et al. [68], who tested multiple brass and stainless-steel screen layers in Freon 113, ethyl alcohol and water; Liu et al. [69], who published experimental results for a single layer of mesh 16 and mesh 50, both for the working fluids methanol and HFE-7100; and the results of Tsay et al. [70] single layers of mesh 24 and mesh 50 with distilled water as the working fluid.

Several researchers have worked on evaporation from wicks, and selected results are summarised in Fig. 4.30. Abhat and Seban [71] reported on heat transfer measurements on vertical tubes using water, ethanol and acetone as the working fluids. In this series of experiments, results were given for smooth surfaces, immersed wicks and evaporating wicks. The authors concluded that up to heat fluxes of 15 W/cm$^2$, the heat transfer coefficient for a screen or felt wicked tube was similar to that of the bare tube and also not very different from that for the evaporating surface.

**TABLE 4.5 Relevant material parameters, exponents and the fitting factor used in the equations.**

| | All surface finishes |
|---|---|
| $m$ | 1 |
| $n$ | 0.69 |
| $r$ | 1.8 |
| $u$ | 1.2 |
| $v$ | 0.12 |
| $K$ | 2.0E + 07 |

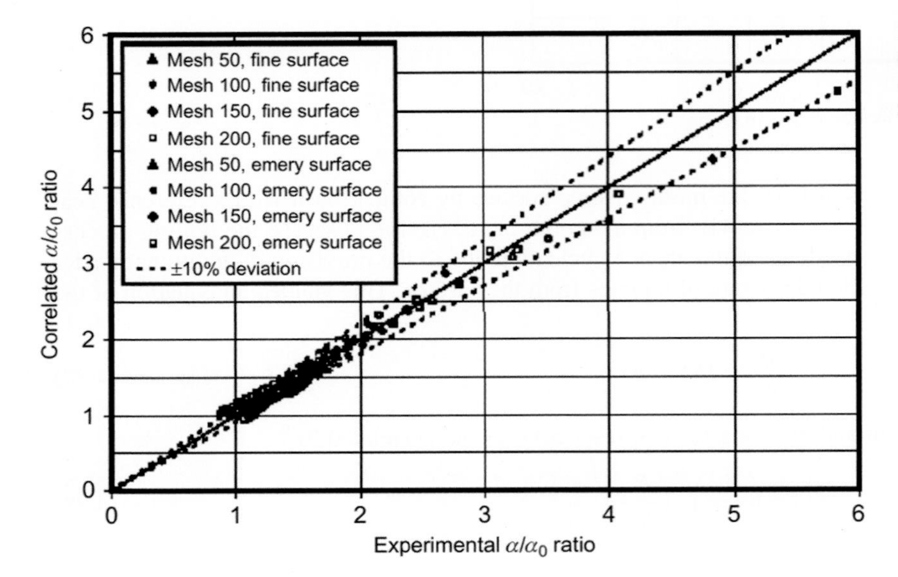


**FIGURE 4.29** Comparison between correlated and experimental heat transfer coefficient ratio for different mesh sizes on different heater surface roughnesses [65].

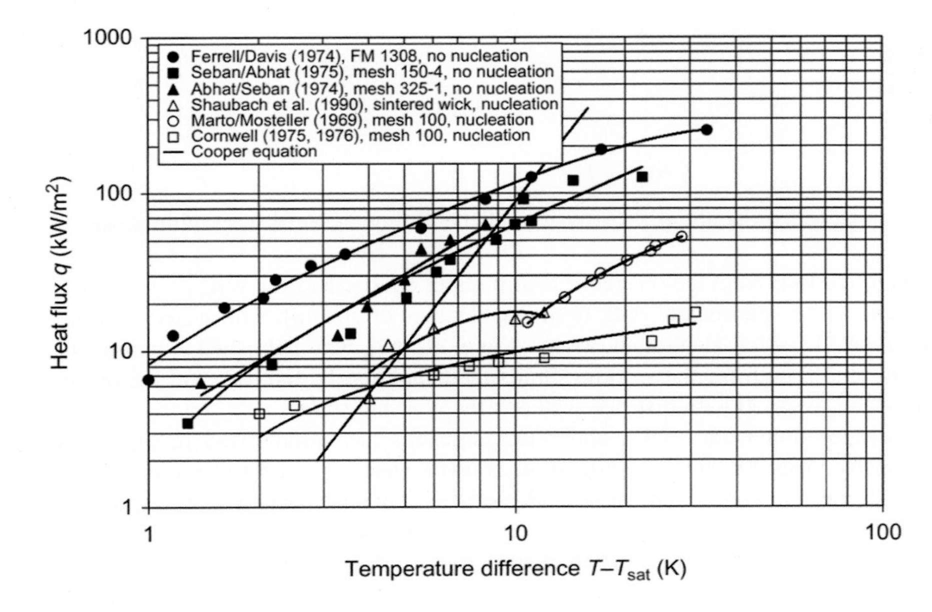


**FIGURE 4.30** Results from researchers measuring boiling and convection in the wick material, working fluid distilled [66]. *Data from Refs [71–76].*

Marto and Mosteller [72] used a horizontal tube surrounded by four layers of 100-mesh stainless-steel screen and water as the working fluid. The outer container was of glass so it was possible to observe the evaporating wick surface. They measured the superheat as a function of heat flux. They found that the critical radius was .013 mm compared to the effective capillary radius of 0.6 mm.

There is some evidence that the critical heat flux for wicked surfaces may be greater than that for smooth surfaces; for example, a report by Costello and Frea [77] suggests that the critical heat flux could be increased by about 20% over that for a smooth tube.

Reiss and Schreitzman [78] report very high values of critical heat flux for sodium in grooved heat pipes. They report values from 2 to 10 times the critical values reported by Balzieser [79] in the temperature region around 550°C. The authors observed the grooves as dark stripes on the outer side of the heat pipe and concluded that evaporation was from the grooves only. Their results are plotted in Fig. 4.31, and the heat flux is calculated on the assumption that evaporation is from the groove only. Moss and Kelley [80] constructed a planar evaporator using a stainless-steel wick ¼ in. thick and water as the working fluid. The authors employed a neutron radiographic technique to measure the water thickness. In order to explain their results, they proposed two models; in the first, it was assumed that vaporisation occurs at the liquid–vapour interface; in the second model, a vapour blanket was assumed to occur at the base of the wick.

Both the Moss and Kelley models can be used to explain most of the published data. These models are discussed in a paper by Ferrell et al. [73] who describe experiments aimed at differentiating between them. The two models are described by the authors as follows:

1. A layer of liquid-filled wick adjacent to the heated surface with conduction across this layer and liquid vaporisation at the edge of the layer. In this model, liquid must be drawn into the adjacent surface by capillary forces.
2. A thin layer of vapour-filled wick adjacent to the heated surface with conduction across this layer to the saturated liquid within the wick. In this model, the liquid is vaporised at the edge of this layer and the resulting vapour finds its way out of the wick along the surface and through large pore size passages in the wick. This model is analogous to conventional film boiling.

Experiments were carried out using a stainless-steel felt wick FM1308 with both water and potassium as the working fluids. Though the results were not entirely conclusive, the authors believed that the second model is the most likely

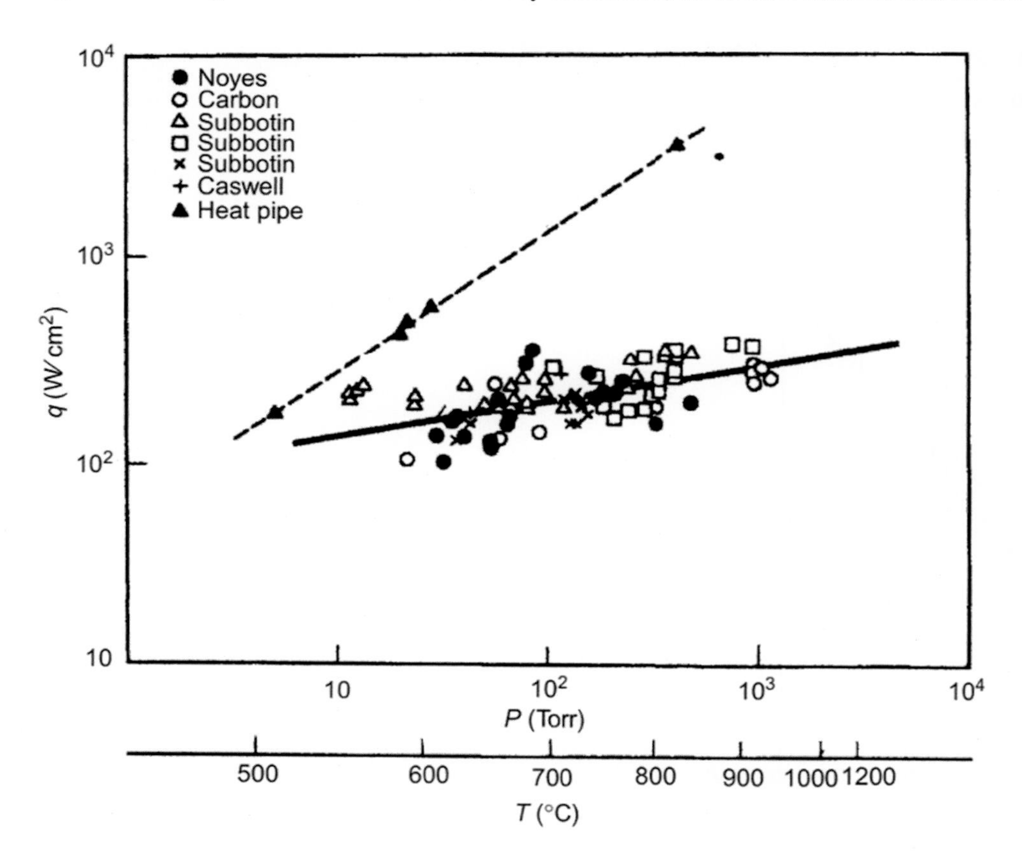


**FIGURE 4.31** Critical flux for sodium pool boiling compared to critical flux in grooved sodium heat pipes [78].

mechanism. Davis and Ferrell [81] report further work using potassium as the working fluid and both a stainless-steel felt wick FM1308 and a steel sintered powder wick Lamipore 7.4. The properties of these wicks are given in Table 4.6.

The data for the vertical heat pipe were different from the heat pipe in the horizontal positions for both types of wick. Figs 4.32 and 4.33 give results for the FM1308 wick in the vertical and horizontal positions. For the vertical position, the heat transfer coefficient for FM1308 increased with increasing flux becoming constant at a value 10.2 kW/$m^2$ K. For Lamipore 7.4, the coefficient decreases with increasing heat flux from 18.2 to 14.8 kW/$m^2$ K.

The effective thermal conductivities of the two wick structures have been calculated using the parallel model Eq. (4.95) and the series model Eq. (4.96). These results are given in Table 4.7 together with the experimentally measured heat transfer coefficients.

The agreement between the limits of the two models and the measured value for FM1309 is close. The agreement for Lamipore is good at low heat fluxes, and there was evidence for the development of a poor thermal contact during the experiment, which may explain the discrepancy at high flux values.

The heat transfer coefficients for the heat pipe in the horizontal position were much lower than for the vertical case. (1.1–5 kW/$m^2$ K) and similar to results obtained for a bare horizontal surface with no boiling. It is suggested that in the horizontal case, further temperature drop occurs by conduction through an excess liquid layer above the wick surface (this would not arise in normal heat pipe operation).

In the case of water, the heat transfer coefficient is 11.3 kW/$m^2$ K for both the vertical and horizontal cases, as shown in Fig. 4.34. The authors conclude that the heat transfer mechanism is different for liquid metals from the mechanism which applies to water and other nonmetallic fluids. In the case of nonmetallic fluids, they suggest that the vaporisation process occurs within the porous media. This vaporisation is probably initiated by inert gas trapped in the porous media or nucleation on active sites on the heated surface. Once initiated, the vapour phase spreads out to form a stable layer.

The data for water show that high fluxes are drawn for quite low values of temperature difference, whereas much larger values of temperature difference might be expected if the mechanism is one of conventional film boiling. The mechanism may be of nucleate boiling at activation sites on the heating surface and wick adjacent to it. This was confirmed by the observations of Brausch [66].

**TABLE 4.6** Dimensions and properties of wick materials.

| Material | Thickness (m) | Porosity | Permeability $m^2 \times 10^{10}$ | Capillary rise m |
|---|---|---|---|---|
| FM 1308 | $3.2 \times 10^{-3}$ | 0.58 | 0.55 | 0.26 |
| Lamipore 7.4 | $1.5 \times 10^{-3}$ | 0.61 | 0.47 | 0.35 |

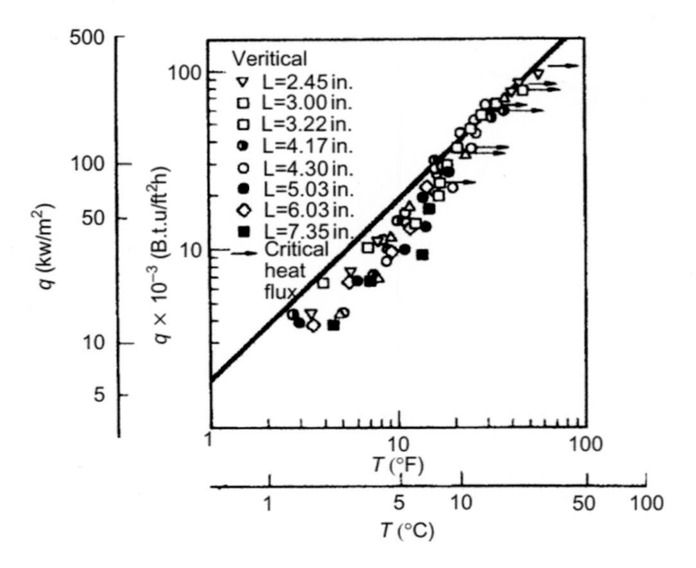


**FIGURE 4.32** Potassium boiling from vertical wick FM1308 (*L* is the vertical height above pool) [81].

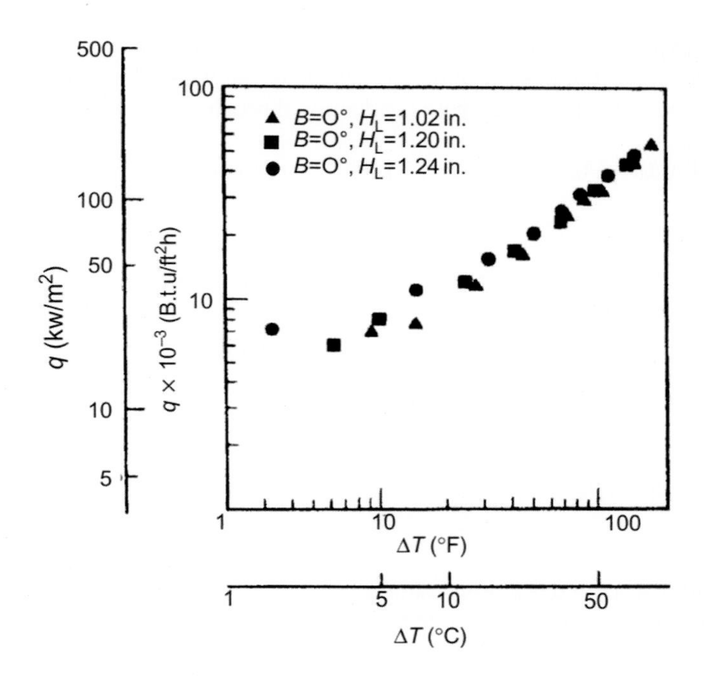


**FIGURE 4.33** Potassium boiling from horizontal wick FM1308 [81].

**TABLE 4.7 Heat transfer coefficients for potassium in vertical heat pipe (α in kW/m² K).**

| Wick | Parallel model, $k_w$ | Series model, $k_w$ | Experimental result, α |
|---|---|---|---|
| FM 1308 | 9.6 | 8.5 | 10.2 |
| Lamipore 7.4 | 19 | 18 | 18.2–14.8 |

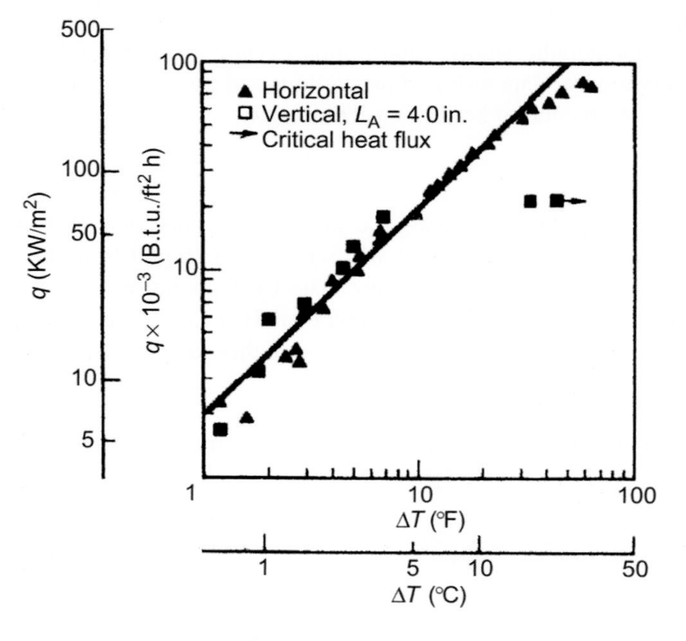


**FIGURE 4.34** Water boiling from FM1308 wick. *Data for vertical and horizontal wick is taken from Ref. [79].*

It is difficult to initiate bubbles in liquid metals and the experimental results of Ref. [66] strongly suggested that for these fluids the wick is saturated and that vaporisation occurs at the liquid surface on the outside of the wick. Hence for liquid metals, the heat transfer coefficient can be accurately predicted by considering conduction through the wick using Eq. (4.95) and Eq. (4.96).

One limit to the radial heat transfer to the evaporating fluid will be set by the 'wicking', that is when the capillary forces are unable to feed sufficient liquid to the evaporator.

The limiting heat flux $\dot{q}_{crit}$ will be given by the expression:

$$\dot{q}_{crit} = \frac{\dot{m}L}{\text{area of evaporator}}$$

where the mass flow in the wick is related to the pressure head, $\Delta P$ by an expression such as Eq. (4.32).

Ferrel et al. [82] have derived a relationship for a planar surface with a homogeneous wick.

$$\dot{q}_{\text{crit}} = \frac{g\left(h_{co}\rho_l \frac{\sigma_l}{\sigma_{lo}} - \rho_l l \sin\varphi\right)}{\frac{l_e \mu_l}{L\rho_l kd}\left[\frac{l_e}{2} + l_a\right]}$$

where

$h_{co}$ is the capillary height of the fluid in the wick measured at a reference temperature;
$\sigma_{lo}, \rho_{lo}$ are the fluid surface tension and density measured at the same reference temperature;
$\sigma_l, \rho_l, \mu_l$ are the fluid surface tension, density and viscosity measured at the operating temperature.

All other symbols have their usual significance. Ferrell and Davis [73] extended their equation by including a correction for thermal expansion of the wick.

$$\dot{q}_{\text{crit}} = \frac{g\left(h_{co}\rho_l \frac{\sigma_l}{\sigma_{lo}} - \rho_l l \sin\varphi(1 + \alpha_t \Delta T)\right)}{\frac{l_e \mu_l}{L\rho_l kd(1 + \alpha_t \Delta T)}\left[\frac{l_e}{2} + l_a\right]} \tag{4.91}$$

where $\alpha_t$ is the coefficient of linear expansion of the wick, and $\Delta T$ the difference between the operating and reference temperature.

Fig. 4.35 shows a comparison of measured values of $\dot{q}_{\text{crit}}$ against predicted values from Eq. (4.91) for both water and potassium. The equation successfully predicts heat flux limits for potassium up to a value of 315 $kW/m^2$. It is also in good agreement for water up to 130 $kW/m^2$. Above this value, the experimental values fall below the values predicted by Eq. (4.91) showing that another mechanism is limiting the flux. The limiting factor is probably due to difficulty experienced by water vapour in leaving the heat surface through the wick. The reduction in heat flux below the predicted value for water further supports the view that for nonmetallic fluids, vaporisation occurs within the wick. More recent work on radial heat flux measurements and observation of vapour/liquid proportions and nucleation within wicks has resulted in more data showing that nucleation is not detrimental to heat pipe performance.

#### 4.3.7.5 Liquid–vapour interface temperature drop

Consider a liquid surface, where there will be a continuous flux of molecules leaving the surface by evaporation. If the liquid is in equilibrium with the vapour above its surface, an equal flux of molecules will return to the liquid from the vapour and there will be no net loss or gain of mass. However, when a surface is losing mass by evaporation, clearly the vapour pressure and hence the temperature of the vapour above the surface must be less than the equilibrium value. In the same way, for net condensation to occur, the vapour pressure and temperature must be higher than the equilibrium value.

The magnitude of the temperature drop can be estimated as follows:

First, consider the vapour near the interface. The average velocity $V_{av}$ in a vapour at temperature $T_v$ and having molecular mass $M$ is given by kinetic theory as

$$V_{av} = \sqrt{\frac{8k_B T_v}{\pi M}} \tag{4.92}$$

where $k_B$ is Boltzmann's constant.

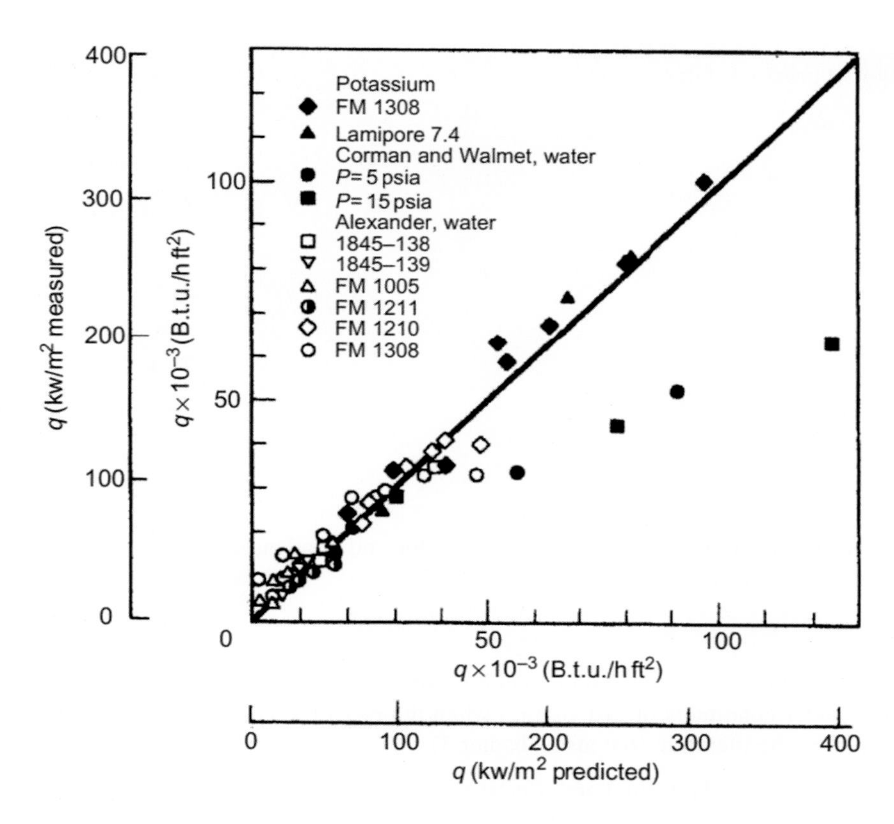


**FIGURE 4.35** Comparison of measured and predicted burnout for water and potassium [81].

The average flow of molecules in any given direction is

$$\frac{nV_{av}}{4}/\text{unit area}$$

and the corresponding flow of heat per unit area is

$$\frac{MLnV_{av}}{4}$$

where $n$ is the number of molecules per unit volume, and $L$ is the latent heat. For a perfect gas,

$$P_v = nk_BT_v$$

hence the heat flux to the liquid surface from the vapour $= P_vL\sqrt{\frac{M}{2\pi kT_v}}$. The heat flux away from the liquid surface is given by $P_lL\sqrt{\frac{M}{2\pi kT_l}}$. Setting $T_v = T_l = T_s$ allows evaluation of the net heat flux across the surface:

$$\dot{q} = (P_l - P_v)L\sqrt{\frac{M}{2\pi k_BT_s}} = \frac{(P_l - P_v)L}{\sqrt{2\pi RT}} \tag{4.93}$$

By using the Clapeyron equation,

$$\frac{dP}{dT} = \frac{L}{(v_v - v_l)T_{sat}}$$

and assuming that the vapour behaves as a perfect gas and the liquid volume is negligible:

$$\frac{\Delta P}{\Delta T} = \frac{PL}{RT^2}$$

Eq. (4.93) may be written:

$$\dot{q} = \frac{\Delta TL^2P}{(2\pi RT_s)^{0.5}}\frac{1}{RT_s^2} \tag{4.94}$$

**TABLE 4.8 Interfacial heat flux as a function of pressure difference [83].**

| Fluid | $T_b$ | $q/(P_l - P_v)$ |
|---|---|---|
| | K | kW/cm$^2$ atm |
| Lithium | 1613 | 55 |
| Zinc | 1180 | 18 |
| Sodium | 1156 | 39 |
| Water | 373 | 21.5 |
| Ethanol | 351 | 13.5 |
| Ammonia | 238 | 15.2 |

thus permitting the calculation of the interfacial temperature change in the evaporator and condenser. Values of $\frac{\dot{q}}{P_l - P_v}$ for liquids near their normal boiling points are given in Table 4.8.

### 4.3.7.6 Wick thermal conductivity

The effective conductivity of the wick saturated with the working fluid is required for calculating the thermal resistance of the wick at the condenser region and also under conditions of evaporation when the evaporation is from the surface for the evaporator region. Two models are used in the literature (see also Chapter 3).

**i.** Parallel case. Here it is assumed that the wick and working fluid are effectively in parallel.

If $k_l$ is the thermal conductivity of the working fluid and $k_s$ is the thermal conductivity of the wick material and

$$\varepsilon = \text{Voidage fraction} = \frac{\text{Volume of working fluid in wick}}{\text{Total volume of wick}}$$

$$k_w = (1 - \varepsilon)k_s + \varepsilon k_l \tag{4.95}$$

**ii.** Series case. If the two materials are assumed to be in series.

$$k_w = \frac{1}{\frac{(1-\varepsilon)}{k_s} + \frac{\varepsilon}{k_l}} \tag{4.96}$$

Additionally, convection currents in the wick will tend to increase the effective thermal conductivity.

### 4.3.7.7 Heat transfer in the condenser

Vapour will condense on the liquid surface in the condenser, the mechanism is similar to that discussed in Section 4.3.7.5 on the mechanism of surface evaporation, and there will be a small temperature drop and hence thermal resistance. Further temperature drops will occur in the liquid film and in the saturated wick and in the heat pipe envelope.

Condensation can occur in two forms, either by the condensing vapour forming a continuous liquid surface or by forming a large number of drops. The former, film condensation, occurs in most practical applications, including heat pipes and will be discussed here. Condensation is seriously affected by the presence of a noncondensable gas. However, in the heat pipe, vapour pumping will cause such gas to be concentrated at the end of the condenser. This part of the condenser will be effectively shut off, and this effect is the basis of the gas-buffered heat pipe. The temperature drop through the saturated wick may be treated in the same manner as at the evaporator.

#### 4.3.7.7.1 Nusselt's theory

Film condensation may be analysed using Nusselt's theory. The first analysis of film condensation was due to Nusselt and given in standard textbooks, for example [84]. The theory considers condensation onto a vertical surface, and the resulting condensed liquid film flows down the surface under the action of gravity and is assumed to be laminar.

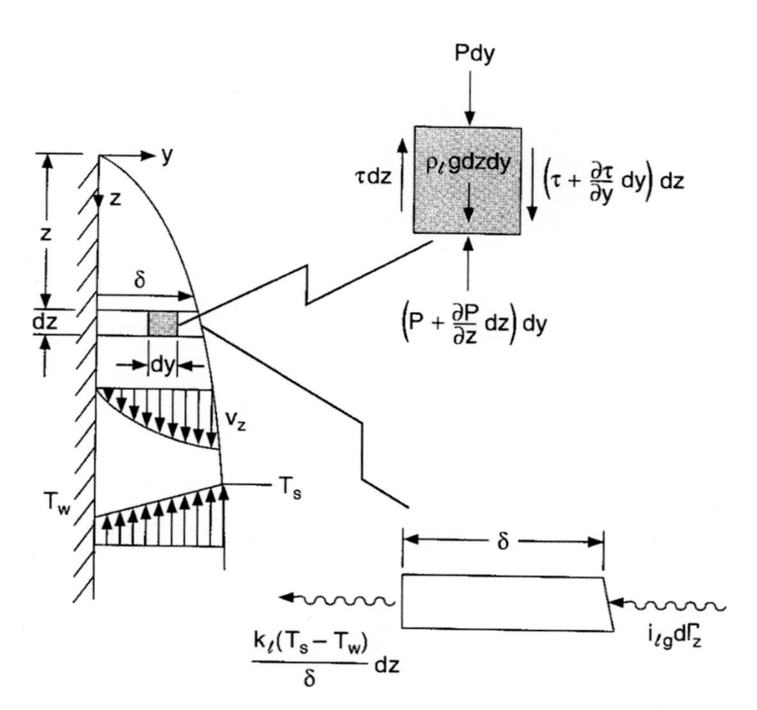


FIGURE 4.36 Film condensation on a vertical surface.

Viscous shear between the vapour and liquid phases is neglected. The mass flow increases with distance from the top, and the flow profile is shown in Fig. 4.36. The average heat transfer coefficient, α, over a distance x from the top is given by

$$\alpha = 0.943 \left[ \frac{L\rho_l^2 g k_l^3}{x\mu_l (T_s - T_w)} \right]^{0.25} \tag{4.97}$$

where $(T_s - T_s)$ is the temperature drop across the liquid film.

The Nusselt condensation heat transfer coefficient is the most commonly used to predict the heat transfer rate by condensation. However, other researchers such as McAdams [85], Chen et al. [86] and Oh and Revankar [87] observed that the experimental heat transfer coefficient of condensation are 15%–20% higher than predicted ones. According to the literature, the under-prediction of the heat transfer coefficient could be attributed to the presence of waves in the condensation film. Furthermore, to account for the potential subcooling of the condensate or the nonlinear temperature distribution in the film, Rohsenow [88] proposed a modified latent heat of vaporisation. In addition, Kapitza [89] reported that the turbulence leads to a reduced average film thickness which increases the heat transfer coefficient; Nusselt made several assumptions in his theory such as considering a laminar flow. Condensation in thermosyphons occurs at the saturation pressure of the vapour. Considering the variety of working fluids used in thermosyphons, in addition to the various shapes and diameters of condensers, condensation heat transfer was widely studied in the literature in order to predict precisely the heat transfer coefficient and validate Nusselt's correlation [90].

#### 4.3.7.7.2 Condensation heat transfer correlations

The recommended correlations for predicting condensation heat transfer coefficients in thermosyphons are listed in Table 4.9.

The applicability of the listed correlations is subject to the working fluid, operating temperature, radial heat flux in the thermosyphon and pipe diameter. For water, methanol, acetone, Dowtherm A and ethanol working fluids, Gross [93] reported that the correlation by Nusselt [90] was in good agreement with the data, whereas the same correlation showed larger deviations for R-11, R-113, R-22 and R-115 working fluids. In many cases, thermosyphon condensers are not sufficiently long to develop a turbulent condensation film, therefore the Nusselt theory is still applied with acceptable accuracy. Fiedler and Auracher [98] reported that the generally recommended condensation correlations are Nusselt [90] for $Re_{f,L_c} < 30$, Kutateladze [91] for $30 < Re_{f,L_c} < 1600$ and Labuntsov [92] for $1600 < Re_{f,L_c}$. These correlations were in good agreement with the data obtained by Fiedler and Auracher [98]. For an inclined thermosyphon,

**TABLE 4.9** Recommended filmwise condensation heat transfer coefficient.

| Author | Frequency of use | Year | Correlation | Valid fluids |
|---|---|---|---|---|
| | | | Laminar falling films ($Re_f \le 20-30$) | |
| Nusselt [90] corrected by Rohsenow [88] | High | 1956 | $\alpha = 0.943\left\{\frac{\rho_l(\rho_l-\rho_v)i_{lv}'gk_l^3}{\mu_l L_c(T_{sat}-T_w)}\right\}^{1/4}$ where, $i_{lv}' = i_{lv} + 0.68c_{pl}(T_{sat} - T_w)$ | All |
| Nusselt [90] | Very high | 1916 | $\alpha_{Nusselt} = 0.943\left\{\frac{\rho_l(\rho_l-\rho_v)i_{lv}gk_l^3}{\mu_l L_c(T_{sat}-T_w)}\right\}^{1/4}$ | All |
| | | | Wavy-laminar falling film ($20-30 \le Re_f \le 600$) | |
| Rohsenow [88] | High | 1956 | $1.51\left(\frac{P_v}{P_{crit}}\right)^{0.14} \times 0.943\left\{\frac{\rho_l(\rho_l-\rho_v)gk_l^3}{\mu_l L_c(T_{sat}-T_w)}\left[i_{lv}+3/8c_{pl}(T_{sat}-T_w)\right]\right\}^{1/4}$ where the fluid properties should be evaluated at a temperature: $T = T_W + 0.31(T_{sat} - T_w)$ | All |
| McAdams [85] | Medium | 1942 | $\alpha = 1.13\left\{\frac{\rho_l(\rho_l-\rho_v)i_{lv}gk_l^3}{\mu_l L_c(T_{sat}-T_w)}\right\}^{1/4}$ | All |
| | | | Wavy falling film ($600 \le Re_f \le 1600$) | |
| Kutateladze [91] | High | 1963 | $\alpha = \frac{Re_{f,Lc}/4}{1.47(Re_{f,Lc}/4)^{1.22}-1.3}k_l\left(\frac{\mu_l^2}{\rho_l(\rho_l-\rho_v)g}\right)^{-1/3}$ where $Re_{f,L_c} = \frac{4\Gamma_{L_c}}{\mu_l}$ | All |
| | | | Turbulent falling film ($1600 \le Re_f \le 3200$) and Highly turbulent falling film ($3200 \le Re_f$) | |
| Labuntsov [92] | Medium | 1957 | $\alpha = 0.0306Re_{f,L_c}{}^{1/4}Pr_l{}^{1/2}k_l\left(\frac{\mu_l^2}{\rho_l(\rho_l-\rho_v)g}\right)^{-1/3}$ where, $Re_{f,L_c} = \frac{4\Gamma_{L_c}}{\mu_l}$ | All |
| Special considerations | | | | |
| | | | For a wide range of Reynolds number (from laminar to highly turbulent) | |
| Gross [93] | Medium | 1992 | $\alpha = \left(\left(0.925f_d Re_{f,max}{}^{-1/3}\right)^2 + \left(0.044Pr_l{}^{2/5}Re_{f,max}{}^{1/6}\right)^2\right)^{1/2}k_l\left(\frac{\mu_l^2}{\rho_l(\rho_l-\rho_v)g}\right)^{-1/3}$ where, $f_d = \left(1-0.63(P_v/P_{crit})^{3.3}\right)^{-1}$ $Re_{f,max} = Re_f = \frac{q}{\pi D_i i_{lv}\mu_l}$ | All |
| Chun and Kim [94] | Low | 1991 | $\alpha = \left[1.33Re_{f,L_c}{}^{-1/3} + 9.56\times10^{-6}Re_{f,L_c}{}^{0.89}Pr_l{}^{0.94} + 8.22\times10^{-2}\right]k_l\left(\frac{\mu_l^2}{\rho_l(\rho_l-\rho_v)g}\right)^{-1/3}$ where $Re_{f,L_c} = \frac{4\Gamma_{L_c}}{\mu_l}$ | Fluids with $1.75 < Pr_l < 5.0$ |

| For low heat flux ($q < 1000$ W) | | | | |
|---|---|---|---|---|
| Jouhara and Robinson [46] | Medium | 2010 | $\alpha = 0.85\mathrm{Re}_f^{0.1}\exp\left(-6.7\times10^{-5}\frac{\rho_l}{\rho_v}-0.14\right)h_{\mathrm{Nusselt}}$<br>where<br>$\mathrm{Re}_f = \frac{4q}{\pi D_i i_{lv}\mu_l}$ | Water, FC-84, FC-77, FC-3283 |
| Hashimoto and Kaminaga [95] | Medium | 2002 | $\alpha = 0.85\mathrm{Re}_f^{0.1}\exp\left(-6.7\times10^{-5}\frac{\rho_l}{\rho_v}-0.6\right)h_{\mathrm{Nusselt}}$<br>where<br>$\mathrm{Re}_f = 4\Gamma/\mu_l$ | All |
| For inclined thermosyphons | | | | |
| Hussein et al. [96] | Low | 2001 | $\alpha = \left(\frac{L_c}{D_i}\right)^{\frac{1}{4}(\cos(\beta))^{0.358}}\left[0.997-0.334(\cos(\beta))^{0.108}\right]h_{\mathrm{Nusselt}}$<br>where,<br>$\beta$: Inclination angle of the thermosyphon (°)<br>$L_c$: Condenser length (m)<br>$D_i$: Internal diameter of the thermosyphon (m) | All |
| Wang and Ma [97] | High | 1991 | $\alpha = \left(\frac{L_c}{r_i}\right)^{\cos(\beta/4)}\left[0.54+(5.68\times10^{-3}\beta)\right]h_{\mathrm{Nusselt}}$<br>where<br>$\beta$: Inclination angle of the thermosyphon (°)<br>$L_c$: Condenser length (m)<br>$r_i$: Internal radius of the thermosyphon (m) | All |

Hussein et al. [96] and Fiedler and Auracher [98] both recommended the correlation by Wang and Ma [97], which was concluded to accurately describe the evolution of the condensation heat transfer with the inclination angle. Based on the research of Wang and Ma [97], Hussein et al. [96] enhanced the previous correlation. The results presented by the authors over a wide range of data from the literature were excellent. At low heat flux ($q < 1000$ W), authors such as Gross [93] and Jouhara and Robinson [46] report that the accuracy of the prediction by Nusselt [90] is doubtful. Jouhara and Robinson [46] showed that, between low and high heat flux, the power law dependency of the condensation heat transfer with the heat flux changes significantly, which confirms a change in the condensation heat transfer mechanisms.

## 4.4 Application of theory to heat pipes and thermosyphons

### 4.4.1 Wicked heat pipes

#### 4.4.1.1 The merit number

It will be shown, with reference to the capillary limit, that if vapour pressure loss and gravitational head can be neglected, then the properties of the working fluid which determine the maximum heat transport can be combined to form a figure of merit, $M$.

$$M = \frac{\rho_l \sigma_l L}{\mu_l} \tag{4.98}$$

The way in which $M$ varies with temperature for a number of working fluids is shown in Fig. 4.37.

#### 4.4.1.2 Operating limits

Upper limits to the heat transport capability of a heat pipe may be set by one or more factors. These limits are illustrated in Fig. 4.1.

##### 4.4.1.2.1 Viscous, or vapour pressure, limit

At low temperatures, viscous forces are dominant in the vapour flow down the pipe. Busse has shown that the axial heat flux increases as the pressure in the condenser is reduced, the maximum heat flux occurring when the pressure is reduced to zero. Busse carried out a two-dimensional analysis, finding that the radial velocity component had a significant effect, and derived the following equation.

$$\dot{q} = \frac{r_v L \rho_v P_v}{16 \mu_v l_{\text{eff}}} \tag{4.99}$$

where $P_v$ and $\rho_v$ refer to the evaporator end of the pipe. This equation agrees well with published data [99]. The vapour pressure limit is described well in the ESDU Data Item on capillary-driven heat pipes [100]. It is stated that when the vapour pressure is very low, the minimum value must occur at the closed end of the condenser section, the vapour pressure drop can be constrained by this very low – effectively zero – pressure, and the low vapour pressure existing at

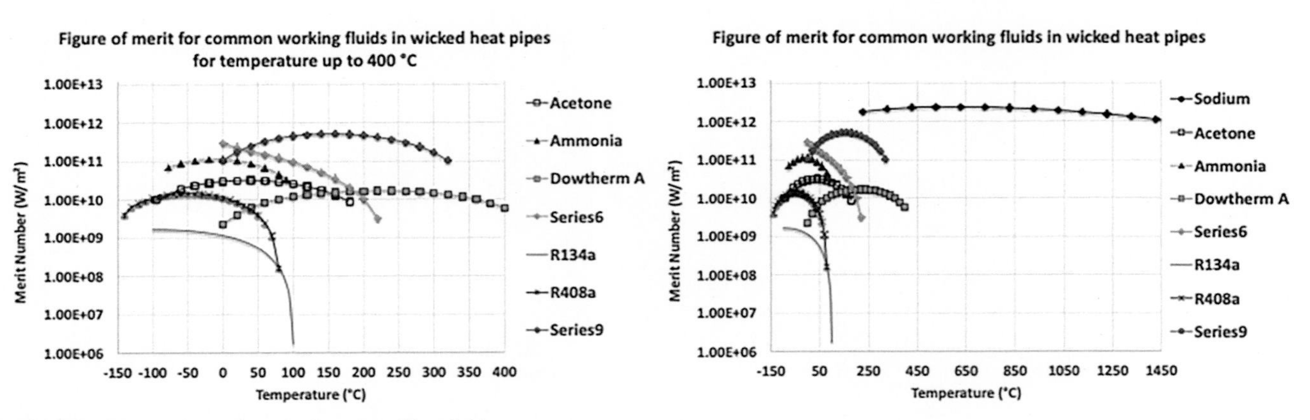


FIGURE 4.37 Merit number of selected working fluids.

the closed end of the evaporator section. The maximum possible pressure drop is therefore equal to the vapour pressure in the evaporator. Because the vapour pressure difference naturally increases as the heat transported by the heat pipe rises, the constraint on this pressure difference thus necessitates that $\dot{q}$ is limited. The limit is generally only of importance in some units during start-up. A criterion for avoidance of this limit is given in reference [100] as

$$\frac{\Delta P_v}{P_v} < 0.1$$

#### 4.4.1.2.2 Sonic limit

At a somewhat higher temperature choking at the evaporator exit may limit the total power-handling capability of the pipe. This problem was discussed in Section 4.3.5. The sonic limit is given by

$$\dot{q} = 0.474L\left(\rho_v P_v\right)^{0.5} \tag{4.100}$$

There is good agreement between this formula and experimental results. Fig. 4.38, due to Busse [99], plots the temperature at which the sonic limit is equal to the viscous limit as a function of $l_{\text{eff}}/d^2$ for some alkali metals, where $d = 2r_v$, the vapour space diameter.

#### 4.4.1.2.3 Entrainment limit

This was discussed in Section 4.3.6, and Eq. (4.65) gave the entrainment limiting flux as

$$\dot{q} = \sqrt{\frac{2\pi\rho_v L^2 \sigma_l}{z}} \tag{4.101}$$

Kemmes experiments suggest a rough correlation of this failure mode with the centre-to-centre wire spacing for $Z$ and that a very fine screen will suppress entrainment.

A number of workers have presented data on entrainment limits for specific cases. Tien and Chung [101] present equations which, it is claimed, enable the user to predict the maximum heat transfer rate due to the entrainment limitation. Some experimental data suggest that the equations give accurate results. Data are presented for 'gravity-assisted wickless heat pipes' (thermosyphons), gravity-assisted wicked heat pipes, horizontal heat pipes, grooved heat pipes and rotating heat pipes. Experiments with gravity-assisted heat pipes employing simple wick structures were carried out by Coyne Prenger and Kemme [102] in order to arrive at a correlation which could be used to predict the entrainment limits for all the data investigated.

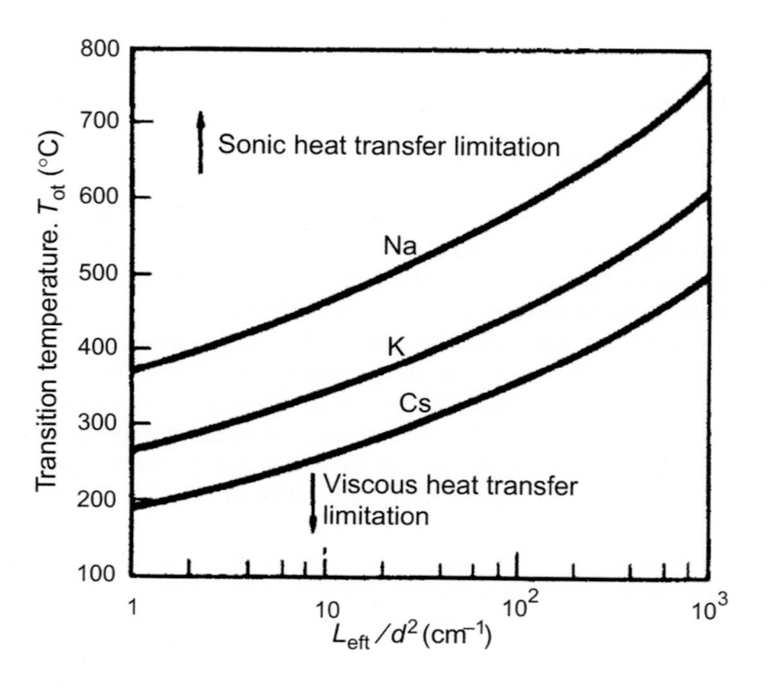


**FIGURE 4.38** Transition from viscous to sonic limit.

This work led to the conclusions that the entrainment limits in the cases studied were vapour dominated and were essentially independent of the fluid inventory. The entrainment limits were successfully correlated with a physical model based on a critical Weber number (see Eq. (4.62)), the expression being

$$\dot{Q} = \sqrt{\frac{2\pi E_t}{\alpha_v}\frac{\delta}{\delta *}} \tag{4.102}$$

where $E_t$ is a dimensionless entrainment parameter defined by $(\sigma_l/\rho_l L\delta)$, $\alpha_v$ the velocity profile correction factor, $\delta$ the surface depth and $\delta*$ a reference surface depth.

The characteristic length in the Weber number formulation was the depth of the wick structure (an example being cited by the authors as the depth of grooves in a circumferentially grooved wick). In the case of mesh wicks, one-half of the wire diameter of the innermost layer is used (this also allowing for mesh composite wicks). The limit is also discussed for gravity-assisted heat pipes by Nguyen-Chi and Groll where it is also called the flooding limit — as the condenser becomes flooded.

#### 4.4.1.2.4 Capillary limit (wicking limit)

In order for the heat pipe to operate, Eq. (4.2) must be satisfied, namely

$$\Delta P_{c,\max} \geq \Delta P_l + \Delta P_v + \Delta P_g \tag{4.2}$$

Expressions to enable these quantities to be calculated are given in Section 4.3.

An expression for the maximum flow rate $m$ may readily be obtained if we assume that

**(i)** the liquid properties do not vary along the pipe
**(ii)** the wick is uniform along the pipe
**(iii)** the pressure drop due to vapour flow can be neglected.

$$\dot{m}_{\max} = \left[\frac{\rho_l\sigma_l}{\mu_l}\right]\left[\frac{KA}{l}\right]\left[\frac{2}{r_e} - \frac{\rho_l g l}{\sigma_l}\sin\varphi\right]_{\max} \tag{4.103}$$

and the corresponding heat transport $\dot{Q} = \dot{m}_{\max}L$, is given by:

$$\dot{Q} = \dot{m}_{\max}L = \dot{m}_{\max} = \left[\frac{\rho_l\sigma_l}{\mu_l}\right]\left[\frac{KA}{l}\right]\left[\frac{2}{r_e} - \frac{\rho_l g l}{\sigma_l}\sin\varphi\right] \tag{4.104}$$

The group defined in Eq. (4.98) as

$$\frac{\rho_l\sigma_l L}{\mu_l}$$

depends only on the properties of the working fluid and is known as the figure of merit, or Merit number, $M$, plotted for several fluids in Fig. 4.37.

#### *4.4.1.3 Burnout*

Burnout will occur at the evaporator at high radial fluxes. A similar limit on peak radial flux will also occur at the condenser. These limits are discussed in Section 4.3.7. At the evaporator, Eq. (4.91) gives a limit which must be satisfied for a homogeneous wick. This equation, which represents a wicking limit, is shown to apply to potassium up to 315 kW/m$^2$ and probably is applicable at higher values for potassium and other liquid metals. For water and other nonmetallics, vapour production in the wick becomes important at lower flux densities (130 kW/m$^2$ for water) and there are no simple correlations. For these fluids, experimental data in Table 3.3 should be used as an indication of flux densities which can be achieved. Tien [103] has summarised the problem of nucleation within the wick and the arguments for possible performance enhancement in the presence of nucleation. Because of phase equilibrium at the interface, liquid within the wick at the evaporator is always superheated to some degree, but it is difficult to specify the degree of superheat needed to initiate bubbles in the wick. Whilst boiling is often taken as a limiting feature in wicks, possibly upsetting the capillary action, Cornwell and Nair [76] found that nucleation reduces the radial temperature difference and increases heat pipe conductance. One factor in support of this is the increased thermal conductivity of liquid-saturated wicks in the evaporator section, compared with that in the condenser.

Tien believed that the boiling limit would become effective only when the bubbles generated within the wick become trapped there, forming a vapour blanket. This was confirmed by Brautsch [66]. Thus some enhancement of the radial heat transfer coefficient in the evaporator section can be obtained by nucleate boiling effects; however, excessive nucleate boiling disrupts the capillary action and reduces the effective area for liquid flow. It is useful to compare the measurements of radial heat flux as a function of the degree of superheat (in terms of the difference between wall temperature $T_w$ and saturation temperature $T_{sat}$) for a number of surfaces and wick forms which have been the subject of recent studies. These are shown in Table 4.10. Abhat and Seban [71], in discussing their results as shown in Table 4.10, stated that for every pool depth, there was a departure from the performance recorded involving a slow increase in the evaporator temperature. In order to guarantee indefinite operation of the heat pipe without dry-out, the authors therefore suggested that the operating flux should be considerably less than that shown in the table. They also found that their measurements of radial flux as a function of $T_w - T_{sat}$ were similar, regardless of whether the surface was flat, contained mesh or a felt. This is partly borne out by the results of Wiebe and Judd [105], also given in Table 4.10. Costello and Frea, however, dispute this [77] (Fig. 4.39).

The work of Cornwell and Nair [75] gives results similar to those of Abhat and Seban [71]. Some of Cornwell and Nair's results are presented in Fig. 4.39 and compared with theory. Cornwell and Nair found that some indication of the $\dot{q}/\Delta T$ curve for boiling within a wick could be obtained by assuming that boiling occurs only on the liquid covered area of the heating surface (measured from observations) and that $\dot{q}$ based on this area may be expressed by a nucleate boiling type correlation:

$$\dot{q} = C\Delta T^a$$

where $C$ is a constant and $a$ is in the range 2–6.

The results of Brautsch [66] followed the same trends as those of Cornwell and Nair. The enhancement due to a mesh wick at low heat fluxes and the reduction in heat transfer due to the wick at higher heat fluxes was explained in terms of the competing effects of additional nucleation sites and vapour blocking the liquid in the wick; the resulting correlation is given as Eq. (4.87). These analyses are restricted to situations where the vapour formed in the wick escapes through the wick surface and not out of the sides or through grooves in the heating surface. A different approach was taken by Nishikawa and Fujita [106], who have investigated the effect of bubble population density on the heat flux limitations and the degree of superheat. Whilst this work is restricted to flat surfaces, the authors' suggested nucleation factor may become relevant in studies where the wick contains channels or consists solely of grooves, possibly flooded, in the heat pipe wall.

**TABLE 4.10 Measured radial heat fluxes in wicks.**

| References | Wick | Working fluid | Superheat (°C) | Flux (W/cm$^2$) |
|---|---|---|---|---|
| Wicbc and Judd | Horizontal flat surface | Water | 3 | 0.4 |
| | | | 6 | 1.2 |
| | | | 11 | 8 |
| | | | 17 | 20 |
| Abhat and Seban [71] | Meshes and Felts | Water | 5 | 1.6 |
| | | | 6 | 5 |
| | | | 11 | 12 |
| | | | 17 | 20 |
| Cornwell and Nair [76] | Foam | Water | 5 | 2 |
| | | | 10 | 10 |
| | | | 20 | 18 |
| | Mesh (100) | water | 5 | 7 |
| | | | 10 | 9 |
| | | | 20 | 13 |
| Abhat and Nguyeachi [104] | Mesh* | Water | 6 | 1 |
| | | | 10 | 8 |

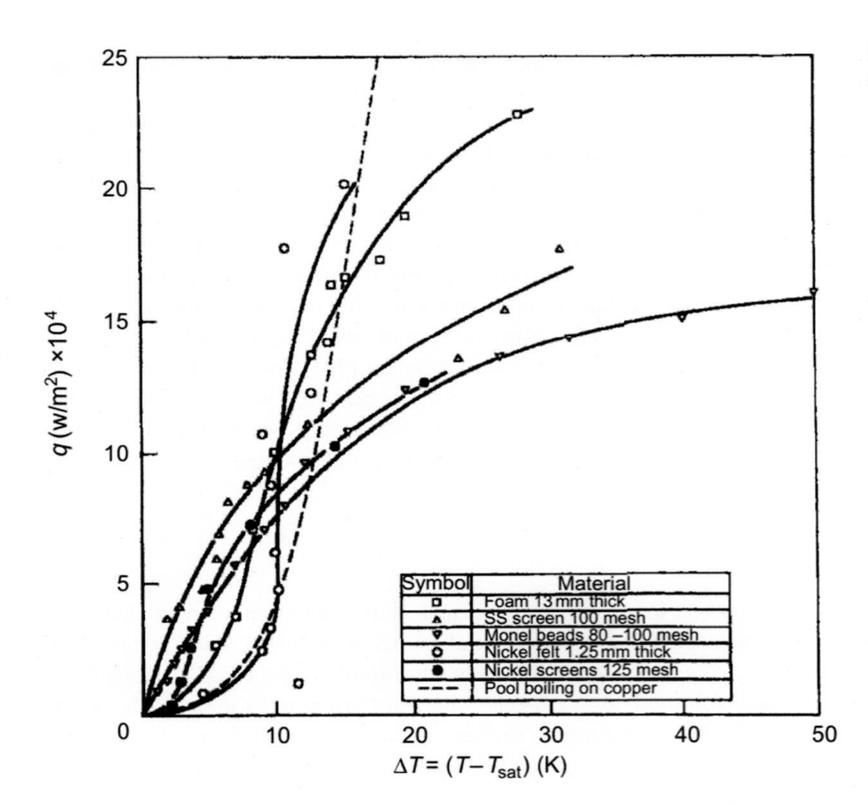


**FIGURE 4.39** Variation of heat flux with superheat for water in porous media [75].

The work of Saaski [107] on an inverted meniscus wick is of considerable interest. This concept, illustrated in Fig. 4.40, embodies the high heat transfer coefficient of the circumferential groove whilst retaining the circumferential fluid transport capability of a thick sinter or wire mesh wick. With ammonia, heat transfer coefficients in the range 2–2.7 $W/m^2$ K were measured at radial heat flux densities of 20 $W/cm^2$. These values were significantly higher than those obtained for other nonboiling evaporative surfaces. Saaski suggested that the heat transfer enhancement may be due to film turbulence generated by vapour shear or surface tension-driven convection. He contemplated an increase in groove density as one way of increasing the heat transfer coefficient, as first results have indicated a direct relationship between groove density and heat transfer coefficient. Vapour resulting from evaporation at the inverted meniscus interface flows along to open ends of the grooves, where it enters the central vapour core.

The main theoretical requirement for inverted meniscus operation, according to Saaski, is the maintenance of a sufficiently low vapour pressure drop in each channel of the wick. Recession of the inverted meniscus, which is the primary means of high evaporative heat transfer, reduces capability considerably. The equation given below describing the heat flux capability contains a term ($j$) which is a function of vapour microchannel pressure drop, being defined as the ratio of this pressure drop to the maximum capillary priming potential.

Thus $j = \Delta P_v/(2\sigma/r_c)$

The heat flux $q_{\max}$ is defined by

$$q_{\max} = \frac{jN^2x^4}{8\pi}\left[\frac{\rho_v L\sigma}{\mu_v M}\right]\frac{1}{N^3 {d_g}^2 r_c} \tag{4.105}$$

$$x = \cos\frac{\Psi}{2}\Big/\left[1 + \sin\frac{\Psi}{2}\right]$$

where $\Psi$ is the groove angle (deg), $M$ the molecular weight of working fluid, $N$ the number of grooves per cm, $d_g$ = heat pipe diameter at inner groove radius ($q_{\max}$) is given in $W/cm^2$ by Eq. (4.105).

Feldman and Berger [108] carried out a theoretical analysis to determine surfaces having potential in a high heat flux water heat pipe evaporator. Following a survey of the literature, circumferential grooves were chosen, these being of rectangular and triangular geometry. The model proposed that:

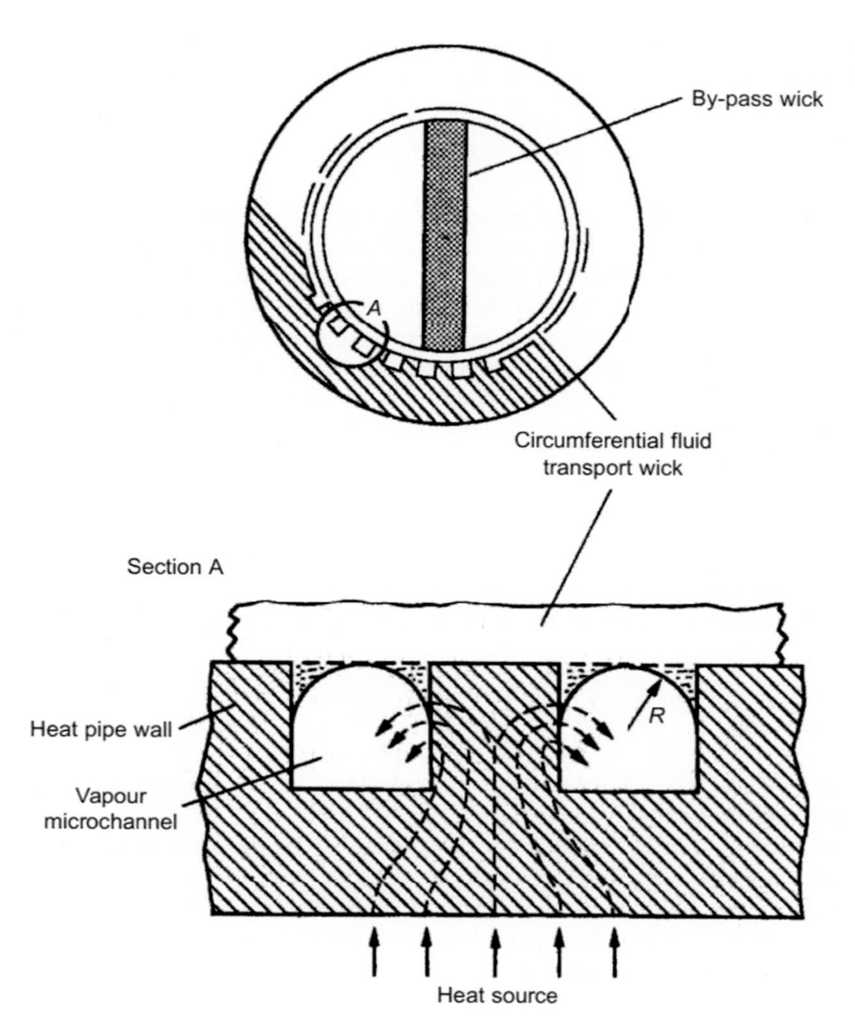


**FIGURE 4.40** High-performance inverted meniscus hybrid wick proposed by Saaski [107].

**(i)** At heat fluxes below nucleate boiling, conduction is the main mode of heat transfer, and vaporisation occurs at the liquid/vapour interface without affecting capillary action.
**(ii)** As nucleation progresses, the bubbles are readily expelled from the liquid in the grooves, with local turbulence, and convection becomes the main mode of heat transfer.
**(iii)** At a critical heat flux, nucleation sites will combine forming a vapour blanket, or the groove will dry out.

Both result (ii) and (iii) in a sharp increase in evaporator temperature. It is claimed that film coefficients measured by other workers supported the computer model predictions, and Feldman stated that evaporator film coefficients as high as 8 W/cm$^2$ K could be obtained with water as the working fluid. Using triangular grooves, it was suggested that radial fluxes of up to 150 W/cm$^2$ could be tolerated. However, such a suggestion appears largely hypothetical, bearing in mind the measured values given in Table 4.10.

Winston et al. [109] reported further progress on the prediction of maximum evaporator heat fluxes, including a useful alternative to Eq. (4.90). Based on modification of an equation first developed by Johnson [110], the relationship given below takes into account vapour friction in the wick evaporator section:

$$\dot{q}_{\text{crit}} = \frac{g\left[h_{\text{co}}\rho_{\text{lo}}\frac{\sigma_1}{\alpha_{\text{lo}}} - \rho_1(l_a + l_e)\sin\Phi\right]}{\frac{l_e\mu_1}{L\rho_1 K\varepsilon(r_\varpi - r_v)}\left[\frac{l_e}{2} + 1_a\right] + \frac{\mu_v}{K\rho_v}\cdot\frac{(r_\varpi - r_v)}{(\varepsilon - \varepsilon_1)L}} \tag{4.106}$$

The solution of the equation was accomplished by arbitrarily varying the porosity for liquid flow, $\varepsilon_l$ from zero to a maximum given by the wick porosity $\varepsilon$. The portion of the wick volume not occupied by liquid was assumed to be filled with vapour. Eq. (4.101) has been used to predict critical heat fluxes, and a comparison with measured values for water has shown closer agreement than that obtained using Eqs (4.90) and (4.91), although the validity of these two latter equations is not disputed for liquid-metal heat pipes.

#### 4.4.1.4 Gravity-assisted heat pipes

The use of gravity-assisted heat pipes, as opposed to reliance on simple thermosyphons, has been given some attention. The main areas of study have centred on the need to optimise the fluid inventory and to develop wicks which will minimise entrainment. Deverall and Keddy [111] used helical arteries, in conjunction with meshes for evaporator liquid distribution, and obtained relatively high axial fluxes, albeit with sodium and potassium as the working fluids. This, together with the work of Kemme [25], led to the recommendation that more effort be devoted to the vapour flow limitations in gravity-assisted heat pipes. The work was restricted to heat pipes having liquid-metal working fluids, and the significance in water heat pipes is uncertain.

However, for liquid-metal heat pipes, Kemme discusses in some detail a number of vapour flow limitations in heat pipes operating with gravity assistance. As well as the pressure gradient limit discussed above, Kemme presents equations for the viscous limit, described by Busse as follows:

$$\dot{q}_v = \frac{A_v D^2 L}{64\mu_v l_{\text{eff}}} \rho_{\text{ve}} P_{\text{ve}} \tag{4.107}$$

where $A_v$ is the vapour space cross-sectional area, $D$ is the vapour passage diameter and suffix e denotes conditions in the evaporator. Kemme also suggested a modified entrainment limitation to cater for additional buoyancy forces during vertical operation:

$$\dot{Q}_{\max} = A_v L \left[ \frac{\rho_v}{A} \left( \frac{2\pi\gamma}{\lambda} + \rho g D \right) \right]^{1/2} \tag{4.108}$$

where $\lambda$ is the characteristic dimension of the liquid/vapour interface and was calculated as $d_w$ plus the distance between the wires of the mesh wick used. $\rho g D$ is the buoyancy force term.

Abhat and Nguyenchi [104] report work carried out at IKE, Stuttgart on gravity-assisted copper/water heat pipes, retaining a simple mesh wick located against the heat pipe wall. Tests were carried out with heat pipes at a number of angles to the horizontal, retaining gravity assistance, with fluid inventories of up to 5 times that required to completely saturate the wick. Thus a pool of liquid was generally present in at least part of the evaporator section. The basis of the analysis carried out by Abhat and Nguyenchi to compare their experimental results with theory was the model proposed by Kaser [112] which assumed a liquid puddle in the heat pipe varying along the evaporator length from zero at the end of the heat pipes to a maximum at the evaporator exit. Results obtained by varying the operating temperature, tilt angle and working fluid inventory (see Fig. 4.41) were compared with Kaser's model, which was, however, found to be inadequate. Kaser believed that the limiting value of heat transport occurred when the puddle commenced receding from the end of the evaporator. However, the results of Abhat and Nguyenchi indicate that the performance is limited by nucleate boiling in the puddle. Whilst these results are of interest, much more work is needed in this area. Strel'tsov [113] carried out a theoretical and experimental study to determine the optimum quantity of fluid to use in gravity-assisted units. Without quantifying the heat fluxes involved, he derived expressions to permit the determination of the fluid inventory. Of particular interest was his observation that film evaporation, rather than nucleation, occurred under all conditions in the evaporator section, using water and several organic fluids, this seeming to contradict the findings of Abbat and Nguyenchi, assuming that Strel'tsov achieved limiting heat fluxes. Strel'tsov derived the following expression for the optimum fluid inventory for a vertical heat pipe:

$$G = (0.8l_c + l_a + 0.8l_e)\sqrt[3]{\frac{3\dot{Q}\mu_1\rho_1\pi^2 D^2}{Lg}} \tag{4.109}$$

where $\dot{Q}$ is the heat transport (W).

For a given heat pipe design and assumed temperature level, the dependence of the optimum quantity of the working fluid is given by the expression:

$$G = K\sqrt[3]{Q}$$

where $K$ is a function of the particular heat pipe under consideration. The predicted performance of a heat pipe using methanol as the working fluid at a vapour temperature of 55°C is compared with measured values in Fig. 4.42.

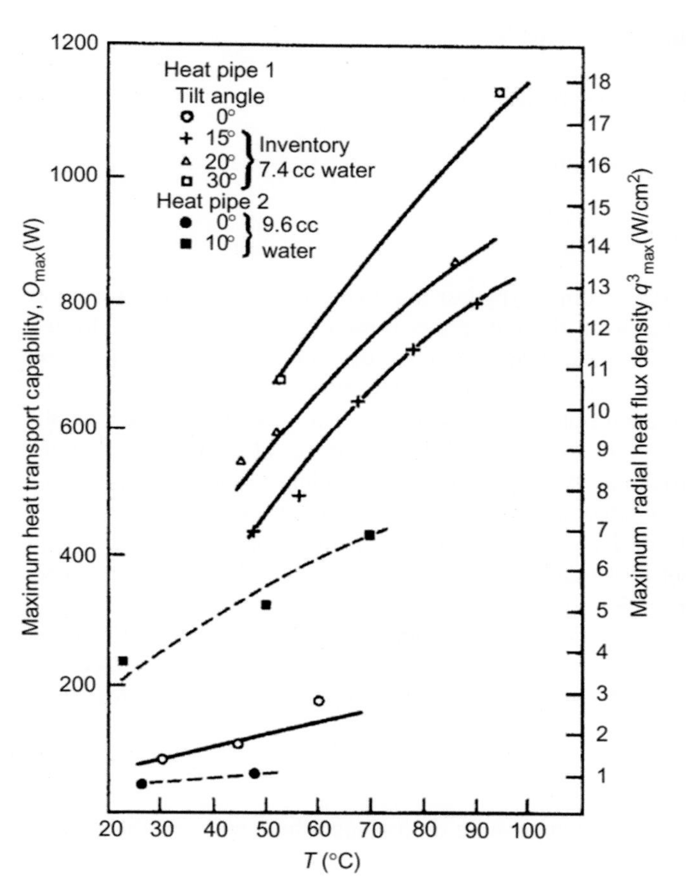


**FIGURE 4.41** The effect of temperature on heat pipe performance for various tilt angles (gravity-assisted angle measured from the horizontal) [105].

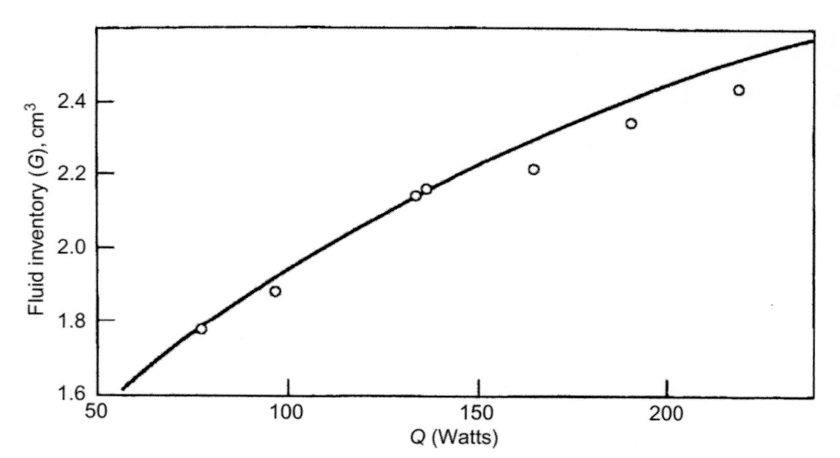


**FIGURE 4.42** The measured effect of fluid inventory on the performance of a methanol heat pipe, and a comparison with results predicted by Eq. (4.109) [113].

The use of arteries in conjunction with grooved evaporator and condenser surfaces has been proposed by Vasiliev and Kiselyov [114]. The artery is used to transfer condensate between the condenser and evaporator, thus providing an entrainment-free path, whilst triangular grooves in the evaporator and condenser wall are used for distribution and collection of the working fluid. It is claimed that this design has a higher effective thermal conductivity than a simple thermal syphon, particularly at high heat fluxes.

The equation developed for the maximum heat flux of such a heat pipe is:

$$\dot{q}_{max}\left(\mathrm{W/m^2}\right) = 3.26 \times 10^{-2} a \cos\frac{\Psi}{2} \cos\theta\theta\,(\Psi)\frac{\sigma\rho_1 L}{\mu_l} \tag{4.110}$$

where $a$ is the groove width, $\Psi$ the groove angle, $\theta$ the contact angle and:

$$f(\Psi) = \left[tg\Psi + c\cdot tg\Psi - \frac{\left(\frac{\pi}{2} - \Psi\right)}{\cos^2\Psi}\right]^3 \cos^2\Psi$$

Other data on gravity-assisted heat pipes, in particular dry-out limits, suggest that two contrasting types of dry-out can be shown to exist. Busse and Kemme [115] state that the 'axial dry-out' arises from a lack of hydrostatic driving force and makes itself evident when the heat flux increases. The so-called 'aximuthal dry-out' is, however, caused by an excess of hydrostatic driving force and was found to be characterised by a concentration of the liquid flow on the lower part of the heat pipe cross-section, without any lack of axial liquid return. It was recommended that graded capillary structures were the best technique for overcoming this limit. Of particular interest is the use of wick-type structures in thermosyphons for enhancing internal heat transfer coefficients. As part of a project leading to the development of improved heat pipe exchangers, UKAEA Harwell inserted longitudinal grooves in the walls of thermosyphons. Using plain bore thermosyphons of near-rectangular cross-section, of 1-m length, with 10% of the vapour space filled with water, a series of measurements of evaporator and condenser heat transfer coefficients was made. Vapour temperatures varied from 40°C to 120°C, with power inputs of 2 and 3 kW [116]. Heat transfer coefficients varied from 15 to 21 kW/m$^2$ K, with a mean for the evaporator of 17,755 W/m$^2$ K and for the condenser of 17,983 W/m$^2$ K. However, with grooves on the inside wall along the complete lm length, tests carried out with 3 kW power input revealed that condenser heat transfer coefficients of 50–100 kW/m$^2$ K were achieved, a mean of 62,061 W/m$^2$ K being recorded.

#### 4.4.1.5 *Total temperature drop*

Referring back to Fig. 4.21, this shows the components of the total temperature drop along a heat pipe and the equivalent thermal resistances.

- $R_1$ and $R_9$ are the normal heat transfer resistances for heating a solid surface and are calculated in the usual way.
- $R_2$ and $R_8$ represent the thermal resistance of the heat pipe wall.
- $R_3$ and $R_7$ take account of the thermal resistance of the wick structure and include any temperature difference between the wall and the liquid together with conduction through the saturated wick. From the discussion in Section 4.3.7.4, it is seen that the calculation of R3 is difficult if boiling occurs. $R_7$ is made up principally from the saturated wick but if there is appreciable excess liquid then a correction must be added.
- $R_4$ and $R_6$ are the thermal resistances corresponding to the vapour–liquid surfaces. They may be calculated from Eq. (4.95) but can usually be neglected.
- $R_5$ is due to the temperature drop $\Delta T_5$ along the vapour column. $\Delta T_5$ is related to the vapour pressure drop $\Delta P$ by the Clapeyron equation.

$$\frac{dP}{dT} = \frac{L}{TV_v}$$

or combining with the gas equation

$$\frac{dP}{dT} = \frac{LP}{RT^2}$$

Hence

$$\Delta T_5 = \frac{RT^2}{LP}\Delta P_v \tag{4.111}$$

where $\Delta P_v$ is obtained from Section 4.3.5, and $\Delta T_5$ can usually be neglected.

- $R_s$ is the thermal resistance of the heat pipe structure, it can normally be neglected but may be important in the start-up of gas-buffered heat pipes.

It is useful to have an indication of the relative magnitude of the various thermal resistances, Table 4.11 (from Asselman and Green [83]) lists some approximate values/cm$^2$ for a water heat pipe.

Expressions for the calculation of thermal resistance are given in Table 4.12.

**TABLE 4.11 Representative values of thermal resistances.**

| Resistance | K (cm$^2$/W) |
|---|---|
| $R_1$ | $10^3$–10 |
| $R_2$ | $10^{-1}$ |
| $R_3$ | 10 |
| $R_4$ | $10^{-5}$ |
| $R_5$ | $10^{-8}$ |
| $R_6$ | $10^{-5}$ |
| $R_7$ | 10 |
| $R_8$ | $10^{-1}$ |
| $R_9$ | $10^3$–10 |

**TABLE 4.12 Summary of Thermal Resistances in a Heat Pipe.**

| Term | Defining relation | Thermal resistance | Comment |
|---|---|---|---|
| 1 | $\dot{Q}_i = \alpha_e A_e \Delta T_1$ | $R_1 = \frac{1}{\alpha_e A_e}$ | |
| 2 | Plane Geometry $\dot{q}_e = \frac{k\Delta T_2}{t}$ Cylindrical Geometry $\dot{q}_e = \frac{k\Delta T_2}{r_2 \ln\frac{r_2}{r_1}}$ | $R_2 = \frac{\Delta T_2}{A_e \dot{q}_e}$ | For thin-walled cylinders $r_2 \ln\frac{r_2}{r_1} = t$ |
| 3 | Eq. (4.95) or (4.96) | $R_3 = \frac{d}{k_w A_e}$ | Correct for liquid metals. Gives upper limit for nonmetallics |
| 4 | $\dot{q}_e = \frac{L^2 P_v \Delta T_4}{(2\pi RT)^{0.5} RT^2}$ | $R_4 = \frac{(2\pi RT)^{0.5} RT^2 L^2 P_v \Delta T_4}{L^2 P_v A_e}$ | Can usually be neglected |
| 5 | $\Delta T_5 = \frac{RT^2 \Delta P_v}{L P_v}$ | $R_5 = \frac{RT^2 \Delta P_v}{\dot{Q} L P_v}$ | Can usually be neglected $\Delta P_v$ from Section 4.3.5 |
| 6 | $\dot{q}_e = \frac{L^2 P_v \Delta T_6}{(2\pi RT)^{0.5} RT^2}$ | $R_6 = \frac{(2\pi RT)^{0.5} RT^2 L^2 P_v \Delta T_4}{L^2 P_v A_c}$ | Can usually be neglected |
| 7 | Eq. (4.95) or (4.96) | $R_7 = \frac{d}{k_w A_c}$ | If excess working fluid is present an allowance should be made for the additional resistance |
| 8 | Plane geometry $\dot{q}_e = \frac{k\Delta T_8}{t}$ Cylindrical Geometry $\dot{q}_e = \frac{k\Delta T_8}{r_2 \ln\frac{r_2}{r_1}}$ | $R_8 = \frac{\Delta T_8}{A_e \dot{q}_e}$ | For thin-walled cylinders $r_2 \ln\frac{r_2}{r_1} = t$ |
| 9 | $\dot{Q}_i = \alpha_e A_e \Delta T_1$ | $R_9 = \frac{1}{\alpha_c A_c}$ | |
| | $\dot{Q}_s = \frac{\Delta T_s}{\left(\frac{l_e + l_a + l_c}{A_w k_w + A_{wall} k}\right)}$ | $R_s = \frac{l_e + l_a + l_c}{A_w k_w + A_{wall} k}$ | Axial heat flow by conduction |

*Notes*: $R_X$ = Thermal Resistance = $\frac{\text{Temperaturedifference}}{\text{Heatflow}} = \left(\frac{\Delta T}{\dot{Q}}\right)_X$, Total heat flow = $\dot{Q}_i = Q + \dot{Q}_s$, $\dot{Q}, \dot{Q}_i$ and $\dot{Q}_s$ are defined in Fig. 4.21. Heat flux = $\dot{q} = \frac{\dot{Q}}{A}$, $\Delta T_X$ and $R_X$ are defined in Fig. 4.21. Other symbols: $\dot{q}_e, \dot{q}_c$ heat flux through the evaporator and condenser walls, $k$ thermal conductivity of the heat pipe wall, $t$ heat pipe wall thickness, $r_1$, $r_2$ inner and outer radii of cylindrical heat pipe, $d$ wick thickness, $k_w$ effective wick thermal conductivity, $P_v$ vapour pressure, $L$ latent heat, $R$ gas constant for vapour, $T$ absolute temperature, $\Delta P_v$ total vapour pressure drop in the heat pipe. $A_w$ cross-sectional area of wick, $A_{wall}$ cross-sectional area of wall.

## 4.4.2 Thermosyphons

### 4.4.2.1 Working fluid selection

Obviously, there is no equivalent of the wicking limit, due to the absence of a wick in a thermosyphon. However, the temperature drop may be appreciable in which case the fluid must be selected to minimise this. A figure of merit $M'$ may be defined for thermosyphon working fluids. This figure of merit, which has the dimensions

$$\frac{\text{kg}}{K^{3/4}s^{5/2}}$$

is defined

$$M' = \left(\frac{Lk_l^3\rho_l^2}{\mu_l}\right)^{\frac{1}{4}} \tag{4.112}$$

where $k_l$ = liquid thermal conductivity.

As with the heat pipe figure of merit, $M'$ should be maximised for optimum performance. Data on $M'$ for a variety of working fluids of use in the vapour temperature range $-60°C$ to 300°C are presented in reference [117] in graphical and tabulated form.

The maximum value of $M'$ for several working fluids together with the corresponding temperature is given in Table 4.13.

The Thermosyphon Merit No is relatively insensitive to temperature, for example, water remains above 4000 for the range 0.01°C–350°C. The figure of merit for the most common working fluids of thermosyphons is shown in Fig. 4.43.

A potential disadvantage of water as a working fluid is that if a thermosyphon may be exposed to temperatures below 0°C prior to operation freezing may impede start-up of the thermosyphon. MacGregor et al. [118] have shown that a water – 5% ethylene glycol gives good performance and further widens the applicability of water as a thermosyphon working fluid. Further data on this work are given in a later section of this chapter.

### 4.4.2.2 Entrainment limit

This limit is described by Nguyen-Chi and Groll [119] where it is also called the flooding limit – as the condenser becomes flooded – and in the ESDU Data Item on thermosyphons [120] – where it is additionally described as the counter-current flow limit. A correlation derived in reference [121] gives good agreement with the available experimental data. For thermosyphons, this gives the maximum axial vapour mass flux as follows:

$$Q_{\max}/AL = f_1 f_2 f_3 (\rho_v)^{0.5} [g(\rho_l - \rho_v)\sigma_l]^{0.25} \tag{4.113}$$

where $f_1$ is a function of the Bond number, defined as follows:

$$Bo = D\frac{g(\rho_\varepsilon - \rho_v)^{0.5}}{\rho_\varepsilon} \tag{4.114}$$

Values of $f_1$ can be obtained from Fig. 4.44, where it is plotted against Bo.

**TABLE 4.13** Maximum Thermosyphon Merit Number for Selected Fluids.

| Fluid | Temperature °C | $M'$ max $\frac{kg}{K^{3/4}s^{5/2}}$ |
|---|---|---|
| Water | 180 | 7542 |
| Ammonia | −40 | 4790 |
| Methanol | 145 | 1948 |
| Acetone | 0 | 1460 |
| Toluene | 50 | 1055 |

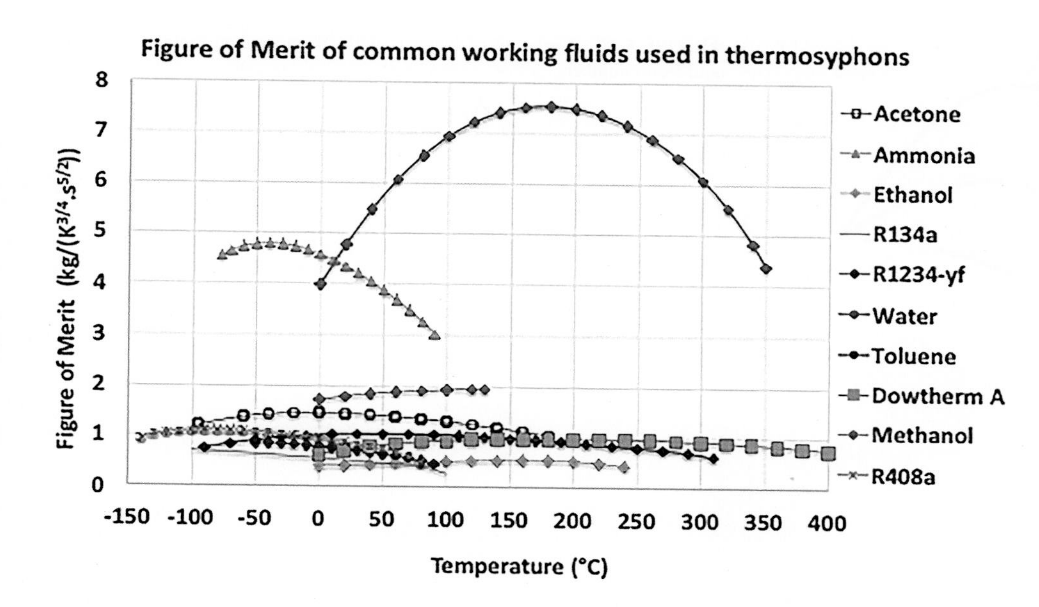


**FIGURE 4.43** Figure of merit of common working fluids used in thermosyphons.

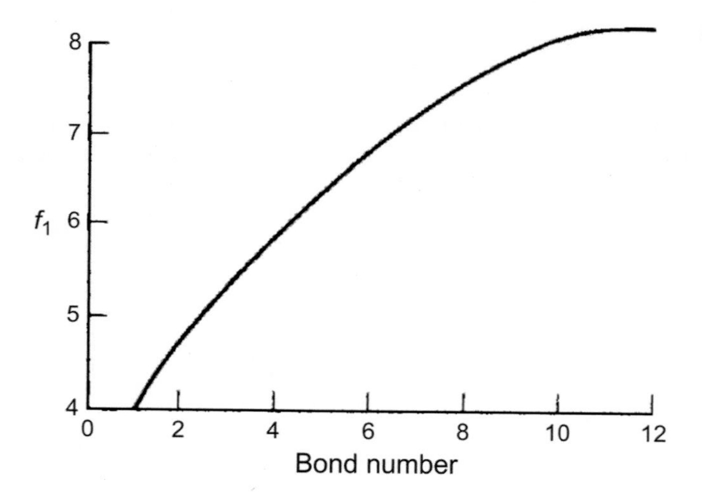


**FIGURE 4.44** Variation of factor $f_1$ with Bond number.

The factor $f_2$ is a function of a dimensionless pressure parameter, $K_p$, where $K_p$ is defined in [119] as follows:

$$K_P = \frac{P_v}{g(\rho_\varepsilon - \rho_v)\delta_1{}^{0.5}} \tag{4.115}$$

$$\text{and } f_2 = K_p^{-0.17} \text{ when } K_P \leq 4 * 10^4$$

$$f_2 = 0.165 \text{ when } K_P > 4 * 10^4$$

The factor $f_3$ is a function of the inclination angle of the thermosyphon. When the thermosyphon is operating vertically, $f_3 = 1$; when the unit is inclined, the value of $f_3$ may be obtained by reading Fig. 4.45, where it is plotted against the inclination angle to the horizontal, Ø, for various values of the Bond number. (Note that the product $f1$, $f2$, $f3$ is sometimes known as the Kutateladze number.)

Note that much of the experimental evidence on the performance of thermosyphons indicates that the maximum heat transport capability occurs when the thermosyphon is at 60 degrees–80 degrees to the horizontal – that is, not vertical. Data are given by Groll [122] and Terdtoon et al. [121] concerning this behaviour.

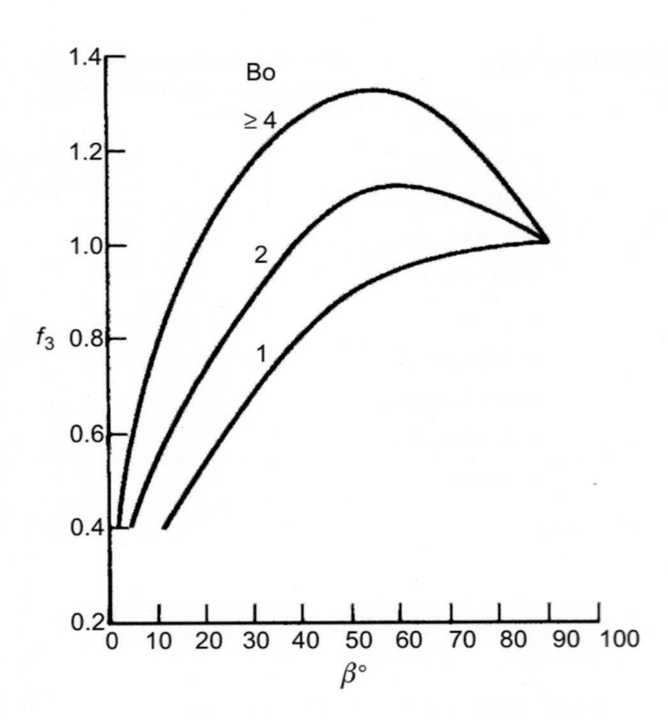


**FIGURE 4.45** Variation of factor $f_3$ with thermosyphon inclination angle and Bond number.

### *4.4.2.3 Thermal resistance and maximum heat flux*

Payakaruk et al. [123] have investigated the influence of inclination angle on thermosyphon performance and have correlated the minimum resistance $R_m$ to the resistance of the thermosyphon, $R_{90}$, in the vertical position. This correlation is shown in equation Eq. (4.116) (Fig. 4.46).

$$R_m/R_{90} = 0.647Ku^{0.0297} \tag{4.116}$$

where

$$Ku = \frac{\dot{q}}{L\rho_v\left[\sigma_l g(\rho_l - \rho_v)/\rho_v^2\right]^{0.25}}$$

and

$$Ku^{**} = Ku \times \frac{l_e}{2r_v} \times \frac{\rho_l}{\rho_v}$$

Golobic and Gaspersic [124] have conducted a review of correlations for the prediction of therosyphon maximum heat flux and developed a correlation based on the principle of corresponding states. This formulation has the advantage of requiring little data regarding fluid properties – only molecular weight, critical temperature, critical pressure and Pitzer acentric factor are needed. The acentric factor may be calculated from

$$w = -1 - \log\left(\frac{p}{p_{0(Tr=0.7)}}\right) \tag{4.117}$$

hence requiring knowledge of the vapour pressure at a reduced pressure of 0.7 (Table 4.14).

## 4.5 Nanofluids

There has been considerable interest in recent years in the use of nanofluids in various heat transfer applications. Nanofluids were first defined by Choi [138] in 1995 as base fluids containing a suspension of nanometre-sized particles. The base fluid is typically a conventional heat transfer fluid such as water, ethylene glycol and oils. The range of nanoparticles used includes pure metals, metal oxides, carbides and nitrides and various forms of carbon. Liu and Li [139] reviewed the published applications of nanofluids to heat pipes and identified some 40 experimental studies relating to

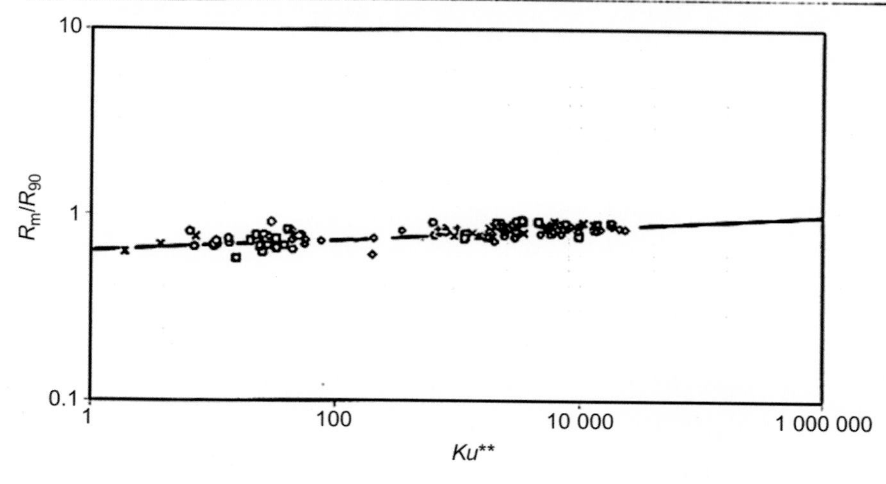


**FIGURE 4.46** $(R_m/R_{90})$ as a function of $Ku^{**}$ for a range of operating conditions compared with Eq. (4.116) [123].

various types of heat pipe including loop and oscillating heat pipes discussed in Chapter 2. The majority of studies show that under at least some conditions the thermal resistance of the heat pipes or thermosyphons tested was lower with the addition of nanoparticles to the base fluid compared to that measured with the base fluid alone.

The reported results are summarised in Table 4.15, together with additional selected papers published after Ref. [139].

The mechanisms through which a small percentage of nanoparticles are added to a fluid are not fully explained and it is not presently possible to predict the enhanced performance. Although there is an increase in thermal conductivity associated with the addition of metallic particles, the improvement in performance which may be achieved is considerably higher than that which would be expected from consideration of the thermal transport properties alone. The presence of nanoparticles in a fluid decreases the contact angle between a surface and the fluid. It has been observed through scanning electron microscopy that after operation the internal surfaces of a heat pipe, including the mesh, become covered with a thin layer comprising deposited nanoparticles. It is suggested that this may further reduce the contact angle. The capillary driving force in a mesh, microgrooved or arterial wick heat pipe is thus increased.

The deposition of nanoparticles on the heat transfer surfaces is generally observed to decrease the thermal resistance within the evaporator and condenser. Putra et al. [156] examined the performance of a range of fluids and concentrations on the performance of a 200 mm long 7.4 mm diameter heat pipe with a four-layer plaited mesh wick and from their results, reproduced in Table 4.16, it can be seen that the overall resistance of the heat pipe is reduced under almost all conditions. Examination of temperature profiles in the heat pipe, shown Fig. 4.47, indicates that the heat transfer coefficients in both evaporator and condenser were significantly increased by the addition of nanoparticles to the water base fluid. The variation of the evaporator heat transfer coefficient is illustrated in Fig. 4.48.

The surface effect will be discussed further in Chapter 2, with reference to the performance of Oscillating heat pipes. On the negative side, increasing the concentration of nanoparticles leads to an increase in density and viscosity; hence, increasing concentration causes the liquid flow resistance to increase. This explains why there appears to be an optimum concentration, above which there is no further improvement in performance. This also explains the cases where the measured performance is observed to decline with the presence of nanoparticles, for example in sintered wicks.

## 4.6 Design guide

### 4.6.1 Introduction

The design of a heat pipe or thermosyphon to fulfil a particular duty involves four broad processes:

i. Selection of appropriate type and geometry
ii. Selection of candidate materials
iii. Evaluation of performance limits

**TABLE 4.14 Correlations for the Maximum Heat Flux in Thermosyphons [125].**

| References | Correlations Kutatladze number $Ku = \frac{q_{co}}{\Delta h_{1V}\rho_V^{0.5}[\sigma g(\rho_1-\rho_v)]^{0.25}}$ |
|---|---|
| Sakhuja [126] | $Ku = \frac{0.725^2}{4} Bo^{1/2} \frac{(d_e/l_e)}{\left[1+\left(\frac{\rho_v}{\rho_l}\right)^{\frac{3}{4}}\right]^2}$ |
| Nejat [127] | $Ku = 0.09 Bo^{1/2} \frac{(d_e/l_e)^{0.9}}{\left[1+\left(\frac{\rho_v}{\rho_l}\right)^{\frac{1}{4}}\right]^2}$ |
| Katto [128] | $Ku = \frac{0.1}{1+0.491(d_e/l_e)Bo^{-0.3}}$ |
| Tien and Chung [129] | $Ku = \frac{3.2}{4} \frac{(d_e/l_e)^{0.9}}{\left[1+\left(\frac{\rho_v}{\rho_l}\right)^{\frac{1}{4}}\right]^2}; Bo \geq 30$ |
| | $Ku = \frac{3.2}{4}\left[\tanh\left(0.5 Bo^{\frac{1}{2}}\right)\right]^2 \frac{(d_e/l_e)^{0.9}}{\left[1+\left(\frac{\rho_v}{\rho_l}\right)^{\frac{1}{4}}\right]^2}$ |
| Harada et al. [130] | $Ku = 9.64(d_e/l_e)\rho_v h_{1V} C\left(\frac{\sigma}{\rho_v}\right)^{\frac{1}{2}}; C\left(\frac{\sigma}{\rho_v}\right)^{\frac{1}{2}} \geq 0.079$ |
| | $Ku = 14.1(d_e/l_e)\rho_v h_{1V}\left[C\left(\frac{\sigma}{\rho_v}\right)^{\frac{1}{2}}\right]^{1.15}; C\left(\frac{\sigma}{\rho_v}\right)^{\frac{1}{2}} < 0.079$ |
| | $C = 1.58\left(\frac{d_e}{\sigma}\right)^{0.4}; (d_e/\sigma) < 0.318 C = 1; (d_e/\sigma) \geq 0.318$ |
| Gorbis and Savchenkov [131] | $Ku = 0.0093(d_e/l_e)^{1.1}(d_e/l_e)^{-0.88} Fe^{-0.74}(1+0.03 Bo)^2; 2 < Bo < 60$ |
| Bezrodnyi [132] | $Ku = 2.55(d_e/l_e)\left\{\frac{\sigma}{P}\left[\frac{g(\rho_l-\rho_v)}{\sigma}\right]^{0.5}\right\}^{0.17} \frac{\sigma}{P}\left[\frac{g(\rho_l-\rho_v)}{\sigma}\right]^{0.5} \geq 2.5 * 10^{-5}$ |
| | $Ku = 0.425(d_e/l_e)\frac{\sigma}{P}\left[\frac{g(\rho_l-\rho_v)}{\sigma}\right]^{0.5} < 2.5 * 10^{-5}$ |
| Groll and Rosler [133] | $Ku = f_{1(Bo)} f_2 f_{3(\varphi,Bo)}(d_e/l_e)$ |
| | $f_2 = \left\{\frac{\sigma}{P}\left[\frac{g(\rho_l-\rho_v)}{\sigma}\right]^{0.5}\right\}^{0.17}; \frac{\sigma}{P}\left[\frac{g(\rho_l-\rho_v)}{\sigma}\right]^{0.5} \geq 2.5 * 10^{-5}$ |
| | $f_2 = 0.165; \frac{\sigma}{P}\left[\frac{g(\rho_l-\rho_v)}{\sigma}\right]^{0.5} < 2.5 * 10^{-5}$ |
| Prenger [134] | $Ku = 0.747(d_e/l_e)[g\sigma(\rho_l-\rho_v)^{0.295}(\rho_v h_{1V})^{-0.045}$ |
| Fukano et al. [135] | $Ku = 2(d_e/l_e)^{0.83} Fe^{0.03}\left\{\frac{[g\sigma(\rho_l-\rho_v)]^{0.5}}{\rho_v h_{1V}}\right\}^{0.2}$ |
| Imura et al. [136] | $Ku = 0.16\{1-\exp[-(d_e/l_e)(\rho_l/\rho_v)^{0.13}]\}$ |
| Pioro and Voroncova [124] | $Ku = 0.131\left\{1-\exp\left[-(d_e/l_e)\left(\frac{\rho_l}{\rho_v}\right)^{0.13}\cos^{1.8}(\varphi-55°)\right]\right\}^{0.8}$ |
| Golobic and Gaspersic [137] | $q_{co} = \frac{0.16 d_e Tc^{1/3} Pc^{11/12} g^{1/4}}{l_e M^{1/4}} \tau \exp(2.530 - 8.233\tau + 1.387\omega + 17.096\tau^2 - 4.944\tau\omega^2 + 15.542\tau^2\omega^2 - 23.989\tau^3 - 19.235\tau^3\omega - 18.071\tau^3\omega^2)$ |
| | $q_{co} = \frac{0.16 d_e Tc^{1/3} Pc^{11/12} g^{1/4}}{l_e M^{1/4}} \tau \exp(2.530 - 13.137\tau^2)$ |

*Notes:* $Bo$ Bond number $= d_e/\left[g\left[\frac{g(\rho_l-\rho_v)}{\sigma}\right]^{0.5}\right.$, $C$ Parameter in Equation (T6), $d_e$ Evaporator diameter, m, $f_1, f_2, f_3$ Parameter in Equation (T9), $F_e$ Fill ratio, liquid fill volume/evaporator volume, $g$ Gravitational acceleration, m/s$^2$, $Ku$ Kutateladze number $= q_{co}/\{\Delta h_{1V}\rho_v^{0.5}[\sigma g(\rho_l-\rho_v)]^{0.25}\}$, $l_e$ Evaporator length, m, $M$ Molecular weight, kg/kmol, $P_c$ Critical pressure, Pa, $q_{co}$ Maximum heat flux, W/m$^2$, $\tau$ Temperature function = 1-$T_\tau$, $T$ Temperature, K, $T_\tau$ Reduced temperature = T/T$_c$, $T_c$, Critical temperature, K, $x_1$ Fluid properties parameter, Eq. (4.2), $x_2$ Density ratio parameter, Eq. (4.3), $\Delta h_{1V}$ Heat of evaporation, J/kg, $\varphi$ Inclination angle, °$\mu$ Dipole moment, debye, $\rho_l$ Liquid density, kg/m$^3$, $\rho_v$ vapor density, kg/m$^3$, $\sigma$ Surface tension, N/m, $\omega$ Pitzer acentric factor.

**iv.** Evaluation of the actual performance

The background to each of these stages is covered in Chapter 3. In this chapter, the theoretical and practical aspects are discussed with reference to sample design calculations.

### 4.6.2 Heat pipes

The design procedure for a heat pipe is outlined in Fig. 4.49. As with any design process, many of the decisions that must be taken are interrelated and the process is iterative. For example, choice of the wick and case material eliminates many candidate working fluids (often including water) due to compatibility constraints. If the design then proves inadequate with the available fluids, it is necessary to reconsider the choice of construction materials.

Two aspects of practical design, which must also be taken into consideration, are the fluid inventory and performance at off design conditions, particularly during the start-up of the heat pipe.

**TABLE 4.15 Summary of Researches of Heat Pipes Using Nanofluids [139].**

| Type and shape of heat pipe | | Researcher | Working liquid type (nanoparticles size and optimal concentration) | Effect |
|---|---|---|---|---|
| Miniature microgrooved heat pipe | Disc shaped | Chien et al. [140] | Au–water (17 nm) | + |
| | Cylindrical | Wei et al. [141] | Ag–water (10 nm, 0.01 wt.%) | + |
| | Cylindrical | Kang et al. [142] | Ag–water (35 nm, 0.01 wt.%) | + |
| | Cylindrical | Yang et al. [143] | CuO–water (50 nm, 1.0 wt.%) | + |
| | Cylindrical | Liu and Lu [144] | CNT–water (diameter: 15 nm, length: 5–15 μm, 2.0 wt.%) | + |
| | Flat | Do and Jang [145] | $Al_2O_3$–water (38.4 nm, 0.8 wt.%) | + |
| | Cylindrical | Shafahi et al. [146] | CuO–water, $Al_2O_3$–water, $TiO_2$–water | + |
| | Disc shaped | Shafahi et al. [147] | CuO–water, $Al_2O_3$–water, $TiO_2$–water | + |
| | Cylindrical | Liu et al. [148] | CuO–water (50 nm, 1.0 wt.%) | + |
| | Cylindrical | Wang et al. [149] | CuO–water (50 nm, 1.0 wt.%) | + |
| Mesh wick heat pipe | Cylindrical | Tsai et al. [150] | Au–water (21 nm) | + |
| | Cylindrical | Liu et al. [151] | CuO–water (50 nm, 1.0 wt.%) | + |
| | Flat | Chen et al. [152] | Ag–water (35 nm, 0.01 wt.%) | + |
| | Cylindrical | Do et al. [153] | $Al_2O_3$–water (30 nm, 2.4 wt.%) | + |
| | Cylindrical | Liu et al. [154] | CuO–water (50 nm, 1.0 wt.%) | + |
| | Cylindrical | Senthikumar et al. [155] | Cu-*n*-aqueous soln *n*-hexanol (40 nm, 0.01 wt.%) | + |
| | Cylindrical | Putra et al. [156] | $Al_2O_3$–water (1–5 vol.%) | + |
| | | | $Al_2O_3$–ethylene glycol (1–5 vol.%) | + |
| | | | $TiO_2$–water (1–5 vol.%) | + |
| | | | $TiO_2$–ethylene glycol (1–5 vol.%) | + |
| | | | ZnO–ethylene glycol (1–5 vol.%) | + |
| | Cylindrical | Wang et al. [157] | | |
| Sintered metal wick heat pipe | Loop heat pipe | Riehl [158] | Ni–water (40 nm, 3.5 wt.%) | – |
| | Cylindrical | Kang et al. [159] | Ag–water (10 nm, 0.01 wt.%) | + |
| | Cylindrical | Morajevi et al. [160] | $Al_2O_3$–water (35 nm, 1–3 wt.%) | + |

(Continued)

**TABLE 4.15 (Continued)**

| Type and shape of heat pipe | | Researcher | Working liquid type (nanoparticles size and optimal concentration) | Effect |
|---|---|---|---|---|
| Oscillating heat pipe | Closed-loop OHP | Ma et al. [161] | Diamond–water (20 and 40 nm, 2.2 wt.%) | + |
| | Closed-loop OHP | Ma et al. [162] | Diamond–water (20 and 40 nm, 2.2 wt.%) | + |
| | Closed-loop OHP | Shang et al. [163] | Cu–water (25 nm, 0.45 wt.%) | + |
| | Closed-loop OHP | Lin et al. [164] | Ag–water (20 nm, 0.1 wt.%) | + |
| | Closed-loop OHP | Park and Ma [165] | CuNi–water (40–150 nm, 8.8 wt.%) | + |
| | Closed-loop OHP | Qu et al. [166] | $Al_2O_3$–water (56 nm, 0.9 wt.%) | + |
| | Closed-loop OHP | Bhuwakietkumjohn and Rittidech [167] | Ag–ethanol | + |
| | Flat plate Closed-loop OHP | Cheng et al. [168] | Diamond–acetone (2 nm, 0.33 wt.%) | + |
| | | | Au–water (3 nm, 0.006 wt.%) | + |
| | | | Diamond–water (2 nm, 2.6 wt.%) | – |
| | Closed-loop OHP | Ji et al. [169] | $Al_2O_3$–water (20–50 μm, 0.5 wt.%) | + |
| | Open-loop PHP | Taslimafar et al. [170] | $Fe_3O_4$. water (25 nm) | + |
| | Open-loop PHP | Riehl et al. [171] | Cu–water (29 nm, 5 wt.%) | + |
| Closed two-phase thermosyphon | Cylindrical | Peng et al. [172] | Al–water (30 nm, 7.8 wt.%) | + |
| | Cylindrical | Xue et al. [173] | CNT–water (15 nm, 2.2 wt.%) | – |
| | Cylindrical | Liu et al. [174] | CuO–water (30 nm, 1.0 wt.%) | + |
| | Flat | Liu et al. [175] | CuO–water (15–50 nm, 1.0 wt.%) | + |
| | Cylindrical | Khandekar et al. [176] | $Al_2O_3$–water (40–47 nm, 1.0 wt.%), CuO–water (8.6–13 nm, 1.0 wt.%) | – |
| | | | Laponite clay-water (25 nm, 1.0 wt.%) | – |
| | Cylindrical | Naphon et al. [177] | Ti–alcohol (21 nm, 0.57 wt.%) | + |
| | Cylindrical | Naphon et al. [178] | Ti-refrigerant R11 (21 nm, 0.31 wt.%) | + |
| | Cylindrical | Noie et al. [179] | $Al_2O_3$–water (20 nm, 3.0 wt.%) | + |
| | Cylindrical | Liu et al. [180] | CNT–water (15 nm, 2.0 wt.%) | + |
| | Cylindrical | Parametthanuwat et al. [181] | Ag–water (<100 nm, 0.5 wt.%) | + |
| | Cylindrical | Parametthanuwat et al. [182] | Ag–water (<100 nm, 0.5 wt.%) | + |
| | Cylindrical | Huminic et al. [183] | Iron oxide–water (4.5 nm, 20 wt.%) | + |
| | Cylindrical | Huminic et al. [184] | Iron oxide–water (4.5 nm, 20 wt.%) | + |
| | Cylindrical | Qu and Wu [185] | $SIO_2$–water, $Al_2O_3$–water (30–56 nm, 0.1–1.2 wt.%) | + |
| | Cylindrical | Hung et al. [186] | $Al_2O_3$–water (10–30 nm, 0.5–3 wt.%) | + |

'+' means heat transfer enhancement and '-' means heat transfer reduction.

**TABLE 4.16** Thermal Resistance of a heat pipe with various working fluids [156].

| No | Working fluids | Thermal resistance $R_{e-a}$(°C/W) | | |
|---|---|---|---|---|
| | | 10 W | 20 W | 30 W |
| 1 | Ethylene glycol | 2.44 | 2.28 | 2.16 |
| 2 | Water | 2.36 | 2.20 | 2.11 |
| 3 | $Al_2O_3$–ethylene glycol | 2.18 | 2.02 | 1.86 |
| 4 | $TiO_2$–ethylene glycol | 2.38 | 2.23 | 2.11 |
| 5 | ZnO–ethylene glycol | 1.82 | 1.67 | 1.55 |
| 6 | $Al_2O_3$–water | 0.51 | 0.36 | 0.26 |
| 7 | $TiO_2$–Water | 0.76 | 0.63 | 0.54 |

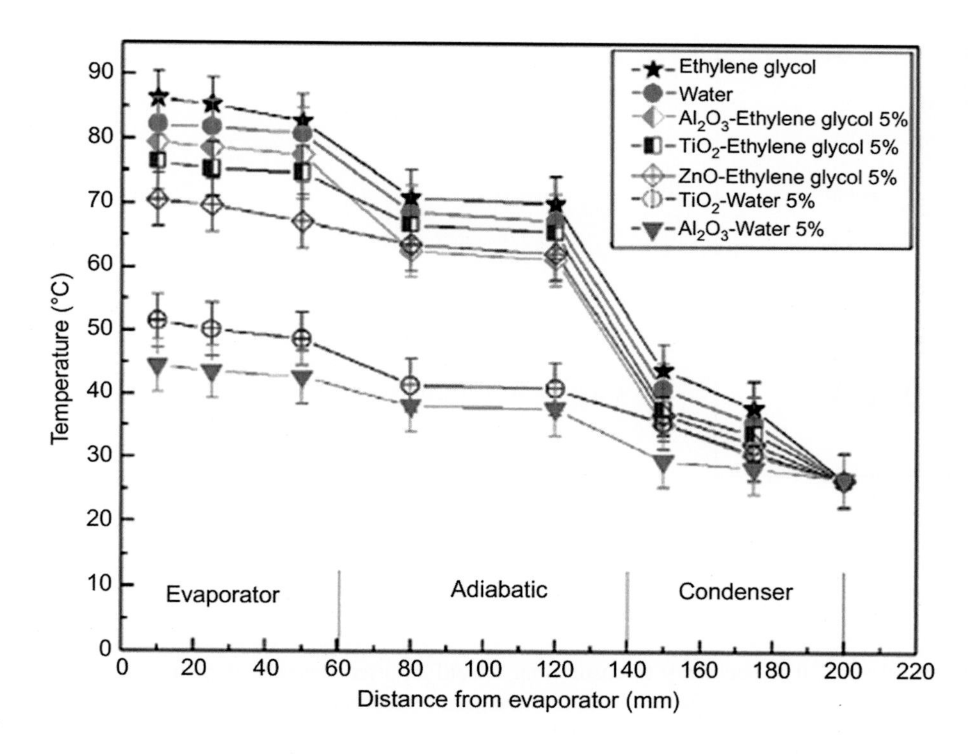


**FIGURE 4.47** Temperature distributions for water, ethylene glycol and ethylene-based nanofluids with 5% volume of different types of nanoparticles of heat load 30 W [156].

### 4.6.2.1 *Fluid inventory*

A feature of heat pipe design, which is important when considering small heat pipes and units for space use, is the working fluid inventory. It is a common practice to include a slight excess of working fluid over and above that required to saturate the wick, but when the vapour space is of small volume a noticeable temperature gradient can exist at the condenser, similar to that which would be observed if a noncondensable gas was present. This reduces the effective length of the condenser, hence impairing heat pipe performance. Another drawback of excess fluid is peculiar to heat pipes in space; in zero gravity the fluid can move about the vapour space, affecting the dynamics of the spacecraft.

If there is a deficiency of working fluid, the heat pipe may fail because of the inability of any artery to fill. This is not so critical with homogeneous wicks as some of the pores will still be able to generate capillary action. Marcus [187] discusses in detail these effects and the difficulties encountered in ensuring whether the correct amount of working fluid is injected into the heat pipe. One way of overcoming the problem is to provide an excess fluid reservoir, which behaves as a sponge, absorbing working fluid that is not required by the primary wick structure.

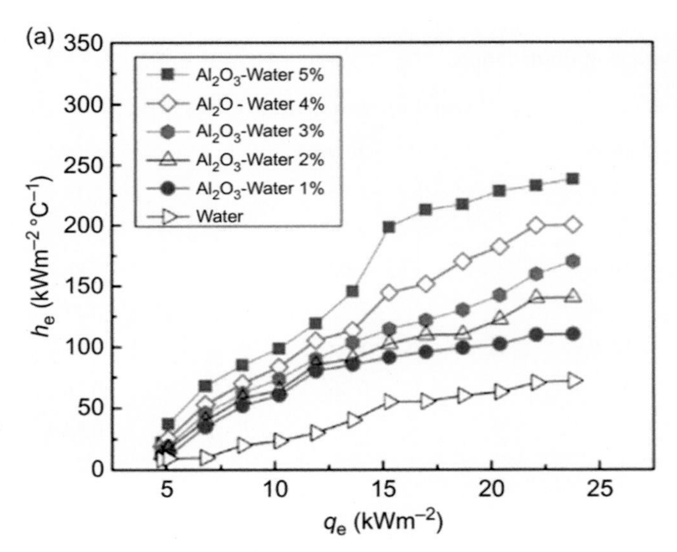


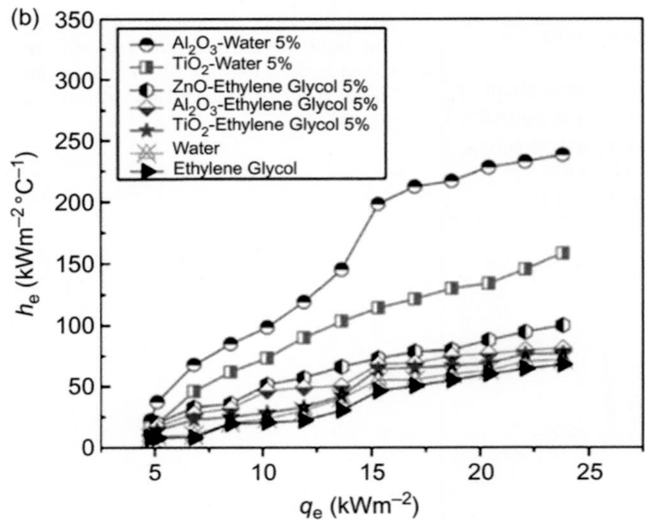


**FIGURE 4.48** (a) Heat transfer coefficients of the evaporator with different concentrations of water-based $Al_2O_3$ nanofluid. (b) Heat transfer coefficients of the evaporator with different type of nanofluids [156].

### 4.6.2.2 *Priming*

With heat pipes having some form of arterial wick, it is necessary to ensure that should an artery become depleted of working fluid, it should be able to refill automatically. It is possible to calculate the maximum diameter of an artery to ensure that it will be able to reprime. The maximum priming height that can be achieved by a capillary is given by the equation.

$$h + h_c = \frac{\sigma_1 \cos\theta}{(\rho_1 - \rho_v)g} \times \left(\frac{1}{r_{p1}} + \frac{1}{r_{p2}}\right) \tag{4.118}$$

where $h$ is the vertical height to the base of the artery, $h_c$ the vertical height to the top of the artery, $r_{p1}$ the first principal radius of curvature of the priming meniscus and $r_{p2}$ is the second principal radius of curvature of the priming meniscus.

For the purpose of priming, the second principal radius of curvature of the meniscus is extremely large (approximately equal to sin $\phi$). For a cylindrical artery

$$h_c = d_a$$

and

$$r_{p1} = \frac{d_a}{2}$$

where $d_a$ is the artery diameter.

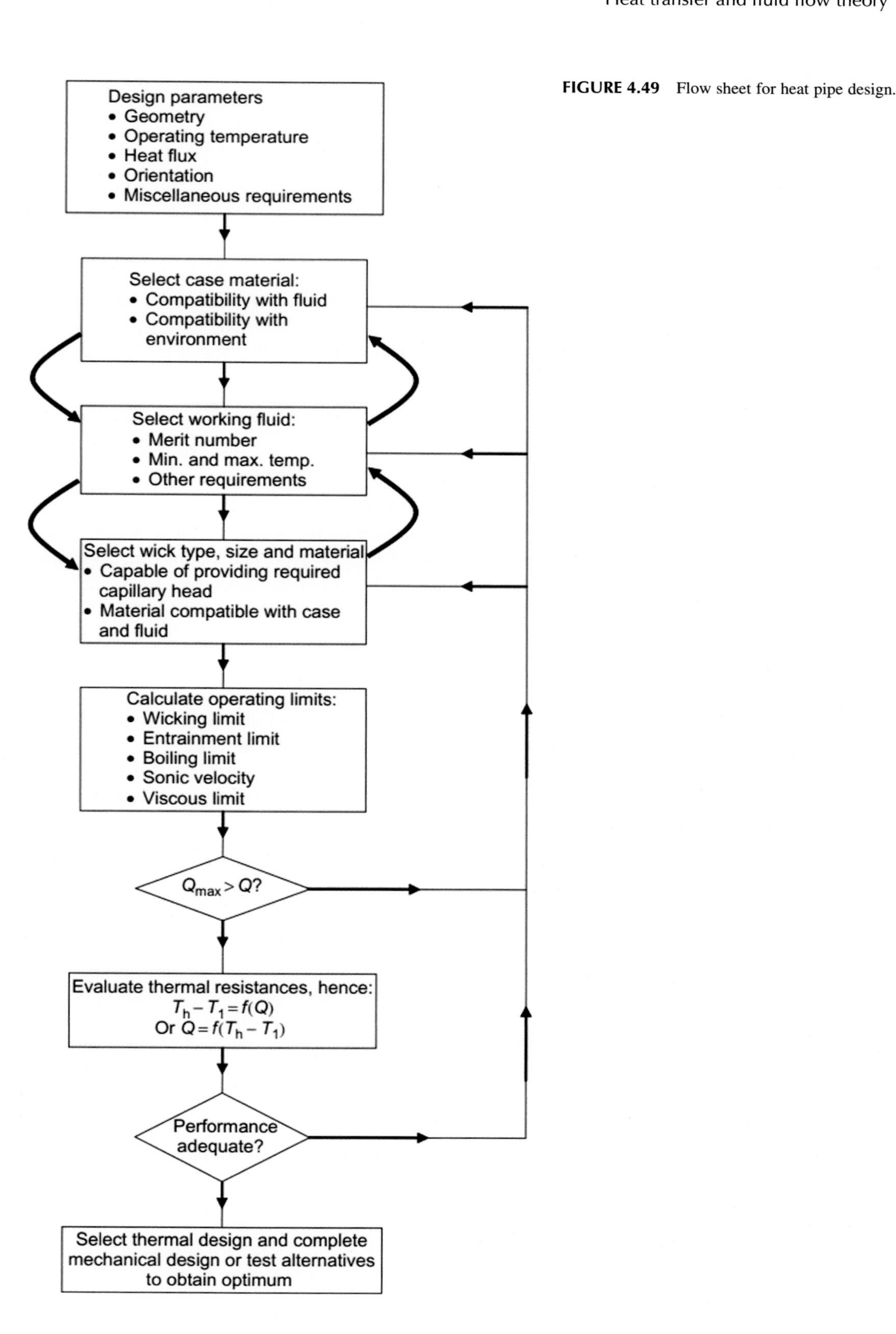


**FIGURE 4.49** Flow sheet for heat pipe design.

Hence the above equation becomes:

$$h + d_a = \frac{2\sigma_1 \cos\theta}{(\rho_1 - \rho_v)g \times d_a} \quad (4.119)$$

which produces a quadratic in $d_a$ that may be solved as:

$$d_a = \frac{1}{2}\left[\left(\sqrt{h^2 + \frac{8\sigma_1 \cos\theta}{(\rho_1 - \rho_v)g}}\right) - h\right] \quad (4.120)$$

An artery can deprime when vapour bubbles become trapped in it. It may be necessary to reduce the heat load in such circumstances, to enable the artery to reprime. It is possible to design a heat pipe incorporating a tapered artery, effectively a derivative of the mono-groove wick system illustrated in Chapter 3. Use of this design [188] facilitated bubble venting into the vapour space and achieved significant performance improvements.

### 4.6.3 Design example 1

#### 4.6.3.1 *Specification*

A heat pipe is required which will be capable of transferring a minimum of 15 W at a vapour temperature between 0°C and 80°C over a distance of 1 m in zero gravity (a satellite application). Restraints on the design are such that the evaporator and condenser sections are each to be 8 cm long, located at each end of the heat pipe, and the maximum permissible temperature drop between the outside wall of the evaporator and the outside wall of the condenser is 6°C. Because of weight and volume limitations, the cross-sectional area of the vapour space should not exceed 0.197 $cm^2$. The heat pipe must also withstand bonding temperatures.

#### 4.6.3.2 *Selection of materials and working fluid*

The operating conditions are contained within the specification. The selection of wick and wall materials is based on the criteria discussed in Chapter 3. As this is an aerospace application, low mass is an important factor.

On this basis, aluminium alloy 6061 (HT30) is chosen for the wall and stainless steel for the wick. If it is assumed that the heat pipe will be of a circular cross-section, the maximum vapour space area of 0.197 $cm^2$ yields a radius of 2.5 mm.

Working fluids compatible with these materials, based on available data, include

- **i.** Freon 11
- **ii.** Freon 113
- **iii.** Acetone
- **iv.** Ammonia

Water must be dismissed at this stage, both on compatibility grounds and because of the requirement for operation at 0°C with the associated risk of freezing. Note that the Freon refrigerants are chlorofluorocarbons (CFCs) and are no longer available but are included in the evaluation in order to demonstrate the properties of a range of fluids. The operating limits for each fluid must now be examined.

##### 4.6.3.2.1 Sonic limit

The minimum axial heat flux due to the sonic limitation will occur at the minimum operating temperature, 0°C, and can be calculated from Equation 2.58 with the Mach number set to unity.

$$\dot{q}_s = \rho_v L \sqrt{\frac{\gamma R T_v}{2(\gamma + 1)}}$$

The gas constant for each fluid may be obtained from

$$R = \frac{R_o}{\text{Molecular weight}} = \frac{8315}{M_w} J/\text{kg } K$$

For ammonia, $\gamma = 1.4$ [189] and its molecular weight is 31, the latent heat and density may be obtained from Appendix 5, therefore $\dot{q}_s$ is given by

$$\dot{q}_s = 3.48 \times 1263 \times \sqrt{\frac{1.4}{2(1.4+1)} \times \frac{8315 \times 273}{31}} = 84 \times 10^7 \mathrm{W/m^2} = 84\ \mathrm{kw/cm^2}$$

Similar calculations may be carried out for the other candidate fluids, yielding

**i.** Freon 11 0.69 kW/cm$^2$
**ii.** Freon 113 3.1 kW/cm$^2$
**iii.** Acetone 1.3 kW/cm$^2$
**iv.** Ammonia 84 kW/cm$^2$

Since the required axial heat flux is 15/0.197 W/cm$^2$ $\equiv$ 0.076 kW/cm$^2$, the sonic limit would not be encountered for any of the candidate fluids.

It is worth noting that the term $\sqrt{\gamma/(\gamma+1)}$ varies from 0.72 to 0.79 for values of $\gamma$ from 1.1 to 1.66; therefore, when the sonic limit is an order of magnitude above the required heat flux it is not essential that $\gamma$ be known precisely for the fluid.

#### 4.6.3.2.2 Entrainment limit

The maximum heat transport due to the entrainment limit may be determined from Eq. (4.65):

$$\dot{q} = \sqrt{\frac{2\pi\rho_v L^2 \sigma_1}{z}}$$

which may be written as follows:

$$\dot{Q}_{\mathrm{ent}} = \pi r_v^2 L \sqrt{\frac{2\pi\rho_v\sigma_1}{z}}$$

where $z$ is the characteristic dimension of the liquid—vapour interface for a fine mesh that may be taken as 0.036 mm. The entrainment limit is evaluated at the highest operating temperature.

The fluid properties of the fluids and the resulting entrainment limits are given in Table 4.17.

A sample calculation for acetone is reproduced below. Particular care must be taken to ensure that the units used are consistent.

$$L = \mathrm{J/kg}\ \sigma_l = \mathrm{N/m}\ \rho_v = \mathrm{kg/s}\ r_v = \mathrm{m}\left(\text{or } A = \mathrm{m}^2\right) z = \mathrm{m}$$

Some useful conversion factors are given in the appendices.

Then we have

$$\mathrm{m}^2 \times \frac{\mathrm{J}}{\mathrm{kg}} \times \sqrt{\frac{\mathrm{kg}}{\mathrm{m}^3}\frac{\mathrm{N}}{\mathrm{m}}\frac{1}{\mathrm{m}}} \equiv \mathrm{m}^2 \times \frac{\mathrm{J}}{\mathrm{kg}} \times \sqrt{\frac{\mathrm{kg}}{\mathrm{m}^3}\frac{\mathrm{kg} \times \mathrm{m}}{\mathrm{m} \times \mathrm{s}^2}\frac{1}{\mathrm{m}}} \equiv \frac{\mathrm{J}}{\mathrm{s}} = \mathrm{W}$$

$$\dot{Q}_{\mathrm{ent,acetone}} = \pi r_v^2 L \sqrt{\frac{2\pi\rho_v\sigma_1}{z}} \pi \times (2.5 \times 10^{-3})^2 \times 495 \times 10^3 \sqrt{\frac{2\pi \times 4.05 \times 0.0162}{0.036 \times 10^{-3}}} = 1040\mathrm{W}$$

**TABLE 4.17 Properties of Candidate Fluids at 80°C.**

| Fluid | $L$ (kJ/kg) | $\sigma_l$ (mN/m) | $\sigma_v$ (kg/m$^3$) | $\dot{Q}_{\mathrm{ent}}$ (kW) |
|---|---|---|---|---|
| Freon 11 | 221 | 10.7 | 27.6 | 0.98 |
| Freon 113 | 132 | 10.6 | 18.5 | 0.48 |
| Acetone | 495 | 16.2 | 4.05 | 1.04 |
| Ammonia | 891 | 7.67 | 34 | 3.75 |

#### 4.6.3.2.3 Wicking limit

At this stage the wick has still to be specified, but a qualitative comparison of the potential performance of the four fluids can be obtained by evaluating the Merit number:

$$\rho_1 \frac{\sigma_1 L}{\mu_1}$$

for each fluid over the temperature range (Fig. 4.50).

#### 4.6.3.2.4 Radial heat flux

Boiling in the wick may result in the vapour blocking the supply of liquid to all parts of the evaporator. In arterial heat pipes, bubbles in the artery itself can create even more serious problems. It is therefore desirable to have a working fluid with a high superheat $\Delta T$ to reduce the chance of nucleation. The degree of superheat to cause nucleation is given by

$$\Delta T = \frac{3.06\sigma T_{\text{sat}}}{\rho_v L \delta}$$

where δ is the thermal layer thickness, and, taking a representative value of 15 μm, allows comparison of the fluids. $\Delta T$ is evaluated at 80°C, as the lowest permissible degree of superheat will occur at the maximum operating temperature. These are:

**i.** Freon 11 0.13 K
**ii.** Freon 113 0.31 K
**iii.** Acetone 0.58 K
**iv.** Ammonia 0.02 K

These figures suggest that the freons and ammonia require only very small superheat temperatures at 80°C to cause boiling. Acetone is the best fluid from this point of view.

#### 4.6.3.2.5 Priming of the wick

A further factor in fluid selection is the priming ability (see Section 4.6.2.2). A comparison of the priming ability of fluids may be obtained from the ratio $\sigma_l/\rho_l$ and this is plotted against vapour temperature in Fig. 4.51.

Acetone and ammonia are shown to be superior to the freons over the whole operating temperature range.

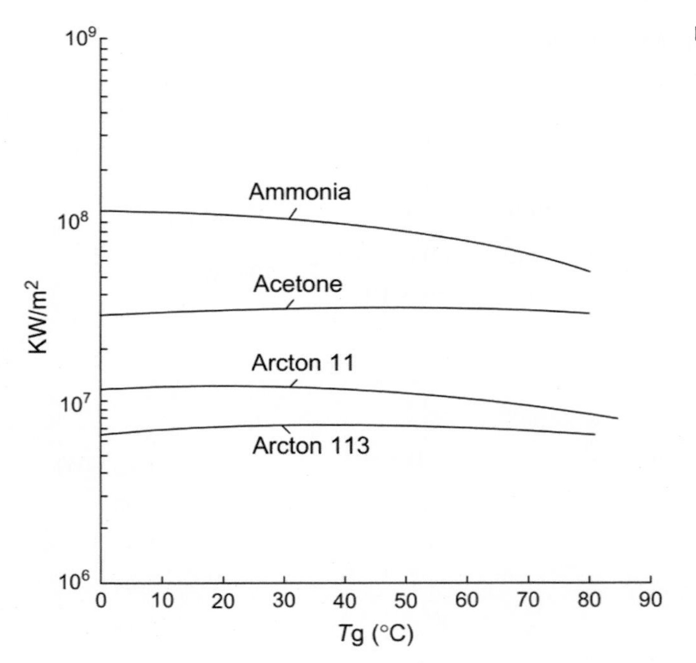


**FIGURE 4.50** Merit number for candidate fluids.

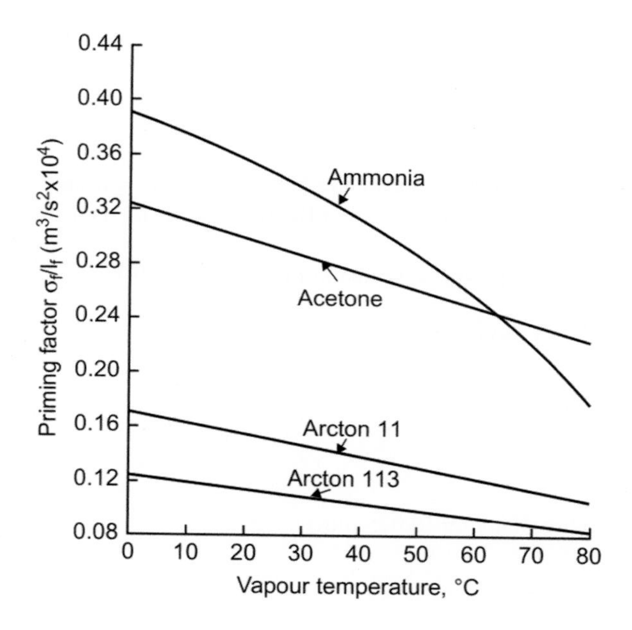


**FIGURE 4.51** Priming factor for selected fluids.

#### 4.6.3.2.6 Wall thickness

The requirement of this heat pipe necessitates the ability to be bonded to a radiator plate. Depending on the type of bonding used, the heat pipe may reach 170°C during bonding, and therefore vapour pressure is important in determining the wall thickness.

At this temperature, the vapour pressures of ammonia and acetone are 113 and 17 bar, respectively. Taking the 0.1% proof stress, $\Omega$, of HT30 aluminium as 46.3 MN/$m^2$ (allowing for some degradation of properties in weld regions), and using the thin cylinder formula,

$$t = \frac{P \times r}{\Omega}$$

the minimum wall thickness for ammonia is 0.65 mm and 0.1 mm for acetone. There is therefore a mass penalty attached to the use of ammonia.

#### 4.6.3.2.7 Conclusions on selection of working fluid

Acetone and ammonia both meet the heat transport requirements, ammonia being superior to acetone in this respect. Nucleation occurs more readily in an ammonia heat pipe, and the pipe may also be heavier. The handling of ammonia to obtain high purity is difficult, and the presence of any water in the working fluid may lead to long-term degradation in performance. Acetone is therefore selected in spite of the somewhat inferior thermal performance.

### *4.6.3.3 Detail design*

#### 4.6.3.3.1 Wick selection

Two types of wick structure are proposed for this heat pipe, homogeneous and arterial types. A homogeneous wick may be a mesh, twill or felt, and arterial types normally incorporate a mesh to distribute liquid circumferentially.

Homogeneous meshes are easy to form but have inferior properties to arterial types. The first question is, therefore, will a homogeneous wick transport the required amount of fluid over 1 m to meet the heat transport specification?

To determine the minimum flow area to transport 15 W, one can equate the maximum capillary pressure to the sum of the liquid and gravitational pressure drops (neglecting vapour $\Delta P$).

$$\Delta P_l + \Delta P_g = \Delta P_c$$

where

$$\Delta P_c = \frac{2\sigma_1 \cos\theta}{r_e}$$

$$\Delta P_1 = \frac{\mu_1 \dot{Q} l_{\text{eff}}}{\rho_1 L A_w K}$$

$$\Delta P_g = \rho_1 g h$$

The gravitational effect is zero in this application, but is included to permit testing of the heat pipe on the ground. The value of $h$ may be taken as 1 cm based on end-to-end tilt plus the tube diameter. The effective length $l_{\text{eff}}$ is 1 m and cos $\theta$ is taken to be unity.

The effective capillary radius for the wick is 0.029 mm; therefore, using the surface tension of acetone at 80°C,

$$\Delta P_c = \frac{2\sigma_1 \cos\theta}{r_e} = \frac{2 \times 0.0162 \times 1}{0.029 \times 10^{-3}} = 1120\text{N/m}^2$$

The wick permeability, $K$, is calculated using the Blake–Kozeny equation

$$K = \frac{d_w^2 (1-\varepsilon)^3}{66.6\varepsilon^2}$$

where $\varepsilon$ is the volume fraction of the solid phase (0.314) and $d_W$ the wire diameter (0.025 mm).

$$K = \frac{(2.5 \times 10^{-5})^2 (1-0.314)^3}{66.6 \times 0.314^2} = 3 \times 10^{-11}$$

Therefore considering the properties of acetone at 80°C

$$\rho_l = 719\text{kg/m}^3$$

$$\mu_l = 0.192 \text{ cP} = 192 \times 10^{-6}\text{kg/ms or Ns/m}^2$$

$$\Delta P_l = \frac{\mu_l \dot{Q} l_{eff}}{\rho_l L A_w K} = \frac{192 \times 10^{-6} \times 15 \times 1}{719 \times 495 \times 10^3 \times 3 \times 10^{-11} A_w} = \frac{0.027}{A_w}\text{N/m}^2$$

The gravitational pressure is

$$\Delta P_g = \rho_l g h = 719 \times 9.81 \times 0.01 = 70 \text{ N/m}^2$$

Equating the three terms gives

$$\Delta P_l + \Delta P_g = \Delta P_c$$

$$\frac{0.027}{A_w} + 70 = 1120$$

$$A_w = \frac{0.027}{1120 - 70} = 26 \times 10^{-6}\text{m}^2 \equiv 0.26\text{cm}^2$$

Since the required wick area (0.26 cm$^2$) is greater than the available vapour space area (0.197 cm$^2$), it can be concluded that the homogeneous type of wick is not acceptable. An arterial wick must be used.

#### 4.6.3.3.2 Arterial diameter

Eq. (5.3) above, and reproduced below, describes the artery priming capability, setting a maximum value on the size of any arteries:

$$d_a = \frac{1}{2}\left[\sqrt{\left(h^2 + \frac{8\sigma_1 \cos\theta}{(\rho_1 - \rho_v)g}\right)} - h\right]$$

Using this equation, $d_a$ is evaluated at a vapour temperature of 20°C (for convenience, priming ability may be demonstrated at room temperature), and $h$ is taken as 1 cm to cater for arteries near the top of the vapour space.

$$d_a = \frac{1}{2}\left[\sqrt{\left(0.01^2 + \frac{8 \times 0.0237 \times 2}{(790 - 0.64) \times 9.81}\right)} - 0.01\right] = 0.58 \times 10^{-3} m$$

Thus the maximum permitted value is 0.58 mm. To allow for uncertainties in fluid properties, wetting ($\theta$ assumed 0 degree) and manufacturing tolerances, a practical limit is 0.5 mm.

#### 4.6.3.3.3 Circumferential liquid distribution and temperature difference

The circumferential wick is the most significant thermal resistance in this heat pipe, and its thickness is limited by the fact that the temperature drop between the vapour space and the outside surface of the heat pipe and vice versa should be 3°C maximum. Assuming that the temperature drop through the aluminium wall is negligible, the thermal conductivity of the wick may be determined and used in the steady-state conduction equation.

$$k_{\text{wick}} = \left(\frac{\beta - \varepsilon}{\beta + \varepsilon}\right) k_1$$

where

$$\beta = \left(1 + \frac{k_s}{k_1}\right) / \left(1 - \frac{k_s}{k_1}\right)$$

$$k_s = 16\text{W/m}^\circ\text{C (steel)}$$

$$k_l = 0.165\text{W/m}^\circ\text{C (acetone)}$$

$$\therefore \beta = \frac{1 + 97}{1 - 97} = -1.02$$

The volume fraction $\varepsilon$ of the solid phase is approximately 0.3:

$$\therefore k_{\text{wick}} = \left(\frac{-1.02 - 0.3}{-1.02 + 0.3}\right) \times 0.165 = 0.3\ W/m^\circ C$$

Using the basic conduction equation:

$$\dot{Q} = kA_e \frac{\Delta T}{t}$$

$$t = \frac{kA_e \Delta T}{\dot{Q}}$$

where $A_e$ is the area of the evaporator (8 cm long, 0.5 cm diameter) and $\dot{Q}$ is the required heat load.

$$t = \frac{0.3 \times 3 \times \pi \times 5 \times 10^{-3} \times 80 \times 10^{-3}}{15} = 75 \times 10^{-6}\ \text{m} \equiv 0.075\ \text{mm}$$

Thus the circumferential wick must be 400 mesh, which has a thickness of 0.05 mm. Coarser meshes are too thick, resulting in unacceptable temperature differences across the wick in the condenser and evaporator.

#### 4.6.3.3.4 Arterial wick

Returning to the artery, the penultimate section revealed that the maximum artery depth permissible was 0.5 mm. In order to prevent nucleation in the arteries, they should be kept away from the heat pipe wall and formed of low-conductivity material. It is also necessary to cover the arteries with a fine pore structure, and 400-mesh stainless steel is selected. It is desirable to have several arteries to give a degree of redundancy, and two proposed configurations are considered, one having six arteries as shown in Fig. 4.52 and the other having four arteries. In the former case, each groove is nominally 1.0 mm wide and, in the latter case, 1.5 mm.

#### 4.6.3.3.5 Final analysis

It is now possible to predict the overall capability of the heat pipe, to check that it meets the specification.

We have already shown that entrainment and sonic limitations will not be exceeded and that the radial heat flux is acceptable. The heat pipe should also meet the overall temperature drop requirement, and the arteries are sufficiently small to allow repriming at 20°C. The wall thickness requirement for structural integrity (0.1 mm minimum) can easily be satisfied. The wicking limitation will therefore determine the maximum performance.

$$\text{i.e. } \Delta P_{\text{la}} + \Delta P_{\text{lm}} + \Delta P_g + \Delta P_v = \Delta P_c$$

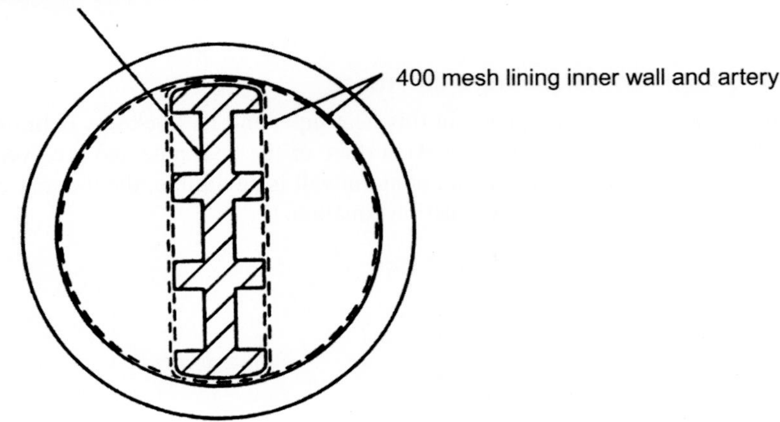


**FIGURE 4.52** Design of wick.

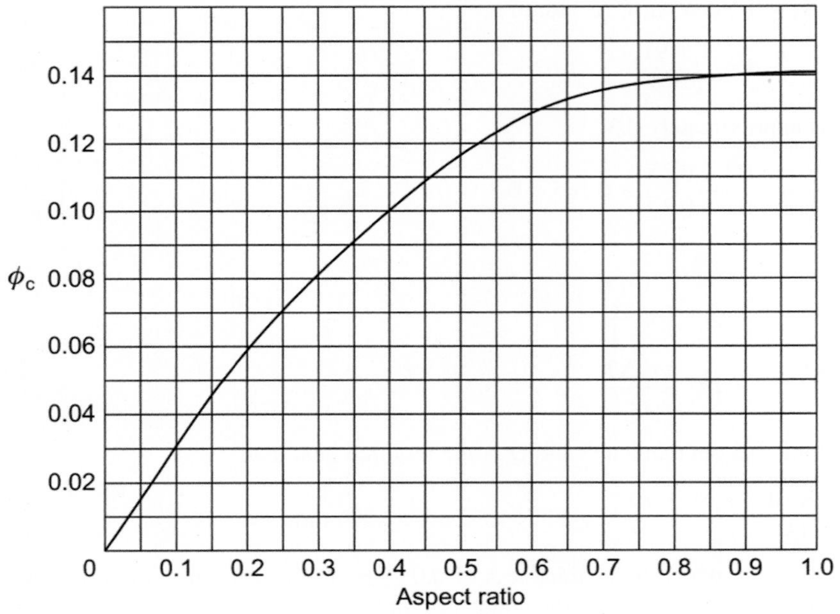

**FIGURE 4.53** Channel aspect ratio factor.

where $\Delta P_{la}$ is the pressure drop in the artery, $\Delta P_{lm}$ the loss in the circumferential wick. The axial flow in the mesh will have little effect and can be neglected. McAdams [190] presents an equation for the pressure loss, assuming laminar flow, in a rectangular duct, and shows that the equation is in good agreement with experiment for streamline flow in rectangular ducts having depth/width ratios. $a_a/b_a$ of 0.05–1.0. This equation may be written as

$$\Delta P_{la} = \frac{4K_1 \times 1_{eff} Q}{a_a^2 b_a^2 \theta_c N}$$

where $N$ is the number of channels, $\theta_c$ a function of channel aspect ratio and is given in Fig. 4.53 and

$$K_1 = \frac{\mu_1}{\pi_1 L}$$

The summed pressure loss in the condenser and evaporator is given by

$$\Delta P_{la} = \frac{K_1 \times 1_{effc} Q}{2KA_c}$$

where $l_{\text{eff c}}$ is the effective circumferential flow length approximately equal to

$$\frac{\pi r_w}{4}\text{carrying}\frac{\dot{m}}{4}$$

where $\dot{m}$ is the liquid mass flow, $A_c$ the circumferential flow area (mesh thickness × cond. or evap. length) and $K$ the permeability of 400 mesh.

The circumferential flow area for 400 mesh with two layers is

$$A_c = 8\times 10^{-2}\times 0.1\times 10^{-3}\text{m}^2 = 8\times 10^{-6}\text{m}^2$$

A resistance occurs in both the evaporator and the condenser; therefore, substituting in the above equation:

$$\Delta P_{lm} = \frac{\pi\times 2.5\times 10^{-3}}{2\times 4}\frac{1}{8\times 10^{-6}}\frac{K_l Q}{0.314\times 10^{-10}}$$

$$= 4.00\times 10^{12}\ K_l Q \text{ for each section}$$

$$\Delta P_{la} = \frac{4\times 0.92\times K_l Q\times 10^{12}}{(0.5)^2(1)^2\times 0.115\times 6}$$

$$= 21.3\times 10^{12} K_l Q \text{ for six channels}$$

$$= \frac{4\times 0.92\times K_l Q\times 10^{12}}{(0.5)^2(1.5)^2\times 0.088\times 4}$$

$$= 18.59\times 10^{12} K_l Q \text{ for four channels}$$

The vapour pressure loss, which occurs in two near-semicircular ducts, can be obtained using the Hagen–Poiseuille equation if the hydraulic radius is used:

$$\Delta P_v = \frac{1}{2}\left\{\frac{8K_v l_{\text{eff}} Q}{\pi r_H^4}\right\}$$

where

$$K_v = \frac{\mu_v}{\rho_v L}$$

Now the axial Reynolds number $Re_z$ is given by:

$$\text{Re}_z = \frac{Q}{\pi r_H \mu_1 L}$$

The transitional heat load at which the flow becomes turbulent can be calculated assuming that transition from laminar to turbulent flow occurs at $Re_z = 1000$, based on hydraulic radius (corresponding to 2000 if the Reynolds number is based on the diameter). Restricting the width of the stainless-steel former in Fig. 4.52 to 1.5 mm, $r_H = 1.07$ mm.

$\dot{Q}$ may be evaluated at the transition point using values of $\mu_v$ and $L$ at several temperatures between 0°C and 80°C, as given below:

| Vapour temperatures (°C) | Transitional load (W) |
|---|---|
| 0 | 31.1 |
| 20 | 31.2 |
| 40 | 30.6 |
| 60 | 30.2 |
| 80 | 30.0 |

The transitional load is always greater than the design load of 15 W, but as the heat pipe may be capable of operating in excess of the design load, it is necessary to investigate the turbulent regime.

For $Re_z > 1000$ and for two ducts

$$\Delta P_v = \frac{0.00896\mu_v^{0.25}Q^{1.75}l_{\text{eff}}}{2\rho_v r_H^{4.75}L^{1.75}}$$

This is the empirical Blasius equation.

$$\Delta P_v(laminar) = \frac{1}{2}\left(\frac{8 \times 0.92K_v \times Q \times 10^{12}}{p \times 1.07^4}\right) = 0.9 \times 10^{12}\ K_v \times Q$$

$$\Delta P_v(turbulent) = \frac{0.00896 \times 0.92}{2 \times \left(1.07 \times 10^{-3}\right)^{4.75}} Q^{1.75}\left(\frac{m_v{}^{0.25}}{L^{1.75}r_v}\right) = 0.53 \times 10^{12}Q^{1.75}\left(\frac{m_v{}^{0.25}}{L^{1.75}r_v}\right)$$

The gravitational pressure drop is

$$\Delta P_g = \rho_l g l \sin\varphi\ = 0.0981\rho_{1'}$$

taking $l \sin \phi = 1$ cm

The capillary pressure generated by the arteries is given by

$$\Delta P_c = \frac{2\sigma_l \cos\theta}{r_c} = \frac{2}{0.003 \times 10^{-2}}\sigma_l \cos\theta = 0.667 \times 10^{-5}\sigma_l$$

Summarising

$$\Delta P_c = 0.667 \times 10^5\sigma_l$$

$$\Delta P_g = 0.0981\rho_l$$

$$\Delta P_{vl} = 0.9 \times 10^{12}K_v \times Q$$

$$\Delta P_{vt} = 0.53 \times 10^{12}Q^{1.75}\left(\frac{m_v{}^{0.25}}{L^{1.75}r_v}\right)$$

$$\Delta P_{lm} = 4 \times 10^{12}K_l \times Q$$

$$\Delta P_{la} = 21.3 \times 10^{12}K_l \times Q \text{ (six channels)}$$

$$= 18.59 \times 10^{12}K_l \times Q \text{ (four channels)}$$

These equations involve $\dot{Q}$ and the properties of the working fluid. Using properties at each temperature (in 20°C increments) over the operating range, the total capability can be determined as

$$\Delta P_c = \Delta P_{\text{lm}} + \Delta P_{\text{la}}\left\{\begin{array}{c}\text{six channels}\\ \text{four channels}\end{array}\right\} + \Delta P_g + \Delta P_v\left\{\begin{array}{c}\text{laminar}\\ \text{turbulent}\end{array}\right\}$$

This yields the following results:

| Vapour temperatures (°C) | Q (W) | | | |
|---|---|---|---|---|
| | **Laminar our channels** | **Laminar six channels** | **Turbulent four channels** | **Turbulent six channels** |
| 0 | 21.6 | 20.9 | – | – |
| 20 | 34.0 | 32.5 | 22.6 | 22.0 |
| 40 | 42.6 | 40.2 | 27.9 | 27.0 |
| 60 | 49.1 | 45.8 | 33.0 | 32.0 |
| 80 | 51.4 | 47.6 | 36.4 | 35.0 |

In this example, it is assumed that the maximum resistance to heat transfer is across the wick; the other thermal resistances have not been calculated explicitly.

This heat pipe was constructed with six grooves in the artery structure and met the specification.

### 4.6.4 Design example 2

#### 4.6.4.1 *Problem*

Estimate the liquid flow rate and heat transport capability of a simple water heat pipe operating at 100°C having a wick of two layers of 250 mesh against the inside wall. The heat pipe is 30 cm long and has a bore of 1 cm diameter. It is operating at an inclination to the horizontal of 30 degrees, with the evaporator above the condenser.

It will be shown that the capability of the above heat pipe is low. What improvement will be made if two layers of 100 mesh are added to the 250-mesh wick to increase liquid flow capability?

#### 4.6.4.2 *Solution – original design*

The maximum heat transport in a heat pipe at a given vapour temperature, if governed by the wicking limit, may be obtained from the equation:

$$Q_{\max} = \dot{m}_{\max} L$$

where $\dot{m}_{\max}$ is the maximum liquid flow rate in the wick.

Using the standard pressure balance equation:

$$\Delta P_c = \Delta P_v + \Delta P_l + \Delta P_g$$

and neglecting, for the purposes of a first approximation, the vapour pressure drop $\Delta P_v$, we can substitute for the pressure terms and obtain:

$$\frac{2\sigma_1 \cos\theta}{r_c} = \frac{\mu_1}{\rho_1 L} \times \frac{Q l_{\text{eff}}}{A_w K} + \rho_1 g l \sin\varphi$$

Rearranging and substituting for $\dot{m}$, we obtain:

$$\dot{m} = \frac{\rho_l K A_w}{\mu_l l_{\text{eff}}} \left\{ \frac{2\sigma_1}{r_c} \cos\theta - \rho_l g l_{\text{eff}} \sin\varphi \right\}$$

The wire diameter of 250 mesh is typically 0.0045 cm, and therefore the thickness of two layers of 250 mesh is $4 \times 0.0045$ cm or 0.0180 cm.

The bore of the heat pipe is 1 cm.

$$\begin{aligned} \therefore A_{\text{wick}} &= 0.018 \times \pi \times 1 \\ &= 0.057\ \text{cm}^2 \\ &= 0.057 \times 10^{-4}\ \text{m}^2 \end{aligned}$$

From Table 3.4, the pore radius $r$ and permeability $K$ of 250 mesh are 0.002 cm and $0.302 \times 10^{-10}\ \text{m}^2$, respectively. Assuming perfect wetting ($\theta = 0°$), the mass flow $\dot{m}$ may be calculated using the properties of water at 100°C.

$$L = 2258\ \text{J/kg}$$

$$\rho_1 = 958\ \text{kg/m}^3$$

$$\mu_1 = 0.283\ \text{mN s/m}^2$$

$$\tau_1 = 58.9\text{mN/m}$$

Converting all terms to the base units (kg, J, N, m and s)

$$\dot{m}_{\max} = \frac{958 \times 0.302 \times 10^{-10} \times 0.057 \times 10^{-4}}{0.283 \times 10^{-3} \times 0.3} \times \left( \frac{2 \times 58.9 \times 10^{-3}}{0.2 \times 10^{-4}} - 958 \times 9.810 \times 0.3 \times 0.5 \right)$$

$$= 1.95 \times 10^{-9} \times (5885 - 1410)$$

$$= 8.636 \times 10^{-6}\text{kg/s}$$

$$\dot{Q}_{\max} = \dot{m}_{\max} \times L$$

$$= 8.636 \times 10^{-6} \times 2.258 \times 10^6$$

$$= 19.5\text{W}$$

#### 4.6.4.3 Solution – revised design

Consider the addition of two layers of 100-mesh wick below the original 250 mesh.

The wire diameter of 100 mesh is 0.010 cm

$\therefore$ thickness of two layers is 0.040 cm

Total wick thickness = 0.040 + 0.018 cm

= 0.058 cm

$$\therefore A_{\text{wick}} \approx 0.058 \times \pi \times 1 \approx 0.182\ \text{cm}^2$$

The capillary pressure is still governed by the 250 mesh, and $r_c = 0.002$ cm. The permeability of 100 mesh is used, Langston and Kunz giving a value of $1.52 \times 10^{-10}\ \text{m}^2$. The mass flow may now be calculated as

$$\dot{m}_{\text{max}} = \frac{958 \times 1.52 \times 10^{-10} \times 0.057 \times 10^{-4}}{0.283 \times 10^{-3} \times 0.3} \times \left(\frac{2 \times 58.9 \times 10^{-3}}{0.2 \times 10^{-4}} - 958 \times 9.810 \times 0.3 \times 0.5\right)$$

$$= 31 \times 10^{-9} \times (5885 - 1410)$$

$$= 139 \times 10^{-6}\text{kg/s}$$

$$\dot{Q}_{\text{max}} = \dot{m}_{\text{max}} \times L$$

$$= 139 \times 10^{-6} \times 2.258 \times 10^{6}$$

$$= 314\text{W}$$

The modified wick structure has resulted in an estimated increase in capacity from 19.5 to 314 W, that is, the limiting performance has been improved by more than an order of magnitude. A disadvantage of the additional layers lies in the additional thermal resistance introduced at the evaporator and the condenser.

### 4.6.5 Thermosyphons

The design process for a thermosyphon is similar to that for a wicked heat pipe in that it requires identification of suitable case material and working fluid, followed by evaluation of the performance of the unit using the techniques discussed earlier in this chapter. The wicking limitation is not relevant to thermosyphon performance. The thermal resistance of any wick used to distribute liquid through the evaporator must, however, be taken into account.

#### 4.6.5.1 Fluid inventory

In the case of thermosyphons, the fluid inventory is based on different considerations from wicked heat pipes [120]. The amount of liquid is governed by two considerations; too small a quantity can lead to dry-out, whilst an excess of liquid can lead to quantities being carried up to the condenser, where blockage of surface for condensation can result. Bezrodnyi and Alekseenko [191] recommended that the liquid fill should be at least 50% of the volume of the evaporator and also that the volume of liquid, $V_l$ should be related to the thermosyphon dimensions as follows:

$$V_l > 0.001D(l_e + l_a + l_c)$$

where $D$ is the pipe internal diameter.

When a wick is fitted to the evaporator section, ESDU [192] recommends:

$$V_l > 0.001D(l_a + l_c)A_w l_e \varepsilon$$

where $\varepsilon$ is the wick porosity.

Comprehensive work by Groll et al. at IKE, Stuttgart, investigated the effect of fluid inventory on the performance of thermosyphons incorporating a variety of novel liquid–vapour flow separators (to minimise interaction between these components). It was found, for several working fluids, [193], that the fluid inventory, expressed as a percentage of the evaporator volume occupied by the liquid working fluid, had a rather flat optimum between about 20% and 80%.

The constraints on filling for a closed two-phase thermosyphon are illustrated by El-Genk and Saber [194]. Low initial filling ratio results in dry-out of the evaporator, whilst excess working fluid results in liquid filling the entire evaporator when bubbles form due to boiling.

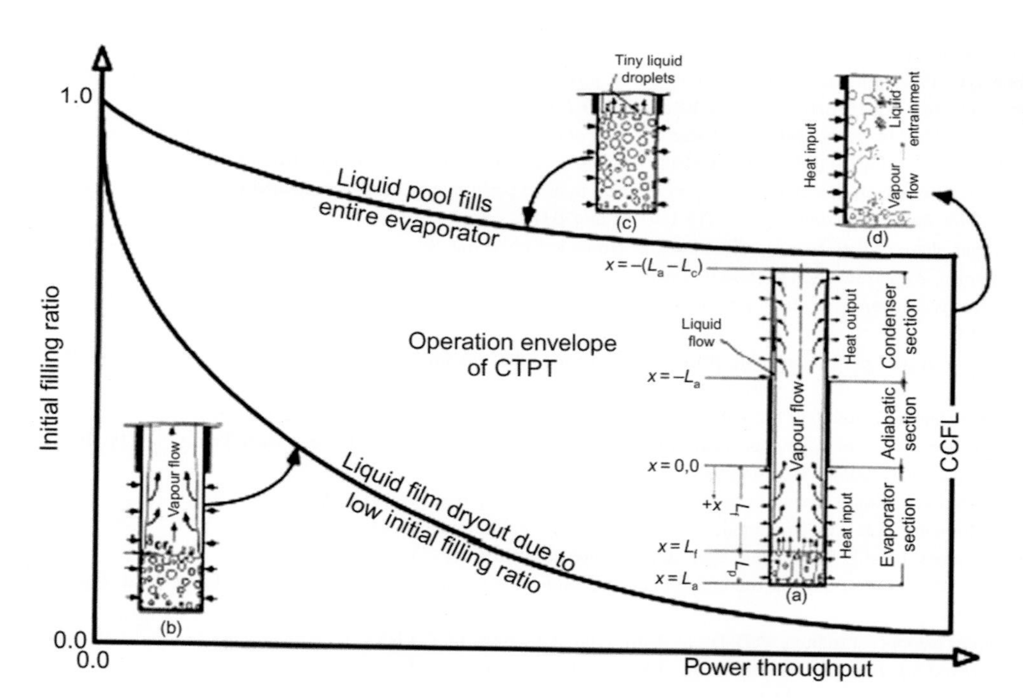


FIGURE 4.54 Operating envelope of closed two-phase thermosyphons (CTPT) [194].

### 4.6.5.2 *Entrainment limit*

The third limit shown in Fig. 4.54 is the entrainment or counter-current flooding limit discussed in Section 4.4.2.

## 4.7 Summary

The basic principles of fluid mechanics and heat transfer relevant to the design and performance evaluation of heat pipes have been outlined and these have been related to studies of heat pipes and thermosyphons. Particular attention has been paid to identifying the operating limits of heat pipes and thermosyphons. Prediction methods give good results with conventional fluids. Nanofluids are currently being investigated for a wide range of heat transfer applications and have been shown to yield significant improvements in performance, however, whilst this has been demonstrated empirically, there is insufficient theoretical background to reliably predict performance with these fluids.

In this chapter, the theory presented in Chapter 4 and the practical considerations of material and working fluid choice described in Chapter 3 have been used to illustrate the process of designing simple wicked heat pipes. The design of thermosyphons uses similar techniques, but the important differences have been highlighted.

## References


[1] T.P. Cotter, Theory of heat pipes, LA-3246-MS, March 26, 1965.
[2] D. Shaw, Introduction to Colloid and Surface Chemistry, second ed., Butterworth, 1970.
[3] V. Semenchenke, Surface phenomena in metals and alloys, Pergamon Press, 1961.
[4] J. Bohdansky, H.E.J. Schins, The surface tension of the alkali metals, J. Inorg. Nucl. Chem. 29 (9) (1967) 2173–2179.
[5] J.J. Jasper, The surface tension of pure liquid compounds, J. Phys. Chem. Ref. data 1 (4) (1972) 841–1010.
[6] R.B. Dooley, Release on surface tension of ordinary water substance, International Association for the Properties of Water and Steam (IAPWS), Electric Power Research Institute, Palo Alto, CA, 1994.
[7] J.S.H.E. Bohdansky, The temperature dependence of liquid metals, Chem. Phys. (1968).
[8] J.K. Fink, L. Leibowitz, Thermodynamic and transport properties of sodium liquid and vapor. ANI/RE-95/2 Argonne National Laboratory, January 1995.
[9] T. Iida, R.I.L. Guthrie.The Physical Properties of Liquid Metals, Clarendon Press, Oxford, 1988.
[10] P. Walden, Uber den zusammenhang der kapillaritaskonstanten mit der latenten verdampfungswarme der losungsmittel, Phys. Chem. 65 (1909) 267–288.
[11] B. Samuel Sugden, A relation between surface tension, density, and chemical composition, J. Chem. Soc. D. 125 (1924) 1177–1189.
[12] O.R. Quale, The parachors of organic compounds, Chem. Rev. 53 (1953) 439–586.

[13] G.F. Hewitt, Handbook of heat exchanger design, Begell House 2 (5) (1992) 3.
[14] A.B. Solomon, K. Ramachandran, B.C. Pillai, Thermal performance of a heat pipe with nanoparticles coated wick, Appl. Therm. Eng. 36 (1) (2012) 106–112. Available from: https://doi.org/10.1016/J.APPLTHERMALENG.2011.12.004.
[15] Y. Nam, S. Sharratt, C. Byon, S.J. Kim, Y.S. Ju, Fabrication and characterization of the capillary performance of superhydrophilic Cu micropost arrays, J. Microelectromechanical Syst. 19 (3) (2010) 581–588. Available from: https://doi.org/10.1109/JMEMS.2010.2043922.
[16] Y. Nam, S. Sharratt, G. Cha, Y.S. Ju, Characterization and modeling of the heat transfer performance of nanostructured Cu micropost wicks, J. Heat. Transf. 133 (10) (2011). Available from: https://doi.org/10.1115/1.4004168/470318.
[17] C.A. Busse, Pressure drop in the vapor phase of long heat pipes, in: Proc. Thermionic Conversion Specialist Conference, 1967, pp. 391–398.
[18] D.H. Lloyd, J.M. Kay, R.M. Nedderman, An introduction to fluid mechanics and heat transfer, Aeronaut. J. 79 (772) (1974) 183.
[19] G.M. Grover, J.E. Kemmy, E.S. Keddy, Advances in heat pipe technology, in: Second International Conference on Thermionic Electrical Power Generation, 1968, pp. 477–490.
[20] D.M. Ernst, Evaluation of theoretical heat pipe performance, in: IEEE conference record of thermionic conversion specialist conference, 1967, pp. 349–354.
[21] C.S. Bankston, J.H. Smith, Incompressible laminar vapour flow in cylindrical heat pipes, ASME Paper No. 7 1-WAJHT-1 5, 1971.
[22] A.R. Rohani, C.L. Tien, Analysis of the effects of vapour pressure drop on heat pipe performance, Int. J. Heat Mass Transf. 17 (1) (1974) 61–67. Available from: https://doi.org/10.1016/0017-9310(74)90038-6.
[23] J.E. Deverall, J.E. Kemme, L.W. Flarschuetz. Some limitations and startup problems of heat pipes, LA-4518, 1970.
[24] J.E. Kemme, Ultimate heat pipe performance. I.E.E.E. Thermionic Specialist Conference, Framingham, MA, 1968.
[25] J.E. Kemme. High performance heat pipes. I.E.E.E. Thermionic Specialist Conference, Palo Alto, CA, 1967.
[26] R.R. Bertossi, C. Avel, A theoretical study and review on the operational limitations due to vapour flow in heat pipes, Front. Heat. Pipes 3 (2012) 023001. ISSN: 2155-658X.
[27] P. Gagneux, Contribution à l'étude des caloducs à sodium, PhD Thesis, Université de Poitiers, LET, ENSMA, 1979.
[28] L.K. Tower, D.C. Hainley, An improved algorithm for the modelling of vapour flow in heat pipes, Paper 13-8, in: Proceedings of the Seventh International Heat Pipe Conference, Minsk, Byelorussia, May 1990.
[29] B.H. Kim, G.P. Peterson, Analysis of the critical Weber number at the onset of liquid entrainment in capillary-driven heat pipes, Int. J. Heat. Mass. Transf. 38 (8) (1995) 1427–1442. Available from: https://doi.org/10.1016/0017-9310(94)00249-U.
[30] H.A. Cheung, Critical review of heat pipe theory and applications, UCRL50453, 1968.
[31] P.G. Drazin, W.H. Reid, Hydrodynamic stability, Cambridge University Press, Cambridge, 1981.
[32] G.F. Hewitt, G.L. Shires, T.R. Bott, Process Heat Transfer, 113, CRC Press, Boca Raton, FL, 1994.
[33] W.M. Rohsenow et al. (Eds.), Handbook of heat transfer. <http://160592857366.free.fr/joe/ebooks/MechanicalEngineeringBooksCollection/HEATTRANSFER/handbookofHeatTransfer.pdf/>, 1998 (accessed 14.11.18).
[34] Y.Y. Hsu, On the size range of active nucleation cavities on a heating surface, ASME J. Heat Transfer (1962).
[35] V.P. Carey, Liquid-vapor phase-change phenomena, Hemisph. Publ. Corp. DC, 1992.
[36] C.H. Wang, V.K. Dhir, On the prediction of active nucleation sites including the effect of surface wettability, in: Proceedings of the Engineering Foundation Conference on Pool and External Flow Boiling, 1992, pp. 111–118.
[37] K.A. Zhokhov, Number of vapour generating centers, in: Aerodyn. Heat Transf. Work. Elem. Power Equipment, Leningrad, Russ., 1969, pp. 131–135.
[38] B.B. Mikic, W.M. Rohsenow, A new correlation of pool-boiling data including the effect of heating surface characteristics, J. Heat. Transf. 91 (2) (1969) 245–250. Available from: http://doi.org/10.1115/1.3580136.
[39] S. Nukiyama, The maximum and minimum values of the heat Q transmitted from metal to boiling water under atmospheric pressure, Int. J. Heat. Mass. Transf. 9 (12) (1966) 1419–1433.
[40] R.F. Gaertner, Photographic study of nucleate pool boiling on a horizontal surface, J. Heat. Transf. 87 (1) (1965) 17–27.
[41] D. Jafari, P. Di Marco, S. Filippeschi, A. Franco, An experimental investigation on the evaporation and condensation heat transfer of two-phase closed thermosyphons, Exp. Therm. Fluid Sci. 88 (2017) 111–123. Available from: https://doi.org/10.1016/J.EXPTHERMFLUSCI.2017.05.019.
[42] R.L. Mohanty, M.K. Das, A critical review on bubble dynamics parameters influencing boiling heat transfer, Renew. Sustain. Energy Rev. 78 (2017) 466–494. Available from: https://doi.org/10.1016/J.RSER.2017.04.092.
[43] F.M. Chowdhury, F. Kaminaga, Boiling heat transfer characteristics of R-113 in a vertical small diameter tube under natural circulation condition, Int. J. Heat. Mass. Transf. 45 (24) (2002) 4823–4829. Available from: https://doi.org/10.1016/S0017-9310(02)00174.6.
[44] D. Jafari, S. Filippeschi, A. Franco, P. Di Marco, Unsteady experimental and numerical analysis of a two-phase closed thermosyphon at different filling ratios, Exp. Therm. Fluid Sci. 81 (2017) 164–174. Available from: https://doi.org/10.1016/J.EXPTHERMFLUSCI.2016.10.022.
[45] V. Guichet, S. Almahmoud, H. Jouhara, Nucleate pool boiling heat transfer in wickless heat pipes (two-phase closed thermosyphons): a critical review of correlations, Therm. Sci. Eng. Prog. 13 (2019). Available from: https://doi.org/10.1016/j.tsep.2019.100384.
[46] H. Jouhara, A.J. Robinson, Experimental investigation of small diameter two-phase closed thermosyphons charged with water, FC-84, FC-77 and FC-3283, Appl. Therm. Eng. 30 (2–3) (2010) 201–211. Available from: https://doi.org/10.1016/J.APPLTHERMALENG.2009.08.007.
[47] I.L. Pioro, W. Rohsenow, S.S. Doerffer, Nucleate pool-boiling heat transfer. II: assessment of prediction methods, Int. J. Heat. Mass. Transf. 47 (23) (2004) 5045–5057. Available from: https://doi.org/10.1016/J.IJHEATMASSTRANSFER.2004.06.020.
[48] Y.J. Park, H.K. Kang, C.J. Kim, Heat transfer characteristics of a two-phase closed thermosyphon to the fill charge ratio, Int. J. Heat. Mass. Transf. 45 (23) (2002) 4655–4661. Available from: https://doi.org/10.1016/S0017-9310(02)00169-2.

[49] S.H. Noie, Heat transfer characteristics of a two-phase closed thermosyphon, Appl. Therm. Eng. 25 (4) (2005) 495–506. Available from: https://doi.org/10.1016/J.APPLTHERMALENG.2004.06.019.
[50] W. Guo, D.W. Nutter, An experimental study of axial conduction through a thermosyphon pipe wall, Appl. Therm. Eng. 29 (17–18) (2009) 3536–3541. Available from: https://doi.org/10.1016/J.APPLTHERMALENG.2009.06.008.
[51] D. Jafari, A. Franco, S. Filippeschi, P. Di Marco, Two-phase closed thermosyphons: a review of studies and solar applications, Renew. Sustain. Energy Rev. 53 (2016) 575–593. Available from: https://doi.org/10.1016/J.RSER.2015.09.002.
[52] M.S. El-Genk, H.H. Saber, Heat transfer correlations for small, uniformly heated liquid pools, Int. J. Heat. Mass. Transf. 41 (2) (1997) 261–274. Available from: https://doi.org/10.1016/S0017-9310(97)00143-9.
[53] T. Kiatsiriroat, A. Nuntaphan, J. Tiansuwan, Thermal performance enhancement of thermosyphon heat pipe with binary working fluids, Exp. Heat. Transf. 13 (2) (2000) 137–152. Available from: https://doi.org/10.1080/089161500269517.
[54] W.M. Rohsenow, A method of correlating heat-transfer data for surface boiling of liquids. <http://iibraries.mit.edu/docs/>, 1952 (accessed 26.01.19).
[55] H. Imura, H. Kusuda, J.-I. Ogata, T. Miyazaki, N. Sakamoto, Heat transfer in two-phase closed-type thermosyphons, JSME Trans. 45 (1979) 712–722.
[56] K. Stephan, M. Abdelsalam, Heat-transfer correlations for natural convection boiling, Int. J. Heat. Mass. Transf. 23 (1) (1980) 73–87. Available from: https://doi.org/10.1016/0017-9310(80)90140-4.
[57] M. Cooper, Saturated nucleate pool boiling – a simple correlation, in: 1st UK Natl. Heat Transf. Conf. IChemE Symp., 1984.
[58] V.I. Subbotin, Heat transfer in boiling metals by natural convection. USAEC-Tr-72 10, 1972.
[59] O.E. Dwyer, On incipient-boiling wall superheats in liquid metals, Int. J. Heat. Mass. Transf. 12 (11) (1969) 1403–1419. Available from: https://doi.org/10.1016/0017-9310(69)90025-8.
[60] W.M. Rohsenow, P. Griffith, Correlation of maximum heat flux data for boiling of saturated liquids, Massachusetts Institute of Technology, Cambridge, Mass, 1955.
[61] B.F.B. Caswell, The critical heat flux for boiling metal systems, in: Chemical Engineering Progress Symposium Series of Heat Transfer, 1966.
[62] E.C.H.J. Philips, Determination of capillary properties useful in heat pipe design, in: S.M.E. – A.I.Ch.E. Heat Transf. Conf. Minneapolis, MN, 1967.
[63] J.K. Ferrell, J. Alleavitch, Vaporisation heat transfer in capillary wick structures, Chem. Eng. Symp. Series 66 (2) (1970).
[64] J.C.W.C. Corman, Vaporisation from capillary wick structures, in: A.S.M.E. – A.I.Ch.E. Heat Transf. Conf., 1971.
[65] A. Brautsch, P.A. Kew, The effect of surface conditions on boiling heat transfer from mesh wicks, Proceedings of the Twelfth International Heat Transfer Conference, Grenoble, Elsevier SAS, 2002.
[66] A. Brautsch, Heat transfer mechanisms during the evaporation process from mesh screen porous structures, PhD Thesis, Heriot-Watt University, 2002.
[67] A. Brautsch, P.A. Kew, Examination and visualisation of heat transfer processes during evaporation in capillary porous structures, Appl. Therm. Eng. 22 (7) (2002) 815–824. Available from: https://doi.org/10.1016/S1359-4311(02)00027-3.
[68] J.P. Asakavicius, A.A. Zukauskav, Heat transfer from Freon-113, ethyl alcohol and water with screen wicks, 1979.
[69] J.W. Liu, D.J. Lee, A. Su, Boiling of methanol and HFE-7100 on heated surface covered with a layer of mesh, Int. J. Heat. Mass. Transf. 44 (1) (2001) 241–246. Available from: https://doi.org/10.1016/S0017-9310(00)00078-8.
[70] J.Y. Tsay, Y.Y. Yan, T.F. Lin, Enhancement of pool boiling heat transfer in a horizontal water layer through surface roughness and screen coverage, Heat. Mass. Transf. 32 (1996) 17–26.
[71] A. Abhat, R.A. Seban, Boiling and evaporation from heat pipe wicks with water and acetone, J. Heat Transf. (1974).
[72] P.S.M.W. Marto, Effect of nucleate boiling and the operation of low temperature heat pipes, in: A.S.M.E. – A.I.Ch.E. Heat Transf. Conf., 1969.
[73] D.R.W.H. Ferrell J.K., Vaporisation heat transfer in heat pipe wick materials, in: Proc. 1st Int. Heat Pipe Conf., 1973.
[74] D.P.M.B.J. Shaubach R.M., Boiling in heat pipe evaporator wick structures, in: Proc. 7th Int. Heat Pipe Conf., 1990.
[75] K. Cornwell, Boiling in wicks. Proceeding of the Heat Pipe Forum Meeting, National Engineering Laboratory, Department of Industry, Report No. 607, East, Kilbride, Glasgow, 1975.
[76] K. Cornwell, B.G. Nair, T.D. Patten, Observation of boiling in porous media, Int. J. Heat. Mass. Transf. 19 (2) (1976) 236–238. Available from: https://doi.org/10.1016/0017-9310(76)90121-6.
[77] C.P.F. Costello, The roles of capillary wicking and surface deposits in attainment of high pool boiling burnout heat fluxes, A.I.Ch.E.J, 1964.
[78] N. Kockmann, Fortschrittsberichte VDI, Düsseldorf, 1997.
[79] R.E. Balhiser, J.A. Clark, E.E. Hucke, L.R. Smith, C.P. Colver, H. Merte. Investigation of liquid metal boiling heat transfer, Air Force Propulsion Lab, Wright Patterson, AFB, OH, AFAPL-TR 66-85, 1966.
[80] R.A. Moss, A.J. Kelly, Neutron radiographic study of limiting planar heat pipe performance, Int. J. Heat. Mass. Transf. 13 (3) (1970) 491–502. Available from: https://doi.org/10.1016/0017-9310(70)90145-6.
[81] W. Davis, J. Ferrell, Evaporative heat transfer of liquid potassium in porous media, in: Thermophysics and Heat Transfer Conference, 1974, p. 719.
[82] J.K. Ferrell et al., Vaporisation heat transfer in heat pipe wick materials, in: A.I.A.A. Thermophys. Conf., 1972.
[83] G.A.A. Asselman, D.B. Green, Heat pipes, Phillips Tech. Rev. 33 (4) (1973) 104–113.
[84] G.F.C. Rogers, Engineering Thermodynamics Work and Heat Transfer, fourth ed., Longman, 1992.

[85] W.H. McAdams, Heat Transmission - Second Edition, McGRAW, New York, 1942.
[86] S.J. Chen, J.G. Reed, C.L. Tien, Reflux condensation in a two-phase closed thermosyphon, Int. J. Heat. Mass. Transf. 27 (9) (1984) 1587−1594. Available from: https://doi.org/10.1016/0017-9310(84)90271-0.
[87] S. Oh, S.T. Revankar, Complete condensation in a vertical tube passive condenser, Int. Commun. Heat. Mass. Transf. 32 (5) (2005) 593−602. Available from: https://doi.org/10.1016/J.ICHEATMASSTRANSFER.2004.10.017.
[88] W.M. Rohsenow, Heat transfer and temperature distribution in laminar film condensation, Trans. ASME 78 (1956) 1645−1648.
[89] P.L. Kapitza, Wave flow of thin layers of viscous liquid. Part I. Free flow, Zhurnal Eksp. i Teor. Fiz. 18 (1948) 3−18.
[90] W. Nusselt, The condensation of steam on cooled surfaces, Z. Ver. Dtsch. Ing. 60 (1916) 541−546.
[91] S.S. Kutateladze, Heat transfer theory fundamentals, Atomizdat, Moscow./E. Arnold Publishers, London, Academic Press Inc, New York, 1963.
[92] D.A. Labuntsov, Heat transfer in film condensation of pure steam on vertical surfaces and horizontal tubes, Teploenergetika 4 (7) (1957) 72−79.
[93] U. Gross, Reflux condensation heat transfer inside a closed thermosyphon, Int. J. Heat. Mass. Transf. 35 (2) (1992) 279−294. Available from: https://doi.org/10.1016/0017-9310(92)90267-V.
[94] M.H. Chun, K.T. Kim, A natural convection heat transfer correlation for laminar and turbulent film condensation on a vertical surface, in: 3rd ASME/JSME Thermal Engineering Joint Conference, 1991, pp. 459−464.
[95] H. Hashimoto, F. Kaminaga, Heat transfer characteristics in a condenser of closed two-phase thermosyphon: effect of entrainment on heat transfer deterioration, Heat. Transf. Res. 31 (3) (2002) 212−225. Available from: https://doi.org/10.1002/htj.10030.
[96] H.M.S. Hussein, M.A. Mohamad, A.S. El-Asfouri, Theoretical analysis of laminar-film condensation heat transfer inside inclined wickless heat pipes flat-plate solar collector, Renew. Energy 23 (3−4) (2001) 525−535. Available from: https://doi.org/10.1016/S0960-1481(00)00149-X.
[97] J.C.Y. Wang, Y. Ma, Condensation heat transfer inside vertical and inclined thermosyphons, J. Heat. Transf. 113 (3) (1991) 777−780.
[98] S. Fiedler, H. Auracher, Experimental and theoretical investigation of reflux condensation in an inclined small diameter tube, Int. J. Heat. Mass. Transf. 47 (19−20) (2004) 4031−4043. Available from: https://doi.org/10.1016/j.ijheatmasstransfer.2004.06.005.
[99] C.A. Busse, Theory of the ultimate heat transfer limit of cylindrical heat pipes, Int. J. Heat. Mass. Transf. 16 (1) (1973) 169−186. Available from: https://doi.org/10.1016/0017-9310(73)90260-3.
[100] Anon. Heat pipes performance of capillary-driven designs. Data Item No. 70012, Engineering Sciences Data Unit, London, September 1979.
[101] C.L. Tien, K.S. Chung, Entrainment limits in heat pipes, in: 3rd International Heat Pipe Conferences, Palo Allo, California, 1978, pp. 36−39.
[102] F.C. Prenger, J.E. Kemme, Performance limits of gravity-assist heat pipes with simple wick structures, in: Proc. IV International Heat Pipe Conference, 1981.
[103] C.L. Tien, Fluid Mechanics of Heat Pipes, 1975 [Online]. Available from: <http://www.annualreviews.org/>.
[104] A. Abhat, H. Nguyenchi, Investigation of performance of gravity assisted copper-water heat pipes. in: Proceedings of the Second International Heat Pipe Conference, Bologna; ESA Report SP112, vol. 1, 1976.
[105] J.R. Wiebe, R.L. Judd, Superheat layer thickness measurements in saturated and subcooled nucleate boiling, J. Heat Transfer. 93 (4) (1971) 455−461.
[106] K. Nishikawa, Y. Fujita, Correlation of nucleate boiling heat transfer based on bubble population density, Int. J. Heat. Mass. Transf. 20 (3) (1977) 233−245. Available from: https://doi.org/10.1016/0017-9310(77)90210-1.
[107] E.W. Saaski, Investigation of an inverted meniscus heat pipe wick concept, NASA Report CR- 137724, 1975.
[108] K.T. Feldman Jr, M.E. Berger, Analysis of a High-Heat-Flux Water Heat Pipe Evaporator., New Mexico Univ Albuquerque Bureau of Engineering Research, 1973.
[109] H.M. Winston, J.K. Ferrell, R.Davis, The mechanism of heat transfer in the evaporator zone of the heat pipe. in: Proceedings of the Second International Heat Pipe Conference, Bologna; ESA Report SP112, vol. 1, 1976.
[110] J.R. Johnson, J.K. Ferrell, The mechanism of heat transfer in the evaporator zone of the heat pipe, A SME Space Technology and Heat Transfer Conference, Los Angeles, CA, Paper 70-HT/ SpT-12, 1970.
[111] J.E. Deverall, E.S. Keddy, Helical wick structures for gravity-assisted heat pipes. in: Proceedings of the Second International Heat Pipe Conference, Bologna; ESA Report sp112, vol. 1, 1976.
J.E. Deverall, E.S. Keddy, Helical wick structures for gravity-assisted heat pipes 1 (1976).
[112] R. Kaser, Heat pipe operation in a gravity field with liquid pool pumping, McDonnell Douglas Corp., July 1972.
[113] A. Strel'tsov, Theoretical and experimental investigation of optimum filling for heat pipes, Heat. Transf-Soviet Res. 7 (1) (1975).
[114] L. Vasiliev, V. Kiselyov, Simplified analytical model of vertical arterial heat pipes - NASA/ADS [Online]. <https://ui.adsabs.harvard.edu/abs/1974hetr.5.209V/abstract\>, 1974 (accessed 29.07.22).
[115] C.A. Busse, J.E. Kemme, Dry-out phenomena in gravity-assist heat pipes with capillary flow, Int. J. Heat. Mass. Transf. 23 (5) (1980) 643−654. Available from: https://doi.org/10.1016/0017-9310(80)90008-3.
[116] M.J. Davies, G.H. Chaffey, Development and demonstration of improved gas to gas heat pipe heat exchangers for the recovery of residual heat, Final report (Technical Report)|OSTI. GOV, Luxembourg [Online]. <https://www.osti.gov/biblio/5555627-development-demonstration-improved-gas-gas-heat-pipe-heat-exchangers-recovery-residual-heat-final-report\>, 1981 (accessed 29.07.22).
[117] Anon, Thermophysical properties of heat pipe working fluids: operating range between −60°C and 300 °C., London, August 1980.
[118] R.W. MacGregor, P.A. Kew, D.A. Reay, Investigation of low Global Warming Potential working fluids for a closed two-phase thermosyphon, Appl. Therm. Eng. 51 (1−2) (2013) 917−925. Available from: https://doi.org/10.1016/J.APPLTHERMALENG.2012.10.049.

[119] H. Nguyen-Chi, M. Groll, Entrainment or flooding limit in a closed two-phase thermosyphon, Proceedings of the Fourth International Heat Pipe Conference, London, Pergamon Press, Oxford, 1981.
[120] Anon, Heat pipes-performance of two-phase closed thermosyphons, Data Item No. 81038, Engineering Sciences Data Unit, London, October 1981.
[121] P. Terdtoon et al., Investigation of effect of inclination angle on heat transfer characteristics of closed two-phase thermosyphons, Paper B9P, in: Proceedings of the Seventh International Heat Pipe Conference, Minsk, May 1990.
[122] M. Groll, Heat pipe research and development in western Europe, Heat. Recover. Syst. CHP 9 (1) (1989) 19–66. Available from: https://doi.org/10.1016/0890-4332(89)90139-7.
[123] T. Payakaruk, P. Terdtoon, S. Ritthidech, Correlations to predict heat transfer characteristics of an inclined closed two-phase thermosyphon at normal operating conditions, Appl. Therm. Eng. 20 (9) (2000) 781–790. Available from: https://doi.org/10.1016/S1359-4311(99)00047-2.
[124] I. Pioro, M. Voronenva, Rascetnoe opredelenie predelnogo teplovogo potoka pri kipenii zidkostej v dvuhfaznih termosifonah, Inz. fiz. Z. 53 (1987) 376–383.
[125] I. Golobič, B. Gašperšič, Corresponding states correlation for maximum heat flux in two-phase closed thermosyphon, Int. J. Refrig. 20 (6) (1997) 402–410. Available from: https://doi.org/10.1016/S0140-7007(97)00019-4.
[126] R.K. Sakhuja, Flooding constraint in wickless heat pipes, ASME Published Paper, 1973.
[127] Z. Nejat, Effect of density ratio on critical heat flux in closed end vertical tubes, Int. J. Multiph. Flow. 7 (3) (1981) 321–327. Available from: https://doi.org/10.1016/0301-9322(81)90025-2.
[128] Y. Katto, Generalized correlation of critical heat flux in natural convection boiling within confined channels, Trans. Jpn. Soc. Mech. Eng. 44 (387) (1978) 3908–3911. Available from: https://doi.org/10.1299/KIKAI1938.44.3908.
[129] C.L. Tien, K.S. Chung, Entrainment limits in heat pipes, AIAA J. 17 (6) (1979) 643–646. Available from: https://doi.org/10.2514/3.61190.
[130] K. Harada, S. Inoue, J. Fujita, H. Suematsu, Y. Wakiyama, Heat transfer characteristics of large heat pipes, Hitachi Zosen Tech. Rev. 41 (1980) 167–174.
[131] Z. Gorbis, G. Savchenkov, Low temperature two-phase closed thermosyphon investigation, in: 2nd International Heat Pipe Conference, 1976, pp. 37–45.
[132] M. Bezrodnyi, Isledovanie krizisa teplomassoperenosa v nizkotemperaturnih besfiteljinyih teplovih trubah, Teplofiz. Visok. Temp. 15 (1977) 371–376.
[133] M. Groll, S. Rosler, Development of advanced heat transfer components for heat recovery from hot waste gases, Final Report CEC Contract No. EN3E0027D(B), 1989.
[134] F. Prenger, Performance limits of gravity-assisted heat pipes, in: 5th Int. Heat Pipe Conf., 1984, pp. 1–5.
[135] T. Fukano, K. Kadoguchi, C. Tien, Experimental study on the critical heat flux at the operating limit of a closed two-phase thermosyphon, Heat. Transf. – Jan. Res. 17 (1998) 43–60.
[136] H. Imura, K. Sasaguchi, H. Kozai, S. Numata, Critical heat flux in a closed two-phase thermosyphon, Int. J. Heat. Mass. Transf. 26 (8) (1983) 1181–1188. Available from: https://doi.org/10.1016/S0017-9310(83)80172-0.
[137] I. Golobic, B. Gaspersic, Generalized method for maximum heat transfer performance in two-phase closed thermosyphon, in: International Conference CFCs, 1994, pp. 607–616.
[138] B.P. McGrail, P.K. Thallapally, J. Blanchard, S.K. Nune, J.J. Jenks, L.X. Dang, Enhancing thermal conductivity of fluids with nanoparticles, Nano Energy 2 (5) (1995) 845–855. Available from: https://doi.org/10.1016/J.NANOEN.2013.02.007.
[139] Z.H. Liu, Y.Y. Li, A new frontier of nanofluid research – application of nanofluids in heat pipes, Int. J. Heat. Mass. Transf. 55 (23–24) (2012) 6786–6797. Available from: https://doi.org/10.1016/J.IJHEATMASSTRANSFER.2012.06.086.
[140] T.H. Tsai, H.T. Chien, P.H. Chen, Improvement on thermal performance of a disk-shaped miniature heat pipe with nanofluid, Nanoscale Res. Lett. 6 (1) (2011) 590. Available from: https://doi.org/10.1186/1556-276X-6-590.
[141] W.C. Wei, S.H. Tsai, S.Y. Yang, S.W. Kang, Effect of nanofluid concentration on heat pipe thermal performance. [Online]. <http://www.me.tku.edu.tw/> (accessed 05.10.22).
[142] S.W. Kang, W.C. Wei, S.H. Tsai, S.Y. Yang, Experimental investigation of silver nano-fluid on heat pipe thermal performance, Appl. Therm. Eng. 26 (17–18) (2006) 2377–2382. Available from: https://doi.org/10.1016/J.APPLTHERMALENG.2006.02.020.
[143] X.F. Yang, Z.H. Liu, J. Zhao, Heat transfer performance of a horizontal micro-grooved heat pipe using CuO nanofluid, J. Micromech. Microeng. 18 (3) (2008). Available from: https://doi.org/10.1088/0960-1317/18/3/035038.
[144] Z.H. Liu, L. Lu, Thermal performance of axially microgrooved heat pipe using carbon nanotube suspensions, J. Thermophys. Heat. Transf. 23 (1) (2012) 170–175. Available from: https://doi.org/10.2514/1.38190. Available from: https://doi.org/10.2514/1.38190.
[145] K.H. Do, S.P. Jang, Effect of nanofluids on the thermal performance of a flat micro heat pipe with a rectangular grooved wick, Int. J. Heat. Mass. Transf. 53 (9–10) (2010) 2183–2192. Available from: https://doi.org/10.1016/J.IJHEATMASSTRANSFER.2009.12.020.
[146] M. Shafahi, V. Bianco, K. Vafai, O. Manca, Thermal performance of flat-shaped heat pipes using nanofluids, Int. J. Heat. Mass. Transf. 53 (7–8) (2010) 1438–1445. Available from: https://doi.org/10.1016/J.IJHEATMASSTRANSFER.2009.12.007.
[147] M. Shafahi, V. Bianco, K. Vafai, O. Manca, An investigation of the thermal performance of cylindrical heat pipes using nanofluids, Int. J. Heat. Mass. Transf. 53 (1–3) (2010) 376–383. Available from: https://doi.org/10.1016/J.IJHEATMASSTRANSFER.2009.09.019.
[148] Z.H. Liu, Y.Y. Li, R. Bao, Thermal performance of inclined grooved heat pipes using nanofluids, Int. J. Therm. Sci. 49 (9) (2010) 1680–1687. Available from: https://doi.org/10.1016/J.IJTHERMALSCI.2010.03.006.

[149] G.S. Wang, B. Song, Z.H. Liu, Operation characteristics of cylindrical miniature grooved heat pipe using aqueous CuO nanofluids, Exp. Therm. Fluid Sci. 34 (8) (2010) 1415–1421. Available from: https://doi.org/10.1016/J.EXPTHERMFLUSCI.2010.07.004.

[150] C.Y. Tsai, H.T. Chien, P.P. Ding, B. Chan, T.Y. Luh, P.H. Chen, Effect of structural character of gold nanoparticles in nanofluid on heat pipe thermal performance, Mater. Lett. 58 (9) (2004) 1461–1465. Available from: https://doi.org/10.1016/J.MATLET.2003.10.009.

[151] T. Shu, Z.-H. Liu, Application of nanofluid in thermal performance enhancement of horizontal screen heat pipe | Request PDF, Aerosp. Power 23 (2008) 1623–1627 (accessed 05.10.22) [Online]. Available from: https://www.researchgate.net/publication/287008013_Application_of_nanofluid_in_thermal_performance_enhancement_of_horizontal_screen_heat_pipe.

[152] Y.T. Chen, W.C. Wei, S.W. Kang, C.S. Yu, Effect of nanofluid on flat heat pipe thermal performance, in: 2008 Twenty-fourth Annual IEEE Semiconductor Thermal Measurement and Management Symposium, 2008, pp. 16–19, https://doi.org/10.1109/STHERM.2008.4509359.

[153] K.H. Do, H.J. Ha, S.P. Jang, Thermal resistance of screen mesh wick heat pipes using the water-based $Al_2O_3$ nanofluids, Int. J. Heat. Mass. Transf. 53 (25–26) (2010) 5888–5894. Available from: https://doi.org/10.1016/J.IJHEATMASSTRANSFER.2010.07.050.

[154] Z. Liu, Q. Zhu, Application of aqueous nanofluids in a horizontal mesh heat pipe, Energy Convers. Manag. 52 (1) (2011) 292–300. Available from: https://doi.org/10.1016/J.ENCONMAN.2010.07.001.

[155] R. Senthilkumar, S. Vaidyanathan, B. Sivaraman, Experimental analysis of cylindrical heat pipe using copper nanofluid with an aqueous solution of n-hexanol, Front. Heat. Pipes 2 (3) (2012). Available from: https://doi.org/10.5098/FHP.V2.3.3004.

[156] N. Putra, W.N. Septiadi, H. Rahman, R. Irwansyah, Thermal performance of screen mesh wick heat pipes with nanofluids, Exp. Therm. Fluid Sci. 40 (2012) 10–17. Available from: https://doi.org/10.1016/J.EXPTHERMFLUSCI.2012.01.007.

[157] P.Y. Wang, X.J. Chen, Z.H. Liu, Y.P. Liu, Application of nanofluid in an inclined mesh wicked heat pipes, Thermochim. Acta 539 (2012) 100–108. Available from: https://doi.org/10.1016/J.TCA.2012.04.011.

[158] R.R. Riehl. Analysis of loop heat pipe behaviour using nanofluid. in: Proceedings of Heat Powered Cycles International Conference (HPC), 11-14 September 2006, Newcastle, UK, 06102, 2006.

[159] S.W. Kang, W.C. Wei, S.H. Tsai, C.C. Huang, Experimental investigation of nanofluids on sintered heat pipe thermal performance, Appl. Therm. Eng. 29 (5–6) (2009) 973–979. Available from: https://doi.org/10.1016/J.APPLTHERMALENG.2008.05.010.

[160] M. Keshavarz Moraveji, S. Razvarz, Experimental investigation of aluminum oxide nanofluid on heat pipe thermal performance, Int. Commun. Heat. Mass. Transf. 39 (9) (2012) 1444–1448. Available from: https://doi.org/10.1016/J.ICHEATMASSTRANSFER.2012.07.024.

[161] H.B. Ma, et al., Effect of nanofluid on the heat transport capability in an oscillating heat pipe, Appl. Phys. Lett. 88 (14) (2006) 143116. Available from: https://doi.org/10.1063/1.2192971.

[162] H.B. Ma, C. Wilson, Q. Yu, K. Park, U.S. Choi, M. Tirumala, An experimental investigation of heat transport capability in a nanofluid oscillating heat pipe, J. Heat. Transf. 128 (11) (2006) 1213–1216. Available from: https://doi.org/10.1115/1.2352789.

[163] F. Shang, D. Liu, H. Xian, Y. Yang, X. Du, Flow and heat transfer characteristics of different forms of nanometer particles in oscillating heat pipe, CIESC J. 58 (9) (2007) 2200–2204 (accessed 05.10.22) [Online].. Available from: https://hgxb.cip.com.cn/EN/Y2007/V58/I9/2200.

[164] Y.H. Lin, S.W. Kang, H.L. Chen, Effect of silver nano-fluid on pulsating heat pipe thermal performance, Appl. Therm. Eng. 28 (11–12) (2008) 1312–1317. Available from: https://doi.org/10.1016/J.APPLTHERMALENG.2007.10.019.

[165] K. Park, H.B. Ma, Nanofluid effect on the heat transport capability in a well-balanced oscillating heat pipe, J. Thermophys. Heat. Transf. 21 (2) (2012) 443–445. Available from: https://doi.org/10.2514/1.22409. Available from: https://doi.org/10.2514/1.22409.

[166] J. Qu, H. Ying Wu, P. Cheng, Thermal performance of an oscillating heat pipe with $Al_2O_3$–water nanofluids, Int. Commun. Heat. Mass. Transf. 37 (2) (2010) 111–115. Available from: https://doi.org/10.1016/J.ICHEATMASSTRANSFER.2009.10.001.

[167] N. Bhuwakietkumjohn, S. Rittidech, Internal flow patterns on heat transfer characteristics of a closed-loop oscillating heat-pipe with check valves using ethanol and a silver nano-ethanol mixture, Exp. Therm. Fluid Sci. 34 (8) (2010) 1000–1007. Available from: https://doi.org/10.1016/J.EXPTHERMFLUSCI.2010.03.003.

[168] P. Cheng, S. Thompson, J. Boswell, H. Ma, An investigation of flat-plate oscillating heat pipes [Online]. <https://proceedings.asmedigitalcollection.asme.org/> (accessed 05.10.22).

[169] Y. Ji, H. Ma, F. Su, G. Wang, Particle size effect on heat transfer performance in an oscillating heat pipe, Exp. Therm. Fluid Sci. 35 (4) (2011) 724–727. Available from: https://doi.org/10.1016/J.EXPTHERMFLUSCI.2011.01.007.

[170] M. Taslimifar, M. Mohammadi, H. Afshin, M.H. Saidi, M.B. Shafii, Overall thermal performance of ferrofluidic open loop pulsating heat pipes: an experimental approach, Int. J. Therm. Sci. 65 (2013) 234–241. Available from: https://doi.org/10.1016/J.IJTHERMALSCI.2012.10.016.

[171] R.R. Riehl, N. Dos Santos, Water-copper nanofluid application in an open loop pulsating heat pipe, Appl. Therm. Eng. 42 (2012) 6–10. Available from: https://doi.org/10.1016/J.APPLTHERMALENG.2011.01.017.

[172] Y. Peng, S. Huang, K. Huang, Experimental study on thermosiphon by adding nanoparticles to working fluid, J. Chem. Industry Eng. (China) 55 (2004) 1768–1772.

[173] H.S. Xue, J.R. Fan, Y.C. Hu, R.H. Hong, K.F. Cen, The interface effect of carbon nanotube suspension on the thermal performance of a two-phase closed thermosyphon, J. Appl. Phys. 100 (10) (2006) 104909. Available from: https://doi.org/10.1063/1.2357705.

[174] Z.H. Liu, X.F. Yang, G.L. Guo, Effect of nanoparticles in nanofluid on thermal performance in a miniature thermosyphon, J. Appl. Phys. 102 (1) (2007) 013526. Available from: https://doi.org/10.1063/1.2748348.

[175] Z. Hua Liu, J. Guo Xiong, R. Bao, Boiling heat transfer characteristics of nanofluids in a flat heat pipe evaporator with micro-grooved heating surface, Int. J. Multiph. Flow. 33 (12) (2007) 1284–1295. Available from: https://doi.org/10.1016/J.IJMULTIPHASEFLOW.2007.06.009.

[176] S. Khandekar, Y.M. Joshi, B. Mehta, Thermal performance of closed two-phase thermosyphon using nanofluids, Int. J. Therm. Sci. 47 (6) (2008) 659–667. Available from: https://doi.org/10.1016/J.IJTHERMALSCI.2007.06.005.
[177] P. Naphon, P. Assadamongkol, T. Borirak, Experimental investigation of titanium nanofluids on the heat pipe thermal efficiency, Int. Commun. Heat. Mass. Transf. 35 (10) (2008) 1316–1319. Available from: https://doi.org/10.1016/J.ICHEATMASSTRANSFER.2008.07.010.
[178] P. Naphon, D. Thongkum, P. Assadamongkol, Heat pipe efficiency enhancement with refrigerant–nanoparticles mixtures, Energy Convers. Manag. 50 (3) (2009) 772–776. Available from: https://doi.org/10.1016/J.ENCONMAN.2008.09.045.
[179] S.H. Noie, S.Z. Heris, M. Kahani, S.M. Nowee, Heat transfer enhancement using $Al_2O_3$/water nanofluid in a two-phase closed thermosyphon, Int. J. Heat. Fluid Flow. 30 (4) (2009) 700–705. Available from: https://doi.org/10.1016/J.IJHEATFLUIDFLOW.2009.03.001.
[180] Z. Hua Liu, X. Fei Yang, G. San Wang, G. Liang Guo, Influence of carbon nanotube suspension on the thermal performance of a miniature thermosyphon, Int. J. Heat. Mass. Transf. 53 (9–10) (2010) 1914–1920. Available from: https://doi.org/10.1016/J.IJHEATMASSTRANSFER.2009.12.065.
[181] T. Parametthanuwat, S. Rittidech, A. Pattiya, A correlation to predict heat-transfer rates of a two-phase closed thermosyphon (TPCT) using silver nanofluid at normal operating conditions, Int. J. Heat. Mass. Transf. 53 (21–22) (2010) 4960–4965. Available from: https://doi.org/10.1016/J.IJHEATMASSTRANSFER.2010.05.046.
[182] T. Paramatthanuwat, S. Boothaisong, S. Rittidech, K. Booddachan, Heat transfer characteristics of a two-phase closed thermosyphon using de ionized water mixed with silver nano, Heat. Mass. Transf. und Stoffuebertragung 46 (3) (2010) 281–285. Available from: https://doi.org/10.1007/S00231-009-0565-Y/FIGURES/6.
[183] G. Huminic, A. Huminic, I. Morjan, F. Dumitrache, Experimental study of the thermal performance of thermosyphon heat pipe using iron oxide nanoparticles, Int. J. Heat. Mass. Transf. 54 (1–3) (2011) 656–661. Available from: https://doi.org/10.1016/J.IJHEATMASSTRANSFER.2010.09.005.
[184] G. Huminic, A. Huminic, Heat transfer characteristics of a two-phase closed thermosyphons using nanofluids, Exp. Therm. Fluid Sci. 35 (3) (2011) 550–557. Available from: https://doi.org/10.1016/J.EXPTHERMFLUSCI.2010.12.009.
[185] J. Qu, H. Wu, Thermal performance comparison of oscillating heat pipes with $SiO_2$/water and $Al_2O_3$/water nanofluids, Int. J. Therm. Sci. 50 (10) (2011) 1954–1962. Available from: https://doi.org/10.1016/J.IJTHERMALSCI.2011.04.004.
[186] Y.H. Hung, T.P. Teng, B.G. Lin, Evaluation of the thermal performance of a heat pipe using alumina nanofluids, Exp. Therm. Fluid Sci. 44 (2013) 504–511. Available from: https://doi.org/10.1016/J.EXPTHERMFLUSCI.2012.08.012.
[187] B.D. Marcus, Theory and design of variable conductance heat pipes [Online]. <https://ntrs.nasa.gov/citations/19720016303\>, 1972 (accessed 28.07.22).
[188] H.R. Holmes and A.R. Feild, Gas-tolerant high-capacity tapered artery heat pipe, AIAA Paper, 1986, doi: 10.2514/6.1986-1343.
[189] ASHRAE, ASHRAE Handbook of Fundamentals, ASHRAE, 2005.
[190] W.H. McAdams, Heat Transmission, third ed., McGraw-Hill, 1954.
[191] M.K.A. Bezrodnyi, Investigation of the critical region of heat and mass transfer in low-temperature wickless heat pipes - NASA, High. Temp. Sci. 15 (2) (1977) 309–313. Accessed: Jul. 28, 2022. [Online]. Available from: https://ui.adsabs.harvard.edu/abs/1977TepVT.15.370B/abstract.
[192] ESDU, Heat pipes - general information on their use, operation, and design. EDU 80013. London: Engineering Sciences Data Unit, 1980. ISBN 978 0 85679 296 0.
[193] M. Groll, D. Heine, T. Spendel, Heat recovery units employing reflux heat pipes as components [Online]. Available: <https://www.osti.gov/etdeweb/biblio/5724880\> January 1984 (accessed 28.07.22).
[194] M.S. El-Genk, H.H. Saber, Determination of operation envelopes for closed, two-phase thermosyphons, Int. J. Heat. Mass. Transf. 42 (5) (1999) 889–903. Available from: https://doi.org/10.1016/S0017-9310(98)00212-9.

Chapter 5

# Additive manufacturing applied to heat pipes

## 5.1 Introduction

Additive manufacturing (AM) is one of many digital technologies that sits within a larger framework called the 'Fourth Industrial Revolution' – or Industry 4.0 for short. It is considered to be a digital technology because it seamlessly 'closes the gap' between a digital model and its corresponding physical model [1]. AM involves the layer-by-layer construction of complex 3D structures where the resulting part exists as a single manufactured piece. The joint ISO/ASTM 52900–15 (2015) standard formally defines AM as 'the process of joining materials to make parts from 3D model data, usually layer upon layer, as opposed to subtractive manufacturing and formative manufacturing'. There are several distinct technologies, each with its own advantages and disadvantages, and these can be largely divided into liquid-based and solid-based techniques [2]. The ISO/ASTM standard recognises seven distinct technologies: (1) vat polymerisation (liquid), (2) material extrusion (solid filament), (3) material jetting (liquid), (4) powder bed fusion (solid powder), (5) binder jetting (solid powder), (6) directed energy deposition (solid powder or filament) and (7) sheet lamination (solid sheet). The capital costs of these vary dramatically: <$1k for material extrusion, ~$5k for vat polymerisation, ~$10k for plastic powder bed fusion, $50k–$250k for binder jetting and $70k–$700k for metal powder bed fusion [2].

Typical advantages of AM include direct translation of design to prototype; easy and affordable customisation; complex geometries; 'lightweighting'; minimal postprocessing; approaching zero-waste manufacturing; rapid prototyping and production; small manufacturing footprints for large product varieties; and on-demand manufacturing [3]. AM is consequently a fast-growing area and there are many examples of AM applied to the field [4]. AM has also been proposed for the development of more advanced wick structures and heat pipe concepts in general, where AM enables the integration of heat pipe components, heat exchangers and heat sinks into a single chassis or an integrated assembly.

There are numerous potential benefits for using AM to fabricate heat pipes. Principally, AM wick pore structures overcome the limitations of sintered wicks: the porosity, pore shape, pore size and pore size distribution can be more precisely controlled to eliminate randomness and maximise pore interconnectivity [5,6]. This enables the realisation of 'volumetric designs' with few geometric constraints, which overcomes the repeated 2D structures often encountered through the diffusion bonding processes of printed circuit heat exchangers. Additionally, laser-based powder bed fusion enables the construction of hermetically sealed metal structures and the elimination of contact thermal resistances. This reduces the number of individual parts and fabrication steps [7,8], enabling the integration of wick structures with the heat pipe wall and the integration of other structures into the heat pipe such as embedded tubes, hollow volumes, porous sections, reticulated structures, extended surfaces and heat exchangers [5,8,9]. Finally, the inherent multiscale surface features (geometry and roughness) can enhance evaporation and condensation whilst simultaneously promoting capillary action and maximising permeability [5].

Standard wicked heat pipes use capillarity to move a liquid condensate from the condenser to the evaporator sections – but this is often a limiting operational factor (see Chapter 2), and so it is desirable for wicks to simultaneously generate high capillary pressures and high permeabilities to minimise pressure drop. However, in most homogeneous and heterogeneous wick structures, reducing the equivalent pore size to increase the capillary pressure simultaneously increases the flow resistance [10]. Because of this, advanced wick design strategies usually involve combining homogeneous and heterogeneous structures, such as mesh covered grooves and arterial wicks to generate both high capillarity (pumping power) and permeability (described in Chapter 2). In addition, as well as generating a capillary pressure, wicks must also simultaneously dissipate heat to the working fluid to enhance evaporation, provide structural support, provide a flow path for the working fluid and recapture the condensate in the condenser section to close the thermal circuit [5,7]. Complex flows therefore often develop in these pore networks [7]. In contrast, pulsed heat pipes typically do

**Heat Pipes. DOI: https://doi.org/10.1016/B978-0-12-823464-8.00003-6**

not use capillary structures at all, but instead rely on complex thermo-hydraulic coupling around a closed loop to transfer thermal energy, involving the flows of alternating liquid and vapour slugs. Furthermore, loop heat pipes (LHPs) involve an additional complexity by combining vapour−liquid slug flows with a wick structure in the evaporator section. Fortunately, since the wick is only used in the evaporator (unlike conventionally wicked heat pipes) permeability and pressure drop are less critical to performance and much finer pore sizes can be used to maximise evaporation and capillary wicking [11].

Numerous design factors therefore influence heat pipe performance, including: *wick properties* (e.g. porosity, thickness and permeability), *material properties* (e.g. effective thermal conductivity), *working fluid properties* (e.g. wettability, density, surface tension and viscosity), *heat pipe geometry* (e.g. surface area, shape and size) and *operating conditions* (e.g. gravitational acceleration and inclination angle, heat source/sink separation, heat input and heat density and cooling duty). There are therefore many potential optima in this large parametric design space and the main role of AM is to make more of these optima accessible.

Although the use of AM for optimising heat pipe performance is still in its infancy, there is, nonetheless, a clear opportunity to use AM to deliver the next generation of heat pipe designs. The aim of this chapter is therefore to discuss the opportunities and challenges based on the current state of the art. This will include a brief description of the AM process; an overview of the current state of the art of additively manufactured heat pipes and wicks; and a review of recent public domain reports concerning the latest advances in materials and surface science to identify possible design features for improving boiling, condensate flow and condensation, whilst highlighting the various specific challenges and limitations relevant to heat pipe development.

## 5.2 Additive manufacturing considerations for heat pipes

Metals can be additively manufactured using all seven technology areas defined by the joint ISO/ASTM standard, but the laser-driven powder bed fusion (L-PBF) process is by far the most common method for metals, enabling *thin* wall thicknesses with *acceptable* costs. And because this technology enables the fabrication of metal structures with near 100% relative densities, it is the most suitable for current novel heat pipe fabrication.

The complete AM process is an amalgam of several production stages, involving: conceptual design (usually with input from prior simulated or experimental results); generation of the 3D model (either by CAD or 3D scanner); validation of the 3D model ('watertightness' and holes checks); build preparation (modification of the model with 'AM-specifics' such as print orientation, inclusion of support structures and layer height settings); generation of machine instructions; fabrication (an automated process taking hours to days with occasional manual inspection/intervention); postprocessing (powder and support material removal as well as other finishes); and qualification [12,13]. Each of these is complex, and each produces its own challenges including standardisation [4] and health/safety issues.

The selection of a suitable machine and material for a heat pipe will depend upon the constraints of the specific printing process, materials, heat pipe application and the heat pipe itself. The Senvol Database (http://senvol.com/database/) is a good starting point − a free comprehensive database of industrial AM machines and materials − providing information such as supplier names, build envelope size, detailed materials properties and costs (where data is available).

The materials landscape is now quite diverse for metal applications. The most commonly reported metals being steel/iron-based alloys (primarily 316 L), titanium/titanium alloys (dominated by the Ti−6Al−4V, or Ti64, alloy [14]), Inconel/nickel-based alloys (primarily IN625 and IN718, but also Abd900 and Hastelloy X) and aluminium (most often the AlSi10Mg casting alloy) because of its established processability [15]. Of course, a variety of other powders such as bronze (CuSn-10) [16], tantalum [17], tungsten carbide [18], magnesium [19] and many other alloys [20] have also been printed. Increasingly the use of copper for heat transfer applications [4] and neodymium (NdFeB) for permanent magnetic applications are emerging [21], and examples and the challenges of these metals in heat transfer applications are discussed in a recent review [4].

Another significant factor with regards to machine and materials selection is the cost, the four main drivers of which are the machine investment and preparation, materials, labour and risk [12]. Of these, the cost of machine investment often dominates, accounting from anywhere between 47% and 78% of the total [22]. Materials contribute not only a direct component (i.e. the purchasing price), but also an indirect component associated with deposition rate (also a function of the powder fusion process). For instance, some materials require longer to distribute and melt which increases the lead time. Labour is especially relevant to the postprocessing stage, accounting for between 4% and 13% of the overall cost depending on build rate, machine utilisation, material cost and machine investment [22]. Risk is also an interesting factor since build failures can account for as much as 26% of the unit cost [12]. Accessing reliable information remains a challenge for performing accurate cost—benefit analyses because a lot of critical information is missing from the literature [12]. Table 5.1 highlights some approximate prices of different materials whilst Table 5.2 summarises the typical price ranges of common postprocessing operations to serve as a guide. It is stressed that the costs in Table 5.1 are subject to the level of utilisation of the build envelope, where compromises will certainly be needed between the production speed (influenced by the printer settings and part packing densities) and materials limitations (e.g. melting points and the need to minimise thermal gradients).

On the design side, heat pipe cases are simple to create using most CAD software packages, requiring only routine push/pull-type manipulations on standard geometric shapes (circles, polygons, lines and curves/arcs and so on). Using these tools, it is also possible to craft simple/regular lattice structures, for example, by copying and merging grids of cylinders together to form the struts of an octahedral lattice network. However, the creation of more advanced designs (as described in Section 5.4.1) would be extremely difficult and tedious with this manual approach. Instead, a range of

**TABLE 5.1** Cost price per $cm^3$ of different materials and processes, based on the assumption of ideal build envelope usage [12].

| Process | Material | Cost (€/$cm^3$) |
|---|---|---|
| L-PBF | SS 304 L | 8.25 |
| | 17—4 PH | 6.62—8.63 |
| | 17—4 PH | 7.17 |
| | 316 L | 7.03 |
| | AlSi10Mg | 7.97 |
| | Titanium | 5.68 |
| E-PBF | Ti—6Al—4V | 2.77 |
| | Titanium | 4.54 |
| DED — powder | Titanium | 2.11 |
| DED — wire | Titanium | 2.11 |
| Binder Jetting | Titanium | 1.96 |

**TABLE 5.2** Cost ranges of common postprocessing operations [12].

| Technique | Details | Cost range |
|---|---|---|
| Part removal | Removal of the part from the build plate by wire EDM | $200—300 per build plate |
| | Removal of the part from the build plate by band saw | Low cost |
| Stress relief | Removal or reduction of thermal residual stresses | $500—600 per build plate |
| Heat treatment | Improvement of the microstructure and mechanical properties | $500—2,000 per build plate |
| Hot Isostatic pressing | Reduce porosity/increase densification and improve fatigue life | $500—2,000 per build plate |
| Machining | For example, improve the accuracy of joining interfaces and surfaces, removal supports/burrs, addition of threads | Depends on geometry, materials and fixture needs |
| Surface treatments | Improve surface finish, quality and/or roughness | $200—2,000 per build plate |
| Inspection/ testing | Process qualification, part validation and certification | 10%—20% of the total part cost |

dedicated automatic lattice generation tools now exist on the market. Autodesk Within and nTopology are both examples described as 'generative design' tools that use algorithms to create structures that would be 'practically impossible' to draw manually. Autodesk Within also enables optimisation of the lattice distribution and includes a stress/structural analyser to gain further insight of the potential vulnerabilities and weaknesses (which may be useful when it comes to cutting the final heat pipe from the build plate for example). Simpler tools such as STL Lattice Generator (runs in Matlab), Meshify (a browser-based solution), Simpleware and Fabpilot (a Cloud-based software package) are also reported. However, often these commercial packages only provided limited control over the lattice properties [12].

In addition to build speed and materials constraints, two other constraints on heat pipe design are the build volume size (limiting the overall length of a linear heat pipe) and the resolution/accuracy (where poor surface finish, pore closing and thermally induced warpage are the main concerns). Currently, the majority of the octahedral lattice structures intended for heat pipes are designed by drawing a single unit cell, using this to populate the volume of interest with repeating octahedral elements (Fig. 5.1). In these cases, the final lattice parameters (such as the pore size and porosity) are influenced by the chosen machine settings such as beam size, layer thickness, laser power and hatch distance, rather than through the CAD file itself. This requires optimisation of the settings for each bespoke lattice design. However, this is not necessarily a disadvantage because, for example, Jafari et al. were able to 3D print a lattice with a mean pore size 6% lower than their CAD model by optimising the laser settings [23]. Further, the inherent surface roughness of the lattice struts is believed to contribute to improved wettability [5], making lower accuracy an asset rather than a liability for heat pipes. For reference, Feng et al. summarise some further elements of generic lattice design principles for AM, including topology, truss, thermal and material considerations [24].

Once the machine and materials have been selected and the heat pipe model generated, subject to the constraints of the particular technology/materials, the build process can proceed. The powder bed fusion process starts by evenly distributing a thin layer of metal powder over the build platform using a rolling arm (also called rake, or recoating arm). A laser beam (or multiple laser beams) driven by the CAD model then scans the powder to selectively melt the powder. This rapidly cools and solidifies leaving the unmelted powder in place. The build platform is then lowered a small amount (typically 20–100 μm to strike a balance between build time, resolution and powder flowability [26]), and a new layer of powder is added with the rolling arm. This new layer of powder is then melted into the first layer with the laser creating an indistinguishable join between the two and the process is repeated. In terms of product purity, typically an inert atmosphere (such as argon gas) fills the print chamber to prevent oxidation. In addition, with the selective laser melting (SLM) technique, the powder is fully melted and nonsintered, and this enables relative densities close to 100%. Each layer can also be slightly offset from the previous one, facilitating the creation of overhangs. So together with the unmelted powder acting as a support material, nonporous tubular structures can be easily produced, and this is critical for the manufacture of heat exchanger and heat pipe devices. A basic schematic of the process is shown in Fig. 5.2.

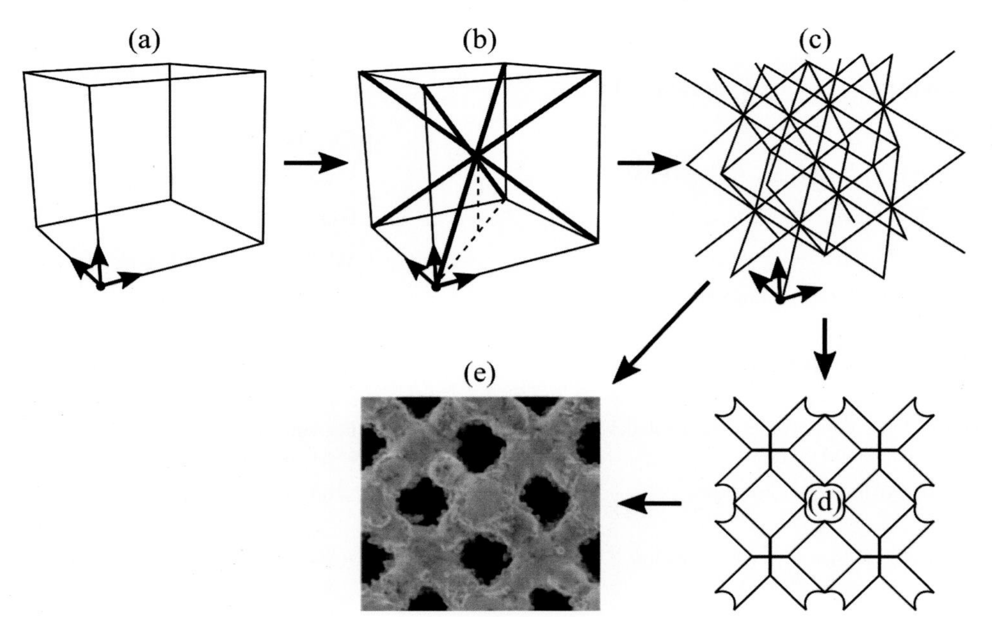


**FIGURE 5.1** Typical procedure for generating octahedral lattice wick structures for heat pipes (redrawn from [25]) | (a) unit cell defined with Cartesian coordinates, (b) unit cell transformed into octahedral structure, (c) replication of the octahedral lattice structure to map out the full wick shape (defines the coordinates for the laser), (d) addition of thickness to the octahedral lattice struts (optional) and (e) final pore structure following optimisation of the laser settings [5].

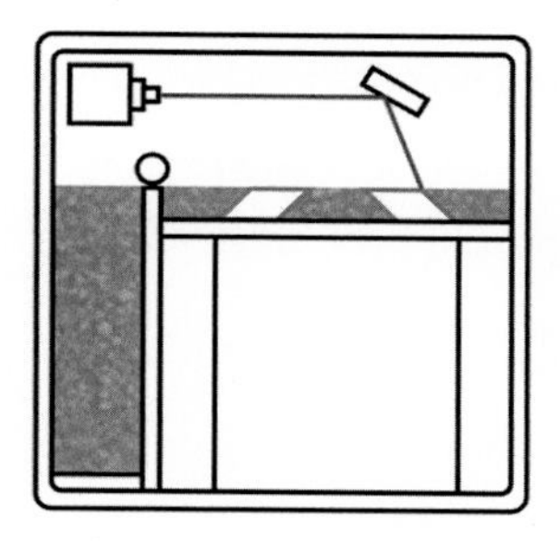
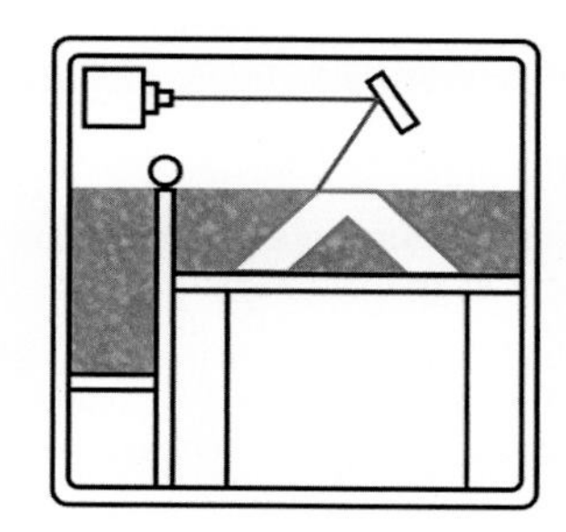
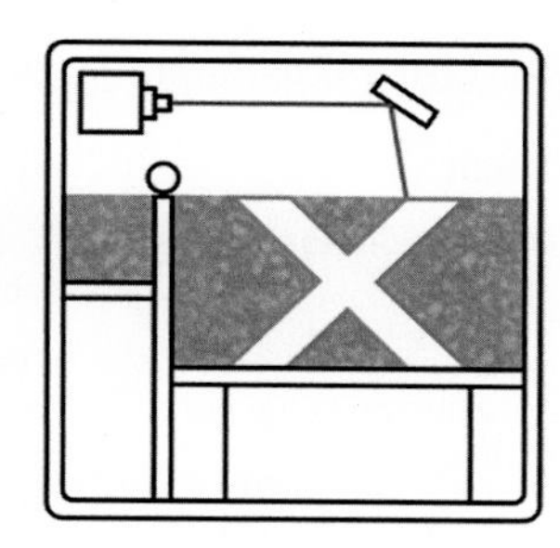

**FIGURE 5.2** Basic schematic of the laser powder bed fusion (PBF) process. *Redrawn from Ref. [4].*

Postprocessing comprises a number of key steps, and these must be planned for in advance. Some of these operations are integral to the AM technology itself (e.g. cutting the part from the build platform in powder bed fusion [PBF] processes), whilst others relate to improving part performance (e.g. modifying the surface finish). In respect to platform separation, parts can be cut from the platform with a band saw or by electric discharge machining (wire cutting). Jafari et al. chose the latter to minimise pore smearing in their stainless steel octahedral lattice structures [7]. All postprocessing activities risk unwanted warpage (especially for thin-walled components), which should be factored in at the design stage.

L-PBF processes typically leave unmelted powder within the internal passages that can be difficult and time consuming to detect and remove. It is important though to remove this powder because it can sinter under high operating temperatures (during postprocessing or in the final application), leading to deviations from the desired dimensions. Apart from negatively affecting the external tolerances, this can also impact performance. For example, the trapped powder in an AlSi10Mg heat exchanger led to underperformance (lower $UA$ and higher $\Delta P$) compared to a 'stock' heat exchanger design [27]. Additionally, partially sintered powder can dislodge under normal operating conditions and then discharge into downstream equipment. For heat pipes, the charging line should serve as a suitable drainage point for internally trapped powder, and usually, a combination of tapping, shaking and compressed air blowing will dislodge the powder, which can then be sieved, sorted and remixed with virgin powder (along with the unused powder remaining in the bed). However, it is noted that blasting might not be effective in channels with length-to-radius ratios greater than 8 [28] and presumably difficult in complex lattice structures as well. Air or grit blasting also risks compactification of the residual powder [29]. Recently, a combination of ultrasonic polishing and vacuum boiling was also found to be effective [30]. These processes are all labour intensive though, contributing to the unit cost.

Generally, the printing process itself will also imbue stresses into the additively manufactured part, as much as 300 MPa in Ti64 [31], especially if the parts are subjected to intense temperature gradients if the bed is not properly managed. These can be minimised somewhat by careful temperature control or preheating of the build platform and atmosphere [32], using support structures [33], through the design of the part itself and careful selection of printing parameters [34], or they can be removed/reduced via the use of postprocessing heat treatments (see Table 5.3). Note, these treatments should be applied to the part in a 'clamped form' (kept in the as-built state) to prevent distortion prior to stress relief [31]. Support structure and burr removal, polishing, surface coatings, drilling/tapping, pore filling and densification are other common examples of postprocessing. As a final thought, the internal surface finish of additively manufactured heat pipes will be 'baked in' at the point of fabrication, so changes to the internal finish will necessitate optimisation of the materials selection and machine settings at the outset.

Finally, the additively manufactured heat pipe must undergo an inspection and/or certification process. Examples include the evaluation of functional performance (i.e. confirmation that the heat pipe performs as expected under the desired heat loads – work that can be contracted out); mechanical performance checks (e.g. strength and/or cyclic loading testing); shape/size tolerance checks; powder removal confirmation (CT scanning remains the most reliable nondestructive approach); demonstration of gas tightness/pressure rating; standards compliance; or some combination thereof. Checks can be performed in situ during the fabrication process or can involve ex situ nondestructive measurements or destructive forensics to evaluate material properties/performance. Everton et al. provide an overview of in situ process monitoring techniques for metal AM [43]. Importantly, these results should feedback to design, either directly (design iteration) or indirectly (sharing of best practices). It has also been suggested that although one of the principal advantages of AM is the ability to easily/rapidly customise parts, the need for bespoke certification might just shift the cost/burden of customisation from fabrication to certification [4]. This could introduce unforeseen time penalties between inception and market delivery. For series-type productions, where customisation is not the focus, validation might only need to involve consistency checks [12]. The joint ISO/ASTM 52900–15 (2015) standard also covers several elements of certification. Recommendations include the use of challenge devices printed alongside the geometry of interest (if

**TABLE 5.3 A selection of postprocessing treatments applied to different additively manufactured metals.**

| Material | Process | Treatment | Results | References |
|---|---|---|---|---|
| 316 L SS | SLM | Low-temperature stress relief | Columnar microstructure and anisotropic corrosion fatigue cracking | [35] |
| | | Hot isostatic pressing (HIP) and solution annealing (full recrystallisation) | The as-built 316 L material produced a good density, so the HIP treatment showed no significant improvement over the low-temperature stress relief method | |
| | | High-temperature solution annealing (partial recrystallisation) | Significant reduction of yield strength observed. The tensile properties remained similar to the HIP-treated samples | |
| 420 SS | SLM | Low-temperature heat treatment | Improved tensile properties and elongation with no change in density | [36] |
| AlSi10Mg | SLM | T6-like heat treatment | Increased tensile elongation at failure, but also decreased nano/microhardness, UTS and yield tensile/compressive strengths | [37] |
| | | Hot isostatic pressing | Caused microstructural changes that strongly affected tensile properties (decreased hardness, fatigue limit and yield strength) | [38] |
| | | Annealing | Effective at removing residual stresses and improving mechanical tensile properties | [39] |
| IN625 | SLM | Annealing | Decreased yield stress observed, but tensile strength remained the same | [40] |
| | | Annealing and polishing | Increased ductility without loss of density or changes in microstructure | [41] |
| Ti–6Al–4V | SLM | Machining and polishing | Improved the endurance limit from 210 MPa for the 'as-built' sample up to 500 MPa for the polished specimen | [42] |
| | SLM | No heat treatment | Tensile ductility of as-built L-PBF around 10% lower than mill-annealed Ti64; tensile strengths are similar | [31] |
| | | Subtransus heat treatment | Heat treatments below 780°C often fail to influence tensile properties. Annealing at or above 940°C can improve tensile elongation | |
| | | Supertransus heat treatment | Can increase tensile elongation over 10% | |
| | EBM | Heat treatments are not usually necessary because of the higher temperatures of this process compared to SLM | | |
| | SLM/EBM | Hot isostatic pressing | Reduced fatigue strength by removing internal porosity and other defects. There is a risk the internal porosity can return | |

the challenge device passes the inspection process, this validates the full batch), using suitable conditions for material storage and following calibration procedures for the machines [12]. It remains to be seen whether the need for bespoke certification will impede the uptake of additively manufactured heat pipes within the various target applications.

## 5.3 State of the art

### 5.3.1 Additive manufacturing wick and heat pipe developments

Several groups have successfully 3D-printed metal lattice structures, some directly intended for heat pipe applications, with Ramirez et al. for example, using electron beam melting (EBM) to fabricate both reticulated meshes and stochastic

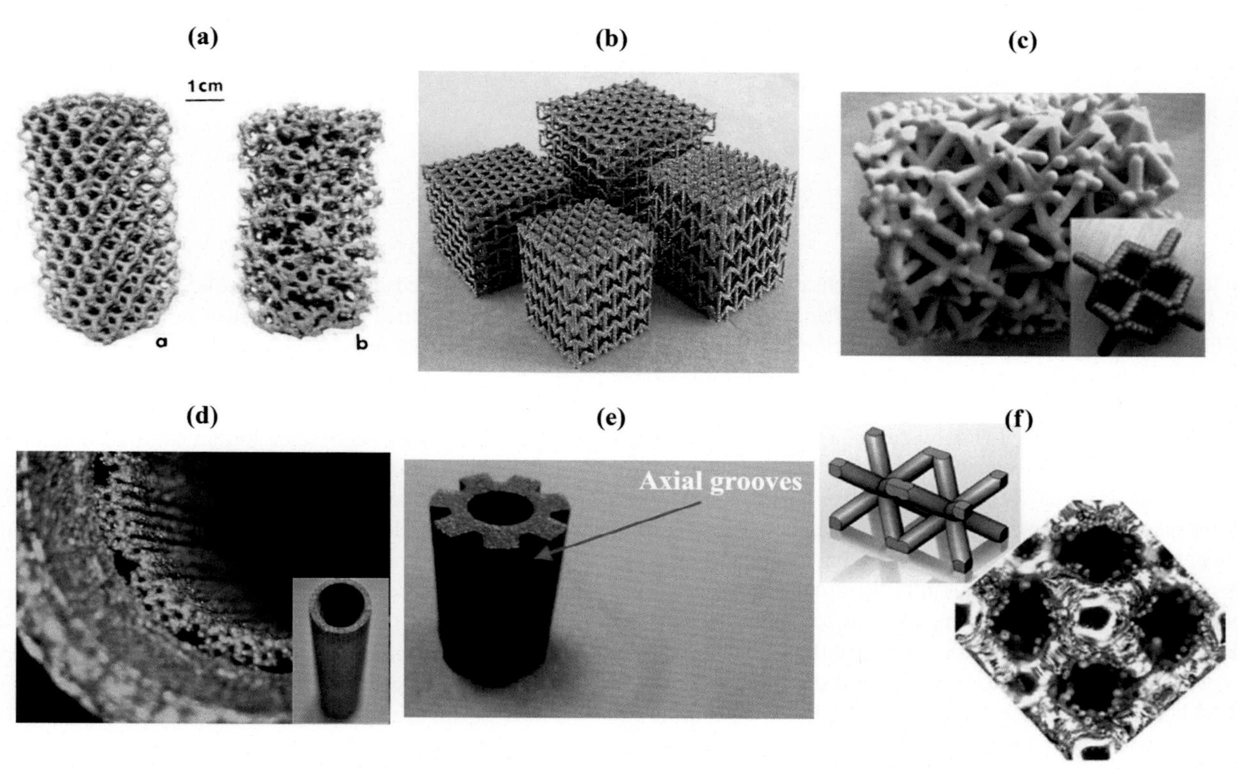


**FIGURE 5.3** Porous/lattice structures fabricated through metal additive manufacturing I (a) reticulated and stochastic copper foams [44], (b) periodic hexagonal cellular structures [47], (c) random and regular (inset) lattice structures [25], (d) arterial wick [25], (e) primary LHP wick with axial grooves [48] and (f) octahedral unit cell and 3D-printed sample [23].

open cellular foams using copper (Fig. 5.3a) [44]. They obtained material densities of between 0.73 and 6.67 g/cm$^3$ with average spacings between the structures of $\sim 2\ \mu m$. Similarly, Kumar et al. [45] report the use of hexagonal lattice structures for air heat exchangers fabricated using a titanium alloy (Fig. 5.3b). One potential issue raised by the authors is the uncontrolled surface roughness using EBM, although for heat pipes this may be advantageous because roughness improves the wetting characteristics [46].

Ameli et al. more comprehensively demonstrated the use of SLM to create randomised and structured aluminium wicks (Fig. 5.3c), including the inclusion of fluid arteries to increase the permeability (Fig. 5.3d) [25]. They obtained porosities of 50%–58% with permeabilities of the order of $10^{-10}\ m^2$ using a 500 μm unit cell, comparable to sintered copper that had a measured porosity of 40%–52% and permeability of $10^{-11}-10^{-12}\ m^2$. Esarte et al. achieved a similar permeability in a dual wick design intended for LHPs [48]. The primary wick was fabricated with SLM (stainless steel AISI 316) and had integrated grooves on the external surface (Fig. 5.3e). Their pore size of 80 μm delivered a permeability of $1.25 \times 10^{-12}$, porosity of 17% and capillary rise height of 13 mm (using water).

Jafari et al. have published several recent papers regarding the use of SLM to fabricate stainless steel heat pipe wicks. In 2017 they characterised the effective thermal conductivities ($k_{eff}$) and contact angles in a CL 20ES octahedral lattice (pore size of 160 μm) containing water, ethylene glycol and no working fluid [7]. They found that $k_{eff}$ was independent of temperature but depended on the working fluid. In order, water produced the highest $k_{eff}$ ($\sim$6 W/m K), followed by ethylene glycol ($\sim$3 W/m K), followed by the empty pores in a vacuum ($\sim$2 W/m K). Water also produced the highest contact angle (56.8 degrees vs 27.4 degrees), demonstrating that it wetted the sample much faster than ethylene glycol. Next, Jafari et al. assessed the capillary performance of stainless steel freeform porous structures created using SLM (Fig. 5.3f) [23]. They obtained a porosity of 46% and a capillary performance ($K/r_{eff}$) of 1.04, which was a factor of 1–6 larger than standard sintered and composite porous wicks, but a factor of 12 lower than grooves. Interestingly, Ameli et al.'s wick design had a capillary performance of $K/r_{eff} = 7.14$ [25] suggesting there is scope for further SLM optimisation of stainless steel wicks. More recently, Jafari et al. validated the thermal performance of an octahedral SS316L lattice where they noted several interesting observations [5]. First, the filling ratio of working fluid mainly affected the evaporator and had minimal effect on the condenser section's performance; the optimal charge ratio that produced the lowest wall temperatures was 110% (slightly above

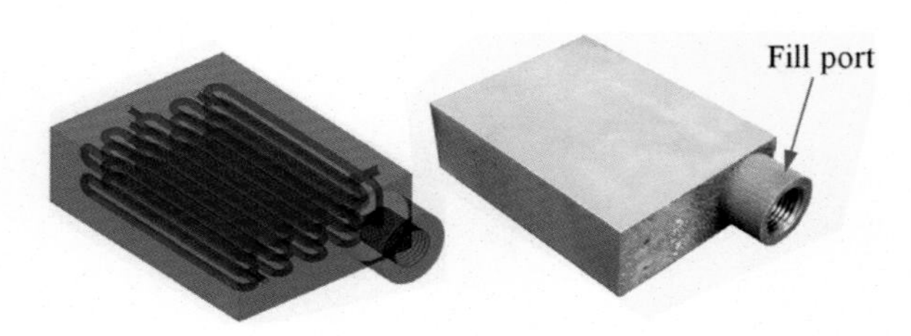


**FIGURE 5.4** Ti64 multilayered oscillating heat pipe [49].

complete wick saturation). Second, condensation in the condenser section occurred at the sintered-like surface, suggesting there is an opportunity to use AM to enhance all parts of the thermal circuit. Finally, the superheat performance of AM SS316L wicks was far superior to conventionally manufactured wicks because of increased meniscus density (high wettability of the rough surface) and high permeability within the pores.

Taking things further, Ozguc et al. fabricated a full stainless steel 316L vapour chamber heat spreader ($54 \times 90$ mm footprint) using deionised (DI) water as a working fluid [8]. Although aluminium possesses a higher thermal conductivity, they chose SS316 because the laser settings for lattice structures was more mature than for aluminium. They observed that the area-weight temperature was the same regardless of whether the device was charged with DI water or not (because of the same heat input and same ambient cooling), but the charged device had a flatter temperature profile. Further, the thermal resistance of the charged chamber was 10 times smaller than the empty vapour chamber and five times smaller than an equivalent solid block of stainless steel, indicating that the device was behaving as a heat pipe/spreader as intended. However, this project did not study the effect of different operating conditions, so device performance under real-world conditions is perhaps questionable. Ozguc et al. also noted that their pores were anisotropic (they were aligned with the vertical print direction), and this may contribute to flow blockage effects [8].

AM has also been used in the development of loop and pulsed heat pipes. Hu et al. used SLM to manufacture LHP evaporator wicks with three different pore radii (100, 200 and 300 μm) and used a transparent cover and liquid transport line to visualise the vapour flow patterns [6]. They observed three start-up stages: (1) initial evaporation, characterised by a rapid increase of the measured evaporator outlet temperature, (2) establishment of circulation, where the condenser inlet temperature rapidly increased, and finally (3) steady-state operation. For applied heat loads of 20–40 W, the 100 μm pore radius produced the lowest operating temperatures, whereas for applied heat loads of >60 W, the 200 μm pore radius produced the lowest temperatures. At low heat loads the 100 μm pore size produces a larger capillary pressure that facilitates surface wetting. However, the 100 μm pore radius also increases the vapour flow friction, trapping vapour in the wick at high heat loads. Additionally, the smaller pore radius produces a higher effective thermal conductivity in general, leading to more severe heat leak. The 300 μm performed worst in all conditions because the higher permeability led to flooding within the evaporator section.

Belfi et al. fabricated a fully operational flat plate pulsed heat pipe for space applications and performed a very basic parametric study [9]. The AlSi10Mg material achieved an absolute roughness of 2.0553 μm and the final dimensions were within the tolerances defined by the CubeSat application limit. They noted that it was difficult to remove powder via a combination of air blasting and ultrasonic washing from the 'standard definition' version which used a 50 μm 'resolution'. Whereas powder could be more easily removed from the 'high definition' 25 μm resolution version. This suggests that higher costs might be incurred (because the cost is a function of the print speed). Their pulsed heat pipe achieved a best relative thermal resistance of 1/10th the equivalent of an empty channel but noticed that gravity assisted the power up process. Several open questions regarding the validity of this particular AM approach remain: the effect of roughness on performance (implications on long-term stability), the cost–benefit of 'standard' versus 'high' definition settings and the lack of performance data under microgravity conditions.

Ibrahim et al. fabricated a Ti64 oscillating heat pipe (Fig. 5.4) and reported 'effective operation' using a variety of working fluids at different orientations [49]. Interestingly, they found that the heat pipe benefited from residual powder sintering on the surface of the channel, which improved the capillary pumping performance of the wick. They also commented that the thermal expansion behaviour of Ti64 is compatible with silicon, implying that this is a suitable solution for electronics cooling, but noted that powder removal must be factored in at the design stage.

## 5.3.2 Commercial examples

The aerospace sector is likely to be one of the earliest adopters of AM heat pipe technology, which will be geared to high-performance but low-volume scales of operation. It is then expected that the technology will see increased deployment in space applications, predominantly telecom satellites and other science-based missions as reliability of the technology becomes more established. Following this, the technology could reach the high-end medium-volume market [50].

Aavid UK are market leaders in high-end heat pipe development, and in collaboration with the University of Liverpool, they have explored the use of AM for heat pipe fabrication in a range of applications, achieving a technology readiness level (TRL) of 4 [50]. They are believed to be the first to be granted a European Patent (No. 2715265) regarding the use of laser powder bed techniques for the creation of capillary devices for heat pipe applications. The key challenge with L-PBF (according to Aavid's early work) is miniaturising the pore size to achieve high capillarity which requires careful optimisation of the laser settings. Their initial demonstration with titanium (pore size of $<700\ \mu m$, Fig. 5.5a) was followed by a European Space Agency (ESA) project to test 'gravity friendly' heat pipes that also function at ground level in addition to microgravity environments (where grooved aluminium heat pipes with ammonia do not). Reducing the pore size increased the capillary pressure at the expense of restricting the mass flow rate, limiting the overall maximum transferable power (Fig. 5.5b). They then fabricated circular Ø8 × 200 mm titanium–ammonia heat pipes using the optimised lattice structure (Fig. 5.6c), which satisfied the requirements of the ECSS-E-ST-31–02C ESA standard (including testing of pressure, vibrations, ageing, noncondensable gases and welds/sealing). These prototype devices enabled the transfer of 30 W of thermal energy at −20 degrees against gravity, far exceeding the −2 degrees of standard screen-mesh wicks. Additionally, Aavid has also started to investigate AM examples in other heat pipe applications. They have manufactured a Ø20 × 45 mm primary evaporator wick for a LHP, but acknowledged that further work is needed to improve the pore size (Fig. 5.5d). They also fabricated a proof-of-concept titanium-water

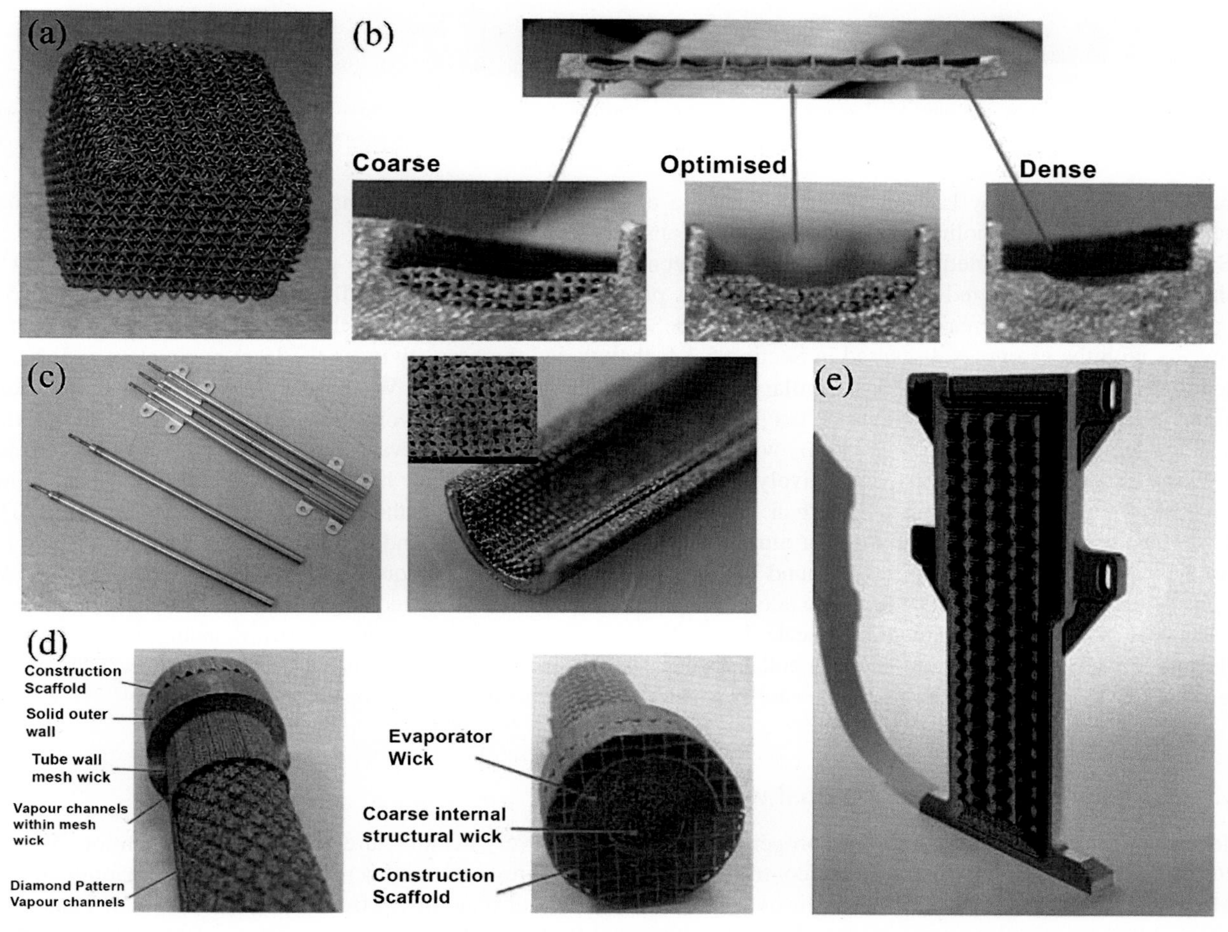


**FIGURE 5.5** Heat pipe development by Aavid, Thermal Division of Boyd Corp. (a) titanium lattice structure, (b) lift height and mass flow rate test pieces with different lattice spacings, (c) AM titanium–ammonia heat pipes for ESA, (d) loop heat pipe evaporator and (e) titanium-water vapour chamber concept [50].

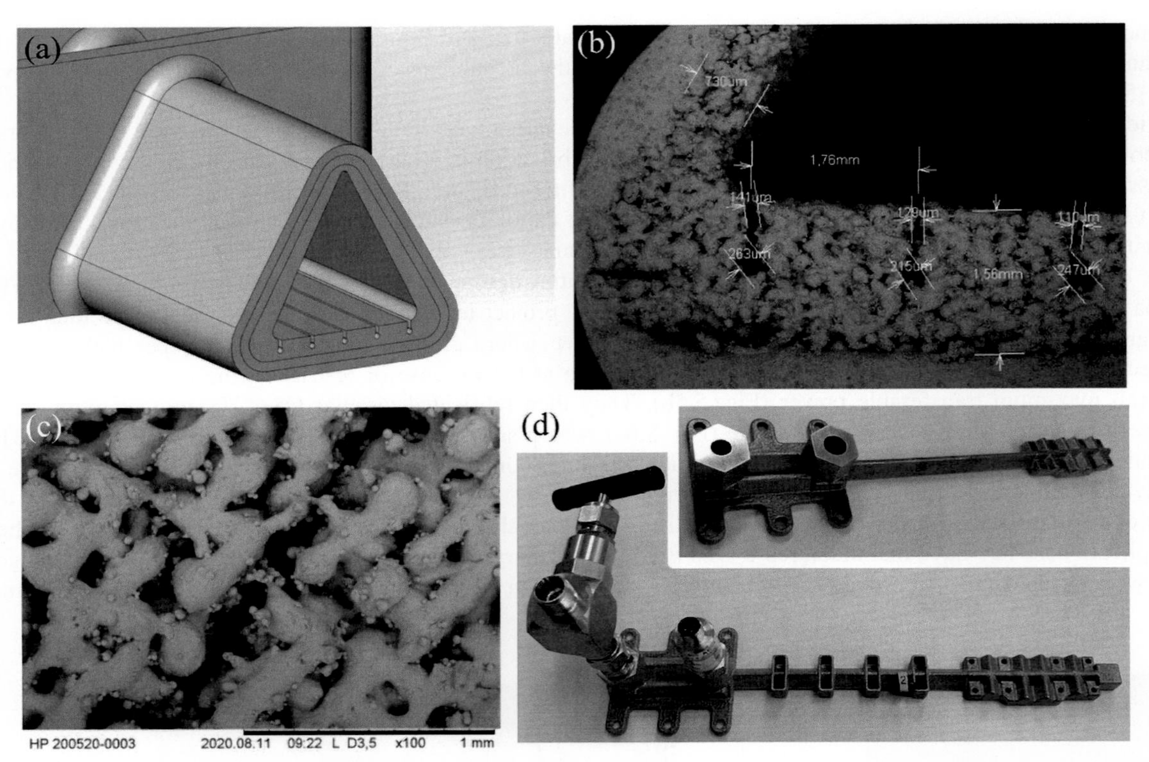


**FIGURE 5.6** FOTEC's ALM heat pipe development for space applications I (a) CAD model of the cross-section, (b) CT scan of the wick showing the slits/grooves, (c) close-up of the pore structures and (d) full test-piece (LHS = condenser, RHS = evaporator) [51].

vapour chamber (funded by Innovate UK), achieving a capillary lift height of 100 mm, that could enable a 20% mass reduction for electronics cooling in aerospace applications (Fig. 5.5e).

Buchner recently reported on Forschungs und Technologietransfer's (FOTEC) progress on the development of Additive Layer Manufactured (ALM) aluminium heat pipes with acetone working fluids for space applications [51]. The goal was to enable increased system power in CubeSat thrusters, which required the dissipation of 15 W of thermal energy. As with the examples discussed in Section 5.3.1, they varied the porosity using the laser parameters and characterised their wick performance through capillary rise height and rise speed tests. With slit-style arteries, they observed a 50 mm rise height in 10 s, and a final rise height between 70 and 90 mm. To avoid the need for internal supports, they used a triangular cross-section (Fig. 5.6a), with additional branches in the evaporator to increase the surface area (Fig. 5.6d). Due to the challenges of additively manufacturing aluminium, their heat pipes stopped working at inclination angles of −2 degrees − much less than Aavid achieved − highlighting the need for further development. This work did acknowledge the opportunities for multifunctional structures though, and they reached a final TRL of 3.

So far, AM heat pipe technology sits around TRL 3−4 [50,51]. It is clear that qualification of both the AM process and heat pipes themselves will be needed prior to adoption by end users, and it is believed that collaboration will be key to progress AM heat pipes from lab through pilot-scale demonstration to full commercialisation. Notwithstanding, issues regarding intellectual property (IP) protection between collaborating partners might introduce bottlenecks into workflows, and, as yet, there is not an obvious solution based on the ease at which physical structures can be copied and 3D-printed [4].

### 5.3.3 3D printed versus conventional wicks

Table 5.4 compares the performance and properties of several additively manufactured wicks with conventionally manufactured wicks. The effective pore radii (controlling the capillary pressure) vary across two orders of magnitude from 0.004 to 0.52 mm, whereas the permeabilities vary over four orders of magnitude (from 0.27 to 3800 $\mu m^2$). Thus it can be inferred that the capillary performance, $K/r_{eff}$ (proportional to the condensate transport rate) is largely influenced by the permeability. The wicks possessing the highest permeabilities, and therefore the largest capillary performance values, were the conventionally manufactured grooved and foam wicks. Interestingly, the aluminium wick 3D-printed

**TABLE 5.4** Comparison of heat pipe wick properties and performance of conventionally manufactured wicks, advanced composite wicks and additively manufactured wicks (using water as the test fluid).

| Wick details | Porosity, $\phi$ | Permeability, $K$ ($\mu m^2$) | Capillary performance, $K/r_{eff}$ ($\mu$m) | Pore radius, $r_{eff}$ (mm) | References |
|---|---|---|---|---|---|
| **3D-printed wicks** | | | | | |
| Aluminium conventional heat pipe wick; apparent octahedral unit cell; $D = 75-108$ $\mu$m | 0.3 | – | – | 0.183[b] | [51] |
| Stainless steel loop heat pipe wick; octahedral unit cell | | | | | [6] |
| $D = 100$ $\mu$m | 0.309 | 5.91 | 0.109 | 0.054 | |
| $D = 200$ $\mu$m | 0.509 | 2.13 | 0.012 | 0.109 | |
| $D = 300$ $\mu$m | 0.614 | 1.34 | 0.008 | 0.162 | |
| Stainless steel vapour Chamber wick; partially sintered powder | 0.39 | – | – | – | [8] |
| Stainless steel conventional heat pipe wick; 500 $\mu$m octahedral unit cell size, $D = 216$ $\mu$m | 0.46 | 130.5 | 1.04 | – | [23] |
| Stainless steel loop heat pipe wick; apparent octahedral unit cell, $D = 70-90$ $\mu$m | 0.17 | 1.25 | 0.016 | 0.08 | [48] |
| Aluminium conventional heat pipe wick; 300 $\mu$m octahedral unit cell size, $D = 120$ $\mu$m | 0.17 | 0.27 | 0.02 | 0.06 | |
| Aluminium conventional heat pipe wick; 500 $\mu$m octahedral unit cell size, $D = 280$ $\mu$m | 0.58 | 270 | 7.14 | 0.14 | [25] |
| **Aluminium wicks** | | | | | |
| Grooved | – | 2049[a] | 8.43[a] | 0.24[a] | [53] |
| Grooved with NaOH wash | | | | | |
| 0.5 mol/L + 5 min contact | – | 2711[a] | 5.16[a] | 0.52[a] | |
| 1.25 mol/L + 5 min contact | – | 2969[a] | 5.96[a] | 0.50[a] | |
| 2.0 mol/L + 5 min contact | – | 2983[a] | 5.89[a] | 0.51[a] | |
| **Copper wicks** | | | | | |
| Sintered powder | | | | | [10] |
| Spherical ($D = 75-110$ $\mu$m) | 0.57 | 4.75 | 0.46 | 0.03[b] | |
| Irregular ($D = 75-110$ $\mu$m) | 0.57 | 6.71 | 0.60 | 0.04[b] | |
| Grooved | 0.50 | 492 | 12.90 | 0.12[b] | |
| Composite: | | | | | |
| Sintered powder/groove | 0.57 | 5.48 | 0.575 | 0.03[b] | |
| Spherical ($D = 75-110$ $\mu$m) | 0.57 | 7.04 | 0.735 | 0.03[b] | |
| Irregular ($D = 75-110$ $\mu$m) | | | | | |
| Foam (Amporcop 220.5) | 0.91 | 1900 | 9.05 | 0.21 | [54] |

(Continued)

**TABLE 5.4 (Continued)**

| Wick details | Porosity, $\phi$ | Permeability, $K$ ($\mu m^2$) | Capillary performance, $K/r_{eff}$ ($\mu m$) | Pore radius, $r_{eff}$ (mm) | References |
|---|---|---|---|---|---|
| Multiartery (sintered spherical powder in square array) | | | | | [52] |
| $D_p = 10\ \mu m$ | – | $0.17^c$ | $0.043^c$ | $0.004^c$ | |
| $D_p = 30\ \mu m$ | – | $1^c$ | $0.098^c$ | $0.010^c$ | |
| $D_p = 50\ \mu m$ | – | $4^c$ | $0.224^c$ | $0.018^c$ | |
| $D_p = 70\ \mu m$ | – | $9^c$ | $0.415^c$ | $0.022^c$ | |
| **Nickel wicks** | | | | | |
| Sintered powder | | | | | [55] |
| $D = 200\ \mu m$ | – | 2.7 | 0.007 | 0.38 | |
| $D = 500\ \mu m$ | – | 8.1 | 0.203 | 0.04 | |
| Fibre ($D = 0.01$ mm) | 0.689 | 0.15 | 0.015 | 0.01 | |
| Felt | 0.89 | 600 | 3.53 | 0.17 | [56] |
| Foam (ampornik 220.5) | 0.96 | 3800 | 16.52 | 0.23 | [54] |
| Monel beads | | | | | |
| 30–40 μm | 0.4 | 415 | 0.80 | 0.52 | |
| 70–80 μm | 0.4 | 78 | 0.41 | 0.19 | [57] |
| 100–140 μm | 0.4 | 33 | 0.25 | 0.13 | |
| 140–200 μm | 0.4 | 11 | 0.12 | 0.09 | |
| **Fibres and weaves** | | | | | |
| Felt metal | | | | | [56] |
| FM1006 | – | 155 | 3.875 | 0.04 | |
| FM1205 | – | 254 | 3.175 | 0.08 | |
| Glass fibre | – | 0.61 | – | – | [55] |
| Stainless steel Dutch twill weave | 0.24 | 0.36 | 0.05 | – | [58] |
| Refrasil ($SiO_2$) fabric | 0.50 | 237 | 14.54 | 0.0163 | [59] |

*D* refers to the pore diameter.
[a] *Values were back-calculated from the performance parameter $\Delta P_c.K$ (N) assuming the properties of water.*
[b] *Effective pore radius calculated from capillary pressure or rise height assuming $\sigma_l = 72$ mN/m (water).*
[c] *Values correspond to wetting regime and were predicted by modelling; $r_{eff}$ calculated from capillary pressure.*

by Ameli et al. using a 500 μm unit cell was able to somewhat match the highest performing conventional wicks [25]. However, these high-permeability wick structures are only likely to be useful in microgravity environments because high capillary pumping pressures are needed to overcome gravitational effects to realise practical heat transfer performance. Here, it can be seen that current 3D-printed wicks produce effective pore radii around 1 order of magnitude higher than the best performing sintered powder reported by Kim and Kaviany [52], but comparable pore radii to sintered powders [10].

## 5.4 Opportunities for additive manufacturing

The wick structures and heat pipe examples presented in the previous section clearly show that AM is capable of fabricating functional heat pipes. Whilst current research seems to be focussed on further developing the L-PBF parameters to achieve better and more reliable wick performance, little attention has been given to full optimisation and integration of the technology. Therefore this section will discuss how each section of the heat pipe could be improved by AM in order to deliver 'truly optimal' heat transfer performance. The specific areas considered are: alternative lattice geometries (5.4.1), enhancement of boiling (5.4.2), enhancement of condensation (5.4.3) and miscellaneous and whole heat pipe considerations (5.4.4).

### 5.4.1 Alternative lattice geometries

The pore size in conventional sintered wicks can be adjusted by changing the size of the powder, but there is minimal scope for also accurately controlling the pore shape, interconnectivity and size distribution. In contrast, AM provides a range of possible unit cell configurations for implementing different cellular/lattice structures that each provides a different combination of mechanical strength, space efficiency and pore interconnectivity. Note, although the mechanical strength is not necessarily a primary consideration for normal heat pipe performance, this attribute may be relevant for the detachment of the lattice from the build platform along with any postprocessing requirements (e.g. powder removal).

One of the simplest cell arrangements is the hexagonal honeycomb (Fig. 5.7a) [47], though this anisotropic structure will inevitably impede liquid distribution in the radial and tangential directions, possibly making it less effective for heat pipe applications. Instead, spatially uniform cubic and octahedral lattices (Fig. 5.7b) created from thin trusses have emerged as the 'default' for AM wick fabrication; the results of Ameli et al. [25], Esarte et al. [48] and Jafari et al. [5,23] were discussed previously. More advanced concepts include Voronoi lattices [60], diameter and spatially graded lattices [61] and triply periodic minimal surfaces (TPMS) [62]. Examples of all of these are shown in Fig. 5.8. Tamburrino et al. have produced a recent review detailing the design processes for some of these lattice structures [63].

Cubic and octahedral lattices are spatially uniform, whereas the Voronoi lattice is stochastic. Voronoi lattices (Fig. 5.7c) are designed through volumetric tessellation, where seed points are randomly dispersed in a 3D space and

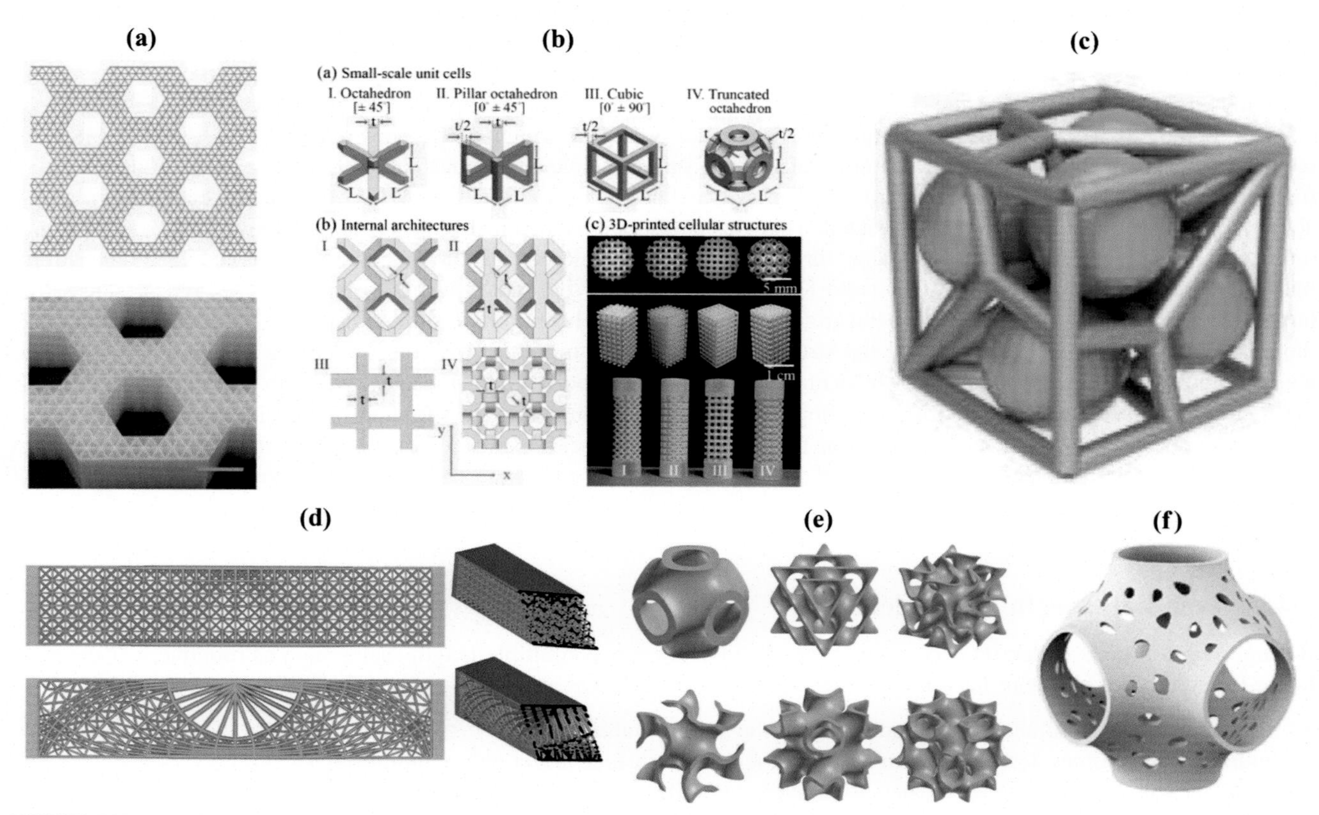


FIGURE 5.7 Potential alternative AM lattice concepts I (a) hierarchical hexagonal honeycomb [64], (b) cubic and octahedral unit cells [65], (c) Voronoi lattice [66], (d) diameter and spatially graded lattices [61], (e) TPMS topologies [67] and (f) hybrid Voronoi-TPMS topology [68].

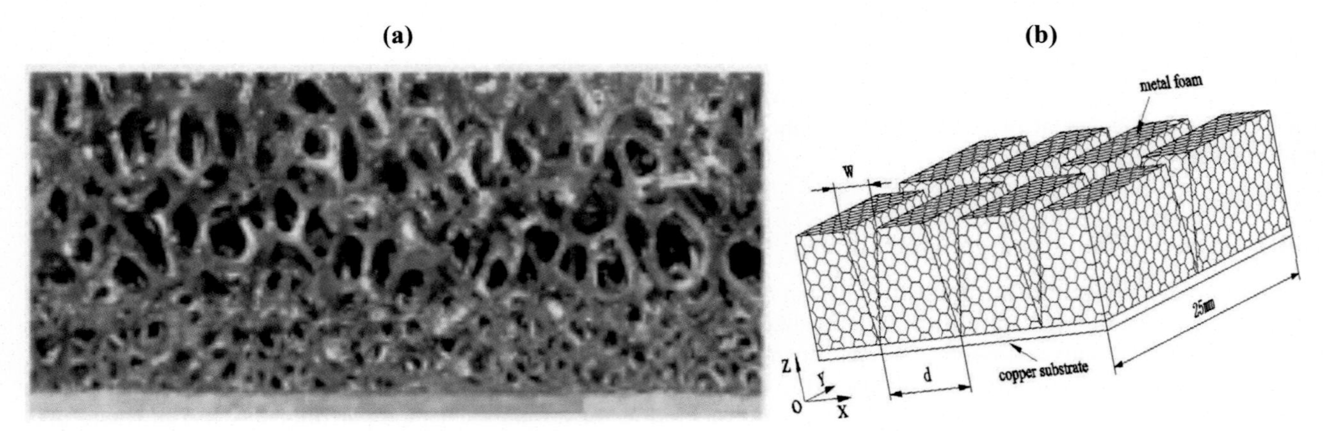


FIGURE 5.8 (a) Dual-porosity foam [69] and (b) sketch of cross-hatched V-shaped grooves in a foam [70].

spheres expanded away from these seed points. At the points, where these expanding spheres meet, boundaries are formed that are equidistant from all local seed points. The porosity can be adjusted by 'dressing' each boundary with a cylinder, and then adjusting the radius of this cylinder [60] similar to the octahedral lattices. The mechanical strength can be substantially improved by adjusting the truss thicknesses without greatly increasing the mass [60]. The principal advantage of the Voronoi lattice is its space-filling properties. They can be easily mapped onto nonuniform heat pipe geometries where it would be otherwise difficult to map more uniform cubic or octahedral lattices.

Daynes et al. proposed an alternative interesting idea. By spatially grading the lattice trusses along lines of isostatic compression and tension (Fig. 5.7d) [61], these geometries are able to reduce bending/deflection against an applied lateral load without increasing the amount of material compared to an ordinary octahedral lattice arrangement. This provides an opportunity to use heat pipes as dual heat transfer and structural components to reduce the physical size of heat pipe-based equipment (e.g. heat pipe heat exchangers/recuperator or heat pipe reactors). A perceived caveat of this approach is that there would be a loss of control over the capillarity and permeability since the pore shape/size would be highly dependent on the calculated applied loads/stresses.

Another class of lattice structure that has received little attention in the heat pipe literature is the TPMS. TPMSs are a class of topologies whose surfaces that traverse a particular boundary within a 3D space have the smallest possible area (some examples are shown in Fig. 5.7e) but because these surfaces have continuous curvature, they minimise the potential for stress concentrations [62] making them mechanically robust. Torquato and Donev [71] have commented that reasonably high fluid permeabilities can be expected. This was confirmed by Femmer et al. [72] who showed that TPMS structures promote higher $CO_2$ transport rates through polydimethylsiloxane (PDMS) compared to similar hollow-fibre membrane configurations. Jung and Torquato [73] have also found that the Schwartz primitive (P) design produces the largest fluid permeability compared to Schwartz diamond (D) and Schoen gyroid minimum surfaces. They conjecture that the highest permeabilities correspond to the structures that globally minimise the specific surface in a given space. The ability to adjust the surface curvature whilst maintaining the same porosity [74] also provides an opportunity to control the meniscus radii/wetting characteristics to improve the capillary pumping pressure and liquid distribution in the wick. Here, it has been shown in nanochannels that multicurvature menisci experience extra positive fluid pressures near the walls because of intermolecular forces near the wall [75]. Also of relevance here, Guest and Prévost [76] proposed a framework for designing optimised TPMS structures for simultaneously maximising the stiffness and permeability.

## 5.4.2 Evaporator section considerations

Boiling within porous materials is complex because it occurs across multiple length scales such as roughness at the surfaces, isolated pores and pore networks, and each influences different aspects of evaporation [77]:

- Surface roughness influences the nucleation of the vapour bubbles.
- Intermolecular forces acting between the liquid and solid surface affect phase equilibria.
- Capillary forces at meso/macroscales affect phase equilibria and surface replenishment.
- Supersaturation levels within the pores, interactions between nucleating clusters and the dissolution of unsaturated vapour bubbles, all influence vapour bubble growth kinetics.

- Curvature effects at the liquid–solid–vapour boundaries affect the liquid/vapour pressures.
- Bulk heterogeneous properties of a lattice influence larger-scale behaviours such as the expulsion of vapour from the porous medium.

There are clearly many ways in which the wick structure could be adapted through AM to improve the aforementioned evaporation characteristics: from increasing nucleation rates, to improving vapour bubble growth, to vapour expulsion from the lattice. This section of the report will describe some of these potential AM opportunities.

In principle, vapour bubbles within the wick impede the capillary-driven liquid flows at the surface whilst also increasing the overall thermal resistance. For this reason, Hanlon and Ma [78] propose that nucleate boiling within the porous wick must be suppressed such that only thin-film evaporation at the top surface takes place. Their numerical and experimental analysis further implies an optimum wick thickness depending on the sintered particle size, porosity and evaporator entry length; generally, smaller particle radii and smaller porosities both increase the optimal thickness. Conversely though, some studies have observed that nucleate boiling in the wick actually enhances heat pipe performance. Ultimately, it is a combination of the wall material, wick structure (type, thickness, porosity, permeability), working fluid, working fluid charge and applied heat load that determines whether nucleate boiling is desirable or not [5].

Sintering two different porosity foams together has been seen to improve the boiling heat transfer performance (Fig. 5.8a). Smaller pores at the lower heated surface maximise the heat transfer area and promote stronger capillary forces to aid liquid replenishment, whilst larger pores at the upper surface minimise the bubble escape resistance [69]. Additionally, Xu and Zhao [70] show that dividing the foam with cross-hatched V-shaped grooves (Fig. 5.8b) also reduces the bubble release resistance, delaying the critical heat flux. These are modifications that can be readily replicated with current AM technologies.

Bodla et al. [79] provide data on the capillary pressures generated as a function of the liquid contact angle for a range of sintered particle sizes. They used X-ray microtomography to recreate computerised models of different commercial sintered wicks (from particle size ranges of 45–60 μm, 106–150 μm and 250–355 μm) and applied a volume of fluid (VOF) approach to model the menisci and study thin-film evaporation rates. From these results they were able to establish that the heat transfer coefficient can be improved by increasing the meniscus surface area per unit volume independently of the contact angle by reducing the pore spacing. Thus smaller AM pores should enhance the evaporation rates. Ranjan et al. [80] however, also note that there will inevitably be a trade-off between the capillary pressure and permeability when reducing the pore size that could potentially offset the evaporation performance.

This matches the observations of Wong and Leong [81] who studied pool boiling inside octahedral lattice structures of varying unit cell sizes fabricated through SLM. Fig. 5.9 shows their unit cell design, along with three SLM samples with different cell sizes. These lattice structures significantly enhanced nucleate boiling heat transfer coefficients and increased the critical heat flux (CHF) compared to plain/flat surfaces. The improvement was attributed to three phenomena: (1) increased surface area, (2) increased nucleation site density and (3) capillary-assisted wetting of the surfaces (a fortuitous consequence of the high surface roughness integral to the L-PBF process). The latter effect in particular improves the liquid replenishing rate at the surface, so delaying the CHF point by eliminating hydrodynamic choking within the porous structure. Otherwise, a 'major liquid–vapour counter flow' establishes and this disrupts the liquid replenishment rate at the surface. They also noted that the bubble evacuation resistance plays a dominant role in the performance of these lattices, so there is a compromise between maximising the lattice surface area by increasing the thickness and minimising bubble drag by decreasing the thickness. A possible caveat is that these results apply to atmospheric pressure conditions only, not the reduced pressure environments of typical heat pipes.

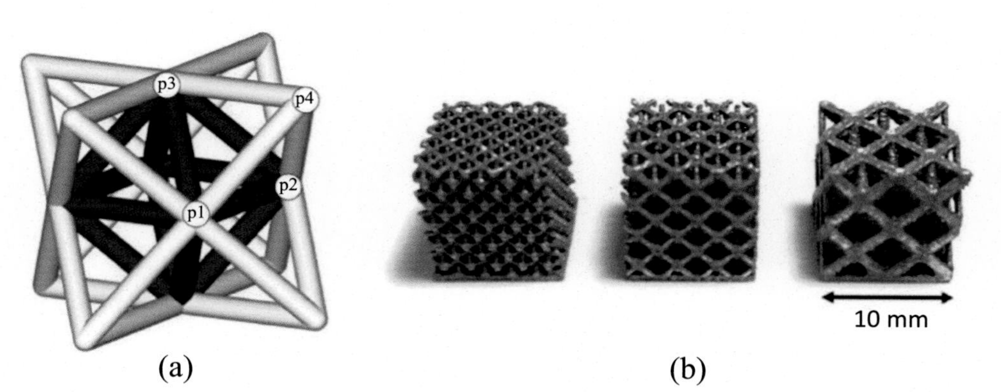


**FIGURE 5.9** (a) Unit cell representation of the 'octet-truss' lattice and (b) SLM-fabricated substrates with 2-mm, 3-mm and 5-mm unit cell sizes (from left to right) [81].

The work of Garimella et al. has also concentrated for many years on understanding the boiling characteristics inside heat pipe wick structures. For instance, Weibel and Garimella [82] visualised vapour formation in capillary-fed sintered wick structures and produced different boiling curves to quantify these observations. For small heat input areas, they found that the conduction resistance is eliminated at the incipience of boiling which then reduces the overall thermal resistance; this also occurs without producing any capillary limitations that would otherwise cause dry-out of the wick. By adding additional patterns into the sintered structure (examples in Fig. 5.10), they also found that they could increase the permeability of the vapour through the wick, which further reduced the thermal resistance during boiling by minimising countercurrent flow between the vapour and liquid. These patterns could conceivably be replicated with current AM technologies for controlling/directing the outflow of the vapour.

Extensions of the patterned-sintered wick concepts proposed by Weibel and Garimella [82] were recently presented by Nasersharifi et al. [83]. They started from the viewpoint described by Wong and Leong [81] that the countercurrent flow of the vapour and liquid is what ultimately limits the boiling heat transfer coefficient and CHF at the heated surface. They subsequently fabricated several 'multilevel modulated wicks' to control the phase separation of the vapour and liquid working fluid; their approaches are shown in Fig. 5.11. Compared to a plain surface, the CHF was increased by 20%, 65% and 87% when using designs B, C and D, respectively. These enhancements were attributed to a reduction of the Rayleigh-Taylor instability wavelength from 14 mm with the plain surface down to 3.5 mm with the columnar posts and mushroom-capped posts. For a plain surface, the Rayleigh-Taylor instability controls the thickness of the falling liquid columns to the surface and rising vapour columns away from the surface (shown schematically in Fig. 5.11a). Therefore reducing the thicknesses of these columns exposes a greater proportion of the heated surface to the liquid, thereby delaying the point of dry-out.

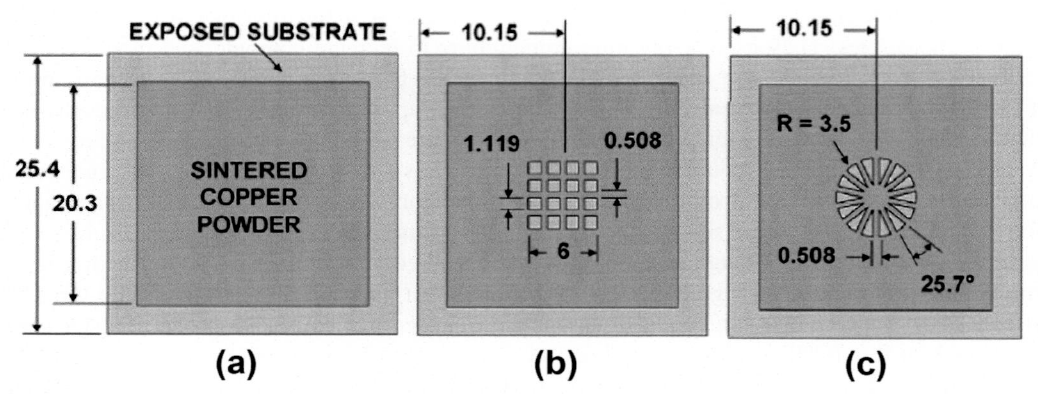


**FIGURE 5.10** Schematics and dimensions of sintered copper powder microwicks: (a) no vapour channels, (b) grid-patterned vapour channels and (c) wedge-patterned vapour channels [82].

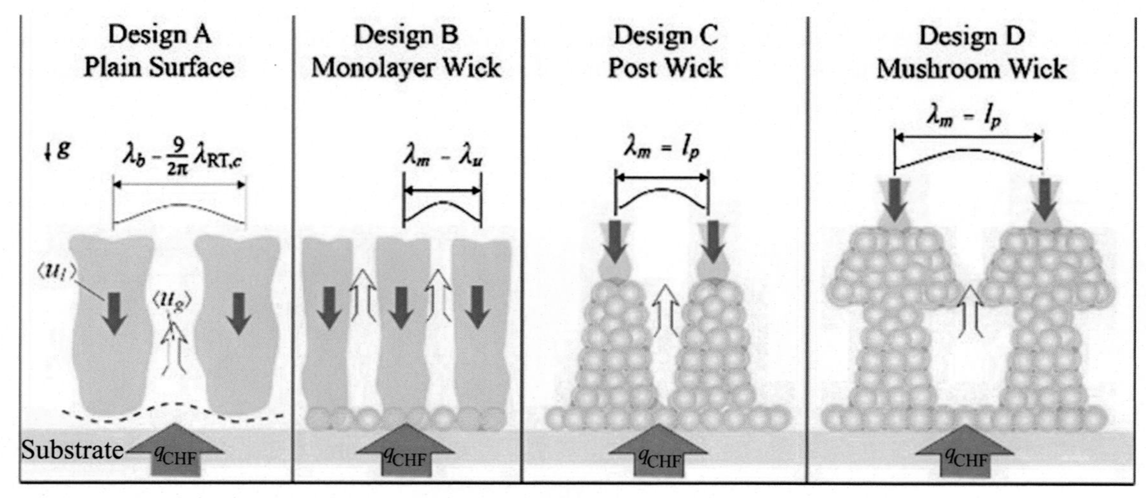


**FIGURE 5.11** Schematics of multilevel modulated wicks for pool boiling enhancement [83].

A 10-fold improvement in the heat transfer coefficient was also observed using designs B and C compared to the plain surface presumably because the surface area was increased. Although these geometries were fabricated through sintering, they could also be made via current AM technologies. The mushroom-cap in particular looks to be rather effective, but comparisons with a plain surface, which would not normally be used in a heat pipe, is perhaps misleading.

An alternative extension of Weibel and Garimella's [82] work is presented by Wen et al. [84]. Conventional mesh wicks provide limited customisability if one is looking to optimise for good capillarity and low flow resistance. Wen et al. [84] subsequently proposed a hybrid structure consisting of a mesh (or micromesh as specifically described) with microchannels interspersed within the mesh layers. Essentially, this is the same concept as a sintered copper wick structure with longitudinal arteries within it formed by burning out polymer filaments during the sintering process. The advantage of this approach is the simple etching process was cost-effective and scalable, although it might be difficult to achieve in an AM-fabricated heat pipe at present. A potential solution would be to 3D print the heat pipe with open ends to allow the interior wick to be exposable to a dip-coating process. Dip coating would then imbue the wick's lattice struts with 'nanograsses' to enhance the surface wettability and simultaneously increase the surface area, both of which would increase the heat transfer coefficient. Fig. 5.12 visualises the hierarchical boiling processes taking place on these multilayer copper meshes. Additionally, Table 5.5 compares the evaporative performance of these multilayer copper meshes with other 'representative' studies of capillary evaporation.

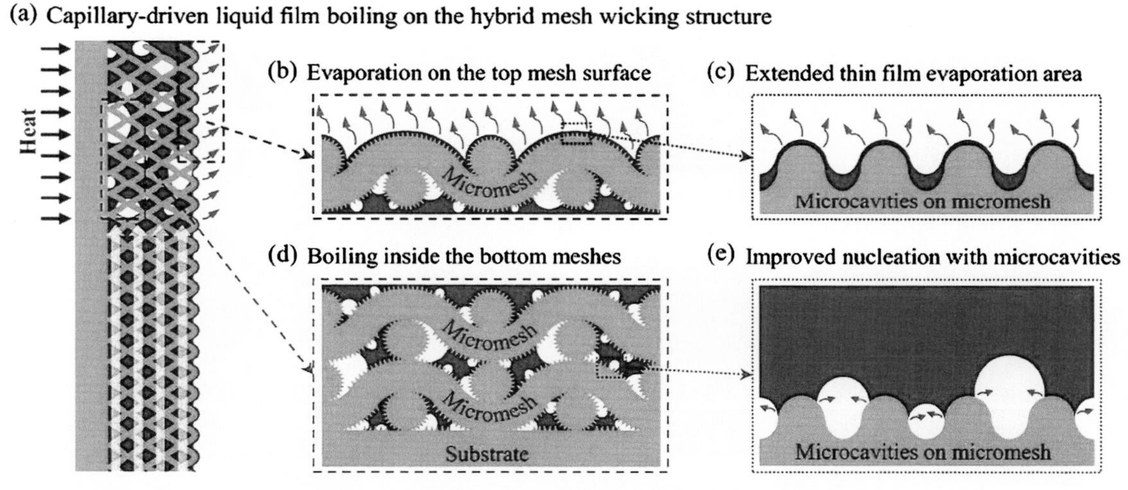


FIGURE 5.12 Simultaneous capillary evaporation and nucleate boiling in multilayer copper woven meshes [84].

**TABLE 5.5 Comparison of experimental studies of capillary evaporation/boiling heat transfer [84].**

| References | Wick structure | Heater area ($mm^2$) | Superheat @ ONB (K) | CHF ($W/cm^2$) | HTC @ CHF ($kW/m^2 K$) |
|---|---|---|---|---|---|
| [85] | Silicon micropillars | 10 × 10 | | 20–50 | 25 |
| [86] | Hybrid silicon micropillars | 10 × 10 | 7–13 | 120–275 | 200 |
| [87] | Carbon nanotubes | 10 × 10 | >10 | 130–140 | 60–80 |
| [88] | Micromembrane on microchannels | 10 × 10 | | 152.2 | 91 |
| [89] | Nanoporous membranes on microchannels | 63 | 12–14 | 40–48 | 10–15 |
| [90] | Metal foam on nanostructured micropillars | 4 × 4 | >20 | 150–420 | 20-60 |
| [84] | Hybrid copper mesh | 10 × 10 | 4.2–8 | 146–199 | 138 |

Table 5.5 refers to micropillars as wick forms. The work of An et al. [91] was similarly directed at improving pool boiling using textured copper pillars of various arrangements. Specifically, frustum pyramids of varying sizes were manufactured by supersonically spraying copper microparticles through a wire mesh; conversion of the kinetic energy upon impact into thermal energy facilitated the bonding process [92]. The concept and resulting pyramids are highlighted in Fig. 5.13. An et al. [91] report that the optimal base size for maximising the heat transfer coefficient and CHF is $D = 0.91$ mm, with a spacing of $d = 0.4$ mm and height of $L = 0.13$ mm (see Fig. 5.14c). Like the mushroom-capped pillars presented by Nasersharifi et al. [83] therefore, these pyramids could be constructed with current L-PBF resolutions. Their optimised pyramids achieved a CHF of 28.5 W/cm$^2$ and HTC of 20 kW/m$^2$ K for a heated area of 27 mm$^2$.

Another set of experiments looking at enhanced surface structures for pool boiling was also recently described [93]. They used a nanosecond fibre laser to create various laser-textured structures on boiling surfaces with features that were spaced apart by 2.5 mm; they note that the optimal spacing is equal to the capillary length of the particular working fluid. Fig. 5.15 shows the arrangements they considered; dimension B in these figures was varied from 0.5 to 2 mm. They observed a maximum fourfold increase of the heat transfer coefficient compared to an untreated surface using hexagonally arranged circular posts ('2D cB' in Fig. 5.14) using B = 2 mm. The interesting part of this work relates to microcavities on the surface. In particular, the scale of these structures is also within the current capabilities of metal

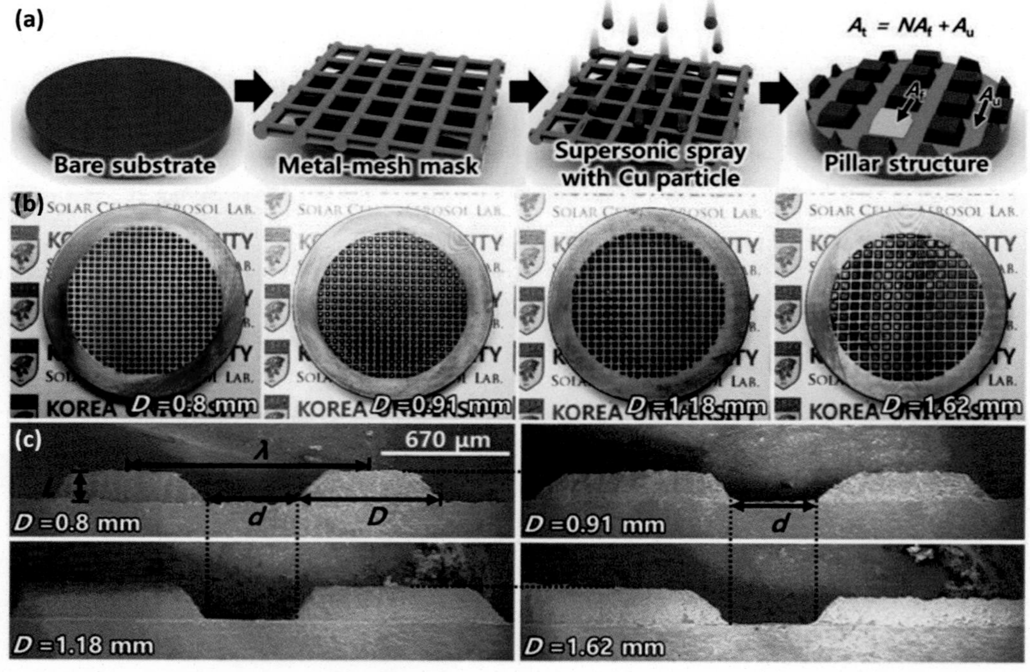


**FIGURE 5.13** (a) Schematic of supersonic cold-spraying process [these pyramids could be readily 3D printed], (b) images of coated substrates and (c) SEM images of frustum-pyramid pillars ($D = 0.8$, 0.91, 1.18 and 1.62 mm) [91].

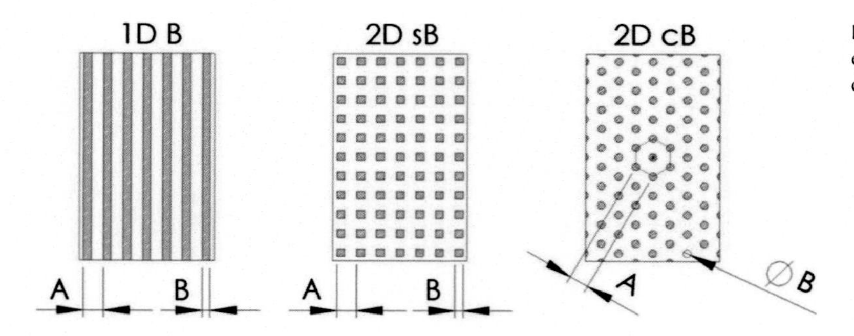


**FIGURE 5.14** Schematics of three designs for boiling enhancement: parallel strips (1D), square posts (2Ds) and circular posts (2Dc) [93].

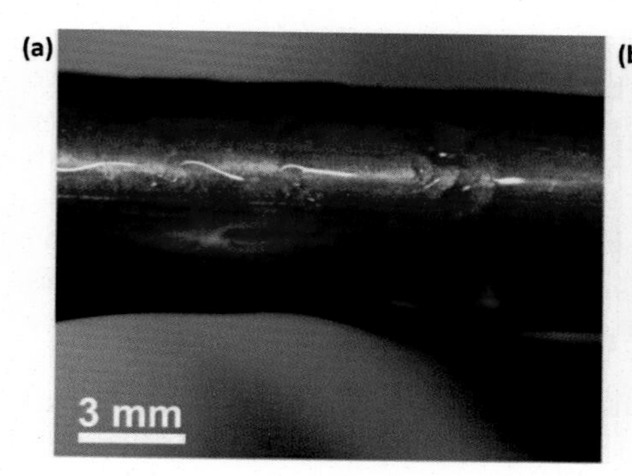

**FIGURE 5.15** Filmwise (left) and dropwise (right) condensation on a copper surface [94].

AM techniques, though it is not known at present if these structures used in conjunction with a lattice structure would provide any additional heat transfer enhancement.

Something that may instead be effectively applied in an AM process is the inclusion of a nanostructure onto a porous structure of larger particles to improve the hydrophilicity/capillary wicking performance which is necessary to improve the surface replenishing rate to improve the CHF. Li et al. [95] achieved this in two stages for a LHP. Firstly sintering of the copper powder, then secondly, applying an alloying-dealloying treatment to produce porous nanostructures on the copper surface. However, this 'multiscale composite porous wick (MCPW)' requires optimisation of the coating size and distribution and is likely to be a trial-and-error process and so may not justify the effort.

Alternatively, although by no means unique, the manufacture of a wick having different thermal conductivities is also of interest for optimising the wick structure, most commonly in the radial direction, as determined by the local heat input (or output) from the heat pipe. For instance, Xin et al. [96] proposed hybrid Ni–Cu sintered wicks to achieve high thermal conductivity on the vapour side of the wick and low thermal conductivity on the side of the wick closest to the heated wall in a LHP. LHPs have their own features that may benefit from such an arrangement, but conventional tubular heat pipes could also use such a design feature. For instance, controlling the presence of the nucleation sites using a lower thermal conductivity metal near the heated wall and higher thermal conductivity closer to the vapour channel might force the stable boiling arrangement described by Wong and Leong [81].

### 5.4.3 Condenser section considerations

Condensation occurs when a saturated vapour contacts a surface below the vapour's saturation temperature and preferentially occurs on a surface instead of within the vapour because heterogeneous nucleation has a lower activation energy than homogeneous nucleation [97]. Filmwise condensation (FWC) involves the formation of a condensate film on the surface and can be considered as the 'default' mode because it does not require any special surface modifications. The resulting condensate film increases the thermal resistance between the surface and vapour because heat must be conducted through the film once it forms. In contrast, dropwise condensation (DWC) can produce much higher heat transfer coefficients than FWC because the condensate film resistance is removed, though Rose [98] acknowledges that erroneous comparisons are often made between FWC and DWC when data is compared across different heat fluxes and operating pressures. Fig. 5.15 shows examples of FWC and DWC on a copper surface.

Generally, higher nucleation site densities produce larger condensing heat transfer coefficients [99]. Additionally, materials with high surface energies produce strong attractive forces that exceed the surface tension of the liquid condensate, causing liquids to wet the surface resulting in the formation of liquid films. Conventional heat transfer materials (such as aluminium, copper, titanium and stainless steel) have high surface energies, meaning either functionalised coatings or other surface modifications are needed to reduce this surface energy so that DWC can be realised.

Another consideration for fabricating an optimised wick structure for the condenser section is the reabsorption of liquid condensate into the wick. Although DWC would promote faster condensation (helping to increase the 'speed' of heat transfer from the source to the sink), there must also be a mechanism for the rewetting of the expelled droplets into the wick structure. This suggests that the presence of a liquid film at some point will be unavoidable to avoid any new bottlenecks in the working fluid circuit, for example, from the transfer of condensate droplets back into the wick structure.

Surface coatings can be thought of as a 'top-down' approach for the enhancement of condensation heat transfer. These reduce the surface energy which promotes the formation and maintenance of dropwise condensation on high surface energy surfaces such as aluminium, copper, stainless steel and titanium. The current challenge is the identification of functional coatings that are simultaneously durable, low cost and impose minimal extra thermal resistance so that the presence of the coating does not offset any performance gain. Enright et al. [99] recently reviewed the main surface functionalisation methods reported in the literature and compiled them (as shown) in Table 5.6. As described in the opportunities for evaporation, these might be accessible to additively manufactured heat pipes by printing them with open ends to expose the interior wick structures to different coating methods.

However, there are also several ‘bottom-up’ surface modification approaches for the enhancement of condensation heat transfer rates that exploit micro- and nanostructures to produce superhydrophilic, superhydrophobic and/or ambiphilic/omniphilic responses that may be compatible with current L-PBF capabilities. These will be the focus of the remaining part of this section.

The wetting characteristics of a surface can be tuned by modifying the surface geometry, provided these surface features are smaller than the capillary length scale (defined by Bond numbers less than 1). The Bond number (Eq. 5.1) characterises the relative importance of gravity compared to surface tension, where $\Delta\rho$ is the density difference between the condensate and solid surface, $g$ is the gravitational acceleration constant (9.81 $m/s^2$), $L$ is a characteristic length scale (often taken to be the height of the surface roughness/geometric features) and $\sigma$ is the surface tension.

$$Bo = \frac{\Delta\rho g L^2}{\sigma} < 1 \tag{5.1}$$

Micropillar grids artificially produce hydrophobicity by lowering the surface energy. In these configurations, depending on the pillar density, two extreme droplet states will occur upon condensation. Wenzel (W) droplet morphologies occur when the condensate nucleates within the pillared surface, spreading out into the structures [100]. In contrast, the mobile Cassie (C) droplets are produced when the structured surface behaves superhydrophobically [101]. A partial-Wenzel (PW) state occurs between the two extremes when the condensate initially nucleates within the pillars before ballooning away. Fig. 5.16 compares these three droplet morphologies.

**TABLE 5.6 Summary of surface functionalisation coatings for DWC [99].**

| Surface functionalisation method | Brief description | Advantages | Disadvantages |
|---|---|---|---|
| Self-assembled monolayers (SAM) | ~1 nm thick molecular films formed on condensing surfaces; hydrophobic tails point outward from the surface whilst a ligand attaches to the surface | Low thermal resistance | Poor or not well-understood durability and long-term stability |
| Polymer coatings | Thin to thick coatings of various polymers (e.g. PTFE, parylene and silicones) applied on the condensing surface | Promotes DWC, excellent heat transfer with thin coatings | Robust thin coatings still need to be developed |
| Noble metal (NM) coatings | DWC observed with the combination of NMs in the presence of physisorbed contaminants. Applied as a coating on the condensing surface | Promotes DWC if hydrocarbon is physisorbed on surface | Costly, requires contaminant source to maintain DWC |
| Ion implantation | Carbon, nitrogen and oxygen ions implanted on metals create nanoscale roughness and generate chemical heterogeneities | Good heat transfer at low subcooling | Flooding occurs at high subcooling, high cost, scalability not demonstrated |
| Rare earth oxides (REO) | Ceramic material deposited on the condensing surface | Intrinsically hydrophobic, potentially more durable than SAMs and polymers | Thin coatings required due to low thermal conductivity; heat transfer data not yet available |
| Lubricant-infused surfaces (LIS)[a] | Low-energy structured surface infused with immiscible low-energy lubricant | Low hysteresis, low contact angles, nucleation density control, omniphobic | Effect of lubricant drainage still needs to be studied |

*[a]LIS is often integrated with the surface modifications described in Section 6.3, but has been included in this table because it would be difficult to implement with the current state-of-the-art metal additive manufacturing technologies.*

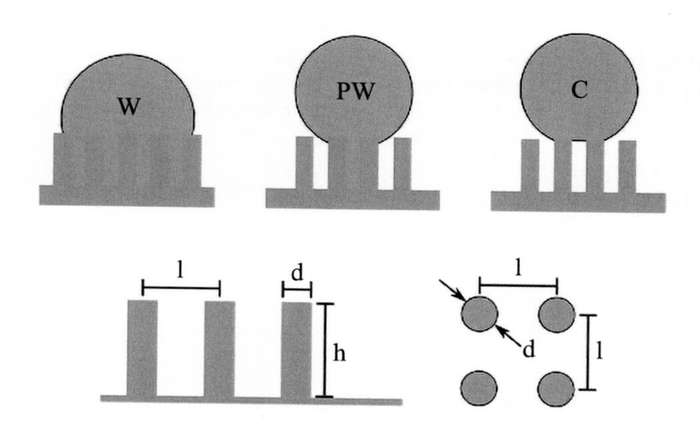


**FIGURE 5.16** Wenzel (W), partial-Wenzel (PW) and Cassie (C) droplets produced on a micropillared surface during condensation (redrawn from [102]).

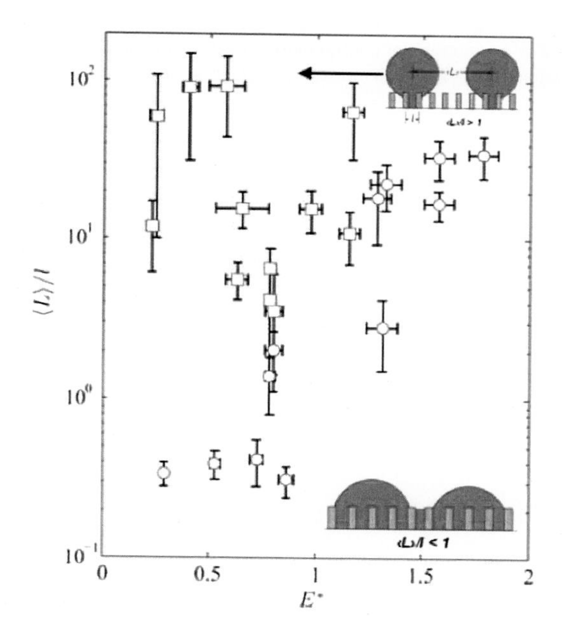


**FIGURE 5.17** Condensation regime map [103].

Enright et al. [103] defined a dimensionless energy ratio, $E^*$, to differentiate the Wenzel and Cassie droplet states. They approximately showed that W droplets occur when $E^* > 1$, whilst Cassie droplet states occur when $E^* < 1$. Note, in this equation, $r$ is the roughness factor (ratio of the total surface area of the pillared surface to the surface area of an equivalent featureless surface) and $\theta_a$ is the advancing contact angle (contact angle at the front of a moving droplet on a surface). Enright et al. [103] also proposed that the ratio of the average spacing between droplets, $L$, and the characteristic separation distance of the pillars (or other surface features), $l$, should be in the range of 2–5 to avoid flooding. They subsequently produced the condensation regime map in Fig. 5.17.

$$E^* = \frac{-1}{r\cos\theta_a} \tag{5.2}$$

$$\frac{L}{l} \sim 2 \text{ to } 5 \tag{5.3}$$

Although the Cassie droplet state enhances the condensation heat transfer coefficient, to complete the thermal circuit in the heat pipe the wick still needs to 'rewet'. One possibility is to exploit the self-propulsion of Cassie droplets away from the surface upon their coalescence. This self-propulsion has been proposed as an antiicing strategy [104] and could transport the liquid condensate independently of gravity and capillary forces [105]. The kinetic energy is provided from the release of the surface energy upon coalescence, whilst the propulsion from the surface occurs when the coalescing liquid bridge strikes the surface [106] (Fig. 5.18). Fig. 5.19 shows a proposed configuration. The main caveat is the

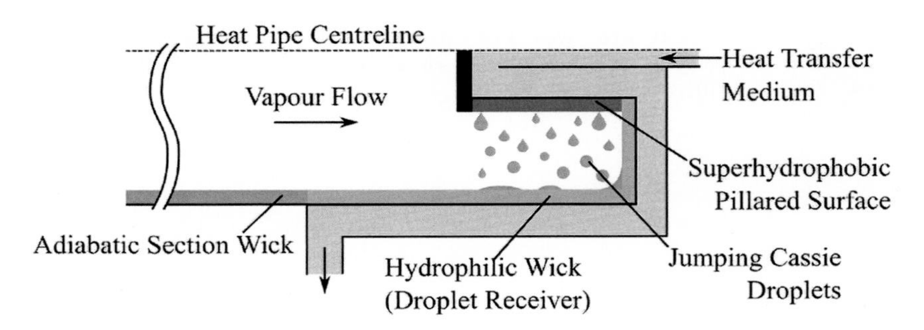


**FIGURE 5.18** Potential heat pipe configuration for realising DWC I Cassie droplets form and are then expelled from a surface covered in micropillars and recaptured by a hydrophilic (unmodified) wick structure.

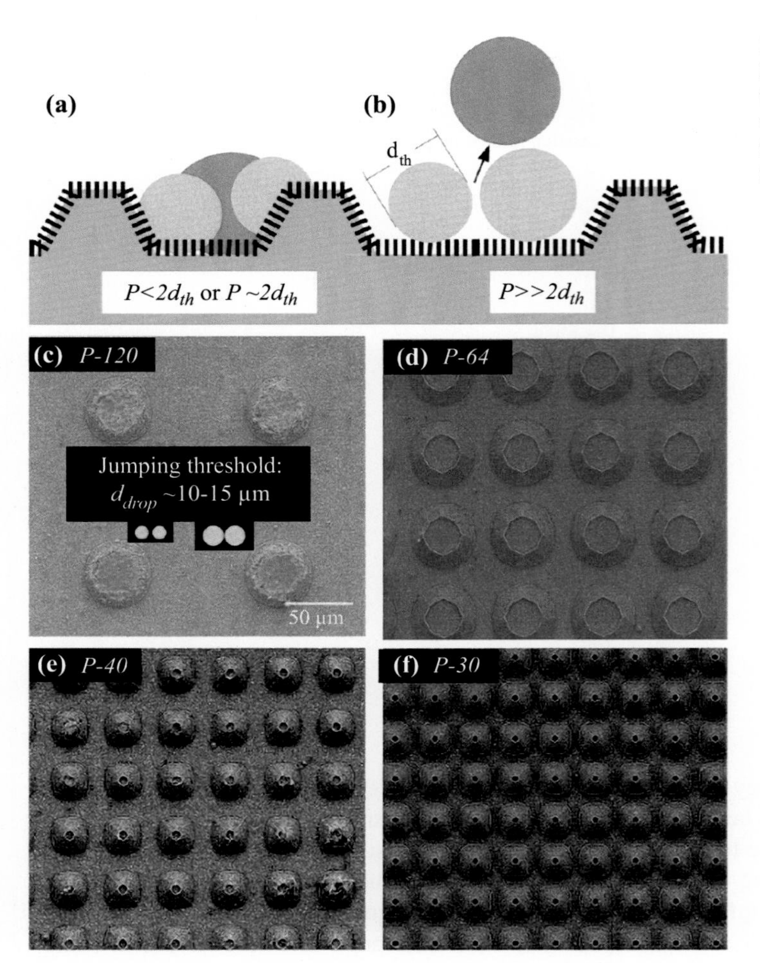


**FIGURE 5.19** (a) Adhesion of droplet to the surface after coalescence when the micropillar spacing $P$ is less than or equal to the threshold droplet diameter $d_{th}$, (b) expelled coalesced droplet when $P$ is greater than twice $d_{th}$, (c) P-120 micropillar configuration, (d) P-64 micropillar configuration, (e) P-40 micropillar configuration and (f) P-30 micropillar configuration [107].

lack of fundamental understanding for this type of design [99]. The main areas still requiring attention are: modelling of the propulsion velocity (defines the maximum separation distance), understanding of the radius of the propelled droplets, and further study of the effects of surface imperfections/texture effects [99]. However, Enright et al. [103] have produced different design maps that correlate the appearance of the jumping phenomena with the dimensions $l$, $d$ and $h$.

In the condenser, one of the limits to the condensation rate is the condensate removal rate from the surface [98]. The capillary length scale of a condensing droplet is defined in Eq. 5.4, where $\gamma$ is the surface tension, $\rho_l$ is the droplet density and $g$ is the gravitational acceleration constant. On 'normal' surfaces, surface tension stabilises the droplet when the droplet diameter is below $L_c$, so the diameter must exceed $L_c$ for it to shed from the surface [107]. However, multiple studies have observed condensate droplet shedding in the DWC regime with diameters below this capillary threshold using nanotextured surfaces [108,109]. This occurs because the nanostructures pin the base of the nucleating droplets so that the contact angle increases as they grow [103,110,111]. Droplet coalescence then releases the excess surface energy leading to increased droplet mobility [105]. Wang et al. [112] report a critical droplet diameter of 10 μm in order for the release of excess surface energy to overcome viscous dissipation and gravity. Interestingly, the use of needle-like CuO nanostructures formed on copper meshes (albeit for promoting superhydrophilicity) have recently been successfully used for enhancing the performance of flexible heat pipes [113]. This suggests that these nanostructures are somewhat mechanically robust.

$$L_c = \sqrt{\frac{\gamma}{\rho_l g}} \tag{5.4}$$

In addition to micropillared and nanotextured surfaces, the literature reports biologically inspired hierarchical surfaces that contain roughness at two length scales. Rykaczewski et al. [107] investigated condensation on micro-truncated cones coated by nanotrees and identified an optimal spacing for the microscale structure that promoted the most microdroplet departures in a given timeframe. The P-64 geometry in Fig. 5.19 provided the best trade-off between forcing droplets to coalesce, whilst being sufficiently spaced apart to minimise impedance to their mobility following coalescence. Feng et al. [114] alternatively focussed on the attributes of nanoribbon structures on a copper surface, which were fabricated using wet chemical oxidation followed by fluorisation. They reported higher droplet shedding rates using narrower spacings and higher aspect ratios (controlled by increasing the fluorisation time). Chen et al. [115] and Preston et al. [116] both report the alternative use of chemical vapour deposition (CVD) for the creation of carbon nanotube (CNT) and ultrathin graphene coatings respectively. Preston et al. [116] commented that the CVD method is robust and scalable, meaning it could be applicable for coating the interiors of heat pipes.

A potential limitation of these hierarchical superhydrophobic surfaces (SHSs) is flooding. Miljkovic et al. [117] observed that the nanostructures flood under high supersaturation conditions (defined as the ratio of the vapour to saturation pressures) and that this reduces the heat transfer coefficient by as much as 40% compared to a 'smooth' dropwise condensing surface. Cheng et al. [118] similarly observed flooding across a broad range of operating conditions on a two-tier surface, negating the observation of any heat transfer enhancement. They propose the use of mechanical agitation to purge the surface of accumulated condensate. Boreyko et al. [119] subsequently studied the reversibility of surface wetting and found that only surfaces with two-tier hierarchical roughness could recover from a flooded state when applying vibration, whereas single-tier micro- or nanoroughened surfaces would be irreversibly flooded.

Another caveat of superhydrophobic surfaces is the requirement for a higher degree of supersaturation before condensation occurs compared to hydrophilic surfaces [99]. Thus the advantages of DWC on hydrophobic surfaces could be offset by any delay in the onset of condensation in the first place. Consequently, biphilic surfaces have been proposed, where hydrophilic regions reduce the energy barrier for vapour condensation and hydrophobic regions promote droplet shedding. This also provides spatial control over the nucleation of the droplets [120], which might also benefit the heat pipe wick. The role of the condensing wick is twofold: (1) to facilitate condensation and (2) to distribute the condensate into the wick structure for transport. Hydrophobic DWC-inducing surfaces benefit the former, whilst hydrophilic (wetting) surfaces benefit the latter.

A multifunctional beetle wing surface (elytra) naturally harvests moisture by exploiting a biphilic surface [121]. Fig. 5.20 shows several bio-inspired biphilic surface approaches reported in the literature. These include hydrophobic/hydrophilic stripes with widths below the capillary length [122]; hydrophobic/hydrophilic dots on hydrophilic/hydrophobic backgrounds [123]; hydrophobic/hydrophilic patterning with structured roughness [124]; and superhydrophobic pores with hydrophilic coatings at the base of the pores [125].

Interestingly, the mixed wettability biphilic surfaces impose spatial ordering that delays the flooding point. For example, Ölçeroğlu and McCarthy [120] examined the condensation performance of various arrays of superhydrophilic 'islands' on a superhydrophobic nanostructured surface using Environmental Scanning Electron Microscopy (ESEM). They found that they could optimise the spacing of the hydrophilic regions to increase the nucleation site density whilst promoting a desired spatial ordering because of preferential nucleation, for example, producing condensate droplets in rectangular arrays [126]. Further, they found that microdroplets (diameters of $<25$ μm) would 'self-organise' through coalescence, supposedly enabling a transition from droplet jumping to droplet shedding when increasing the supersaturation without inducing flooding. More recent innovations include applying this 'ambiphilic' concept to hierarchical surfaces to avoid excess flooding via droplet 'burst sites' to maintain thin liquid films [127] and coating hydrophilic microgroove tips with hydrophobic coatings to promote self-cleaning [128]. This latter concept appears to be suitable for grooved heat pipes Fig. 5.21.

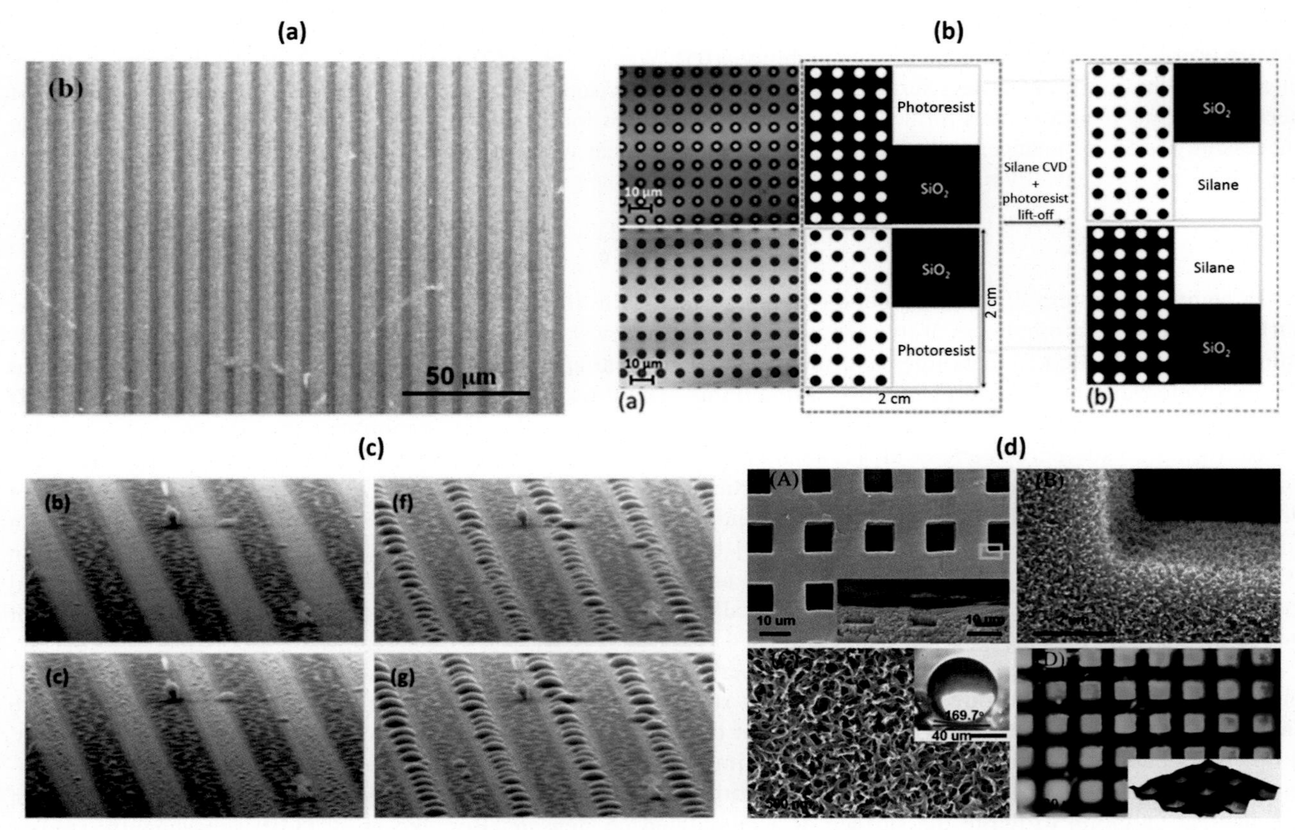


FIGURE 5.20 Examples of biphilic surface configurations for droplet spatial control | (a) stripes below the capillary length scale [122], (b) dots [123], (c) stripes with structured roughness [124] and (d) biphilic pores [125].

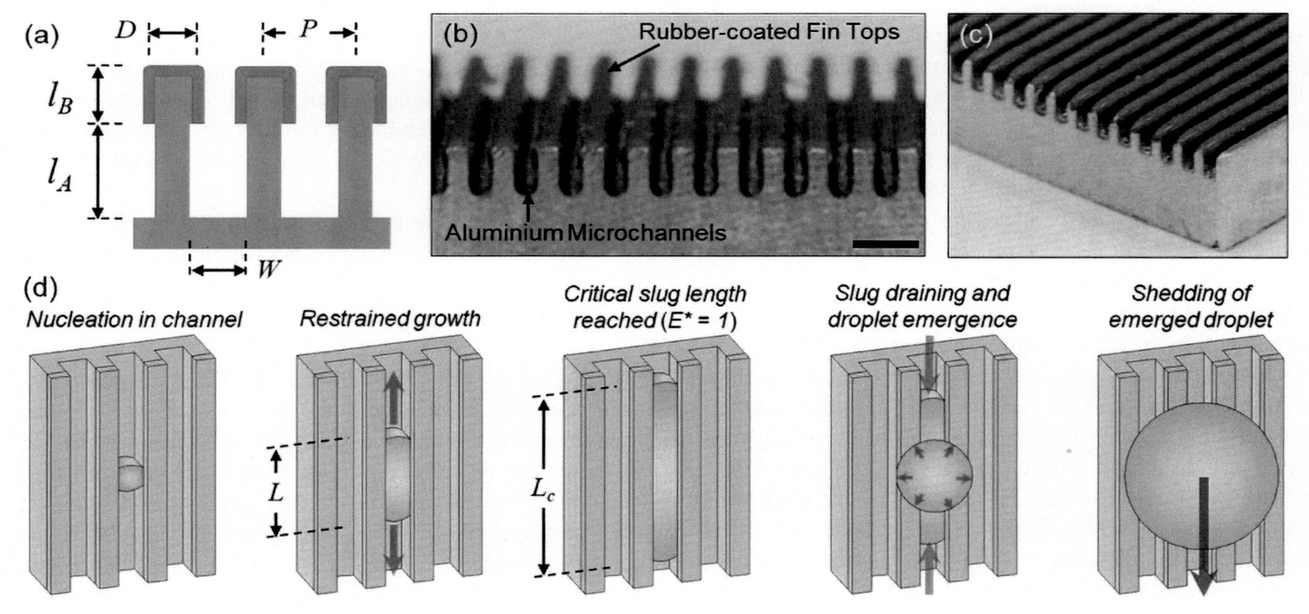


FIGURE 5.21 Illustration of ambiphilic microchannels for enhanced condensation via droplet shedding | (a) microchannel geometry, (b,c) aluminium microchannels with silicone-coated fin tips (1 mm scale bar) and (d) nucleation and growth, dewetting, emergence and shedding of condensate droplets [128].

Enright et al. [99] argue that structured SHSs can only be used for heat fluxes below 8 $W/cm^2$ because beyond this point flooding of the micro- and nanostructures is likely to occur. They propose five design principles for structured SHSs that should be considered at the AM design stage:

1. Promotion of the PW droplet morphology is preferred.
2. Surface structure length scales should be minimised (ideally $<2\ \mu m$).
3. High surface structure densities should be used.
4. Thermal conductivity should be maximised.
5. Hydrophobic coatings must be ultrathin ($<40$ nm) and highly conformal.

To overcome the 8 $W/cm^2$ limits, finer surface structures/features are required, but this is not possible with the current metal AM technologies available. Considering that the maximum heat flux for electronic devices (a common heat pipe application) is 190 $W/cm^2$ [23], smooth DWC surfaces are likely to provide more robust operation in the short term. These will likely require surface functionalisation via a suitable coating method (which will be dependent on the specific application) but might be achievable by fabricating the heat pipe with open ends so that the interior wick surfaces can be accessed for treatment.

### 5.4.4 Whole heat pipe and miscellaneous considerations

Gradient wettability surfaces have recently emerged as an option for enhancing the capillary performance of micro heat pipes. Hu et al. [129] adjusted the contact angle of water on copper grooves between 20 degrees and 85 degrees by immersing the surfaces in aqueous solutions of 1 M NaOH and 0.1 M $(NH_4)S_2O_8$ (see Table 5.7 for further details). They studied five different wetting configurations: uniform and high wetting (20 degrees), uniform and medium wetting (45 degrees), uniform and low wetting (85 degrees), step change in wetting from condenser to adiabatic to evaporator (85 degrees, 45 degrees and 20 degrees, respectively) and gradient wetting in the adiabatic section from 85 degrees to 20 degrees from the condenser to evaporator. They found that the latter gradient configuration produced the lowest thermal resistance of all designs, 33% lower than the resistance of the uniform low wetting configuration. Cheng et al. [130] also found an optimal wetting arrangement using a low wetting condenser (85 degrees), high wetting evaporator (20 degrees) and continuous axial variation of the wetting angle in the adiabatic section with the result that a maximum reduction of the thermal resistance of 92.6% was obtained. Singh et al. [131] have more recently identified the true optimal wetting gradient profiles through formal mathematical optimisation. Fig. 5.22a shows the optimal gradient. Fig. 5.22b shows that the maximum heat transfer rate is improved at all operating temperatures. Singh et al. [131] noted that the shapes of the optimal gradients are largely invariant to the working fluid charge (at least between 2.4 and 4.8 mg). They also showed that the linear gradient wetting surfaces were favourable for water, ammonia, methanol, acetone and heptane working fluids.

Fig. 5.23 illustrates the operation of a microheat pipe with trapezoidal grooves. Singh et al. [131] propose that larger contact angles decrease the liquid flow area in the groove compared to low contact angles because the wetted groove length (Fig. 5.23c) must be smaller if the meniscus curvature remains constant. This means larger contact angles increase the liquid pressure drop relative to smaller contact angles. In the heat pipe, the radius of curvature of the meniscus gradually decreases from the condenser to the evaporator. Consequently, if a step change in the contact angle is introduced, the flow area will suddenly change resulting in a sudden change of liquid pressure drop that causes a change in the distribution of the liquid. That is, by suddenly decreasing the contact angle in the evaporator, the flow area will suddenly increase, decreasing the pressure, resulting in the redistribution of the working fluid to the higher wetting evaporator. Optimal thermal performance requires a trade-off between higher liquid flow resistance in the condenser, and high liquid mass in the evaporator; this is achieved using the continuous wettability gradient opposed to step change in wettability.

Table 5.7 summarises several methodologies for engineering surfaces with wettability gradients. These all require the use of some form of functionalised coating, meaning these methods might be difficult to integrate with an additively manufactured heat pipe. One option might be to print the heat pipe with designated openings and employ a soaking and withdrawal process. The method used by Faustini et al. [136] appears to be the most universally compatible for metallic heat pipes. Alternatively, it might be possible to engineer the wettability through AM without the need for additional surface coatings. For example, increasing the density of micropillars decreases the wettability compared to smooth metal surfaces [137].

A related concept to the gradient wettability surface is the use of variable wick thickness (Fig. 5.24). Wang et al. [138] proposed this biomimicking structure to attempt to achieve optimal performance in each section of the heat pipe.

**TABLE 5.7 Reported methodologies for varying the wettability of a surface.**

| Method | Surface | Description | References |
|---|---|---|---|
| Alkali-assisted surface oxidation | Copper | Copper substrate is vertically immersed in aqueous solution of NaOH and $(NH_4)_2S_2O_8$ and slowly drawn out. Additional NaOH and $(NH_4)_2S_2O_8$ are added over time increasing the concentration at the surface of the solution. The immersion time gradually increases moving from the upper to lower part of the substrate, creating different amounts of $Cu(OH)_2$ nanoribbon growth, creating a controllable wettability gradient. | [129,132] |
| Laser ablation then $H_2O_2$ wash | Copper | A pulsed 20 μm focal laser beam (20 W, 200 mm/s scanning speed, 100 ns pulse @ 40 kHz) is used to create 140 μm deep and 40 μm wide grooves. Following 2-week exposure to air, the surfaces transform from initially hydrophilic to hydrophobic because the outer layer of copper is converted to $Cu_2O$. Varying the groove interval from 50 to 200 μm decreases the contact angle from 145 degrees to 125 degrees, due to the disappearance of microdebris. Wettability gradients are achieved by sitting the grooves in a 15% $H_2O_2$ aqueous solution and adjusting the soak time. | [133] |
| Silanisation | Silicon | A small drop (~2 μL) of alkyltrichlorosilane is suspended above a clean silicon wafer. The silane evaporates from the drop and diffuses to, then reacts with, the silicon ($Si/SiO_2$) surface, creating a self-assembled monolayer with normally distributed concentration. This creates a corresponding normal distribution of the contact angle with the maximum contact angle of water (100 degrees) occurring at the centre of the wafer and near zero contact angle occurring at the periphery. | [134] |
| Contact printing | Silicon | Different gradient arrangements can be created by coating differently shaped elastomeric stamps in octadecyltrichlorosilane (OTS). Hemispherical, quarter-cylinder and half-cylinder stamps produce radial, linear and bilinear gradients, respectively. Application of stepwise or continuous force to the stamps results in different exposure times of the OTS to the surface as the stamp deforms. | [135] |
| Dip coating | Ceramic, metallic | The substrate is first soaked in a precursor solution (alkoxide in alcohol solvent) and is then drawn out at a constant rate. The alcohol solvent evaporates leaving a thin layer of alkoxide on the surface. Longer soak times increase the thickness of the alkoxide coating, meaning the withdraw rate can be used to tune the contact angle gradient. | [136] |

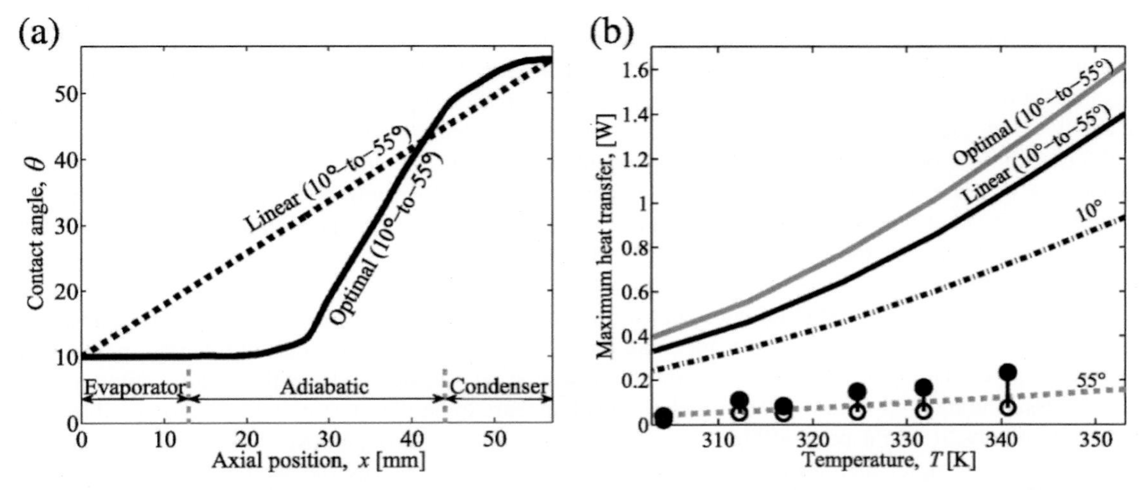


**FIGURE 5.22** (a) Optimal vs linear wettability gradients, (b) Comparison of maximum heat transfer capacities using the optimal wettability gradient, linear wettability gradient, uniform and high wetting and uniform and low wetting surfaces | each configuration used 3.2 mg of water as the working fluid [131].

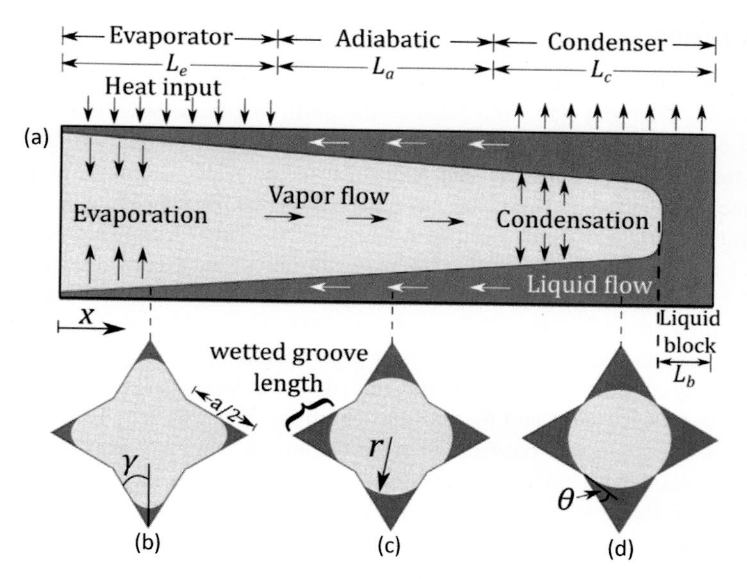


**FIGURE 5.23** Schematic illustration of trapezoidally grooved microheat pipe. Evaporation of the working fluid in the evaporator section causes the radius of curvature of the meniscus to recede, whilst condensation of the working fluid in the condenser section increases the radius of curvature of the meniscus; an axial capillary pressure gradient consequently develops; (a) liquid distribution in longitudinal cross-section, (b) small radius of curvature in evaporator, (c) medium radius of curvature in adiabatic section, (d) large radius of curvature in condensor [131].

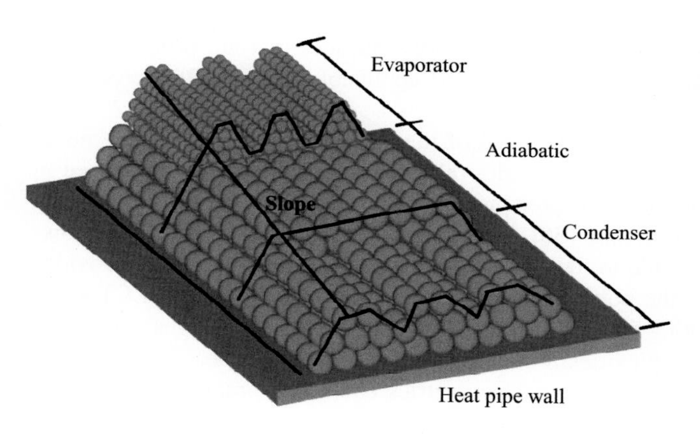


**FIGURE 5.24** Simple schematic of a variable thickness wick design with additional grooves in the evaporator and condenser to improve the surface area [138].

Here, the wick thickness controls the distribution of working fluid within the thermal circuit. A thicker wick in the evaporator section could delay dry-out in high heat flux applications although this might overburden the condenser section without proper testing/validation. This approach could also be applied in addition to, or in place of, the gradient wettability surface.

LHPs are not limited to metal wick structures. Low thermal conductivity materials such as polymers and ceramics are purported to be advantageous in some LHP applications because they reduce heat leak. For example, Wu et al. [139] report the use of sintered polytetrafluoroethylene (PTFE) as the primary LHP wick material, which increased the CHF to 600 W compared to 500 W for nickel when operating at a comparably low temperature (85°C) because of the reduced heat leak. The use of these materials is an interesting prospect. Limitations in conventional shaping and finishing technologies previously restricted the development of ceramic devices [140]. Desktop-scale stereolithography (SLA) has now advanced to the point where high-temperature polymers and even ceramics can be printed at low costs which could massively increase the accessibility of LHP development research. Ceramic SLA involves the fabrication of 'green parts' via the photopolymerisation of resin slurries containing micro and/or nanoparticles. Following a secondary sintering stage (to remove the polymer matrix) fully ceramic parts can be achieved. The method is explained in more detail in a recent review [141], whilst other ceramic AM methods are covered in [142]. The key challenge as explained is the optimisation of the rheology of the ceramic-resin slurry [141]. Nevertheless, several commercial examples already exist, removing the burden from the heat pipe researcher (e.g. FormLabs' RS-F2-CEWH-01 Ceramic Resin product). Sixel et al. [143] used ceramic SLA to fabricate a ceramic heat exchanger for thermally managing direct windings, validating this option.

## 5.5 General challenges areas for heat pipes

The allowable tolerance between adjacent structures before pore closing occurs remains one of the main limitations that reduces the effectiveness of current-generation AM wicks. This gap size limit negates the ability of these structures to generate capillary pumping effects as large as the best conventional wick designs. Additionally, the wall thickness limit (around 0.15–0.2 mm) also places limits on the heat transfer rate via the thermal resistance of the wall. The minimum wall thickness arises from the need to avoid leaks (since the relative densities are not quite 100%) whilst simultaneously ensuring the printed components are mechanically robust enough to survive removal from the platform. AlSi10Mg has been found to highly distort for a wall thickness of 0.5 mm [144], whilst 100%–300% deviations in dimensional accuracy have been observed when printing Ti64 with 0.1 mm wall thickness [145]. Ameli et al. achieved even small wall thicknesses of 0.07–0.13 mm in their aluminium heat pipe structures (see Section 5.3.1) [25]. These different results demonstrate that optimisation of the specific L-PBF process may be required in the heat pipe development process, providing another potential for the introduction of time delays. Other factors affecting the wall thickness selection and wick pore size include the reliability of the powder fusion process itself and the build orientation.

On the latter subject, although AM removes many of the geometric constraints of conventional manufacturing processes, the need to orientate the parts in 3D space combined with different resolutions in the $X$, $Y$ and $Z$ directions (depending on the AM technology) places an upper limit on the design freedom. Build orientation affects the final part quality (such as surface roughness), but optimisation is largely based on a combination of trial and error and user experience [146]. This might introduce further bottlenecks or unforeseen time delays in the development of novel heat pipe geometries although these will be somewhat offset by the rapid build times. Zwier and Wits [146] proposed a method for optimising the orientation of objects manufactured by fused deposition modelling (FDM) by minimising the number of overhangs, the logic being that this would minimise the total area of lower-quality surfaces that are produced from contact with the support structures/build platform. This same approach may also be applicable to the L-PBF process to maximise the quality of the external surface of complex heat pipe geometries.

The challenge of powder removal comes in two parts – the actual removal of the powder and reliable postremoval inspection. The consequences of ineffective powder removal and some strategies relevant to heat pipes were described in Section 5.2. On the inspection side, CT scanning remains the most comprehensive and nondestructive method, enabling the precise characterisation of trapped powder quantity and positioning [30]. Fig. 5.25 shows an example of trapped powder within a heat pipe wick observed using a CT scan. Being able to locate the position of this trapped powder provides feedback for the design of the wick and the effectiveness of the powder removal method. However, although CT scanning is effective, its cost and accessibility might be a mitigating factor. Table 5.8 summarises the advantages and disadvantages of a variety of inspection methods of differing complexities. For heat pipes, the use of 'cleaning challenge devices' may prove to be somewhat useful. These are objects with internal geometries equal to or more complex than the heat pipe wick that would be printed in the same batch. Confirming the removal of all powder from the cleaning challenge devices using comprehensive destructive methods then validates the batch of heat pipes [147].

There are several potential challenges on the materials side of heat pipe development. The design of metal powder blends is often by trial and error, requiring significant time and financial investment to get right [3]. For this reason, most heat pipe development projects will likely be restricted to using proprietary materials already available at market, which may impose limits on higher-end performance. The reader is again advised to consult the Senvol Database for further details (http://senvol.com/database/). Another potential issue is a lack of a suitable industry standard. Although the ASTM F3049–14 standard exists, it only provides signposting to various characterisation methods of metal powders and was not devised with AM in mind [149]. In Section 5.4.2, the use of multiple metal wicks that exploit multiple

**FIGURE 5.25** Trapped powder inside a porous lattice structure observed via CT scan.

**TABLE 5.8** Advantages and disadvantages of nondestructive residual powder inspection methods [30,147,148].

| Method | Advantages | Disadvantages |
|---|---|---|
| X-ray CT scanning | • Can locate precise location of trapped powder<br>• Can quantify the amount of trapped powder<br>• The entire device can be inspected, providing feedback on potential problematic design features | • Expensive<br>• Very time consuming<br>• Low voxel accuracy<br>• 2D XCT cannot quantify 3D geometries (e.g., helical coils) effectively |
| Weighing *(comparison of part weight to design spec.)* | • Fast<br>• Low cost<br>• Easy and simple to implement<br>• Widely available apparatus | • Cannot examine individual features, only the device as a whole<br>• Requires careful precleaning<br>• No guarantee of complete powder removal – surface roughness would be of a similar magnitude to trapped powder |
| SEM surface Imaging | • Can observe the layered structures in detail<br>• Support material burrs distinct from the model via the texture | • Time consuming<br>• Limited field of view (impractical for larger parts) |
| Optical microscopy | • Widely accessible<br>• Easy to implement | • Limited to observations of the external surface |
| 'Ruler-Drop' *(comparison of cavity depths to design spec.)* | • Fast and simple to implement<br>• Low cost | • Not suitable for complex internal channels |
| Shadowgraphs *(related to schlieren imaging)* | • Can generate accurate 2D data of channel shapes for quantifying distortion<br>• Can also quantify curvature, roughness, wall thickness and pore closing | • Only applicable to shallow channels (5 mm depths), so this method would need to be paired with a cleaning challenge device |

thermal conductivities was suggested, but this idea is ultimately contingent on the advancement of multimaterial printing. Some predict this will not be readily accessible until 2030, with others even predicting 10 years beyond this [150]. That being said, various researchers have demonstrated different types of multimetal printing. The simplest studies report blends of primary metals acting as a matrix for secondary reinforcement metals. For example, Onuike et al. [151] enhanced the thermal conductivity of IN718 by 300% by adding in GRCop-84 copper alloy. More impressively, Heer and Bandyopadyay [152] managed to achieve a graded transition from SS316 to magnetic SS430L without a weld seam. This latter capability would clearly be beneficial to heat pipes by enabling the functionalisation of different parts of the thermal circuit for example. However, a multitude of roadblocks remain. Primarily, more trial-and-error research is required to get the material blends correct. Part integrity, dimensional accuracy, part size, quality control and repeatability could all be compromised by differences in metal properties such as thermal expansion, thermal conductivities, powder 'flowabilities', melting points and laser absorption [153]. On top of the physical limitations, there are also shortcomings on the software side; there are no packages that support the analysis and modelling, let alone the creation, of multimaterial structures, imposing further burdens on the experimental side.

Scaling up also comes with its own set of challenges, and similar to the issues of materials described above, includes both hardware and software considerations. A detailed discussion of these was provided in a recent review article [4], so a shorter summary relevant to heat pipes is provided here. First, the hardware. The extra time to process larger powder volumes can be offset somewhat using more lasers, and various solutions have been demonstrated by different commercial vendors. For example, the X Line 2000R (800 × 400 × 500 mm) uses dual lasers, the Renishaw AM500Q (250 × 250 × 350 mm) uses quad 500 W lasers, and EOS's 'LaserProFusion' technology purportedly manages 'up to one million diode lasers' to fire simultaneously with a combined 5 kW. However, scaling up a powder bed involves more than increasing the print times. Fundamentally, scale-up changes the surface area to volume ratio making it harder to maintain uniform temperature profiles. And as discussed previously, this can imbue stresses into the parts leading to dimensional deviations, worse reliability and consistency and unwanted pore formation [154].

On the software side, the file size is generally proportional to the surface area (based on the stl file standard) so the file size will roughly quadruple for a doubling of the principal length. Heat pipes are particularly susceptible to this issue because of the large surface areas of the lattice/wick geometries. Although geometric simplifications may alleviate the issue to a small degree (e.g. approximating cylindrical struts as polygons to reduce the number of surfaces), it is likely that more powerful and expensive computers will be needed just to create larger heat pipe CAD models. Another potential solution would be to define the heat pipe lattice as a series of repeated instructions for the laser to follow (where complexity emerges from these simpler instructions), but this then restricts the design freedom. Alternatively, heat pipes could be modularised, which would then remove the need for larger printers in the first place. A secondary advantage of modularisation is it reduces the consequences of print failures that would be 'more inherent' to larger designs. The final 'software' consideration is the design itself. Successful prototyping requires that users have extensive prior knowledge and experience, but few experts exist within an organisation that meet these needs; consequently, there are calls to improve 'AM literacy' to fix this [4]. Being able to predict the 'printability' of new concepts a priori will provide time and cost savings in the development in next-generation heat pipes.

## 5.6 Summary and outlook

AM involves *adding* material instead of *subtracting* material and sits within the 'Industry 4.0' framework as one of many new digital technologies enabling new engineering practices. In particular, AM unlocks new geometric complexity and provides access to new optima within the large potential design space of heat pipe concepts, involving the combination of *heat pipe geometry*, *wick design*, *materials properties*, *working fluid properties* and *operating conditions*. However, the prospect of using AM to fabricate and develop heat pipes is clearly still in its infancy.

One of the current limitations is the allowable tolerance between adjacent structures without pore closing, meaning it is difficult to generate strong capillary pumping effects. Whilst AM wicks are able to match the capillary performance of current-generation grooves and foams, the pore sizes are still 1 order of magnitude higher than the theoretical best sintered wicks meaning the capillary pressures are not as high as the current best performing conventional wick structures. Continued improvement in AM technologies such as SLM will likely overcome this issue.

To unlock the next generation of heat pipe capabilities, each part of the thermal circuit should be considered so that truly optimal performance can be realised. To aid in this development, this chapter has discussed many of the opportunities that AM can potentially deliver now. As a final note, the wick design features proposed in this chapter are still grounded in conventional manufacturing thinking and techniques. Many of these suggested ideas are born from conventional 'top-down' or subtractive manufacturing approaches. Instead, by thinking in 'bottom-up' terms, whereby the design is driven by the fundamental concepts in heat transfer, thermodynamics and fluid mechanics, next-generation heat pipe concepts should become reality.

## References

[1] C. Parra-Cabrera, C. Achille, S. Kuhn, R. Ameloot, 3D printing in chemical engineering and catalytic technology: structured catalysts, mixers and reactors, Chem. Soc. Rev. 47 (2018) 209.

[2] P.N. Nesterenko, 3D printing in analytical chemistry: current state and future, Pure Appl. Chem. (2020). Available from: https://doi.org/10.1515/pac-2020-0206.

[3] S.A.M. Tofail, E.P. Koumoulos, A. Bandyopadhyay, S. Bose, L. O'Donoghue, C. Charitidis, Additive manufacturing: scientific and technological challenges, market uptake and opportunities, Mater. Today 21 (1) (2018) 22–37.

[4] J.R. McDonough, A perspective on the current and future roles of additive manufacturing in process engineering, with an emphasis on heat transfer, Therm. Sci. Eng. Prog. 19 (2020) 100594.

[5] D. Jafari, W.W. Wits, B.J. Geurts, Phase change heat transfer characteristics of an additively manufactured wick for heat pipe applications, Appl. Therm. Eng. 168 (2020) 113890.

[6] Z. Hu, D. Wang, J. Xu, L. Zhang, Development of a loop heat pipe with the 3D printed stainless steel wick in the application of thermal management, Int. J. Heat. Mass. Transf. 161 (2020) 120258.

[7] D. Jafari, W.W. Wits, B.J. Geurts, An investigation of porous structure characteristics of heat pipes made by additive manufacturing, in: 23rd International Workshop on Thermal Investigations of ICs and Systems, Therminic, 2017.

[8] S. Ozguc, S. Pai, L. Pan, P.J. Geoghegan, J.A. Weibel, Experimental demonstration of an additively manufactured vapor chamber heat spreader, In: 18th IEEE ITHERM Conference, 2019.

[9] F. Belfi, F. Iorizzo, C. Galbiati, F. Lepore, Space structures with embedded flat plate pulsating heat pipe built by additive manufacturing technology: development test and performance analysis, J. Heat. Transf. 141 (2019) 095001.

[10] D. Deng, Y. Tang, G. Huang, L. Lu, D. Yuan, Characterisation of capillary performance of composite wicks for two-phase heat transfer devices, Int. J. Heat. Mass. Transf. 56 (2013) 283–293.

[11] Y.F. Maydanik, Loop heat pipes, Appl. Therm. Eng. 25 (2005) 635–657.

[12] T. Vaneker, A. Bernard, G. Moroni, I. Gibson, Y. Zhang, Design for additive manufacturing: framework and methodology, CIRP Ann. – Manuf. Technol. 69 (2020) 578–599.

[13] D.B. Kim, P. Witherell, R. Lipman, S.C. Feng, Streamlining the additive manufacturing digital spectrum: a systems approach, Addit. Manuf. 5 (2015) 20–30.

[14] S.L. Liu, Y.C. Shin, Additive manufacturing of Ti6Al4V alloy: a review, Mater. Des. 164 (2019) 107552.

[15] N.T. Aboulkhair, M. Simonelli, L. Parry, I. Ashcroft, C. Tuck, R. Hague, 3D printing of aluminium alloys: additive manufacturing of aluminium alloys using selective laser melting, Prog. Mater. Sci. 106 (2019) 100578.

[16] C. Deng, J. Kang, T. Feng, Y. Feng, X. Wang, P. Wu, Study on the selective laser melting of $CuSn_{10}$ powder, Materials 11 (4) (2018) 614.

[17] L. Thijs, M.L.M. Sistiaga, R. Wauthle, Q. Xie, J.-P. Kruth, J. Van Humbeeck, Strong morphological and crystallographic texture and resulting yield strength anisotropy in selective laser melted tantalum, Acta Mater 61 (12) (2013) 4657–4668.

[18] N. Kang, W. Ma, L. Heraud, M. El Mansori, F. Li, M. Liu, et al., Selective laser melting of tungsten carbide reinforced maraging steel composite, Addit. Manuf. 22 (2018) 104–110.

[19] M. Gieseke, C. Noelke, S. Kaierle, V. Wesling, H. Haferkamp, Selective laser melting of magnesium and magnesium alloys, Magnesium Technology, Springer, 2013, pp. 65–68.

[20] C.Y. Yap, C.K. Chua, Z.L. Dong, Z.H. Liu, D.Q. Zhang, L.E. Loh, et al., Review of selective laser melting: materials and applications, Appl. Phys. Rev. 2 (2015) 041101.

[21] J. Jacimovic, T. Christen, E. Dénervaud, Self-organised giant magnetic structures via additive manufacturing in NdFeB permanent magnets, Addit. Manuf. 34 (2020) 101288.

[22] D.S. Thomas, S.W. Gilbert, NIST Special Publication 1176, Costs and Cost Effectiveness of Additive Manufacturing: A Literature Review and Discussion, 2014.

[23] M. Ameli, B. Agnew, P.S. Leung, B. Ng, C.J. Sutcliffe, J. Singh, et al., A novel method for manufacturing sintered aluminium heat pipes (SAHP), Appl. Therm. Eng. 52 (2013) 498–504.

[24] D. Jafari, W.W. Wits, B.J. Geurts, Metal 3D-printed wick structures for heat pipe application: capillary performance analysis, Appl. Therm. Eng. 143 (2018) 403–414.

[25] J. Feng, J. Fu, Z. Lin, B. Li, A review of the design methods of complex topology structures for 3D printing, Visual Comput. Ind. Biomed. Art. 1 (2018) 5.

[26] A.T. Clare, P.R. Chalker, S. Davies, C.J. Sutcliffe, S. Tsopanos, Selective laser melting of high aspect ratio 3D-titanium structures two way trained for MEMS applications, Int. J. Mech. Des. 4 (2008) 181–187.

[27] B.J. Hathaway, K. Garde, S.C. Mantell, J.H. Davidson, Design and characterization of an additive manufactured hydraulic oil cooler, Int. J. Heat. Mass. Transf. 117 (2018) 188–200.

[28] G.A.O. Adam, D. Zimmer, On design for additive manufacturing: evaluating geometrical limitations, Rapid Prototyp. J. 21 (6) (2015) 662–670.

[29] H. Hasib, O.L.A. Harrysson, H.A. West II, Powder removal from Ti-6Al-4V cellular structures fabricated via electron beam melting, JOM 67 (3) (2015) 639–646.

[30] L.W. Hunter, D. Brackett, N. Brierley, J. Yang, M.M. Attallah, Assessment of trapped powder removal and inspection strategies for powder bed fusion techniques, Int. J. Adv. Manuf. Technol. 106 (2020) 4521–4532.

[31] M. Qian, W. Xu, M. Brandt, H.P. Tang, Additive manufacturing and postprocessing of Ti-6Al-4V for superior mechanical properties, MRS Bull. 41 (10) (2016) 775–784.

[32] D. Buchbinder, W. Meiners, N. Pirch, K. Wissenbach, J. Schrage, Investigation on reducing distortion by preheating during manufacture of aluminium components using selective laser melting, J. Laser Appl. 26 (2014) 012004.

[33] T. Mishurova, S. Cabeza, T. Thiede, N. Nadammal, A. Kromm, M. Klaus, et al., The influence of the support structure on residual stress and distortion in SLM Inconel 718 Parts, Metall. Mater. Trans. A 49 (2018) 3038–3046.

[34] Renishaw, Design for Metal AM – A Beginner's Guide. 2017. https://www.renishaw.com/en/design-for-metal-am-a-beginners-guide--43333. (Accessed 23 June 2023).

[35] X. Lou, M.A. Othon, R.B. Rebak, Corrosion fatigue crack growth of laser additively-manufactured 316L stainless steel in high temperature water, Corros. Sci. 127 (2017) 120–130.

[36] S.D. Nath, E. Clinning, G. Gupta, V. Wuelfrath-Poirier, G. L'Espérance, O. Gulsoy, et al., Effects of Nb and Mo on the microstructure and properties of 420 stainless steel processed by laser-powder bed fusion, Addit. Manuf. 28 (2019) 682–691.

[37] N.T. Aboulkhar, I. Maskery, C. Tuck, I. Ashcroft, N.M. Everitt, The microstructure and mechanical properties of selectively laser melted AlSi10Mg: The effect of a conventional T6-like heat treatment, Mater. Sci. Eng. A 667 (2016) 139–146.

[38] N.E. Uzan, R. Shneck, O. Yeheskel, N. Frage, Fatigue of AlSi10Mg specimens fabricated by additive manufacturing selective laser melting (AM-SLM), Mater. Sci. Eng. A 704 (2017) 229–337.

[39] L. Zhuo, Z. Wang, H. Zhang, E. Yin, Y. Wang, T. Xu, et al., Effect of post-process heat treatment on microstructure and properties of selective laser melted AlSi10Mg alloy, Mater. Lett. 234 (2019) 196–200.

[40] M.A. Anam, Microstructure and Mechanical Properties of Selective Laser Melted Superalloy Inconel 625 (Electronic theses and dissertations), Paper 3029, 2018.

[41] H. Wong, K. Dawson, G.A. Ravi, L. Howlett, R.O. Jones, C.J. Sutcliffe, Multi-laser powder bed fusion benchmarking – initial trials with Inconel 625. The, Int. J. Adv. Manuf. Technol. 105 (2019) 2891–2906.

[42] E. Wycisk, A. Solback, S. Siddique, D. Herzog, F. Walther, C. Emmelmann, Effects of defects in laser additive manufactured Ti-6Al-4V on fatigue properties, Phys. Proc. 56 (2014) 371–378.

[43] S.K. Everton, M. Hirsch, P. Stravroulakis, R.K. Leach, A.T. Clare, Review of *in-situ* process monitoring and *in-situ* metrology for metal additive manufacturing, Mater. Des. 95 (2016) 431–445.

[44] D.A. Ramirez, L.E. Murr, S.J. Li, Y.X. Tian, E. Martinez, B.I. Machado, et al., Open-cellular copper structures fabricated by additive manufacturing using electron beam melting, Mater. Sci. Eng. A 528 (2011) 5379–5386.

[45] V.P. Kumar, G.P. Manogharan, D. Cormier, Design of periodic cellular structures for heat exchanger applications, in: 20th Annual International Solid Freeform Fabrication Symposium, SFF, 2009.

[46] J. Esarte, J.M. Blanco, A. Bernardini, J.T. San-José, Optimising the design of a two-phase cooling system loop heat pipe: wick manufacturing with 3D selective laser melting printing technique and prototype testing, Appl. Therm. Eng. 111 (2017) 407–419.

[47] L. Yang, O. Harrysson, H. West, D. Cormier, Compressive properties of Ti-6Al-4V auxetic mesh structures made by elecctron beam melting, Acta Mater 60 (8) (2012) 3370–3379.

[48] A. Malijevský, Does surface roughness amplify wetting? J. Chem. Phys. 141 (2014) 184703.

[49] O.T. Ibrahim, J.G. Monroe, S.M. Thompson, N. Shamsaei, H. Bilheux, A. Elwany, et al., An investigation of a multi-layered oscillating heat pipe additively manufactured from Ti-6Al-4V powder, Int. J. Heat. Mass. Transf. 108 (Part A) (2017) 1036–1047.

[50] R.J. McGlen, An introduction to additive manufactured heat pipe technology and advanced thermal management products, Therm. Sci. Eng. Prog. 25 (2021) 100941.

[51] C. Buchner, Development of a 3D-printed heat pipe for thermal management of a high-power CubeSat thruster, in: European Space Thermal Engineering Workshop [Online], 6th October, 2020.

[52] G. Huang, W. Yuan, Y. Tang, B. Zhang, S. Zhang, L. Lu, Enhanced capillary performance in axially grooved aluminium wicks by alkaline corrosion treatment, Exp. Therm. Fluid Sci. 82 (2017) 212–221.

[53] E.C. Phillips, J.D. Hinderman, Determination of properties of capillary media useful in heat pipe design, ASME Paper 69-HT-18, 1969.

[54] M. Kim, M. Kaviany, Multi-artery heat-pipe spreader: monolayer-wick receding meniscus transitions and optimal performance, Int. J. Heat. Mass. Transf. 112 (2017) 343–353.

[55] A.M. Schroff, M. Armand, Le Caloduc. Rev. Tech. Thomson-CSF, vol. 1, no. 4, 1969 (in French).

[56] R.A. Freggens, Experimental determination of wick properties for heat transfer applications, in: 4th Intersociety Energy Conference Engineering Conference, Washington DC, 22–26 September, 1969, pp. 888–897.

[57] J.K. Ferrell, J. Alleavitch, Vaporisation heat transfer in capillary wick structures, Chem. Eng. Progr. Symp. Ser. 66 (102) (1969) 82–91.

[58] N. Fries, K. Odic, M. Conrath, M. Dreyer, The effect of evaporation on the wicking of liquids into a metallic weave, J. Colloid Interface Sci. 321 (1) (2008) 118–129.

[59] A. Faghri, Heat Pipe Science and Technology, Taylor & Francis, US, 1995.

[60] Y. Chen, T. Li, Z. Jia, F. Scarpa, C.-W. Yao, L. Wang, 3D printed hierarchical honeycombs with shape integrity under large compressive deformations, Mater. Des. 137 (2018) 226–234.

[61] S. Limmahakhun, A. Oloyede, K. Sitthiseripratip, Y. Xiao, C. Yan, 3D-printed cellular structures for bone biomimetic implants, Addit. Manuf. 15 (2017) 93–101.

[62] H.-Y. Lei, J.-R. Li, Z.-J. Xu, Parametric design of Voronoi-based lattice porous structures, Mater. Des. 191 (2020) 108607.

[63] S. Daynes, S. Feih, W.F. Lu, J. Wei, Optimisation of functionally graded lattice structures using isostatic lines, Mater. Des. 127 (2017) 215–223.

[64] F. Günther, M. Wagner, S. Pilz, A. Gebert, Design procedure for triply periodic minimal surface based biomimetic scaffolds, J. Mech. Behav. Biomed. Mater. 126 (2022) 104871.

[65] G. Maliaris, E. Sarafis, Mechanical behaviour of 3D printed stochastic lattice structures, In: 8th International Conference on Materials Structures and Micromechanics of Fracture, Brno University of Technology, Czech Republic, 2016.

[66] O. Al-Ketan, M.A. Assad, R.K.A. Al-Rub, Mechanical properties of periodic interpenetrating phase composites with novel architected microstructures, Compos. Struct. 176 (2017) 9–19.

[67] Z.G. Xu, C.Y. Zhao, Experimental study on pool boiling heat transfer in gradient metal foams, Int. J. Heat. Mass. Transf. 85 (2015) 824–829.

[68] Z.G. Xu, C.Y. Zhao, Pool boiling heat transfer of open-celled metal foams with V-shaped grooves for high pore densities, Exp. Therm. Fluid Sci. 52 (2014) 128–138.

[69] F. Tamburrino, S. Graziosi, M. Bordegoni, The design process of additively manufactured mesoscale lattice structures: a review, J. Comput. Inf. Sci. Eng. 18 (2018) 040801.

[70] S. Torquato, A. Donev, Minimal surfaces and multifunctionality, Proc. R. Soc. Lond. A. 460 (2004) 1849–1856.

[71] T. Femmer, A.J.C. Kuehne, J. Torres-Rendon, A. Walther, M. Wessling, Print your membrane: rapid prototyping of complex 3D-PDMS membranes via a sacrificial resist, J. Membr. Sci. 478 (2015) 12–18.

[72] Y. Jung, S. Torquato, Fluid permeabilities of triply periodic minimal surfaces, Phys. Rev. E 72 (2005) 056319.

[73] S.B.G. Blanquer, M. Werner, M. Hannula, S. Sharifi, G.P.R. Lajoinie, D. Eglin, et al., Surface curvature in triply-periodic minimal surface architectures as a distinct design parameter in preparing advanced tissue engineering scaffolds, Biofabrication 9 (2017) 025001.

[74] P. Kim, H.-Y. Kim, J.K. Kim, G. Reiter, K.Y. Suh, Multi-curvature liquid meniscus in a nanochannel: evidence of interplay between intermolecular and surface forces, Lab. Chip 9 (2009) 3255–3260.

[75] J.K. Guest, J.H. Prévost, Optimizing multifunctional materials: design of microstructures for maximised stiffness and fluid permeability, Int. J. Solids Struct. 43 (2006) 7028–7047.
[76] Y.C. Yortsos, A.K. Stubos, Phase change in porous media, Curr. Opin. Colloid Interface Sci. 6 (2001) 208–216.
[77] M.A. Hanlon, H.B. Ma, Evaporation heat transfer in Sintered Porous Media, J. Heat. Transf. 125 (2003) 644–652.
[78] K.K. Bodla, J.Y. Murthy, S.V. Garimella, Evaporation analysis in sintered wick microstructures, Int. J. Heat. Mass. Transf. 61 (2013) 729–741.
[79] R. Ranjan, J.Y. Murthy, S.V. Garimella, Analysis of the wicking and thin-film evaporation characteristics of microstructures, J. Heat. Transf. 131 (10) (2009) 101001.
[80] K.K. Wong, K.C. Leong, Saturated pool boiling enhancement using porous lattice structured produced by selective laser melting, Int. J. Heat. Mass. Transf. 121 (2018) 46–63.
[81] J.A. Weibel, S.V. Garimella, Visualization of vapor formation regimes during capillary-fed boiling in sintered-powder heat pipe wicks, Int. J. Heat. Mass. Transf. 55 (2012) 3498–3510.
[82] Y. Nasersharifi, M. Kaviany, G. Hwang, Pool-boiling enhancement using multilevel modulated wick, Appl. Therm. Eng. 137 (2018) 268–276.
[83] R. Wen, S. Xu, Y.-C. Lee, R. Yang, Capillary-driven liquid film boiling heat transfer on hybrid mesh wicking structures, Nano Energy 51 (2018) 373–382.
[84] S. Adera, D.S. Antao, R. Raj, E.N. Wang, Design of micropillar wicks for thin-film evaporation, Int. J. Heat. Mass. Transf. 101 (2016) 280–294.
[85] D. Ćoso, V. Srinivasan, M.C. Lu, J. Chang, A. Majumdar, Enhanced heat transfer in biporous wicks in the thin liquid film evaporation and boiling regimes, J. Heat. Transf. 134 (2012) 101501.
[86] Q. Cai, A. Bhunia, High heat flux phase change on porous carbon nanotube structures, Int. J. Heat. Mass. Transf. 55 (2012) 5544–5551.
[87] X. Dai, F. Yang, R. Yang, Y. Lee, C. Li, Micromembrane-enhanced capillary evaporation, Int. J. Heat. Mass. Transf. 64 (2013) 1101–1108.
[88] K.L. Wilke, B. Barabadi, Z. Lu, T. Zhang, E.N. Wang, Parametric study of thin film evaporation from nanoporous membranes, Appl. Phys. Lett. 111 (2017) 171603.
[89] S. Ryu, J. Han, J. Kim, C. Lee, Y. Nam, Enhanced heat transfer using metal foam liquid supply layers for micro heat spreaders, Int. J. Heat. Mass. Transf. 108 (2017) 2338–2345.
[90] S. An, B. Joshi, A.L. Yarin, M.T. Swihart, S.S. Yoon, et al., Supersonic cold spraying for energy and environmental applications: one-step scalable coating technology for advanced micro- and nanotextured materials, Adv. Mater. 32 (2020) 1905028.
[91] J.-G. Lee, D.-Y. Kim, B. Kang, D. Kim, H.-E. Song, J. Kim, et al., Nickel-copper hybrid electrodes self-adhered onto silicon wafer by supersonic cold-spray, Acta Mater. 93 (2015) 156–163.
[92] J. Voglar, P. Gregorčič, M. Zupančič, I. Golobič, Boiling performance on surfaces with capillary-length-spaced one- and two-dimensional laser textured patterns, Int. J. Heat. Mass. Transf. 127 (2018) 1188–1196.
[93] H. Li, S. Fu, G. Li, T. Fu, R. Zhou, Y. Tang, et al., Effect of fabrication parameters on capillary pumping performance of multi-scale composite porous wicks for loop heat pipe, Appl. Therm. Eng. 143 (2018) 621–629.
[94] G. Xin, P. Zhang, Y. Chen, L. Cheng, T. Huang, H. Yin, Development of composite wicks having different thermal conductivities for loop heat pipes, Appl. Therm. Eng. 136 (2018) 229–236.
[95] D. Kashchiev, Nucleation: Basic Theory with Applications, Butterworth-Heinemann, Oxford, UK, 2000.
[96] J.W. Rose, Dropwise condensation theory and experiment: a review, Proc. Instn. Mech. Engrs. 216 (Part A) (2002) 115–128.
[97] R. Enright, N. Miljkovic, J.L. Alvarado, K. Kim, J.W. Rose, Dropwise condensation on micro- and nanostructured surfaces, Nanosc. Microsc. Thermophys. Eng. 18 (2014) 223–250.
[98] R.N. Wenzel, Resistance of solid surfaces to wetting by water, Ind. Eng. Chem. 28 (1936) 988–994.
[99] A.B.D. Cassie, S. Baxter, Wettability of porous surfaces, Trans. Faraday Soc. 40 (1944) 546–551.
[100] N. Miljkovic, R. Enright, E.N. Wang, Modeling and optimization of superhydrophobic condensation, J. Heat. Transf. 135 (11) (2013) 111004.
[101] R. Enright, N. Miljkovic, A. Al-Obeidi, C.V. Thompson, E.N. Wang, Condensation on superhydrophobic surfaces: the role of local energy barriers and structure length scale, Langmuir 28 (40) (2012) 14424–14432.
[102] Q. Zhang, M. He, J. Chen, J. Wang, Y. Song, L. Jiang, Anti-icing surfaces based on enhanced self-propelled jumping of condensed water microdroplets, Chem. Commun. 49 (2013) 4516–4518.
[103] J.B. Boreyko, C.H. Chen, Vapor chambers with jumping-drop liquid return from superhydrophobic condensers, Int. J. Heat. Mass. Transf. 61 (2013) 409–418.
[104] J.B. Boreyko, C.H. Chen, Self-propelled dropwise condensate on superhydrophobic surfaces, Phys. Rev. Lett. 103 (2009) 184501.
[105] K. Rykaczewski, A.T. Paxson, S. Anand, X. Chen, Z. Wang, K.K. Varanasi, Multimode multidrop serial coalescence effects during condensation on hierarchical superhydrophobic surfaces, Langmuir 29 (2013) 881–891.
[106] K. Rykaczewski, J. Chinn, L.A. Walker, J.H.J. Scott, A.M. Chinn, W. Jones, et al., How nanorough is rough enough to make a surface superhydrophobic during condensation? Soft Matter 8 (2012) 8786–8794.
[107] C. Dorrer, J. Rühe, Wetting of silicon nanograss: from superhydrophilic to superhydrophobic surfaces, Adv. Mater. 20 (2008) 159–163.
[108] K. Rykaczewski, Microdroplet growth mechanism during water condensation on superhydrophobic surfaces, Langmuir 28 (2012) 7720–7729.
[109] T. Liu, W. Sun, X. Sun, H. Ai, Thermodynamic analysis of the effect of the hierarchical architecture of a superhydrophobic surface on a condensed drop state, Langmuir 26 (2010) 14835–14841.

[110] F.-C. Wang, F. Yang, Y.-P. Zhao, Size effect on the coalescence-induced self-propelled droplet, Appl. Phys. Lett. 98 (2011) 053112.
[111] D. Lee, C. Byon, Fabrication and characterization of pure-metal-based submillimeter-thick flexible flat heat pipe with innovative wick structures, Int. J. Heat. Mass. Transf. 122 (2018) 306–314.
[112] J. Feng, Z. Qin, S. Yao, Factors affecting the spontaneous motion of condensate drops on superhydrophobic copper surfaces, Langmuir 28 (2012) 6067–6075.
[113] C.-H. Chen, Q. Cai, C. Tsai, C.-L. Chen, G. Xiong, Y. Yu, et al., Dropwise condensation on superhydrophobic surfaces with two-tier roughness, Appl. Phys. Lett. 90 (2007) 173108.
[114] D.J. Preston, D.L. Mafra, N. Milijkovic, J. Kong, E.N. Wang, Scalable graphene coatings for enhanced condensation heat transfer, Nano Lett. 15 (2015) 2902–2909.
[115] N. Miljkovic, R. Enright, Y. Nam, K. Lopez, N. Dou, J. Sack, et al., Jumping-droplet-enhanced condensation on scalable superhydrophobic nanostructured surfaces, Nano Lett. 13 (2013) 179–187.
[116] J. Cheng, A. Vandadi, C.-L. Chen, Condensation heat transfer on two-tier superhydrophobic surfaces, Appl. Phys. Lett. 101 (2012) 131909.
[117] J.B. Boreyko, C.H. Baker, C.R. Poley, C.-H. Chen, Wetting and dewetting transitions on hierarchical superhydrophobic surfaces, Langmuir 27 (12) (2011) 7502–7509.
[118] E. Ölçeroğlu, M. McCarthy, Self-organization of microscale condensate for delayed flooding of nanostructured superhydrophobic surfaces, ACS Appl. Mater. Interfaces 8 (2016) 5729–5736.
[119] A.R. Parker, C.R. Lawrence, Water capture by a desert beetle, Nature 414 (2001) 33–34.
[120] M. Morita, T. Koga, H. Otsuka, A. Takahara, Macroscopic-wetting anisotropy on the line-patterned surface of fluoroalkylsilane monolayers, Langmuir 21 (2005) 911–918.
[121] R. Raj, R. Enright, Y. Zhu, S. Adera, E.N. Wang, Unified model for contact angle hysteresis on heterogeneous and superhydrophobic surfaces, Langmuir 28 (2012) 15777–15788.
[122] K.K. Varanasi, M. Hsu, N. Bhate, W. Yang, T. Deng, Spatial control in the heterogeneous nucleation of water, Appl. Phys. Lett. 95 (2009) 094101.
[123] M. He, Q. Zhang, X. Zeng, D. Cui, J. Chen, H. Li, et al., Hierarchical porous surface for efficiently controlling microdroplets' self-removal, Adv. Mater. 25 (16) (2013) 2291–2295.
[124] E. Ölçeroğlu, M. McCarthy, Spatial control of condensate droplets on superhydrophobic surfaces, J. Heat. Transf. 137 (2015) 080905.
[125] E. Ölçeroğlu, C.Y. Hsieh, K.K.S. Lau, M. McCarthy, Thin film condensation supported on amphilic microstructures, Trans. ASME 139 (2017) 020910–020911.
[126] R.L. Winter, M. McCarthy, Dewetting from amphiphilic minichannel surfaces during condensation, ACS Appl. Mater. Interfaces 12 (6) (2020) 7815–7825.
[127] Y. Hu, J. Cheng, W. Zhang, R. Shirakashi, S. Wang, Thermal performance enhancement of grooved heat pipes with inner surface treatment, Int. J. Heat. Mass. Transf. 67 (2013) 416–419.
[128] Z. Huang, J. Zhang, J. Cheng, S. Xu, P. Pi, Z. Cai, et al., Preparation and characterization of gradient wettability surface depending on controlling $Cu(OH)_2$ nanoribbon arrays growth on copper substrate, Appl. Surf. Sci. 259 (2012) 142–146.
[129] X. Xie, Q. Weng, Z. Luo, J. Long, X. Wei, Thermal performance of the flat micro-heat pipe with the wettability gradient surface by laser fabrication, Int. J. Heat. Mass. Transf. 125 (2018) 658–669.
[130] S. Daniel, M.K. Chaudhury, J.C. Chen, Fast drop movements resulting from the phase change on a gradient surface, Science 291 (2001) 633–636.
[131] S.H. Choi, B.Z. Newby, Micrometer-scaled gradient surfaces generated using contact printing of octadecyltrichlorosilane, Langmuir 19 (2003) 7427–7435.
[132] M. Faustini, D.R. Ceratti, B. Louis, M. Boudot, A. Albouy, C. Boissiere, et al., Engineering functionality gradients by dip coating process in acceleration mode, ACS Appl. Mater. Interfaces 6 (2014) 17102–17110.
[133] J. Cheng, G. Wang, Y. Zhang, P. Pi, S. Xu, Enhancement of capillary and thermal performance of grooved copper heat pipe by gradient wettability surface, Int. J. Heat. Mass. Transf. 107 (2017) 586–591.
[134] M. Singh, N.V. Datla, S. Kondaraju, S.S. Bahga, Enhancemed thermal performance of micro heat pipes through optimization of wettability gradient, Appl. Therm. Eng. 143 (2018) 350–357.
[135] X. Zhang, J. Xu, Z. Lian, Z. Yu, H. Yu, Influence of microstructure on surface wettability, in: International Conference on Advanced Mechatronic Systems, Beijing, China, 22–24 August, 2015.
[136] Q. Wang, J. Hong, Y. Yan, Biomimetic capillary inspired heat pipe wicks, J. Bionic Eng. 11 (3) (2014) 469–480.
[137] S.C. Wu, T.W. Gu, D. Wang, Y.M. Chen, Study of PTFE wick structure applied to loop heat pipe, Appl. Therm. Eng. 81 (2015) 51–57.
[138] U. Scheithauer, R. Kordaß, K. Noack, M.F. Eichenauer, M. Hartmann, J. Abel, et al., Potentials and challenges of additive manufacturing technologies for heat exchanger, in: C. Gómez, V.M.V. Flores (Eds.), Advances in Heat Exchangers, IntechOpen, 2018.
[139] J.W. Halloran, Ceramic stereolithography: additive manufacturing for ceramics by photopolymerization, Annu. Rev. Mater. Res. 46 (2016) 19–40.
[140] J. Deckers, J. Vleugels, J.-P. Kruth, Additive manufacturing of ceramics: a review, J. Ceram. Sci. Tech. 05 (04) (2014) 245–260.
[141] W. Sixel, M. Liu, G. Nellis, B. Sarlioglu, Ceramic 3D printed direct winding heat exchangers for improving electric machine thermal management, 2019 IEEE Energy Conversion Congress and Exposition (ECCE), Baltimore, MD, USA (2019) 769–776. Available from: https://doi.org/10.1109/ECCE.2019.8913234.

[142] A. Ahmed, A. Majeed, Z. Atta, G. Jia, Dimensional quality and distortion analysis of thin-walled alloy parts of AlSi10Mg manufactured by selective laser melting, J. Manuf. Mater. Process. 3 (2019) 51.
[143] S. Dobson, Y. Wu, L. Yang, Material characterization for lightweight thin wall structures using laser powder bed fusion additive manufacturing, in: Proceedings of the 29th Annual International Solid Freeform Fabrication Symposium, 2018.
[144] M.P. Zwier, W.W. Wits, Design for additive manufacturing: automated build orientation selection and optimization, Proc. CIRP 55 (2016) 128–133.
[145] B. Verhaagen, T. Zanderink, D.F. Rivas, Ultrasonic cleaning of 3D printed objects and cleaning challenge devices, Appl. Acoust. 103 (Part B) (2016) 172–181.
[146] D. Thomas, The Development of Design Rules for Selective Laser Melting (Ph.D. thesis), Cardiff Metropolitan University, 2009.
[147] J.A. Slotwinski, E.J. Garboczi, Metrology needs for metal additive manufacturing powders, JOM 67 (2015) 538–543.
[148] R. Jiang, R. Kleer, F.T. Piller, Predicting the future of additive manufacturing: a Delphi study on economic and societal implications of 3D printing for 2030, Technol. Forecast. Societal Change 117 (2017) 84–97.
[149] B. Onuike, B. Heer, A. Bandyopadhyay, Additive manufacturing of Inconel 718-Copper alloy bimetallic structure using laser engineering net shaping (LENS™), Addit. Manuf. 21 (2018) 133–140.
[150] B. Heer, A. Bandyopadhyay, Compositionally graded magnetic-nonmagnetic bimetallic structure using laser engineering net shaping, Mater. Lett. 216 (2018) 16–19.
[151] E.W.I., Dissimilar Materials Joining [Online]. https://ewi.org/dissimilar-materials-joining/, 2012.
[152] I. Echeta, X. Feng, B. Dutton, R. Leach, S. Piano, Review of defects in lattice structures manufactured by powder bed fusion, Int. J. Adv. Manuf. Technol. 106 (2020) 2649–2668.
[153] B. Liu, M. Liu, H. Cheng, W. Cao, P. Lu, A new stress-driven composite porous structure design method based on triply periodic minimal surfaces, Thin-Walled Struct. 181 (2022) 109974.
[154] N. Miljkovic, E.N. Wang, Condensation heat transfer on superhydrophobic surfaces, MRS Bull. 38 (2013) 397–406.

Chapter 6

# Heat pipe heat exchangers

## 6.1 Introduction

Although heat pipes were first introduced in the 1830s with the development of the Perkins' tube, because a wicked construction requires quite complex manufacturing techniques (which were not readily obtainable at the time), it was not until 1944 that a fully viable and commercial wicked heat pipe was available (see Chapter 1). As a result, early implementation was unfavourable, limited and lacked some of the later necessary theoretical analysis, and this led to the development of thermosyphons which were reliant on gravity during the interim period. In contrast with today, the features of heat pipe heat exchangers (HPHEs) that are attractive in industrial heat recovery applications are as follows:

- No moving parts and no external power requirements, implying high reliability.
- Cross-contamination is eliminated due to a solid wall between the hot and the cold gas streams.
- The units are generally compact, suitable for various processes and are also available in a wide range of sizes.
- Collection of the formed condensate from the exhaust gases can be arranged, and the flexibility of the technology allows ease of maintenance for both finned and unfinned systems.

Their applications fall into three main categories:

- Heat recovery in air-conditioning systems, normally involving comparatively low temperatures and duties, but including evaporative cooling.
- Recovery of heat from a process exhaust stream to preheat air for space heating.
- Recovery of waste heat from a process for reuse in other processes, for example, preheating of combustion air. This area of application is the most diverse and can involve a wide range of temperatures and duties.

The materials and working fluids used for heat pipe heat recovery units depend to a large extent, on their operating temperature range and, as far as the external tube surface and fins are concerned, on contamination from the environment in which the unit is to operate. Working fluids for air-conditioning and other applications – where operating temperatures are unlikely to exceed 40°C – include hydrofluorocarbons (HFCs) or their replacements such as hydrofluoroolefins (HFO)s. Moving up the temperature range, water is the best fluid to use, and for hot exhausts in furnaces and direct gas-fired air circuits, higher temperature organics can be used. There are also cases where liquid metal has been used as a working fluid, sodium perhaps being the most common – however, industrial examples are few and they are the subject of continuing research.

In most instances, the tube material is copper or aluminium with the same materials being used for the extended surfaces. Where the contamination from the gas is likely to be acidic, or at a higher temperature where a more durable material may be required, carbon steel or stainless steel is generally selected.

Fig. 6.1 shows a selection of heat pipes ready for assembly into an HPHE. The finned units on the left are 8 m long. The lower illustration shows three of the possible options for the heat pipes – (i) sections using two materials (stainless steel [SS] and carbon steel [CS]) where one end of the heat pipe is subject to possible corrosion attack, (ii) a partially finned heat pipe and (iii) an unfinned unit.

The tube bundle may comprise commercially available helically wound finned tubes or may be constructed like a refrigeration coil – the tubes being expanded into plates forming a complete rectangular 'fin' running the depth of the heat exchanger – the latter technique being preferable from a cost viewpoint. The size of unit varies with the air flow, a velocity of about 2–4 m/s being generally acceptable to keep the pressure drop through the bundle to a reasonable level. Small units having a face size of 0.3 m (height) × 0.6 m (length) are available.

This chapter focuses on industries and case studies which highlight the flexibility and versatility of the technology.

Heat Pipes. DOI: https://doi.org/10.1016/B978-0-12-823464-8.00002-4

**FIGURE 6.1** Heat pipe tube lengths.

## 6.2 Heat pipe heat exchangers in buildings

Heat exchangers and cooling coils have wide application within the ventilation industry, but in addition the wraparound loop heat pipes (WLHPs) has also been investigated by Jouhara and Meskimmon [1] and Jouhara and Ezzuddin [2]. A basic feature of a traditional system is the cooling coil which is meant to do the majority of the cooling and dehumidifying. Whereas with a wraparound heat pipe heat exchanger (Fig. 6.2) – in which the cooling coil is located between the evaporator and the condenser of a wickless heat pipe tube – the unit can operate by reversing the temperature difference across the heat pipe, so transferring the heat of the incoming hot air to the heat pipe which then contributes to the dehumidification process and maximises the operation of the cooling coil.

The studies highlighted the problem that traditional systems (involving a cooling coil) are inefficient because the mass transfer is primarily occurring at the cooling coil and not at the surface of a heat exchanger, whereas the outcome of the studies showed that allowing the heat exchanger to act as a precooler for the incoming air prior to the cooling coil improves the overall system efficiency.

Additional studies have also been conducted and applied within the ventilation industry. For example, Mahajan et al. [3] investigated the design of an oscillating heat pipe charged with pentane within a closed ducted air-conditioning system (Fig. 6.3). The system operates by containing both phases within the evaporator and condenser section; the thermal difference and pressures between the two phases causing oscillations within the tube depending on the location of the vapour plug or liquid slug flow. For example, when the hot air is drawn and heated from ambient, this produces vapour plugs which then push liquid slugs further up to the condenser system. The outcome from the study highlighted a maximum recovery rate of 200 W. However, regrettably, the study did not include a long-term evaluation because oscillating heat pipes are prone to dry out after prolonged operation when applied to low-temperature applications.

The potential of heat pipe application to ventilation has also been investigated by Diao et al. [4] who examined the use of flat microheat pipe heat exchangers for waste heat recovery from residential buildings. The feature of the technology (as shown in Fig. 6.4) is to allow warm air to pass through the evaporator section where the microfins on both the evaporator and condenser sections increase their effective surface area so improving heat transfer. Subsequently, a developed panel was investigated on a laboratory-scale basis for multiple flow rates, and the outcome of the study highlighted that with the addition of extra panels, the maximised heat recovery could reach approximately 78% efficiency when operating under summer conditions.

Jouhara [5] has also presented a patented thermal transfer loop which is an innovative heat pipe heat exchanger for ventilation (Fig. 6.5). The novel heat pipe heat exchanger (which can be retrofitted in buildings) transfers the heat from the evaporator to the condenser in an arrangement where the evaporator is located at a level higher than the condenser,

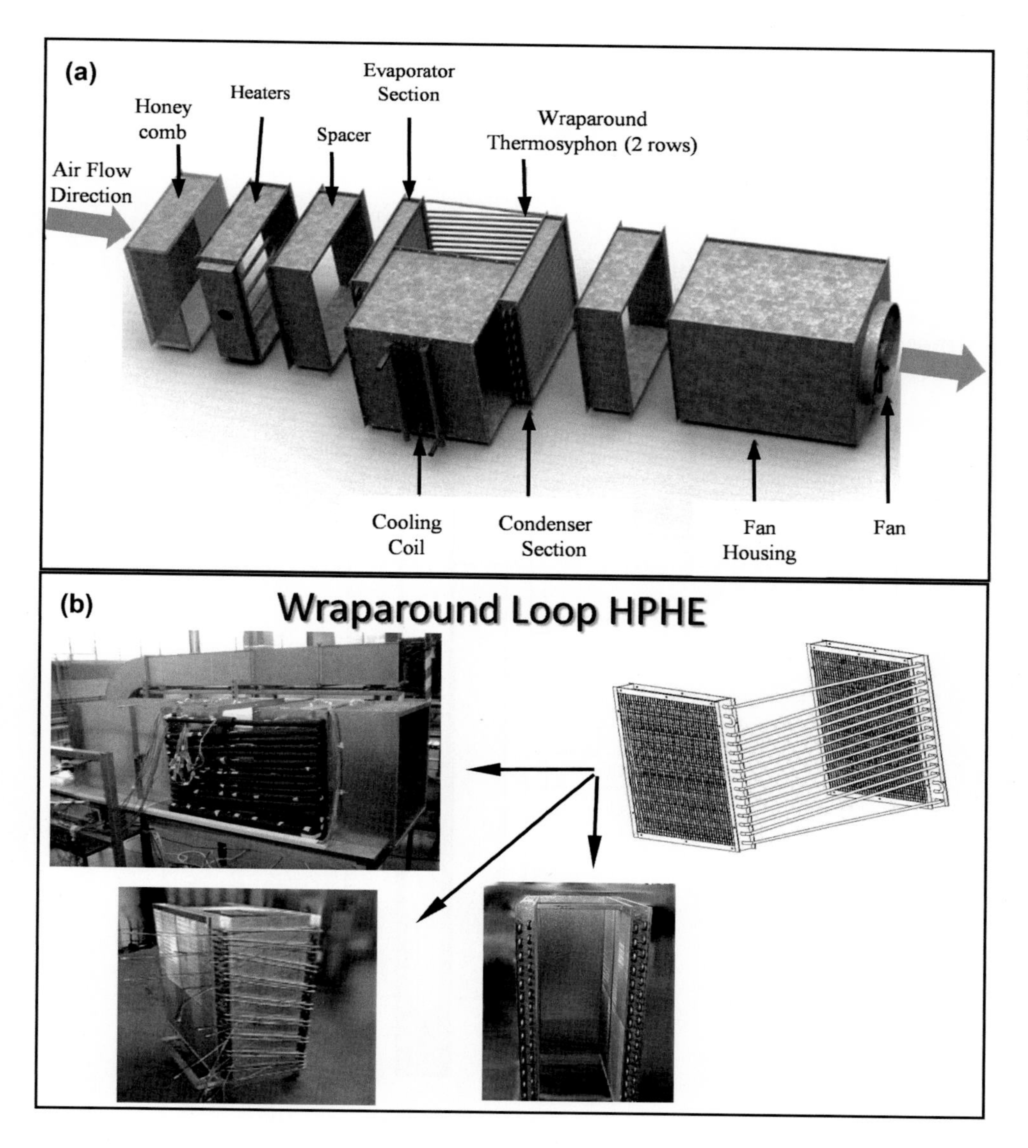


**FIGURE 6.2** Wraparound loop heat pipe (WLHP): (a) exploded 3D model of the system; (b) mechanical design of the system.

the condensate returning to the evaporator by the use of a pump rather than needing a wick. The feature and reason for this placement (i.e. the evaporator being higher than the condenser) is that it allows the system to recover heat from a building during natural ventilation – the exhaust warm air flowing through the evaporator at the top whilst fresh air flows through the condenser heat exchanger at the bottom. Moreover, the system is also provided with a liquid reservoir allowing the pump to work intermittently so reducing energy consumption.

## 6.3 Heat pipe heat exchangers in food processing

Currently, the food industry (specifically food production) offers one of the biggest potentials for waste heat recovery – where the addition of an HPHE allows heat recovery from medium to high-temperature exhaust streams, which can then be redirected into other processes and so reduce the overall consumption of fossil fuels. For example, following a study, Jouhara et al. [6] developed the HPHE that was installed at the Erva Mate tea drying plantation in Brazil. Traditionally, the drying process involves exposing the tea leaves to combustion gases to dry them, and although this recycles the combustion fumes it also creates a hazardous end product as the leaves are exposed to polycyclic aromatic hydrocarbons (PAHs) which are known carcinogens. To eliminate this risk, a heat pipe heat exchanger was designed and installed (shown in Fig. 6.6) to recover heat from the exhaust gases (with temperatures up to 450°C) from a biomass burner as they pass through the heat pipe evaporator section. Meanwhile, the condenser section draws in fresh air

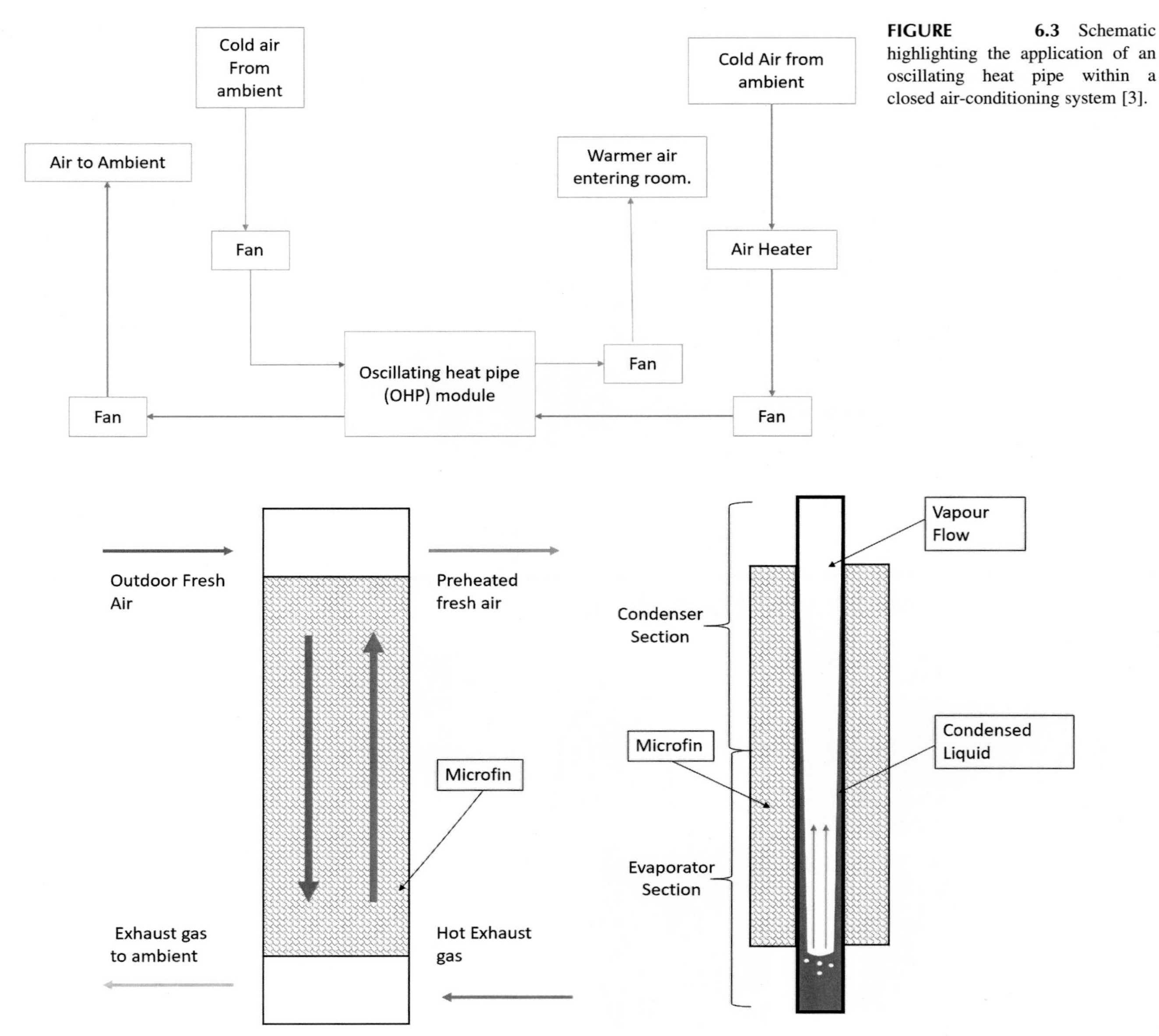


**FIGURE 6.3** Schematic highlighting the application of an oscillating heat pipe within a closed air-conditioning system [3].

**FIGURE 6.4** Schematic and structure of a microfinned heat pipe [4].

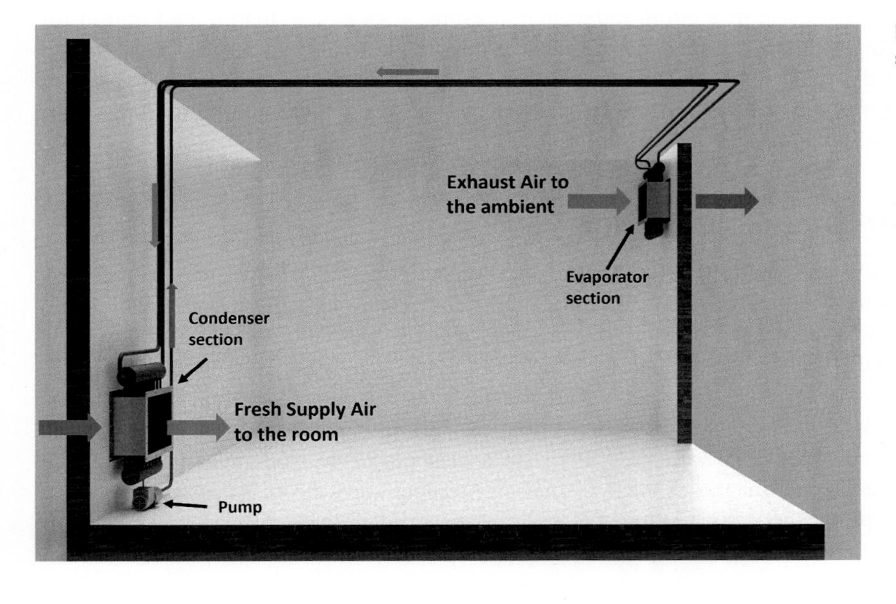


**FIGURE 6.5** Thermal transfer loop heat pipe system by Jouhara [5].

**FIGURE 6.6** Schematic of HPHE installed for Erva Mate tea drying [6].

and leaves the unit as clean hot air which can be used for drying. By implementing this method, the risk of cross-contamination is eliminated resulting in a safe high-quality product.

Also, within the food processing industry, a study conducted by Lukitobudi et al. [7] investigated the role of HPHEs for recovering medium-grade heat from bakery processes. The study involved the construction and the application of a cross-flow, air to air, HPHE to study the addition of fins and fin geometry. The study (as shown in Fig. 6.7) tested the HPHE under laboratory-scale conditions, cold air being drawn into the system through the condenser section (acting as a heat sink) and the resultant warm air being further heated and recirculated to the evaporator section to mimic the temperatures seen in the bakery industry,

The study further investigated the applicability of finned copper, finned steel and smooth copper tubes at various set point temperatures and velocities (1–5 m/s), and the outcomes of the study suggested a heat exchanger effectiveness ranging from 17.8% to 63% for finned copper units as the set point temperature and the inlet velocity increase. Similarly, when finned stainless steel was tested under identical conditions the heat exchanger effectiveness ranged from 6.2% to 48.5%, which in comparison against the smooth tubes, was still an improvement over its tube bundle which only showed a heat exchanger effectiveness between 3.2% and 19.7%. For its application within a bakery, the

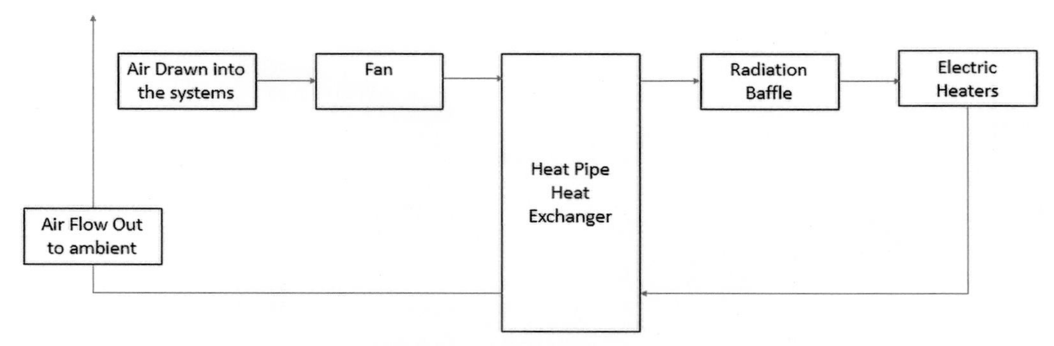


**FIGURE 6.7** Schematic of HPHE developed for waste heat recovery in bakery processes [7].

**FIGURE 6.8** Convective HPHE used within the ceramic industry [8].

unit was then installed in the exhaust of a baking oven, the reclaimed heat from which was then used as the heat source for a proving oven – previously heated by waste steam – and the addition of this HPHE itself indicating a heat exchanger effectiveness of 65%.

## 6.4 Heat pipe heat exchangers for the ceramics sector

### 6.4.1 Cross-flow heat pipe heat exchanger

Heat pipe heat exchanger technology has been applied to the ceramics industry by Jouhara et al. [8] (as a further example of an energy-saving application) with the installation of a cross-flow flue, the gas-to-air heat exchanger in the exhaust stack of the cooling section of a roller hearth kiln. The heat recovered by the system is used within the dryers of the plant to reduce their burner gas consumption by preheating the incoming combustion air. The system was installed on a platform next to the kiln as shown in Fig. 6.8.

The system was monitored using pressure sensors located at the outlet of each stream, with pitot probes for each stream and thermocouples at the inlet and outlet of the heat pipe heat exchanger. The hot flue gas entering the heat pipe was set at 210°C and a flow rate of 4300 $Nm^3/h$ whilst the condenser section parameters were 25°C (season dependent) for a flow rate of 2000 $Nm^3/h$. The results of the installed unit can be seen in Fig. 6.9 and Fig. 6.10, and these reflect the significant variations that occurred over to the operation cycle of the ceramic kiln.,

The heat pipe heat exchanger recovered up to 100 kW at an outlet temperature for the condenser section of 164°C and an evaporator temperature of 145°C, and the system managed to recover up to 876 MWh per year. The total return on investment (ROI) of the system (based on this value) was estimated at 16 months.

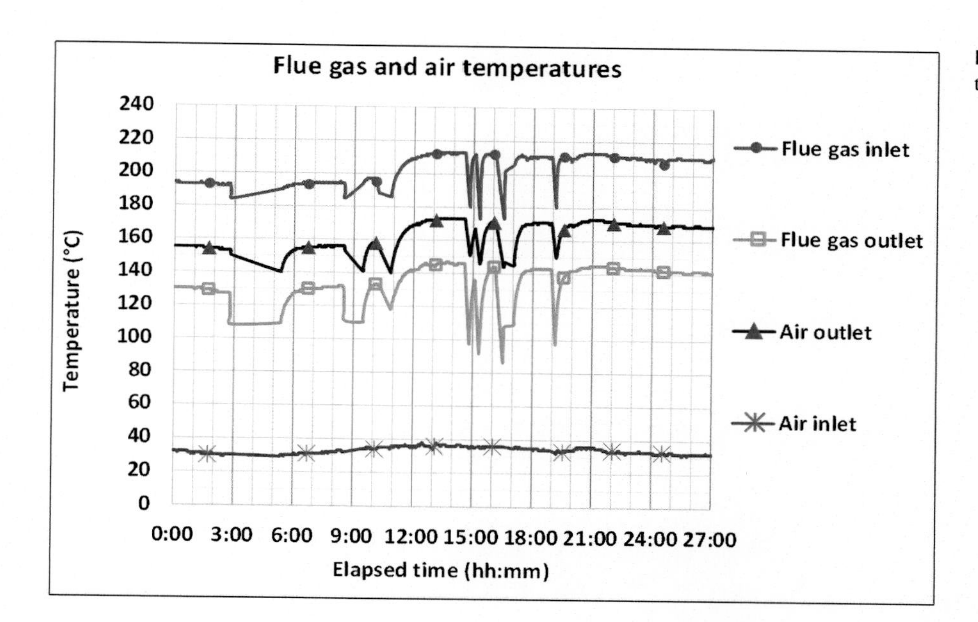


**FIGURE 6.9** Heat pipe heat exchanger temperatures during 27 working hours [8].

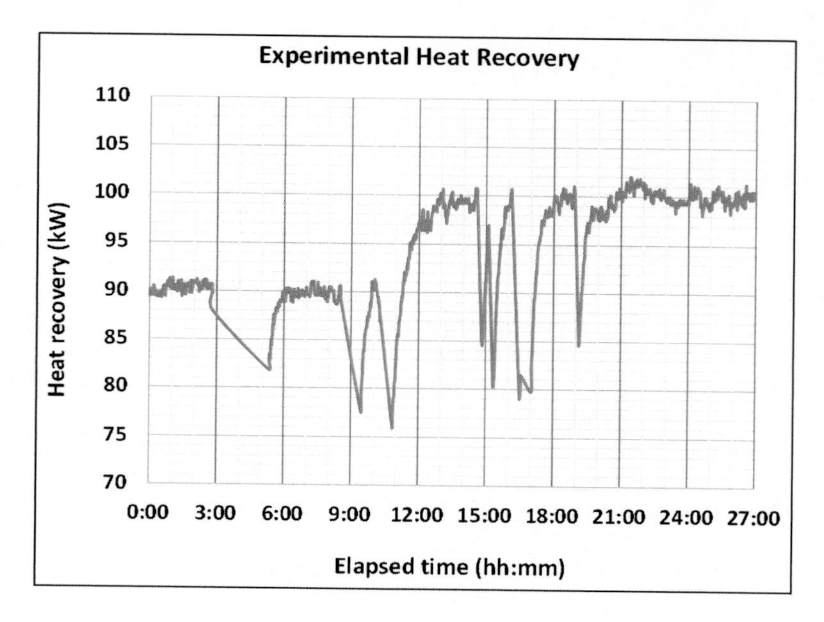


**FIGURE 6.10** Heat recovery during 27 working hours test under variable production levels [8].

Computational fluid dynamics (CFD) was used by Delpech et al. [9] to determine the predicted heat transfer value of a heat pipe heat exchanger operating under similar conditions, and their recovered heat was estimated at 863 MWh per year – similar to the experimental value determined above. The total saving in terms of natural gas was estimated at 110,600 $Sm^3$ per year whilst the $CO_2$ emission was reduced by 164 tonnes per year. This system was installed within the DREAM Project (Design for Resource and Energy efficiency in cerAMic kilns) – funded by the European Commission.

## 6.4.2 Radiative heat pipe heat exchanger

In order to reduce tile wastage and recover waste heat, a radiative heat pipe heat exchanger or radiative heat pipe module was investigated by Delpech et al. [10] (within the DREAM project) to actively cool tiles during the slow cooling section of a roller hearth kiln. As the tile cools within the kiln, a temperature difference between the sides and the centre of the tile occurs and this sets up stresses in the tile which can, on occasions, lead to cracking. To tackle this issue, a

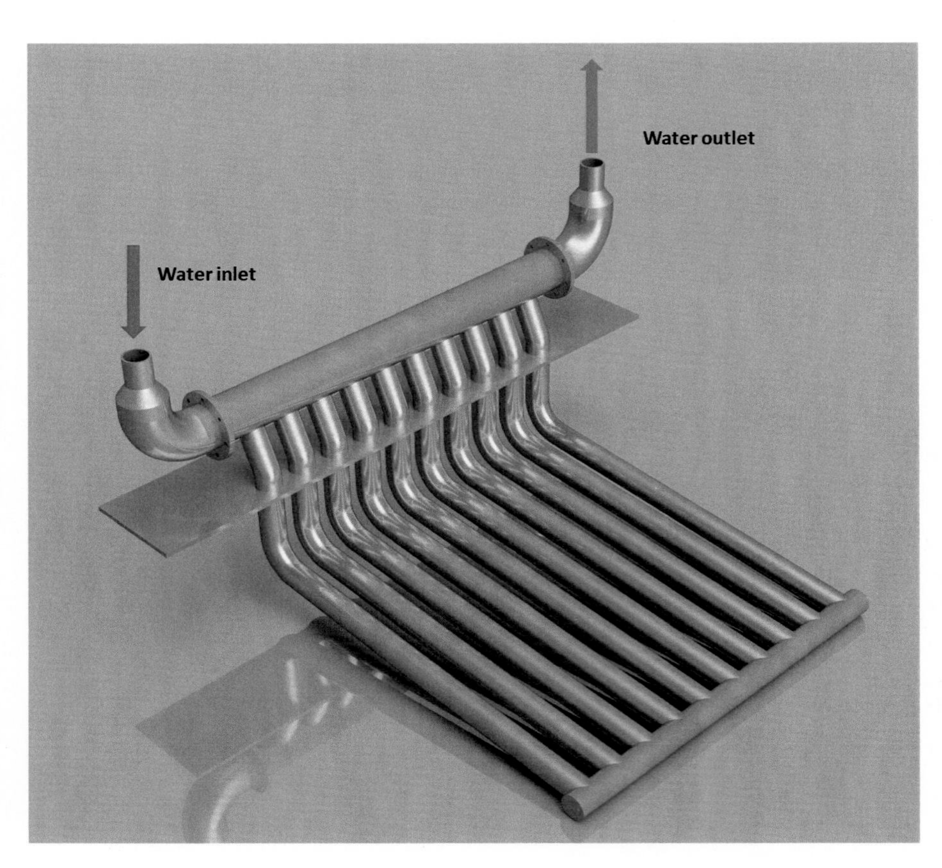


**FIGURE 6.11** Radiative heat pipe [10].

radiative heat pipe was designed (as shown in Fig. 6.11) which then acts as an active cooling device to achieve a more uniform temperature over the tile during this cooling period.

As shown in Fig. 6.12, the radiative heat pipe described in this paper embodied 10 parallel pipes, each of 28 mm diameter and 2 mm wall thickness connected by a bottom collector (also of 28 mm diameter) and a condenser section of diameter 50 mm and wall thickness of 2 mm. The condenser section is, itself, a shell and tube type heat exchanger with rows of 10 mm tubes comprising a total of nine tubes. The system was entirely monitored by thermocouples welded on the surface of the pipes to reduce the effect on the readings of natural convection and radiation, and the system was tested in a ceramics pad heater kiln at a range of temperatures from 200°C to 500°C and flow rates between 5 and 20 L/min. A theoretical model was also developed to predict the temperature and heat recovery of the system under different conditions.

The system was able to recover up to 3700 W at 500°C with a flow rate of 20 L/min, but it was also determined that the radiation effect takes place at higher temperatures whilst at lower temperatures, natural convection is the main mode of heat transfer. It was therefore concluded that this configuration would be more efficient at higher temperatures. The theoretical model developed for this system was in good agreement with the experimental data with an error of $\pm 15\%$; the model also allowed scalability to an industrial size of plant. The system also presents a good solution to help control the temperature of any tile as the surface area of the heat pipe (exposed to the tile) can be modified to suit the temperature requirements. Additionally (and a further advantage for the system), the temperature control of the tile can be facilitated using the uniform temperature of the heat pipe evaporator section. Plus, the system can be cooled by water without any risk of steam generation or explosion resulting from the high temperatures occurring within the kiln. Finally, it should also be noted that this design of heat pipe can be used for many other applications where radiative heat is available.

### 6.4.3 Multipass heat pipe heat exchanger

Still in the field of ceramics production, an experimental and computational study on a vertical multipass heat pipe heat exchanger was conducted by Brough et al. [11] to investigate the performance of a heat pipe heat exchanger on a

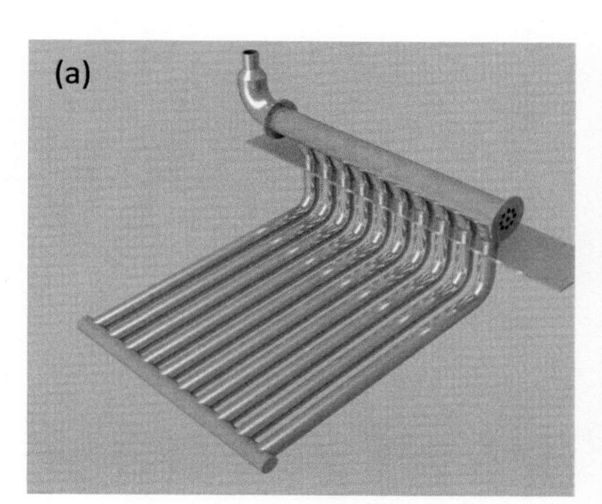


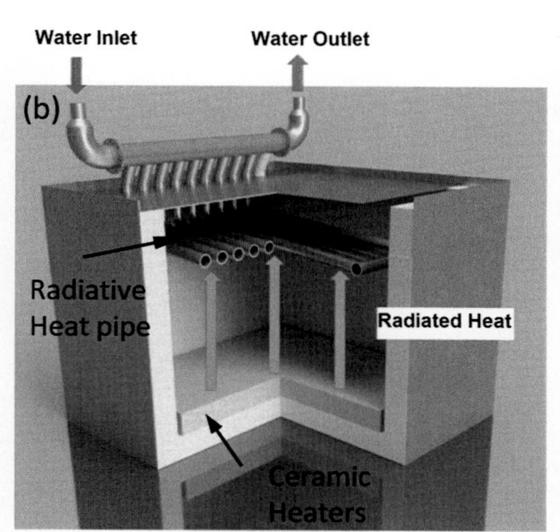


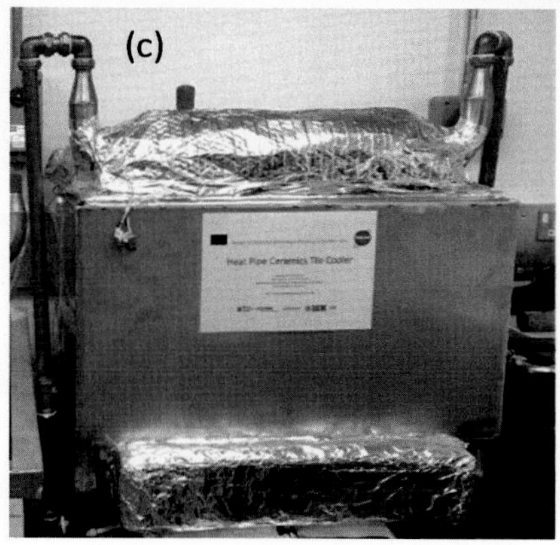


**FIGURE 6.12** (a) Developed radiative heat pipe design; (b) schematic highlighting the operation of the radiative panel; (c) developed lab scale unit [10].

laboratory-scale ceramic firing kiln. The heat exchanger was composed of 100, 28 mm diameter heat pipes in a staggered arrangement with a 1210 mm length evaporator section fitted with baffles to maximise the available heat transfer area. The condenser section was 250 mm long (also fitted with baffles) and cooled by water. An illustration of the system can be seen in Fig. 6.13.

The flue gas supplied by the kiln was varied in temperature over the range 135°C–265°C with the hot air being delivered to the heat pipe using a bypass to control the outlet temperature of water supplied at 20°C by a centrifugal pump. The piping and instrument diagram (P&ID) for the system can be seen in Fig. 6.14.

The installed system was able to recover up to 64 kW at an exhaust temperature and flow rate of 270°C and 1298 kg/h, respectively, and a condenser section flow rate of 1320 kg/h, and the developed model (for the prediction of the energy generated from the heat pipe) was in good agreement with the experimental results within ± 15%. The payback for a full-scale installation was estimated at 33 months and would certainly decrease the boiler load used for the generation of hot water for a facility; the system's main advantage (as developed) is its vertical multipass effect which not only greatly improves efficiency, but at the same time reduces the footprint required for such a unit.

## 6.5 Heat pipe heat exchangers waste heat boiler

Waste heat boilers are typically used in industrial plants to recover waste heat from furnaces and other such plants where the exhaust gases they produce contain considerable amounts of thermal energy. As such, the application of heat pipe heat exchangers to waste heat boilers has been widely researched, and over the past decades, their installation has been favoured because of their passive nature and high flexibility. For example, Littwin and McCurley [12] recognised the potential for the use of heat pipes within waste heat boilers in the 1980s, their study proposing the use of a bank of heat pipes within a pressure vessel, with a finned evaporator section being placed in the exhaust stream of a process and a condenser section within the boiler water where the steam is to be generated (as shown in Fig. 6.15).

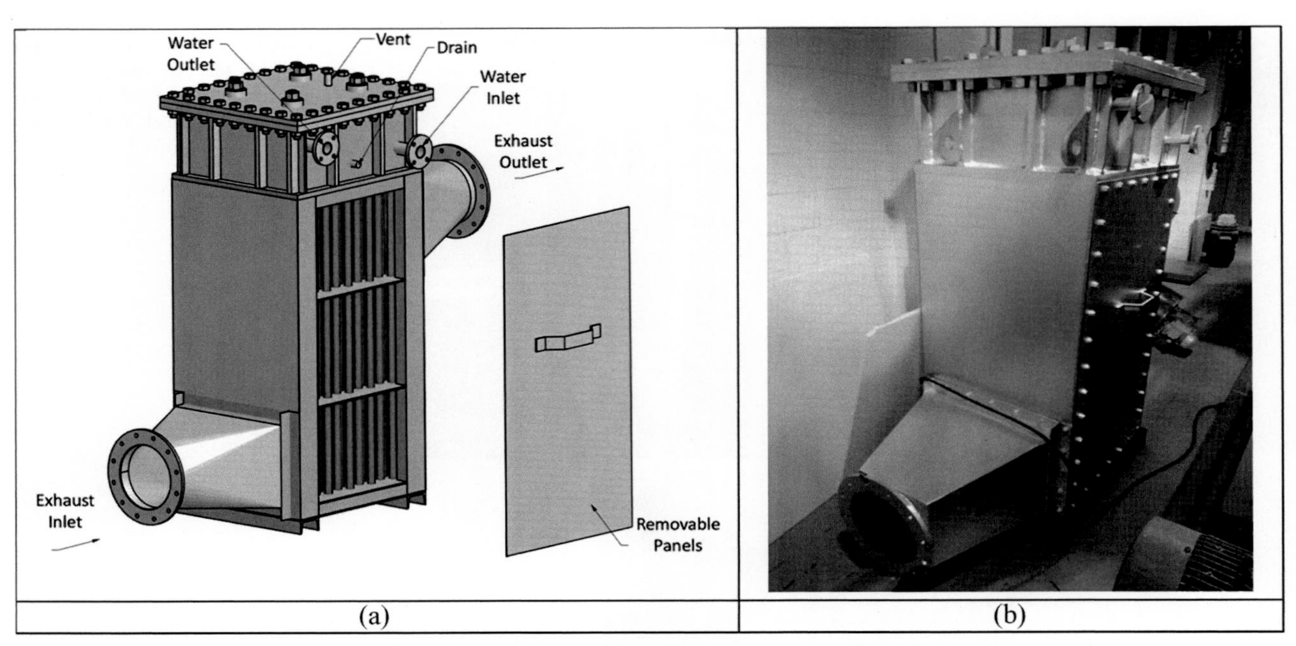


**FIGURE 6.13** (a) 3D model of the HPHE. (b) HPHE installed [11].

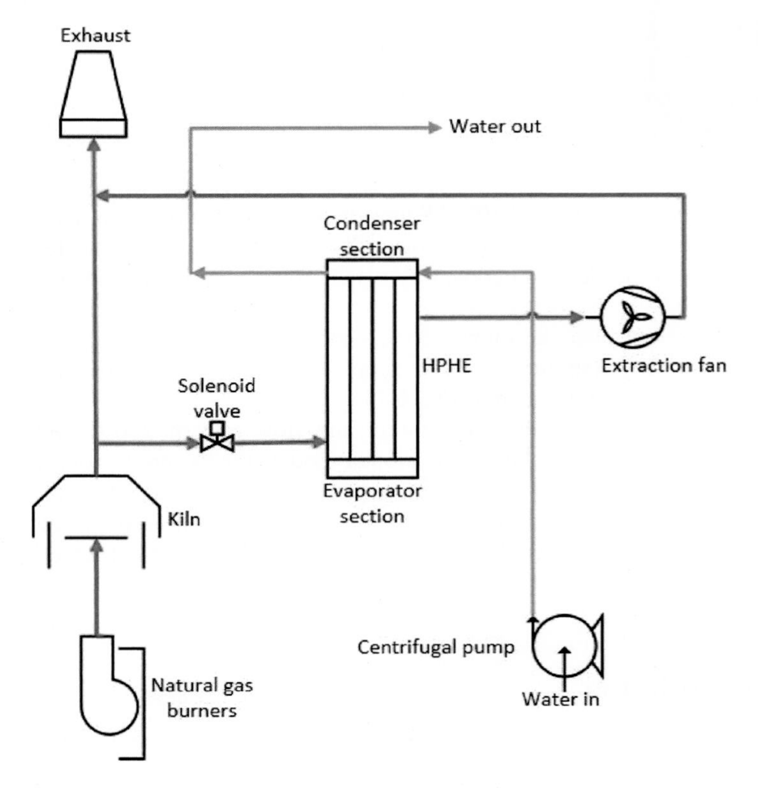


**FIGURE 6.14** P&ID of the installation [11].

As a beginning, their investigation formed an initial introduction and highlighted the applicability of heat pipes within the sector. More recently, multiple studies have been conducted to optimise their use, for instance, studies conducted by Ma et al. [13] investigated methods to optimise heat exchanger copper tube bundles used within waste heat boilers. By ensuring that the evaporator system was placed within the hot stream and the condenser within a cold water

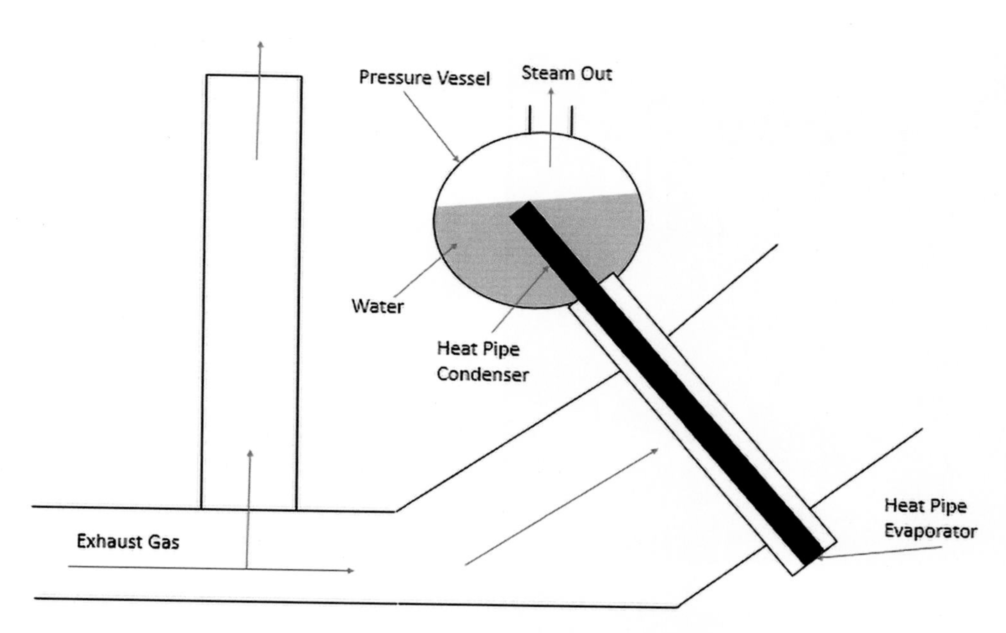


**FIGURE 6.15** Proposed heat pipe application in waste heat boilers [12].

stream, the outcome of their laboratory-scale unit showed that the effectiveness of the system improved with an increase in the evaporator flow rate and temperature. A similar study was also conducted by Elmas et al. [14] to develop and optimise a heat pipe tube bundle for applications within waste heat boilers – the system being developed to recover heat from exhaust gases which could then be reused in steam generators. The study developed a wicked tube bundle and, by replicating the conditions within a steam generation plant, recorded the recovery of 1 kW for a laboratory-scale unit.

## 6.6 Flat heat pipe heat exchangers

### 6.6.1 Flat heat pipes for solar applications

Photovoltaic (PV) solar systems are a vital technology in the transition to green energy production. However, when exposed to heat PV cells lose efficiency and degrade faster over time. To tackle this issue, Jouhara et al. [15] developed a new type of heat pipe called the heat mat. The heat mat itself is a specifically designed heat pipe, constructed from aluminium extrusion, and is similar to a thermosyphon in that the heat mat is also charged and hermetically sealed. The heat mat, by being directly attached to the PVs, allowed the heat generated within the PV cells (due to the solar radiation) to be absorbed and then discarded to a manifold located at the top. The main advantage of this system is that the surface of the heat mat is flat and uniform in temperature, allowing the PV cells to be cooled evenly. An illustration of the heat mat/solar roof can be seen in Fig. 6.16.

Use of the heat mat increased the efficiency of the PV electrical production by 16% compared to uncooled PV cells, but in addition the design allows for the rejected heat to be employed within a central heating system. The heat mat therefore offers a complete solution for electrical and thermal solar power harvesting, and because of the nature of the heat mat/PV cell construction, it also benefits from being a finished modular structural element for the building. Modelling of the PV/T system was conducted by Khordegah et al. [16] in order to investigate the potential of the heat mat on a residential building in London. The heat mat system was modelled using transient system simulation (TRNSYS) and an illustration of the developed model can be seen in Fig. 6.17.

The study concluded (using a flat heat mat with a surface area of 6.4 $m^2$) that the system increased the efficiency of PV cells by 12% and in addition, the thermal energy needs of a boiler would be reduced.

Gang et al. [17] developed a heat pipe to be used in a similar way, their aim being to reduce the thermal energy consumption of a building whilst increasing the power output and life span of the PVs attached to it. The heat pipe collector comprised $9 \times 8$ mm diameter copper heat pipes for the evaporator section, 24 pipes for the condenser and an aluminium plate placed on top of the evaporator section to accommodate the PV cells. An illustration of the system can be seen in Fig. 6.18.

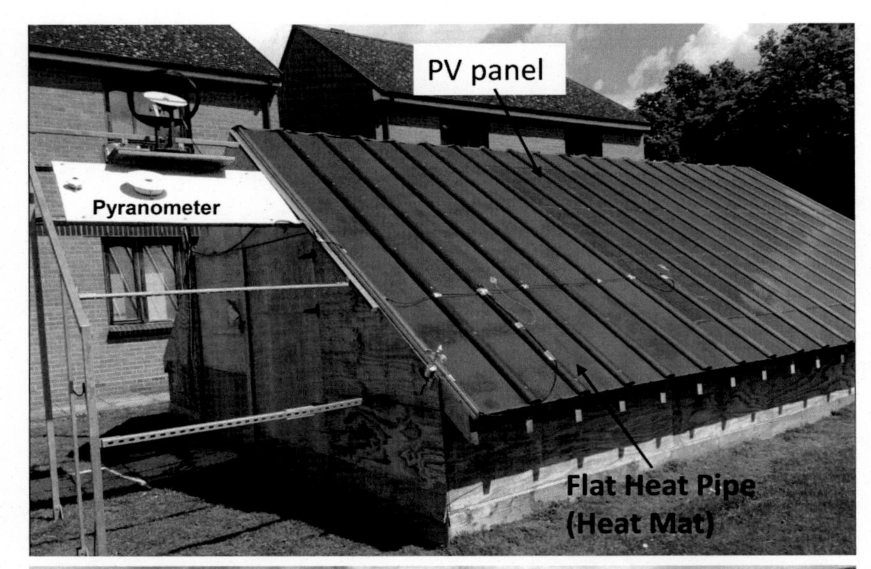

FIGURE 6.16 Heat mat solar roof [15].



FIGURE 6.17 Schematic of the PV/T system in TRNSYS simulation platform [16].

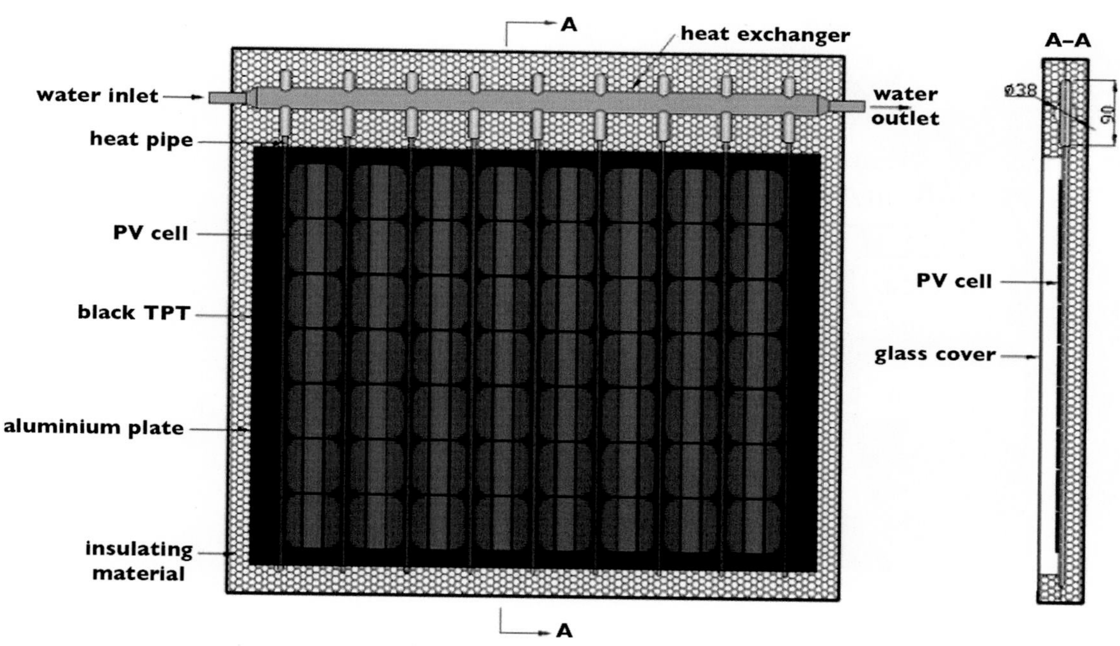


FIGURE 6.18 HP-PV/T solar collector. [17].

The work showed that the PV efficiency increased by around 10% whilst also providing thermal energy to a storage tank – the thermal heat gain varying from 363 to 550.6 W/m$^2$ and the electrical gain from 34.8 to 36.5 W/m$^2$. The study also concluded that the pipe separation in the evaporator section had a strong impact on the energy gain, but that reducing the pipe separation below 80 mm has a cost impact that is not mitigated by the increase in thermal and electrical production.

## 6.6.2 Heat pipe thermal collector

Heat pipes can also be used in an array to solely collect thermal solar radiation, and Alshukri et al. [18] proposed an evacuated tube thermal heat pipe solar collector to harvest solar radiation using various phase change materials as storage. The heat pipe was located at the focus of a mirror (to increase the temperature and solar radiation collected) and the condenser section was connected to a paraffin tank. An illustration of the rig can be seen in Fig. 6.19.

The system was then tested by exposure to solar radiation whilst monitoring the temperature of the heat pipe and the phase change material (PCM) tanks. The research showed that by using a heat pipe as the heat transfer device the efficiency of a solar collector can be considerably increased – to around 80% for flow rates varying from 1 to 2 L/h. The arrangement therefore provides a fast, reliable supply of energy for locations with no direct power supply.

## 6.6.3 Battery thermal management using heat mat

Developing battery technologies and their thermal management solutions are amongst the main challenges for electric-powered vehicles, because both decrease the charging time and increase the life span of the battery a careful battery thermal management system needs to be put in place. In addition, to ensure that batteries are operating at their optimum conditions, the cells should not reach temperatures above 35°C and the temperature difference between the cells in a pack should not exceed 2°C. To facilitate this, Jouhara et al. [19] proposed a further use for the heat mat technology to extract the heat produced by a battery pack during charge and discharge cycles – the study replicating the conditions of a bus trip to determine the thermal behaviour of the heat mat. An illustration of the system can be seen in Fig. 6.20.

The tested system comprised a battery pack of 16 prismatic lithium-titanium-oxide (LTO) cells together with an EA-PS 8080-340 DC power supply and an EA-EL 9080-400 electronic load to charge and discharge the pack, and the heat mat then connected to either a refrigerant or a water-cooling system. The test results from system charge/discharge cycles showed that the heat mat was able to maintain a temperature difference of ± 1.5°C across the battery pack whilst also keeping the temperature of the cells below 35°C. Indicating that the heat mat approach would not only greatly

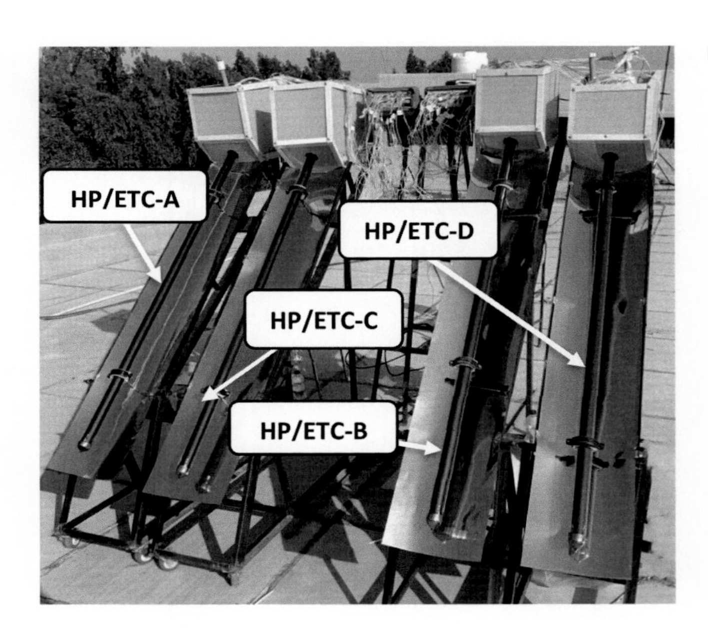


**FIGURE 6.19** Heat pipe solar collector [18].

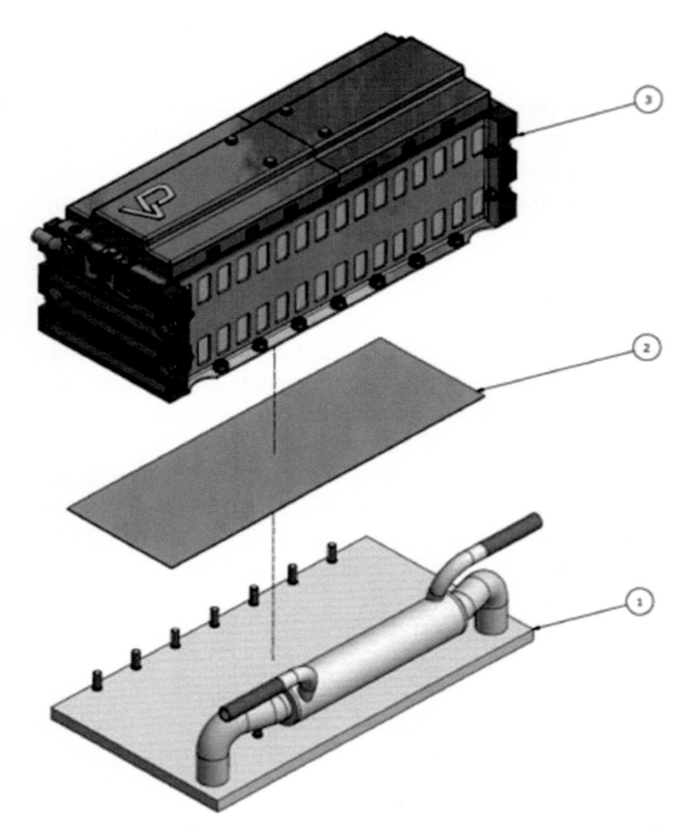


**FIGURE 6.20** (1) Heat mat, (2) thermal paste, (3) battery pack.

benefit the thermal management of vehicle battery pack, but also aid construction in that the heat mat can be used as a structural element allowing battery cells to be directly attached to its surface. In comparison to other systems, the heat mat is compact and provides a barrier between batteries and coolant avoiding any problems due to leakages from the pack.

### 6.6.4 Flat heat pipe for high temperatures

The steel manufacturing and processing industry is one of the energy-intensive industries, and large amounts of heat are emitted during cooling following high-temperature manufacturing and production processes. There is therefore a huge potential to enhance metal production efficiency by recovering some of this heat and reusing it within the plant [20]. As an example, a flat heat pipe (FHP) – using a single row of tubes instead of a flat surface – was developed by Jouhara et al. [21] to recover heat from a hot conveyor in a steel wire manufacturing process. Traditionally, after the wires are shaped, they are exposed to ambient air (and forced convection where necessary) and so the radiated heat from the wire is wasted to the environment. To recover this heat, an FHP was developed (as shown in Fig. 6.21) which was composed of 14 stainless-steel pipes linked at the bottom by a collector and at the top by a condenser section. The condenser section – itself a shell and tube style heat exchanger – consisted of rows of tubes cooling the vapour generated by the evaporator section with a total of eight pipes of 10 mm diameter cooling the heat pipe using water circulating in the shell. The total area of the evaporator section is 1 m$^2$ and a back panel was also used to reflect thermal radiation between the pipes.

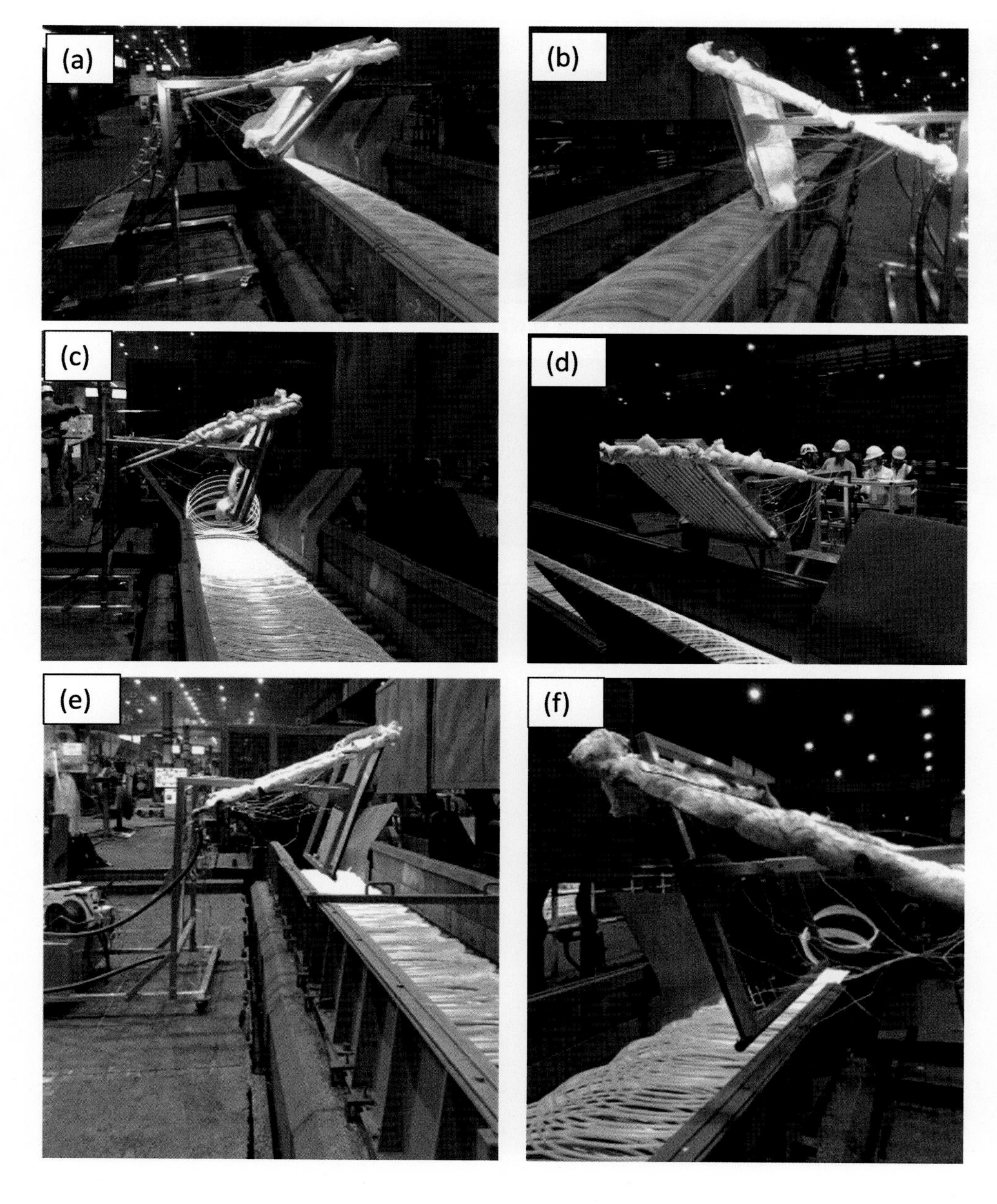


**FIGURE 6.21** Flat heat pipe (FHP) Industrial testing at different positions: (a) At 25 degrees inclination angle from vertical, (b) at 12.5 degrees inclination angle from vertical, (c) Positioning the FHP above the hot steel, (d) At 25 degrees inclination angle from vertical, (e) black-painted FHP and (f) black FHP without a back panel.

The test heat pipe, sited above hot wire rods which emerged at a temperature of 450°C, reached temperatures varying between 60°C and 80°C whilst the maximum temperature of the external heat pipe surface reached 160°C. The flow rate at the heat sink was fixed at 0.38 kg/s at which the heat pipe managed to recover up to 15.6 kW during the test.

### 6.6.5 Flat heat pipes within refrigeration

The retail sale and distribution industry is also another high consumer of energy, particularly with respect to the refrigeration of products that use display cabinets and other refrigeration systems which typically rely on forced convection to maintain items at their optimum temperature. Unfortunately, this technique can also lead to hot spots within the refrigerator which then leads to a variation in product temperatures. To overcome this and obtain a uniform temperature right across the refrigerator for all the products, Jouhara et al. [6] proposed using the heat mat technology as a way of providing a uniform, regulated and efficient method of cooling and storing products. Because by replacing the current conventional shelves with active cooling heat mat shelves, the cabinet is able to maintain isothermal surfaces to the benefit of the goods being displayed. An illustration of a retrofitted cabinet can be seen in Fig. 6.22.

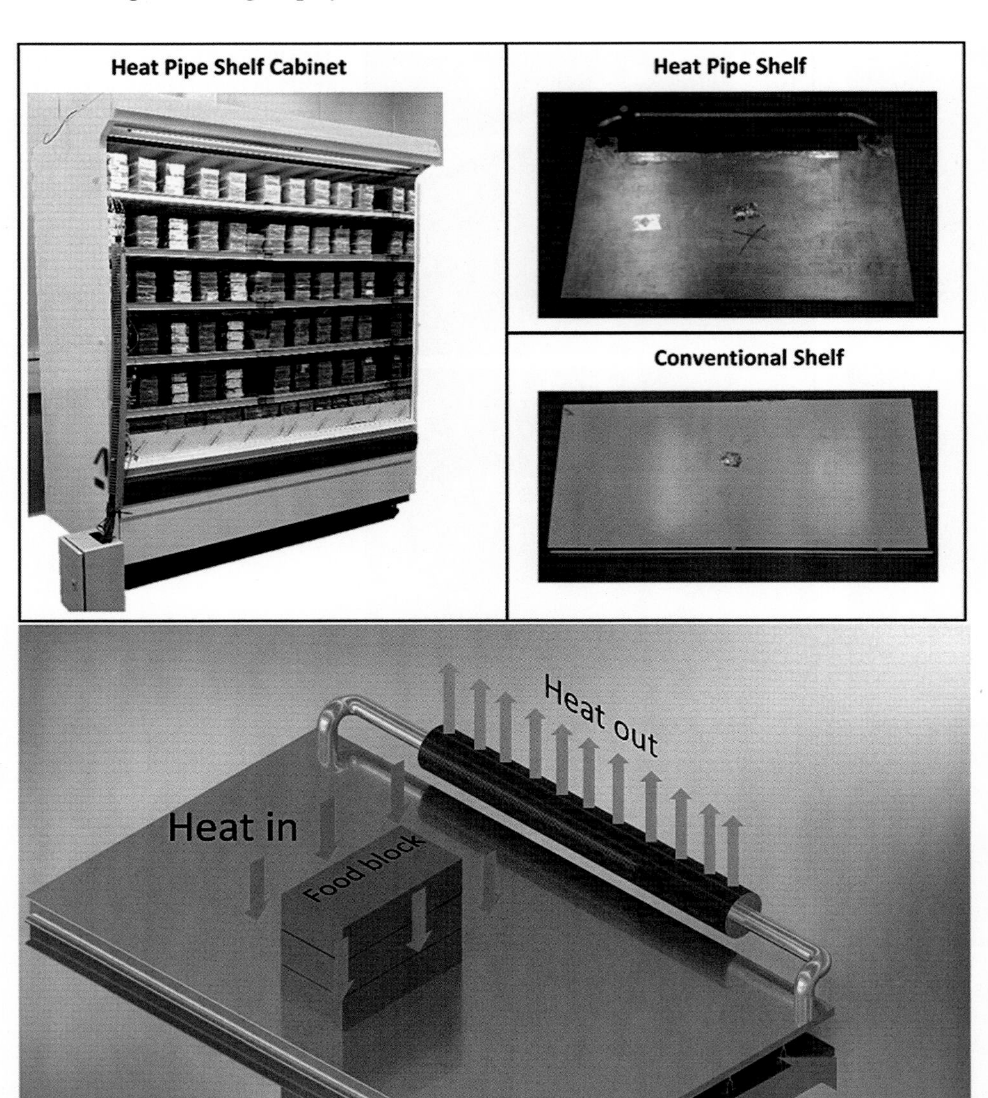


FIGURE 6.22 Heat pipe refrigeration shelf.

To test the concept, blocks (representing food products) were fitted with thermocouples and placed on the shelves, the thermocouples being used to measure the temperature at the centre of each block. The heat mats were placed inside a cabinet with a back panel to separate the flat surface and the condenser section and cold air was supplied (from the compressor) to extract the heat from the mats. The system was then tested in a fully controlled chamber, with different configurations being checked to ensure that the system is able perform in any situation. For a representative comparison, a conventional cabinet was also tested under similar conditions, and it was determined that in any situation the uniformity of temperatures between all the *'food product blocks'* in the retrofitted cabinet outperformed those in the conventional cabinet. The study also concluded that retrofitting commercial refrigerated displays with heat mat shelves would lead to a reduction in electricity consumption of between 20% and 35% compared to their conventional cabinet equivalents.

## 6.7 Heat pipe units for waste management

Waste management mainly comprises three aspects: recycling, incineration, and landfill, and all of these require large logistical frameworks to collect domestic waste from homes and for its transportation for processing at an industrial site. This generates a considerable amount of $CO_2$, especially in the transportation and incineration stages, and so to help tackle the challenge of the transportation issue Jouhara et al. [22] developed a domestic-size pyroliser relying on heat pipe technology to provide a uniform breakdown of the waste. The system consists of an insulated chamber and a heat pipe composed of legs connected at the bottom with a large collector (Fig. 6.23).

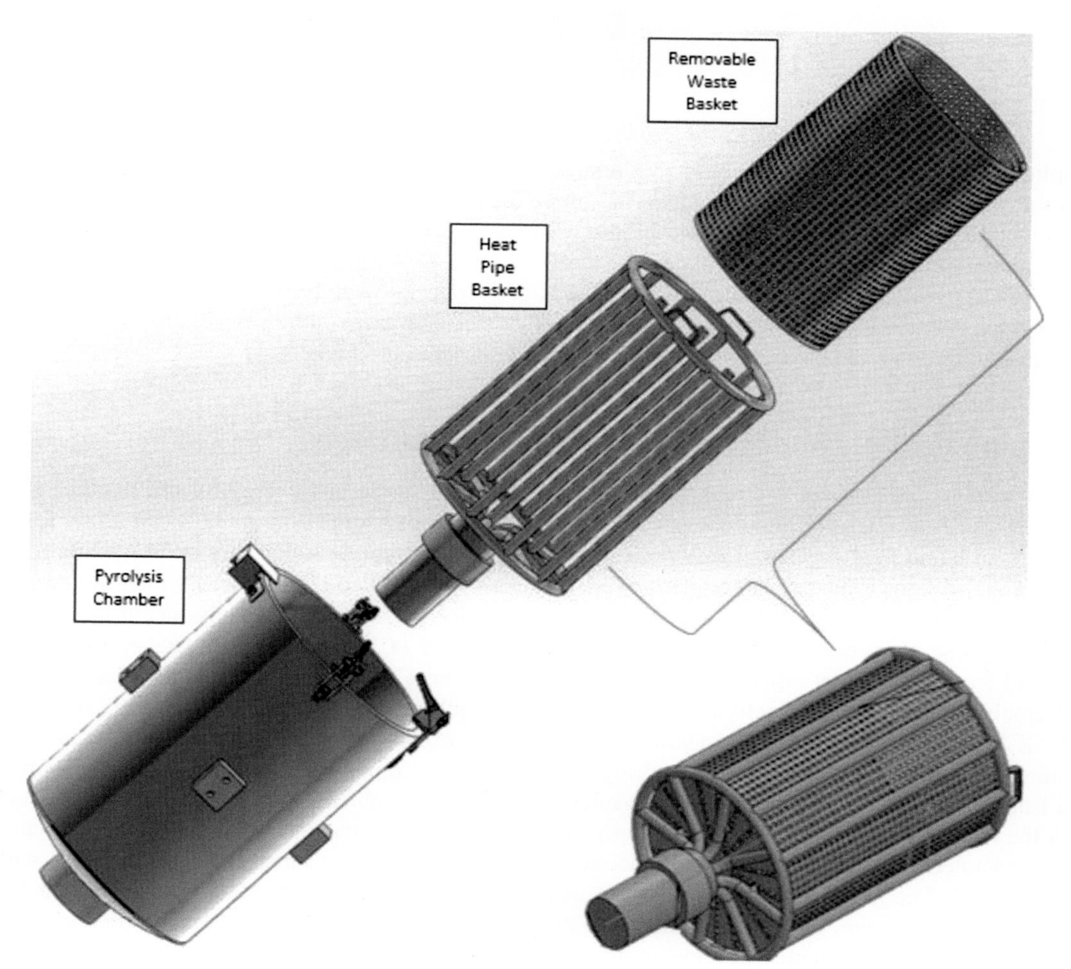


**FIGURE 6.23** Design features of home energy recovery unit (HERU).

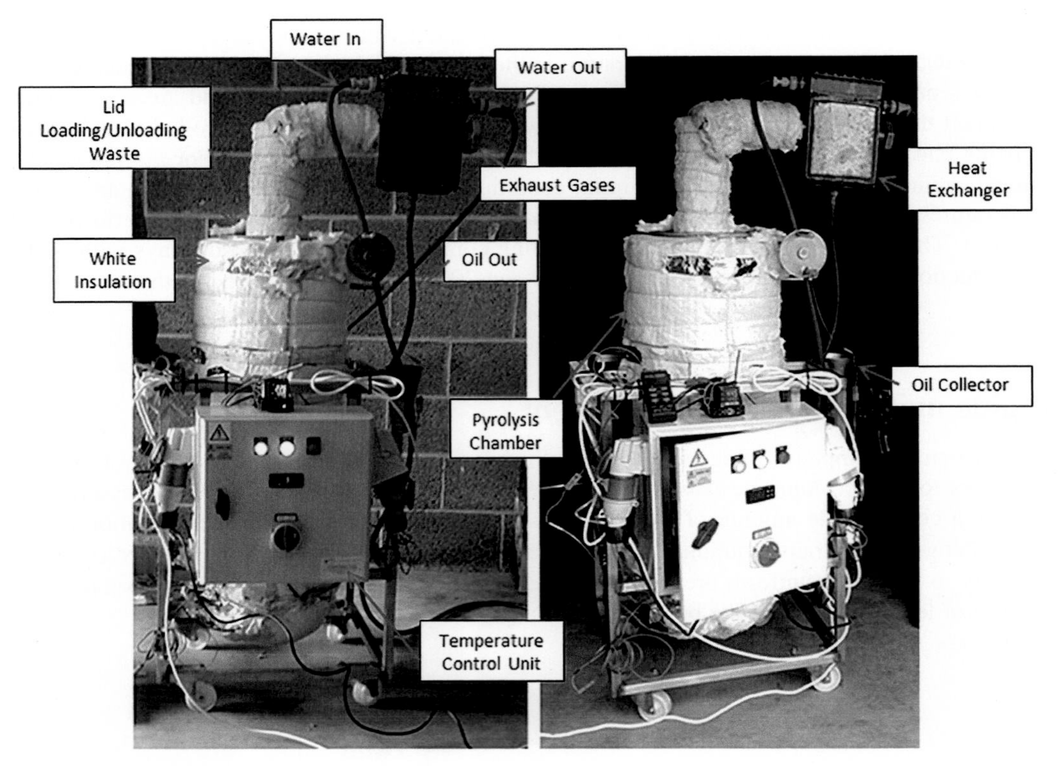


**FIGURE 6.24** HERU test rig.

Heat supplied by an electric heater increases the temperature of the waste to 300°C, and this is maintained until all moisture is removed. Once this stage is complete, the air is then injected into the chamber to disintegrate the waste. During the incineration phase, the heat generated by the waste is recovered and can be used for domestic purposes. The system tested can be seen in Fig. 6.24.

At the end of the pyrolysis cycle: 75% of the waste was converted into solid waste (biochar), 5% was transformed to oil and 20% was vapourised as pyrolysis gases, showing that this compact solution is able to both reduce the necessity for waste transportation and beneficially generate hot water for household use

## 6.8 Heat pipe heat exchangers in thermal energy storage

Thermal energy storage (TES) is applied in various sectors to accumulate thermal energy for use in other process applications such as waste heat recovery and domestic and commercial heating and cooling. TES can allow for the storage of both sensible and latent heat – where sensible heat allows energy storage in a liquid or solid medium by increasing the temperature but without a phase change. However, sensible TES has two major flaws which are: the capacity of the system relative to its volume, and the inconsistency of temperature control. Conversely, latent heat storage utilises the application of phase change materials (PCMs) allowing the melted PCM to transfer its energy into heat exchanger tube bundles and from there to an adjacent fluid. But PCM latent heat storage also has its drawbacks, one key issue being its extremely low thermal conductivity. So with a view to help improve this, Amini et al. [23] investigated the use of a PLUSICE S89 (hydrated salt) PCM within a heat pipe heat exchanger as shown in Figs 6.25 and 6.26.

The test unit consisted of a stainless-steel working fluid coil, a heat pipe tube bundle (used to transport heat to the PCM) and a final section containing the PCM. The inlet and outlet highlighted in the figure shows the condenser section in which the PCM is located along with the supply of steam and liquid condensate. The evaporator section removes both the heat generated by the condensation of the steam flow (through the evaporator side) together with the heat released into the PCM. The justification for the use of this PCM material is based on several parameters such as its thermal conductivity, melting range, chemical stability and cost. PLUSICE S89 is paraffin based, which means its cost is low and it is highly compatible with other materials, so issues such as corrosivity and chemical stability are greatly improved. During the discharge process, the water temperature was 24°C with a flow rate of 5.5 L/min, and the overall

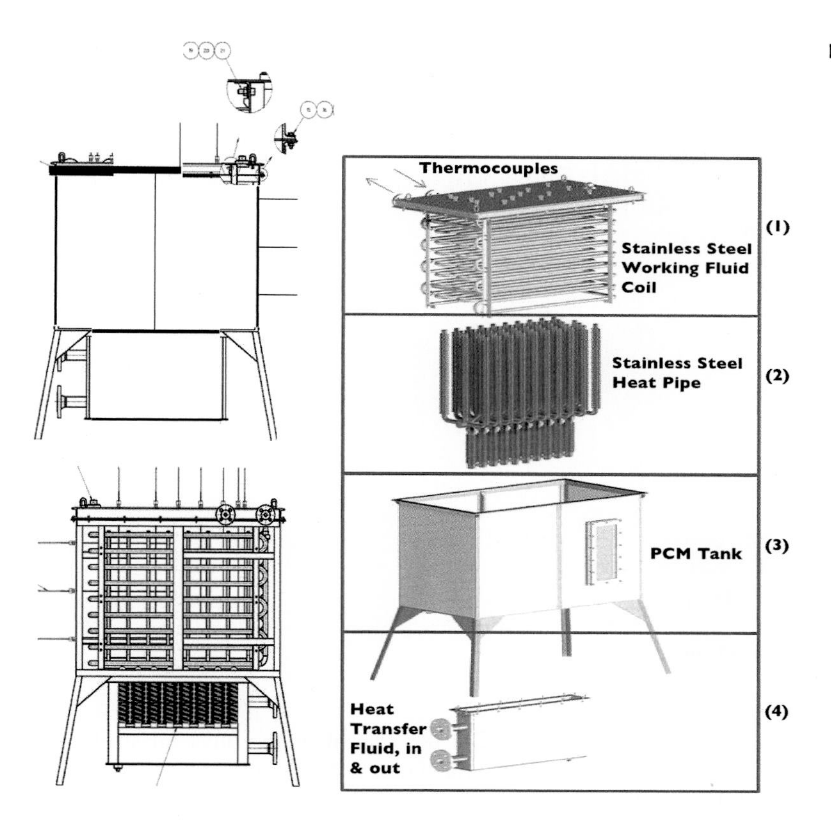


**FIGURE 6.25** Heat pipe thermal storage schematic.

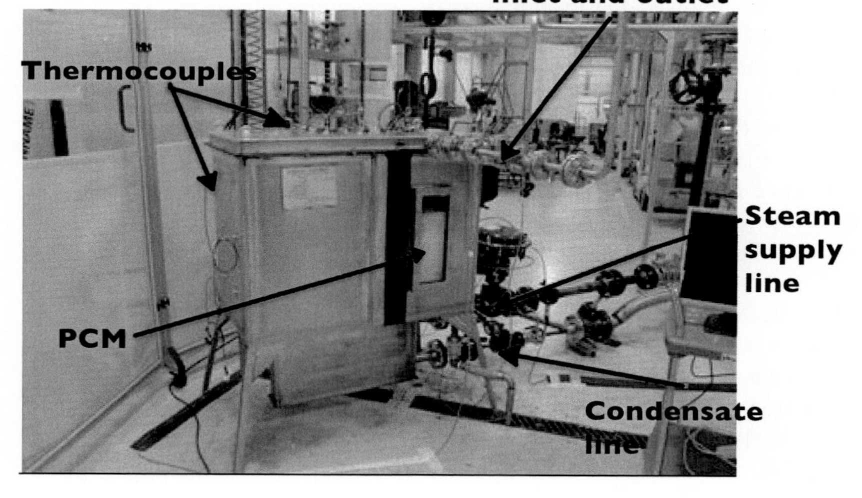


**FIGURE 6.26** Heat pipe thermal storage experimental test rig.

ratio between the stored energy within the PCM that was discharged by water, to the energy that was supplied to the heat pipe was approximately 27%. However, the author also indicated that by optimising the tube bundle this ratio could be improved to 50%. The outcome from the study thereby demonstrates the feasibility of the PCM-heat pipe unit as an effective method for the recovery and reuse of waste heat.

In practice, the combination of both heat pipe technology with PCM technologies has been widely investigated, and a study conducted by Robak et al. [24] looked at the application and influence of fins within a hybrid heat pipe system combined with a PCM-based energy storage method. Using a laboratory-scale unit, a comparative study was conducted for a tube bundle of five heat pipes using solid stainless-steel tubes charged with water, and similar to the study

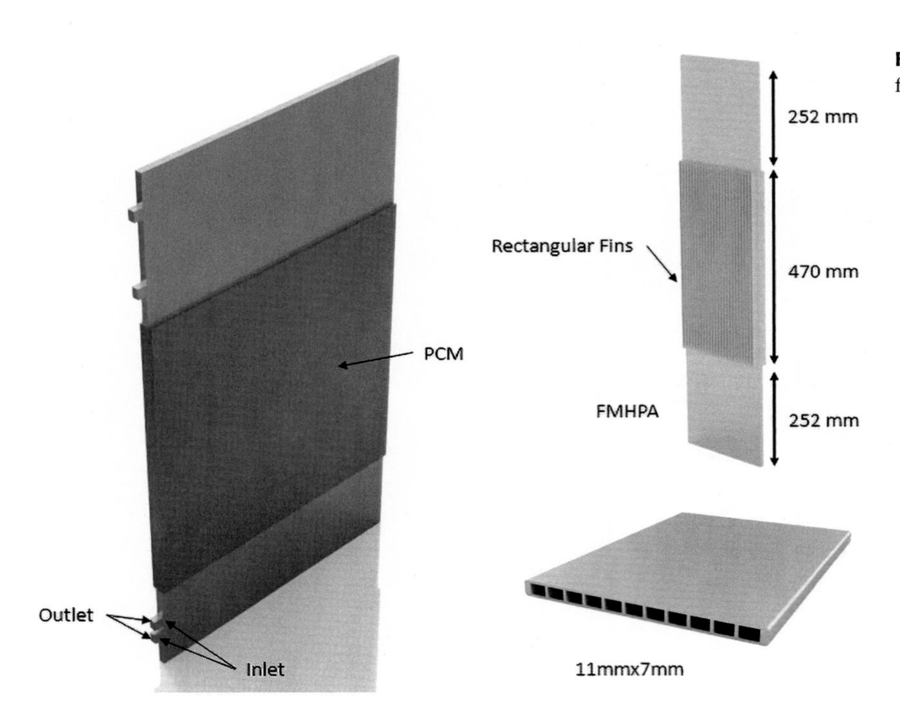


FIGURE 6.27 PCM combined with rectangular finned microheat pipes.

presented by Amini et al. [23], this comparative study also used a paraffin PCM because of its the physical attributes. The outcome highlighted that by virtue of their higher thermal conductivity, the PCM solidification rate was almost doubled when using heat pipes compared to using solid rods. This result was also supported in a study conducted by Sharifi et al. [25] who investigated the differing effects of a stainless-steel heat pipe (charged with potassium) as against a solid rod and a hollow tube during the melting phase of a sodium nitrate PCM. The outcome of the study again highlighted that the heat pipe recorded the best performance in terms of melting rate. The use of fins has also been investigated in a study by Diao et al. [26], and this involved the examination of rectangular finned microheat pipes within a hybrid system. The system proposed and developed (shown in Fig. 6.27) maximises the available heat transfer area which then offers a greater probability of providing a uniform heat flux during the charging process, minimising any potential risk of uneven melts.

Similar to the other studies presented throughout this chapter, the study by Diao et al. [26] showed that an increased heat transfer area significantly improves charging time and allows for a more homogenous thermal distribution across the PCM. Their study also investigated the effect of fin spacing and highlighted that an increase in fin spacing significantly altered the solidification of the PCM – which then formed mushy zones (representing two phases) between the fins – at which point, natural convection becomes the dominant method of heat transfer leading to improper and inefficient melts.

Extended heat pipe networks with latent heat storage have also been investigated by Tiari et al. [27], and their study involved the development of an experimental unit where different modes of operation were investigated. Their study (as shown in Fig. 6.28), consisted of a primary heat pipe combined with a secondary heat pipe, where the role of the primary heat pipe was to transfer heat to the heat sink located in the condenser section, and the role of the secondary heat pipe to transport any surplus heat to the PCM during charging and to support energy retrieval during discharging. Similar to the study conducted by Amini et al. [23], this study also utilised RUBITHERM RT55 – a mixture of paraffins and waxes, ensuring that the PCM is not only extremely chemically stable but is also cost-efficient and temperature stable. To allow for a correct characterisation of the primary and secondary heat pipes, the PCM was first tested without the heat pipe network. The outcome of this initial study highlighted that when the bottom of the PCM tank was heated the resultant data indicated significant temperature differences throughout the PCM. However, when the heat pipe network was tested under similar conditions the heat pipe significantly improved heat transfer during the charging process. This occurs because the rate of heat transfer is directly proportional to the fluid flow rate, so an increase in the flow rate results in an increase in the charging rate. And at the start of the changing process, as conduction is (at that point) the main heat transfer process, the melting rate remains high with a significant amount of solidified PCM existing until its melting point is reached. After which, natural convective effects start to influence the heat transfer mode and the

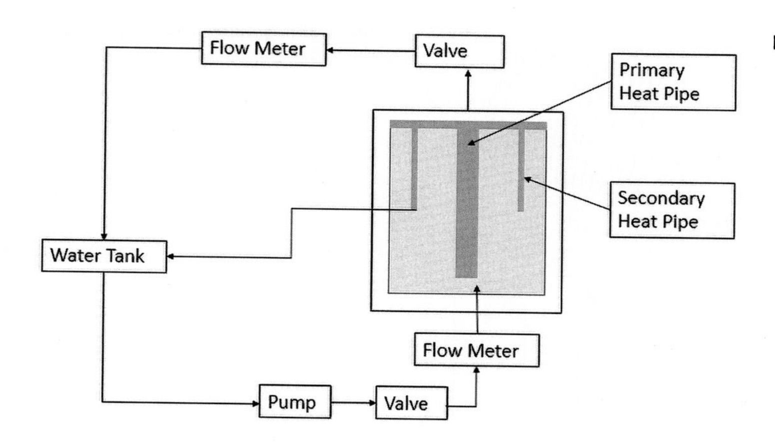


FIGURE 6.28 Experimental setup.

rate falls. During the discharge process, cold water at approximately 20°C was fed through the system at various flow rates and circulated throughout the heat pipe network, and because the discharge rate is dependent on the heat conduction rate, the low thermal conductivity of the PCM significantly effects this.

This difference between the charging and the discharging rates is one of the key operational parameters of a PCM, and throughout this chapter the presented studies have shown that higher thermal conductivity can be achieved by the inclusion of heat pipes, thereby allowing such systems to outperform traditional PCM units. Nonetheless, both traditional and hybrid systems still suffer from prolonged discharge times.

## 6.9 Other applications and case studies

### 6.9.1 Variable conductance heat pipe for automotive thermal management

The compact nature of heat pipes allows the technology to be applied for vehicle thermal management. For example, Leriche et al. [28] tested a variable conductance heat pipe (VCHP) to cool the engine oil of an internal combustion engine during a vehicle cold start and to reduce consumption during this period. The VCHP in this experiment was connected to resistive heaters on the evaporator section and a fan in the cooling section and the VCHP was inclined to allow the two-phase flow to occur by gravity. The VCHP was made of copper, with an evaporator length of 120 mm, an adiabatic section of 80 mm and a condenser section of 150 mm. The reservoir for the inert gas (nitrogen) was 68.9 mm and water was used as the working fluid. An illustration of the system can be seen in Fig. 6.29.

The tests showed that the VCHP was able to reduce the variation of the oil temperature during cold starting and the warm-up period of the car by acting as a thermal switch so that the VCHP only starts to remove heat from the oil once it reaches 80°C – so offering better control of the oil temperatures. The work also showed that by varying the condenser area with the nitrogen, the conductance of the heat pipe could be adapted to suit the required oil temperature.

### 6.9.2 Heat pipe radiator unit for space nuclear power reactor

Safety is amongst the main advantages of a heat pipe because it can operate passively, and so this aspect is well suited to the thermal management of space nuclear power reactors; Wang et al. [29] use a heat pipe as a heat rejection system. Their system connects: the reactor core to the primary coolant, a thermoelectric power converter and a cooling loop – the cooling loop then being connected to a radiative heat pipe system. The radiator unit was composed of rows of high-temperature potassium heat pipes covered with fins to increase heat transfer. An illustration of the system can be seen in Fig. 6.30.

The work showed that this type of heat pipe radiator could be a good solution for rejecting the heat from a space nuclear reactor. However, it was also noted that the start-up period of the heat pipe (from frozen to full operation) could take at least 5 min. Due mainly to the wick, the sonic vapour velocity limit of the heat pipe is rapidly reached in the second phase of its start-up, which could then cause issues with the cooling loop. Nonetheless, the work concluded, that at operating temperature, a radiative heat pipe is a good solution for cooling a space nuclear power reactor, rejecting the required heat load with minimal size and weight.

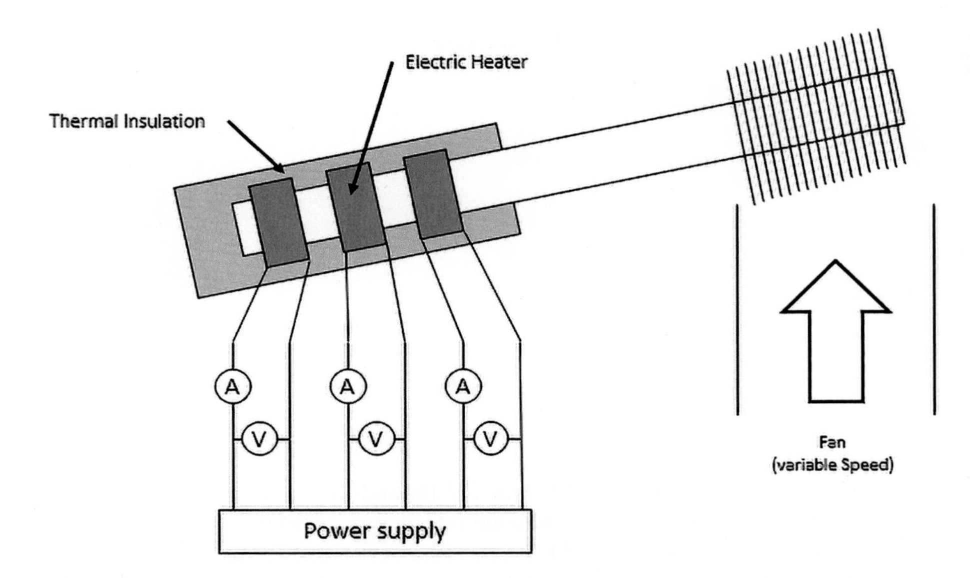


**FIGURE 6.29** Variable conductance heat pipe test rig.

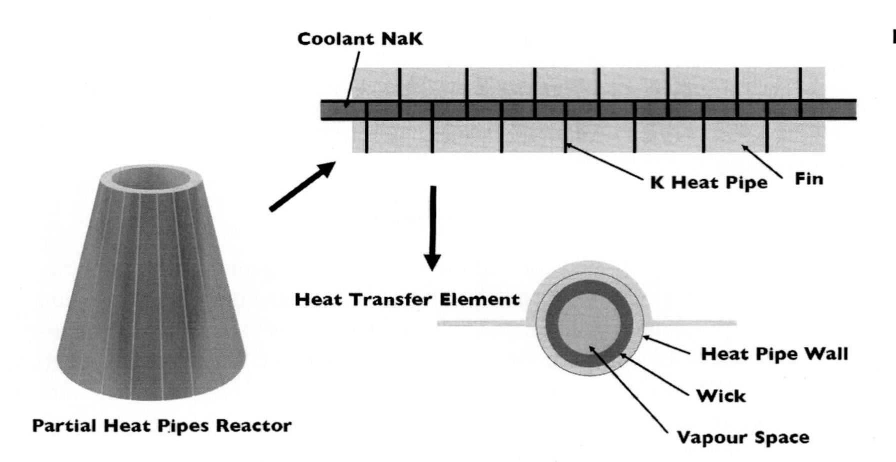


**FIGURE 6.30** Heat pipe radiator.

## 6.9.3 Hybrid heat pipes for nuclear applications

Current technologies for dry nuclear fuel storage rely mainly on natural convection to remove heat produced by the decay of spent fuel located in a casket. Unfortunately, because these technologies have heat transfer limitations and cannot prevent an accident in the case of flooding, Jeong et al. [30] suggested applying heat pipe technology to cool the casket. Their design would also recover heat and consist of a thermoelectrical module, a Stirling engine and/or a phase change material – a form of design considered to be a hybrid heat pipe. The shell contains the working fluid (water) and a neutron-absorbing material to control the reactivity of the spent fuel, and an illustration of the system can be seen in Fig. 6.31.

The normal temperature of spent fuel in dry storage depends on the environmental conditions and is expected to be 290°C, but by applying heat pipe technology for each assembly, the author was able to reduce the temperature of the spent fuel rods to 165°C. Indicating that a hybrid heat pipe system can, by decreasing the temperature of the fuel, significantly increase the safety of spent fuel storage and all with no moving parts.

Jeong et al. [31] also investigated the use of the hybrid heat pipe for the control and cooling of a nuclear reactor core, the temperature of which is usually controlled by inserting neutron-absorbing rods into the core to slow the nuclear fission process. However, the control rods are not able to remove heat from the reactor during an emergency shutdown. So in this regard, applying hybrid heat pipes to replace the control rods could provide a good solution for temperature control during an emergency. Even so, amongst the main challenges of applying conventional heat pipes technologies for this application is the extreme pressure of the reactor during an emergency shutdown. Under such

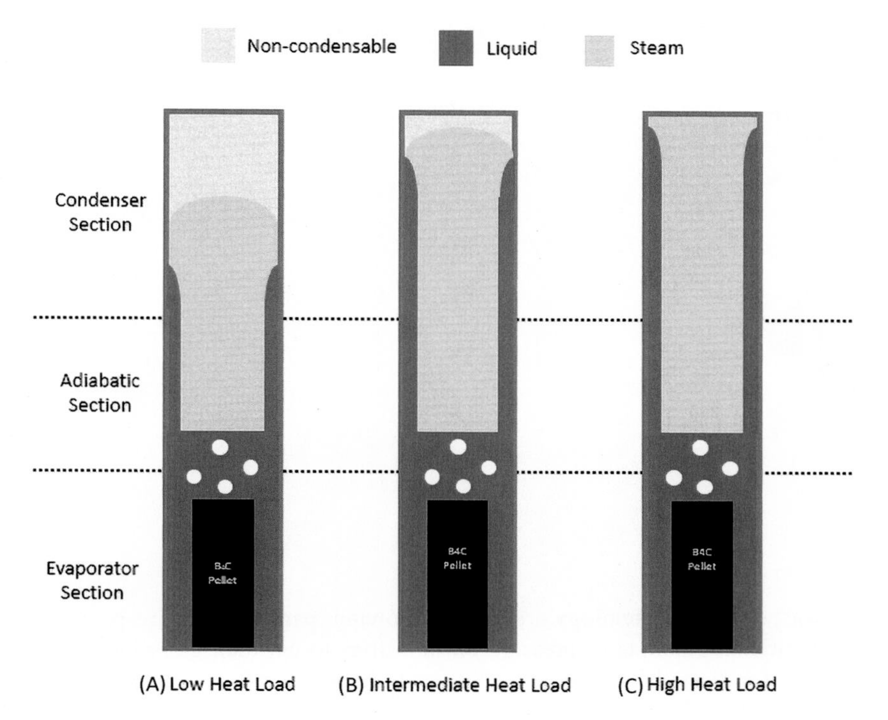


**FIGURE 6.31** Hybrid heat pipe.

conditions, the external pressure applied to the heat pipes can reach 155 bar producing large stresses in the heat pipe shells which, because these heat pipes rely on wicks to move the working fluid, need minimal thickness. Nonetheless, the study determined that for situations when the pressure in the reactor is too extreme for more cooling water to be injected into the primary circuit, hybrid heat pipes could remove decay heat and help slow the reactor fission using neutron absorbers. To consider this, the research determined that a single hybrid heat pipe could extract up to 18 kW from the reactor and so a transient analysis was also undertaken on a full-scale reactor to investigate the benefits of hybrid heat pipes. From this work, it was determined (in the case of loss of the primary fluid coolant), that the time required to boil the water in the reactor would be increased by 13 min and the fuel rod uncovering time by 5.4 h, allowing more time for safety measures to be put in place. Thereby leading to the conclusion that if hybrid heat pipes are applied to the entire reactor core, they could prevent the core uncovering for a day.

## 6.9.4 Hybrid pump-assisted loop heat pipe

Loop heat pipes are used for various thermal management applications but their heat transfer capabilities are constrained by capillary force limitations. To overcome the issue of dry-out at the evaporator and increase the maximum heat flux, Setyawan [32] presented a hybrid loop heat pipe (HLHP) integrated with a diaphragm pump. The condensate in the HLHP returns to the evaporator by wick capillary forces during normal condition, but a mechanical pump can be activated to circulate the working fluid when the capillary force is not sufficient for the heat load. The tested HLHP system comprised an evaporator, condenser, reservoir and diaphragm pump as shown in Fig. 6.32 – the reservoir stores the liquid so that it could be used when needed. The experimental results indicated that, with the assistance of the pump, an HLHP was able to overcome the issue of dry-out and transfer larger amounts of heat compared to the case of the wick alone.

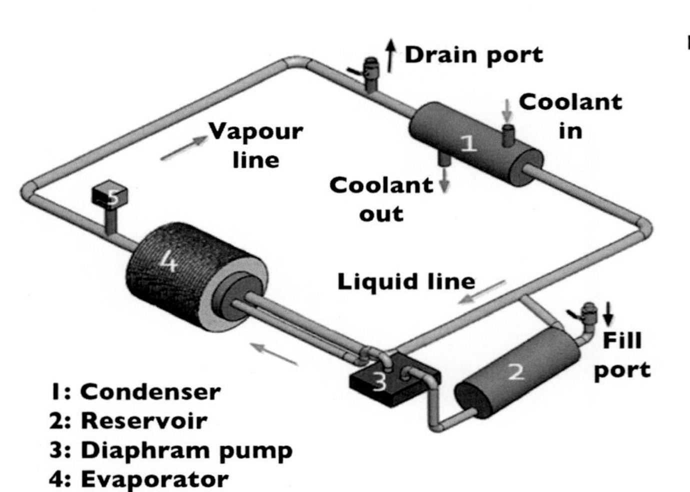


**FIGURE 6.32** Hybrid loop heat pipe.

## 6.10 Concluding remarks

Looking towards the longer term, applications for heat pipe technology are rapidly growing, particularly in response to climate change regulations which are putting significant amounts of pressure on industries to cut their emissions. And this has resulted in many industries/sectors looking for waste heat recovery solutions to help lower their dependency on fossil fuels. Overall, the versatility of the technology is evident because the same technology can be used multiple times for many diverse industries such as the case with the heat mat described earlier in this chapter. Commercial applications have also been widely considered with real-life industrial applications within the food processing and ceramic industries being discussed, highlighting an attractive reduction in emissions and a short return on investment period as a result of installing this technology.

## References

[1] H. Jouhara, R. Meskimmon, Experimental investigation of wraparound loop heat pipe heat exchanger used in energy efficient air handling units, Energy 35 (12) (2010) 4592–4599. Available from: https://doi.org/10.1016/j.energy.2010.03.056. Dec.

[2] H. Jouhara, H. Ezzuddin, Thermal performance characteristics of a wraparound loop heat pipe (WLHP) charged with R134A, Energy 61 (2013) 128–138. Available from: https://doi.org/10.1016/j.energy.2012.10.016. Nov.

[3] G. Mahajan, S.M. Thompson, H. Cho, Experimental characterization of an n-pentane oscillating heat pipe for waste heat recovery in ventilation systems, Heliyon 4 (11) (2018) e00922.

[4] Y.H. Diao, L. Liang, Y.M. Kang, Y.H. Zhao, Z.Y. Wang, T.T. Zhu, Experimental study on the heat recovery characteristic of a heat exchanger based on a flat micro-heat pipe array for the ventilation of residential buildings, Energy Build. 152 (2017) 448–457.

[5] H. Jouhara, Thermal transfer loop, WO/2019/234423, Mar. 2021 [Online]. Available: https://patentscope.wipo.int/search/en/detail.jsf?docId = WO2019234423.

[6] H. Jouhara, Heat pipes (gravity assisted and capillary-driven), heat exchanger design handbook, Begell House (2019).

[7] A.R. Lukitobudi, A. Akbarzadeh, P.W. Johnson, P. Hendy, Design, construction and testing of a thermosyphon heat exchanger for medium temperature heat recovery in bakeries, Heat. Recover. Syst. CHP 15 (5) (1995) 481–491. Available from: https://doi.org/10.1016/0890-4332(95)90057-8.

[8] H. Jouhara, et al., Investigation on a full-scale heat pipe heat exchanger in the ceramics industry for waste heat recovery, Energy 223 (2021) 120037. Available from: https://doi.org/10.1016/J.ENERGY.2021.120037. May.

[9] B. Delpech, et al., Energy efficiency enhancement and waste heat recovery in industrial processes by means of the heat pipe technology: case of the ceramic industry, Energy 158 (2018) 656–665. Available from: https://doi.org/10.1016/j.energy.2018.06.041.

[10] B. Delpech, B. Axcell, H. Jouhara, Experimental investigation of a radiative heat pipe for waste heat recovery in a ceramics kiln, Energy 170 (2019) 636–651. Available from: https://doi.org/10.1016/j.energy.2018.12.133.

[11] D. Brough, et al., An experimental study and computational validation of waste heat recovery from a lab scale ceramic kiln using a vertical multi-pass heat pipe heat exchanger, Energy (2020) 118325. Available from: https://doi.org/10.1016/j.energy.2020.118325.

[12] D. Littwin, J. Mccurley, Heat pipe waste heat recovery boilers, J. Heat. Recover. Syst. I (4) (1981) 339–348.

[13] G. Ma, Y. Zhang, M. Yue, Y. Shi, Thermal economy study on the waste heat utilization of a double reheat unit under coupled steam turbine and boiler, Appl. Therm. Eng. 175 (2020) 115112.

[14] E.Taner Elmas, Design and production of high temperature heat pipe heat recovery units, J. Mol. Struct. 1212 (2020) 127927. Available from: https://doi.org/10.1016/j.molstruc.2020.127927.

[15] H. Jouhara et al., The performance of a novel flat heat pipe based thermal and PV/T (photovoltaic and thermal systems) solar collector that can be used as an energy-active building envelope material, 2015, doi: 10.1016/j.energy.2015.07.063.

[16] N. Khordehgah, V. Guichet, S.P. Lester, H. Jouhara, Computational study and experimental validation of a solar photovoltaics and thermal technology, Renew. Energy 143 (2019) 1348−1356. Available from: https://doi.org/10.1016/j.renene.2019.05.108.

[17] P. Gang, F. Huide, Z. Huijuan, J. Jie, Performance study and parametric analysis of a novel heat pipe PV/T system, Energy 37 (1) (2012) 384−395. Available from: https://doi.org/10.1016/j.energy.2011.11.017.

[18] M.J. Alshukri, A.A. Eidan, S.I. Najim, Thermal performance of heat pipe evacuated tube solar collector integrated with different types of phase change materials at various location, Renew. Energy 171 (2021) 635−646. Available from: https://doi.org/10.1016/j.renene.2021.02.143.

[19] H. Jouhara, N. Serey, N. Khordehgah, R. Bennett, S. Almahmoud, S.P. Lester, Investigation, development and experimental analyses of a heat pipe based battery thermal management system, Int. J. Thermofluids 1 (2) (2020) 100004. Available from: https://doi.org/10.1016/j.ijft.2019.100004.

[20] H. Jouhara, Waste Heat Recovery in Process Industries, Wiley, Weinheim, 2022.

[21] H. Jouhara, et al., Experimental investigation on a flat heat pipe heat exchanger for waste heat recovery in steel industry, Energy Procedia 123 (2017) 329−334. Available from: https://doi.org/10.1016/j.egypro.2017.07.262.

[22] H. Jouhara, T.K. Nannou, L. Anguilano, H. Ghazal, N. Spencer, Heat pipe based municipal waste treatment unit for home energy recovery, Energy (2017). Available from: https://doi.org/10.1016/j.energy.2017.02.044.

[23] A. Amini, J. Miller, H. Jouhara, An investigation into the use of the heat pipe technology in thermal energy storage heat exchangers, Energy 136 (2017) 163−172. Available from: https://doi.org/10.1016/j.energy.2016.02.089.

[24] C.W. Robak, T.L. Bergman, A. Faghri, Enhancement of latent heat energy storage using embedded heat pipes, Int. J. Heat. Mass. Transf. 54 (15) (2011) 3476−3484. Available from: https://doi.org/10.1016/j.ijheatmasstransfer.2011.03.038.

[25] N. Sharifi, S. Wang, T.L. Bergman, A. Faghri, Heat pipe-assisted melting of a phase change material, Int. J. Heat. Mass. Transf. 55 (13) (2012) 3458−3469. Available from: https://doi.org/10.1016/j.ijheatmasstransfer.2012.03.023.

[26] Y.H. Diao, L. Liang, Y.H. Zhao, Z.Y. Wang, F.W. Bai, Numerical investigation of the thermal performance enhancement of latent heat thermal energy storage using longitudinal rectangular fins and flat micro-heat pipe arrays, Appl. Energy 233 (234) (2019) 894−905. Available from: https://doi.org/10.1016/j.apenergy.2018.10.024.

[27] S. Tiari, M. Mahdavi, S. Qiu, Experimental study of a latent heat thermal energy storage system assisted by a heat pipe network, Energy Convers. Manag. 153 (2017) 362−373. Available from: https://doi.org/10.1016/j.enconman.2017.10.019.

[28] M. Leriche, S. Harmand, M. Lippert, B. Desmet, An experimental and analytical study of a variable conductance heat pipe: application to vehicle thermal management, Appl. Therm. Eng. 38 (2012) 48−57. Available from: https://doi.org/10.1016/j.applthermaleng.2012.01.019.

[29] C. Wang, J. Chen, S. Qiu, W. Tian, D. Zhang, G.H. Su, Performance analysis of heat pipe radiator unit for space nuclear power reactor, Ann. Nucl. Energy 103 (2017) 74−84. Available from: https://doi.org/10.1016/j.anucene.2017.01.015.

[30] Y.S. Jeong, I.C. Bang, Hybrid heat pipe based passive cooling device for spent nuclear fuel dry storage cask, Appl. Therm. Eng. 96 (2016) 277−285 [Online]. Available. Available from: http://www.sciencedirect.com/science/article/pii/S1359431115013320.

[31] Y.S. Jeong, K.M. Kim, I.G. Kim, I.C. Bang, Hybrid heat pipe based passive in-core cooling system for advanced nuclear power plant, Appl. Therm. Eng. 90 (2015) 609−618. Available from: https://doi.org/10.1016/j.applthermaleng.2015.07.045.

[32] I. Setyawan, N. Putra, I.I. Hakim, Experimental investigation of the operating characteristics of a hybrid loop heat pipe using pump assistance, Appl. Therm. Eng. 130 (2018) 10−16. Available from: https://doi.org/10.1016/j.applthermaleng.2017.11.007.

Chapter 7

# Cooling of electronic components

As the deployment of heat pipes in electronics cooling is dominant, this chapter is dedicated to electronic cooling applications and emerging heat pipe technologies that aim to meet the future needs of electronic devices.

As pointed out by Boyd Corporation [1] – a global leader and manufacturer of heat pipe technologies for electronics thermal control applications – the electronics sector is constantly striving towards miniaturisation of devices, which inevitably results in greater power densities, resulting in the requirement for enhanced heat dissipation technology. The challenge to develop efficient thermal management technologies for heat removal from high flux devices is growing, both in the heat dissipation challenge and volume of the deployment across the electronics sectors. Microelectronics sectors, such as the telecommunications, data centre and power electronics sectors, have now been joined by the emerging e-mobility, aerospace and space sectors as major deployers of electronics in applications with challenging thermal management requirements. The wearable electronics market and applications internal to the body are expected to grow in the future.

The performance of microprocessors has progressed significantly, with servers achieving 10 million I/O operations per second and 80 million has been demonstrated [2]. The inception of dual- and quad-core processors around a decade ago eased the thermal challenge by dividing the overall chip power between the cores, reducing the local junction to case thermal resistance and smoothing the heat flux profile across the component surface that interfaces with the cooling system. Multicore processors, such as the Intel Xeon Platinum 8380 Processor, have progressed to 40 cores with maximum case temperatures of 83°C and thermal dissipation powers of 270 W, versus 136 W reported previously [3]. Furthermore, progression of telecom server heat dissipation from 120 W card level heat dissipation to 120 W per $35 \times 35$ mm chip (heat flux $\sim 10$ W/cm$^2$), is now predicted to exceed 350 W card level dissipation, with multiple miniaturised 80 W microprocessors ($10 \times 20$ mm), with surface heat flux of $\sim 40$ W/cm$^2$, deployed in emerging applications.

Traditionally electronics housings and printed circuit boards (PCBs) were maintained within industry standards. Although this is still true, radical new design approaches, that aim to maximise the efficiency of thermal management systems, are now being implemented. Optimising PCB layouts to enhance thermal management and integration of the thermal components with the electronics chassis/housing are now a key focus and are at the forefront of the system architects' design methodology.

Previously, DARPA, the Defence Advanced Research Projects Agency in the United States, identified the requirements for disruptive thermal technologies that mitigate thermal limitations on the operation of military electronic systems, whilst significantly reducing size, weight and power consumption, by integrating thermal management techniques into the chip layout, substrate structure and/or packaging design [4]. DARPA's ICECool activity had the target to achieve a significant reduction in the cooling system size and demonstrate chip-level heat removal in excess of 1 kW/cm$^2$ heat flux and 1 kW/cm$^3$ heat density. In the space sector, the challenges of miniaturisation have similar implications. In 2003 NASA [5] had summarised that: 'While advanced two-phase technology such as capillary pumped loops and loop heat pipes (LHPs) offer major advantages over traditional thermal control technology, it is clear that this technology alone will not meet the needs of all future scientific spacecraft ....'

Over the last decade, geostationary telecom satellites (GEO) have seen an increase in heat dissipation from 5 to $>20$ kW. The absence of convective heat transfer in the space environment requires new and novel thermal management technology, and the role of heat pipes is becoming more significant, to overcome the limitations of conduction heat transfer through PCBs and aluminium chassis material.

Heat pipe technology is emerging that addresses thermal management requirements of future electronics systems, by incorporating multifunctionality, increased level of integration with the heat-dissipating components and architecture, enhancing thermal performance. Examples of high-tech heat pipe applications and future heat pipe technologies are presented in this chapter.

Heat Pipes. DOI: https://doi.org/10.1016/B978-0-12-823464-8.00006-1

In addition to the established electronics cooling markets, rapidly growing markets include portable electronics, wearable electronics and e-mobility applications such as electric vehicle (EV) battery cells and advanced, more efficient electric motors, which are essential for electric aircraft and electric vehicles. A review by Rafal Wrobel and an author of this book provides a good overview of heat pipe technology deployed in motor machines [6]. In the case of automotive applications, the increased sophistication of electronics deployed in vehicles has led to the description 'smartphones on wheels', where semiconductors are utilised in multiple applications from fuel sensors, digital speedometers, through to reversing cameras and navigation systems [7]. Artificial intelligence enables driver assists such as proximity sensors to detect lane drift and reversing sensors. A connected autonomous road network is predicted for the future. Automotive electronics now deploy the same level of microprocessor technology as tablet computers and game consoles. The impact of the COVID-19 pandemic on reduced EV sales and increased electronic entertainment system sales has led to a severe shortage in EV microprocessors, which were predicted to have global production of 7–8 million units in 2021.

Extending the range of electric vehicles requires the development of battery cells with increased capacity. The thermal management of battery cells is a major challenge to address, to maintain cells within optimal temperature ranges and with isothermal surface temperature profiles, which offers opportunities for the development of advanced heat pipe technologies [8]. New technologies such as lightweight ultra-thin heat pipes and high-performance, low-cost heat pipe loops are required.

## 7.1 Features of the heat pipe

As outlined at the beginning of this chapter, it is convenient to describe in broad terms how the heat pipe might be used in the area of electronics, based on its main properties. Three are particularly important here.

1. Separation of heat source and heat sink
2. Temperature flattening
3. Temperature control

The system geometry may be divided into five major categories, each representing a different type of heat pipe system.

1. Tubular (including individual microheat pipes)
2. Flat plate (including vapour chambers)
3. Microheat pipes and arrays
4. LHPs
5. Direct contact systems

The following lists emerging heat pipe technologies, which offer enhanced integration and multifunctionality, in addition to improved thermal performance. The list is expected to expand in the future:

1. Advanced three-dimensional loop thermosyphon heat exchanger chassis
2. Ultra-thin vapour chambers
3. Additive-manufactured integrated heat pipes and vapour chambers

### 7.1.1 Tubular heat pipes

In its tubular form, with round, oval, rectangular or other cross-sections, two prime functions of the heat pipe may be identified:

1. Transfer of heat to a remote location
2. Production of a compact heat sink

It is possible to add heat flux transformation to this, although in electronics thermal management it would be difficult to find applications that do not involve taking a high heat flux and dissipating it in the form of a low heat flux, over a larger area.

By using the heat pipe as a heat transfer medium between two isolated locations (and as shown in Fig. 7.1, these may be isolated electrically as well as physically), recognisable applications become evident. It becomes possible to connect the heat pipe condenser to any of the following:

**FIGURE 7.1** An electrically isolated heat pipe assembly. *Courtesy: Boyd.*

- a solid heat sink (including finned heat sinks),
- immersion in another cooling medium,
- a separate part of the component or component array,
- another heat pipe,
- the wall of the module containing the component(s) being cooled.

### *7.1.1.1 Electrically isolated heat pipes*

Heat pipes can be used for applications requiring electrical isolation. Fig. 7.1 shows a heat pipe assembly that incorporates ceramic isolated inserts in the adiabatic section of the heat pipe. These inserts can resist breakdown with kilovolts of electricity. This particular assembly was deployed in a high-speed train application to cool power electronics (insulated-gate bipolar transistors, IGBTs). The internal working fluid is a dielectric to prevent the electrical bypass of the ceramic.[1]

In applications where size and weight need to be kept to a minimum (this applies to most applications in the 21st century), the near-isothermal operation of the heat pipe may be used to raise the temperature of fins or other forms of extended surface. This leads to higher heat transfer to the ultimate heat sink (e.g. air in the case of laptop computers) (Fig. 7.2), and the advantage may be used to uprate the device or reduce the size and weight of the metallic heat sink.

There are two possible ways of using the heat pipe here:

- mount the component to be cooled directly onto the heat pipe,
- mount the component onto a plate into which the heat pipes are inserted.

### 7.1.2 Flat plate heat pipes

The second of the five categories of heat pipes likely to be the most useful in electronics cooling is the flat plate unit. It is often used for 'heat spreading' or 'temperature flattening' and is highly effective in this role. The applications of the flat plate unit include, but are not limited to:

- multicomponent array temperature flattening,
- multicomponent array cooling,
- doubling as a wall of a module or mounting plate,
- single component temperature flattening.

The ability of flat plate heat pipes (and vapour chambers) to double as structural components opens up many possibilities and of course not just related to electronics thermal management. The example shown in Fig. 7.3, discussed in the next section, although not strictly a flat plate heat pipe, rather heat pipes in a flat plate, is the first step towards producing the near-isothermal surface.

#### *7.1.2.1 Embedded heat pipes*

It is very common to take the standard heat pipes and embed them into the base of a heat sink as shown in Fig. 7.3 [9]. This allows the use of high-volume, lower-cost heat pipes in a heat sink design. The heat pipes are flattened, soldered or epoxied into the heat sink and then machined to create a flat mounting surface for the electronics being cooled.

Fig. 7.4 shows a more complex embedded heat sink with multiple bent heat pipes soldered into the heat sink base to spread the heat more uniformly across the heat sink base. In this application, 1.5 kW of heat is dissipated from metal

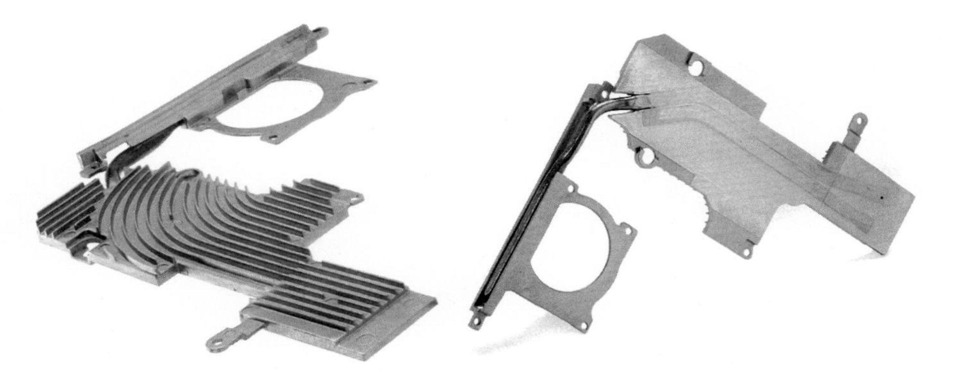

**FIGURE 7.2** Laptop computer heat pipe assembly. *Courtesy: Boyd.*

oxide semiconductor field effect transistor (MOSFETS) grouped towards the bottom of the sink. The heat pipes are

1. Electrically insulated heat pipes have been examined in a number of electrical engineering areas too. One of the authors was involved many years ago with the A. Reyrolle Switchgear Company, testing heat pipes for use at up to 33 kV potential, examining the suitable dielectric fluids and wall materials. As in many application areas, including cooling buried cable junctions, judicious use of heat pipes to remove hotspots can allow upgrading of equipment at minimum extra cost.

**FIGURE 7.3** Heat pipes embedded in a Hi-Contact heat sink base, with enhanced thermal contact by flattening and machining of the heat pipe vessel material. *Courtesy: Boyd.*

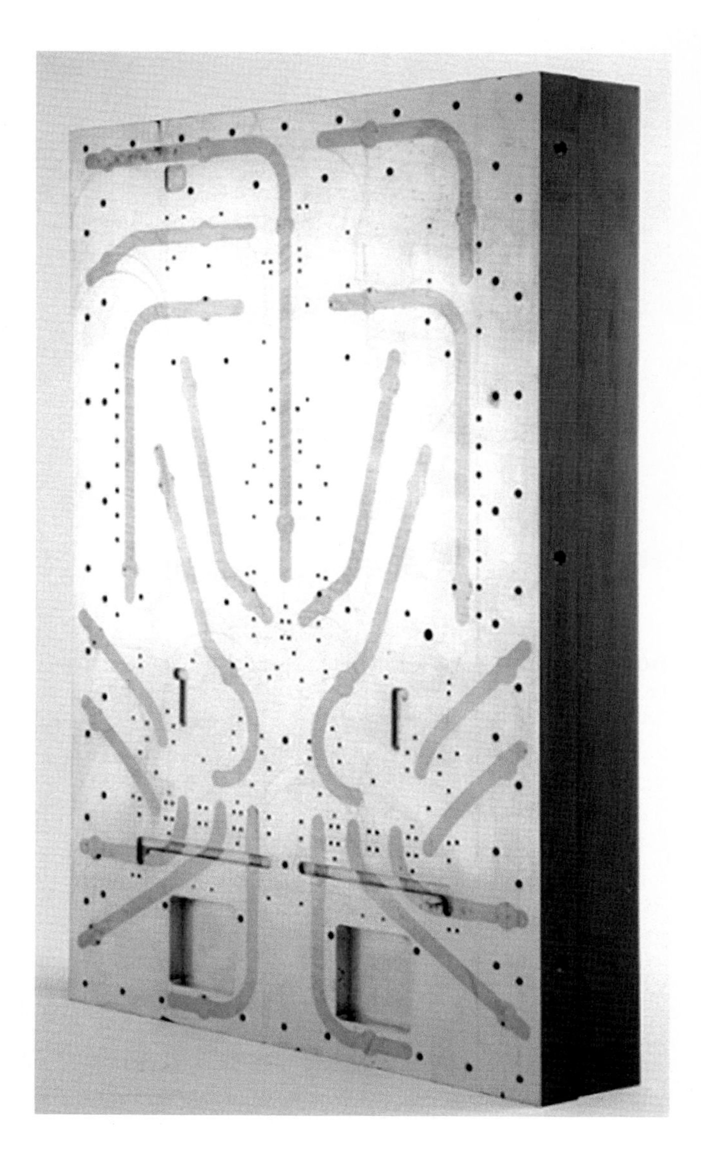

**FIGURE 7.4** Embedded heat pipe heat sink. *Courtesy: Boyd.*

routed to transport this heat to regions of the sink with lower heat input, allowing a 45% reduction in size, whilst maintaining the maximum operating temperature below 120°C. The addition of heat pipes delivered a reduction in the maximum MOSFET temperature from 118.8°C to 86.5°C, increasing the reliability and lifetime of the components.

#### *7.1.2.2 Bent and flattened heat pipes*

The most common application for heat pipes in electronics cooling is to bend and flatten heat pipes, to route them around electronic components along the heat pipe path between the heat source and the heat sink. Conventionally, this type of heat pipe assembly would deploy a metal evaporator block over the evaporator section of the heat pipe which interfaces with the component to collect the waste heat and either a second block at the condenser region or a direct connection into a heat sink or fin stack. It should be noted that a distinction should be made between flattened conventional tubular heat pipes, such as illustrated in Fig. 7.5 and vapour chamber heat pipes as described below. Fig. 7.6 shows a heat pipe assembly with four copper–water heat pipes soldered to an aluminium evaporator block. The heat pipes transfer the heat to a forced convection fin stack. Note the leading edges of the upper and lower fins are folded to form a duct to minimise bypass airflow over the protruding pipe ends and enable attachment of the fin stack to the aluminium block.

**FIGURE 7.5** Direct cooling of a discrete component utilising a bent and flattened heat pipe assembly. *Courtesy: Boyd.*

**FIGURE 7.6** Heat pipe assembly with zipper fin forced air convection fin stack. *Courtesy: Boyd.*

#### 7.1.2.3 Vapour chamber heat pipe

A flat plate heat pipe (vapour chamber) is shown in Fig. 7.7. This flat plate heat pipe design is hollow on the inside and the vapour is free to travel throughout. The internal structure is custom designed to include support pillars to prevent the collapse of the chamber, thermal vias, and a custom wick with varied thickness to target high flux regions. The vapour chamber is typically soldered or epoxied to the heat sink base.

As shown in Fig. 7.8, through holes and tapped holes may also be incorporated into the vapour chamber design to allow mechanical connections to be made to electronic devices and heat sinks (Patent no. 6302192). The vapour chamber shown in the figure is deployed in a medical polymerase chain reaction (PCR) test application and has contributed towards the testing of tens of millions of samples during the COVID-19 pandemic. Examples of the internal vapour chamber sintered wick structure are shown in Fig. 7.9.

### 7.1.3 Microheat pipes and arrays

Microheat pipes were reported previously as an option for the thermal control of miniaturised components, but potentially they are succeeded by flattened heat pipes and ultra-thin vapour chambers, presented later in this chapter, which offers smaller thicknesses and thermal performance enhancement versus microheat pipes. Examples, [10,11], of

**FIGURE 7.7** A vapour chamber flat plate heat pipe. *Courtesy: Boyd.*

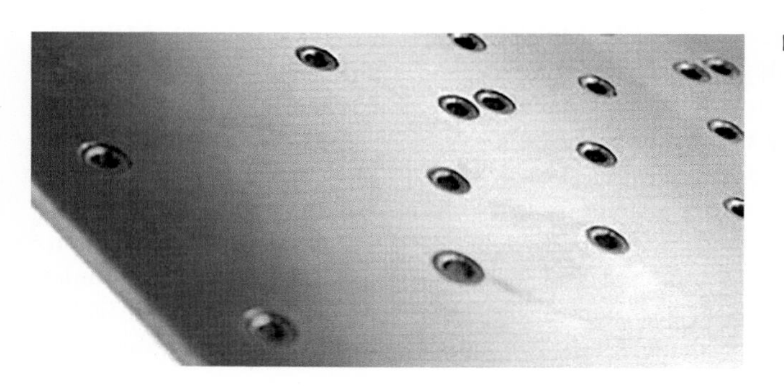

**FIGURE 7.8** Vapour chamber through holes. *Courtesy: Boyd.*

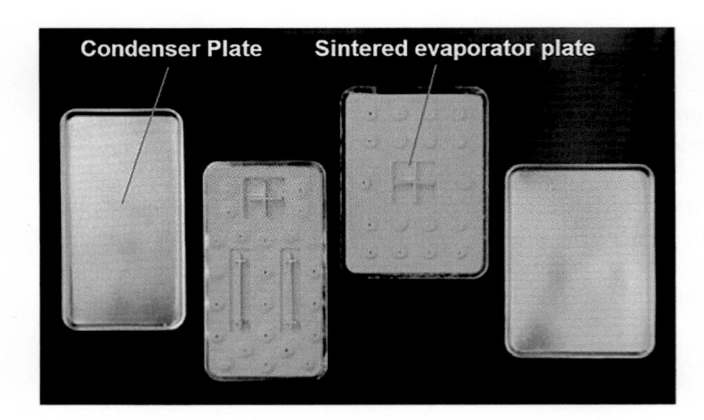


**FIGURE 7.9** Vapour chamber components before assembly, showing the internal wick structure. *Courtesy: Boyd.*

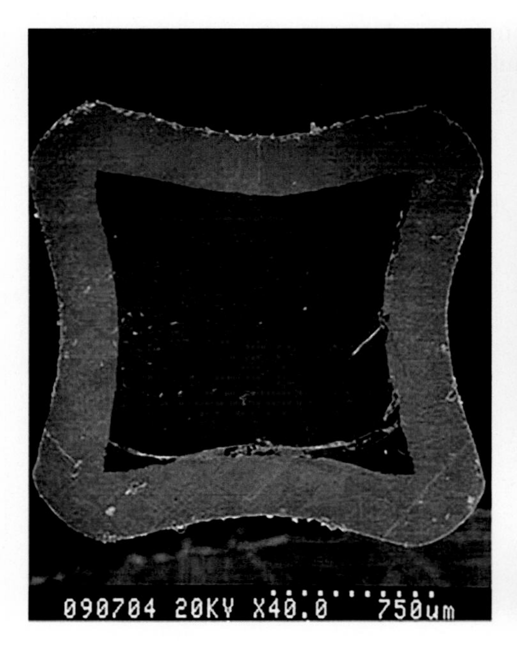

**FIGURE 7.10** Microheat pipe cross-sections. The wall material is copper and working fluid is water [10].

miniature (2–4 mm diameter) microheat pipes with characteristic dimensions down to 1 mm, reported by Moon et al. in South Korea, are shown in Fig. 7.10. The units of triangular cross-section can transport up to 7 W.

### 7.1.4 Loop heat pipes

LHPs are traditionally deployed in hi-tech applications such as space and aerospace (fast jet) applications. The combination of a loop system and primary wick technology offers a combination of high effective heat transfer over distances of metres; against gravity or against high gravitational acceleration loads. Multiple evaporator and condenser LHPs are also possible [12]. In satellite applications, LHPs are deployed to transport heat from specific high-power electronic components to a radiator panel. By minimising the thermal resistance between the heat source and the radiator, as well as effective thermal control of the electronics, the operating temperature and radiative heat capacity of the radiator can be raised. An LHP evaporator with a transport capacity of 1.5 kW in 0-g is shown in Fig. 7.11.

Fig. 7.12 shows an ammonia-charged LHP with 10 years flight heritage in Northrop Grumman's Block 60 F-16 aerospace application. The LHP is capable of functioning under 9*g* acceleration in any direction. The LHP is able to transport up to 250 W from a remote evaporator to the condenser unit and is designed to decouple the electronics mounted onto the evaporator section from the condenser section, to minimise the effect of vibration generated within the condenser air box on the electronics. The complexity of manufacturing the LHP's evaporator assembly limits

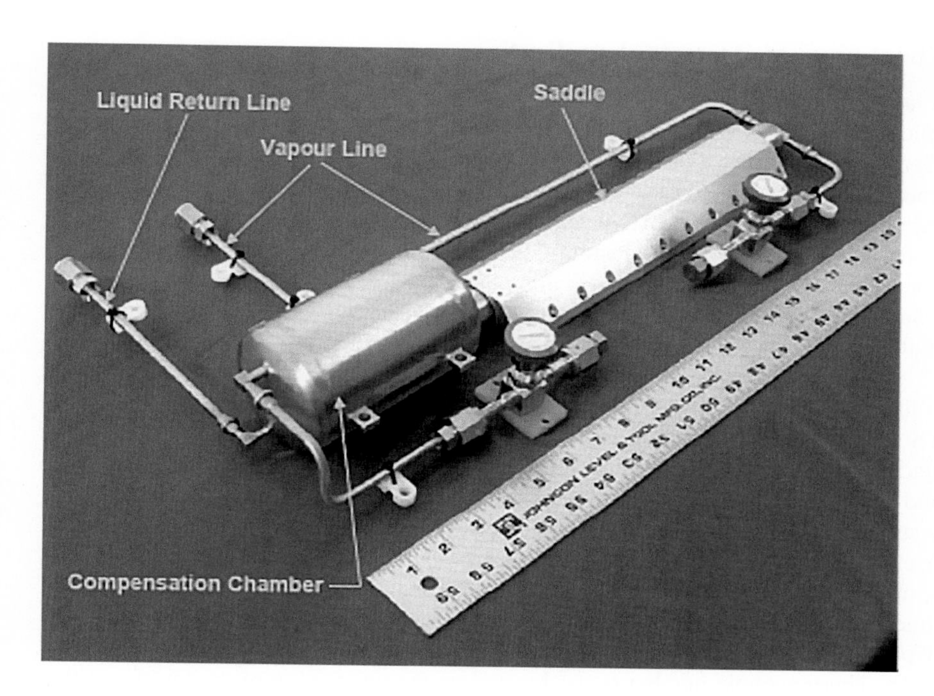


**FIGURE 7.11** LHP evaporator, 1.5 kW stainless steel – ammonia. *Courtesy: Boyd.*

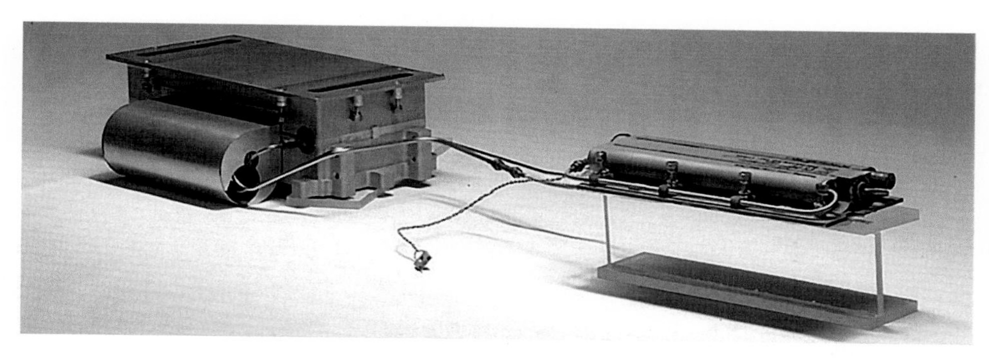

**FIGURE 7.12** Aerospace LHP for jet aircraft application, with more than 10 years of flight heritage. *Courtesy: Boyd.*

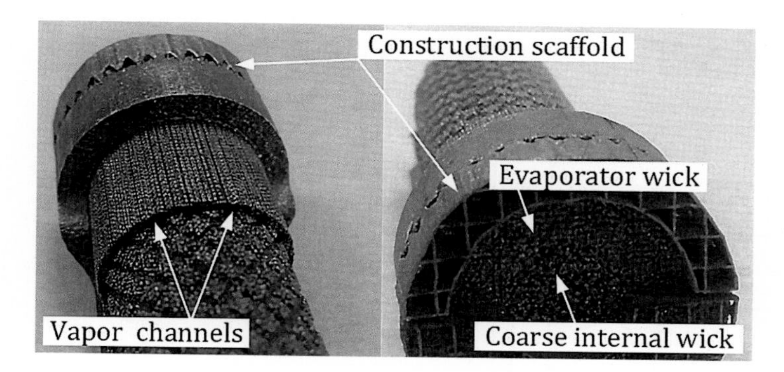


**FIGURE 7.13** Additive manufactures loop heat pipe primary wick sample. *Courtesy: Boyd.*

commercial applications of the technology. Additive manufacturing potentially can address the cost challenge by constructing the LHP as a single component.

Fig. 7.13 shows a proof-of-concept LHP primary wick, developed by an author of this book. The component was manufactured by AM from titanium (Ti6Al4V) powder using laser powder bed fusion (LPBF) [13–15]. The component includes a miniaturised lattice wick structure with integral vapour flow channels, external solid vessel material and solid bulkheads. The bore of the primary wick incorporates a coarse, randomised wick that is representative of an LHP secondary wick.

### 7.1.5 Heat pipe loops and loop heat pipes for terrestrial applications

The engineering complexity and associated high manufacturing cost limit the deployment of high-performance LHPs (as discussed above) in terrestrial applications such as data servers and consumer electronics applications. Therefore there is an opportunity for lower-tech heat pipe loops such as loop thermosyphons and low-cost LHPs to be deployed in consumer electronics and server applications.

The first deployment of LHPs in laptop PCs was reported by Maydanik in 2001 [12], when a copper—water LHP, with a diameter of 6 mm was tested to dissipate 25—30 W from custom CPUs. The target at the time was to dissipate about 70 W from Atholon XP processors targeted at notebook PCs. As this power load is easily transported with conventional copper—water sintered wick heat pipes, low-cost LHPs have not yet emerged into the market. However, with increasing data server power and miniaturised consumer electronics, a need is forseen for new low-cost devices and new technology is emerging.

An early version of the Boyd loop thermosyphon (LTS) with horizontal transport lines is shown in Fig. 7.14 [16]. To reduce the fabrication cost, the LTS was designed using the company's standard vapour chamber parts as building blocks, therefore the primary wick found within hi-tech LHPs is removed and the LTS is an open loop, which requires gravity-aided return of the working fluid to function. The flat copper evaporator with a footprint of $65 \times 90$ mm$^2$ had a 1.5 mm thick sintered copper capillary structure on the inner surfaces with an effective pore radius of about 50 μm. The condenser consisted of two flat copper chambers, $65 \times 90$ mm$^2$ each, without any internal structure. The two chambers were connected using two short tubes, one above the other and were cooled by forced air convection through external fin stacks. The smooth-wall transport lines were 600 mm long.

The LTS was filled with 75 cc of pure ethanol and tested in the horizontal orientation as shown in Fig. 7.14. A selection of test data is shown in Fig. 7.15. Thermocouple number 6 was located on the top portion of the far end of the

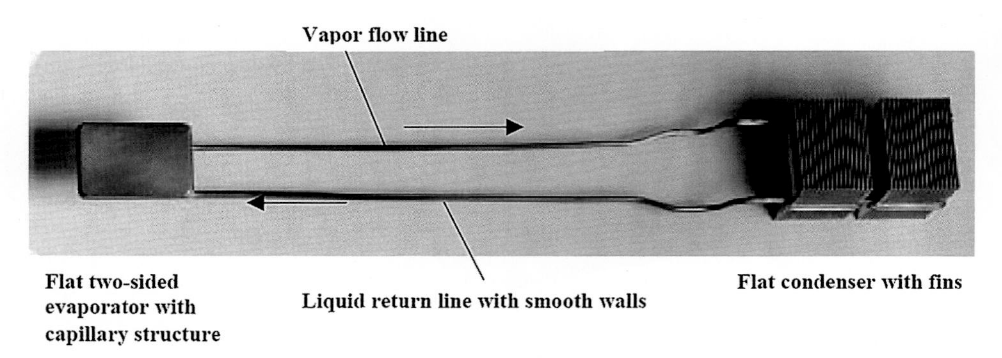


FIGURE 7.14 A Boyd LTS with horizontal transport lines.

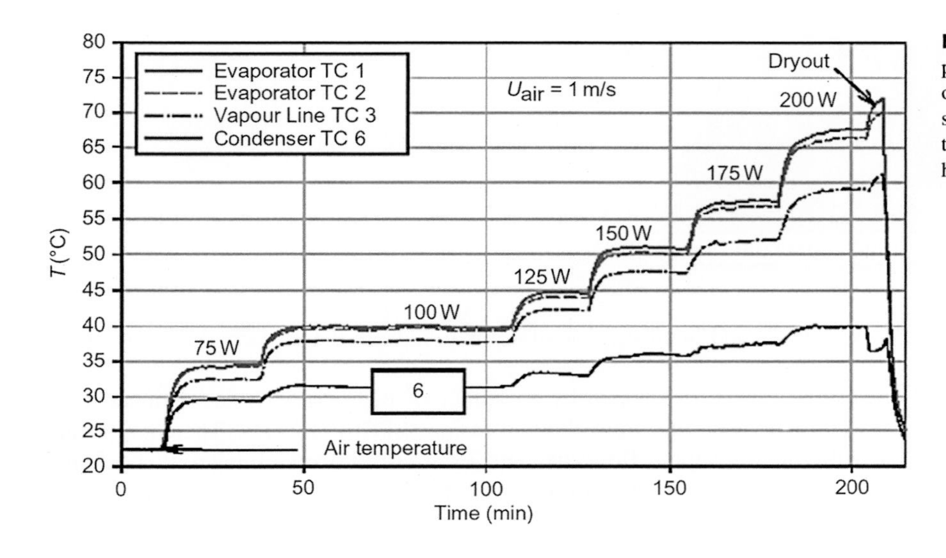


FIGURE 7.15 Test data for the LTS heat pipe. The increase in heat transport, measured over a period of about 200 min, with steady state being achieved at each step, is shown. As the capability is exceeded, dry-out occurs and heat transport breaks down.

condenser. The LTS transported up to 200 W with an air velocity across the condenser of 1 m/s and the heated wall temperature below 70°C.

A current, up-to-date LTS developed for data centre server thermal management applications is shown in Fig. 7.16. The LTS incorporates a dual-evaporator design and an integral forced air convection condenser unit and has a case to the ambient thermal resistance of Rca ~ 0.08 K/W. The LTS is mounted horizontally in the server application as shown in Fig. 7.17, to provide a gravity head, the vapour lines exit the evaporator bodies vertically, and the liquid lines are in-plane with the evaporator bodies. Various single- and dual-evaporator LTS designs are currently being tested with transport capacity ranging from 300 to 600 W.

### 7.1.6 Pulsating heat pipes

Pulsating heat pipes (PHPs) function by vapour bubble growth within a capillary tube, which transports heat from the evaporator to the condenser region. At the condenser, the vapour bubble collapses and condenses to a liquid. The expansion and collapse of vapour bubbles create an oscillatory flow within the transport lines, which drives the flow of heat between the evaporator and condenser regions. Experimental results of an unlooped PHP targeted at the electronic thermal management field within hybrid vehicle applications are given in Ref. [17]. The 2.5 mm inner tube diameter device

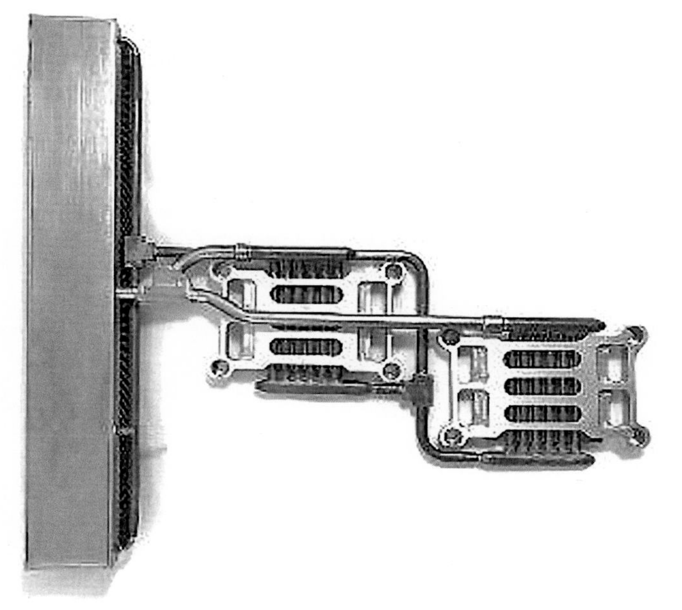

**FIGURE 7.16** Dual-evaporator LTS for data server applications. *Courtesy: Boyd.*

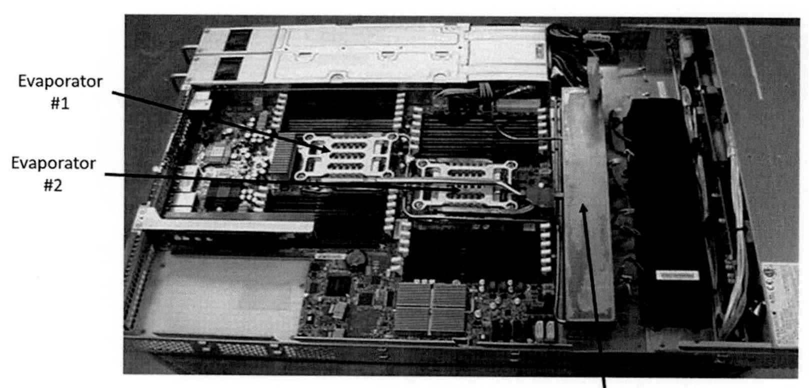


**FIGURE 7.17** Dual-evaporator LTS installed into server application. *Courtesy: Boyd.*

was cooled by an air heat exchanger to replicate the environment of a vehicle. An improvement in performance and functionality is achieved by forming the PHP into a loop system. A closed loop PHP developed by Prof. Marco Marengo and his team for space applications is shown in Fig. 7.18 [18]. Multiple parallel serpentine tubes efficiently transport a 200 W power load, with a thermal resistance of 0.15 K/W. A challenge in developing PHPs for space applications is that the inside diameter Ø3 mm of the tube (Ø5 mm OD) is larger than the critical capillary diameter on Earth, therefore parabolic flight testing is required to prove performance in a 0-g environment. The aluminium PHP vessel was charged with refrigerants and ethanol. Peltier devices or forced convection was used during testing.

### 7.1.7 Direct contact systems

One of the challenges with integrating heat pipes and electronic components is mounting the devices and minimising interface resistances. Two ways for easing this problem have been proposed and patented in the United Kingdom.

The first of these, developed at Marconi, involves using a conformable heat pipe, or more accurately, a plate, which can be pressed into intimate contact with heat-generating components, with a minimal thermal interface between these and the wick. The second method involves the removal of the heat pipe wall altogether and directly integrating the heat pipe evaporator wick across the microprocessor surface. In this case, the module would be a sealed unit with a provision for heat extraction located externally.

A number of challenges have limited the deployment of direct contact heat pipe-related systems in commercial applications, including fluid compatibility, fluid electrical conductivity, encapsulation of the electronics packages, hermetic sealing, in situ heat pipe charging in the application, complexity of installation, maintenance and component replacement.

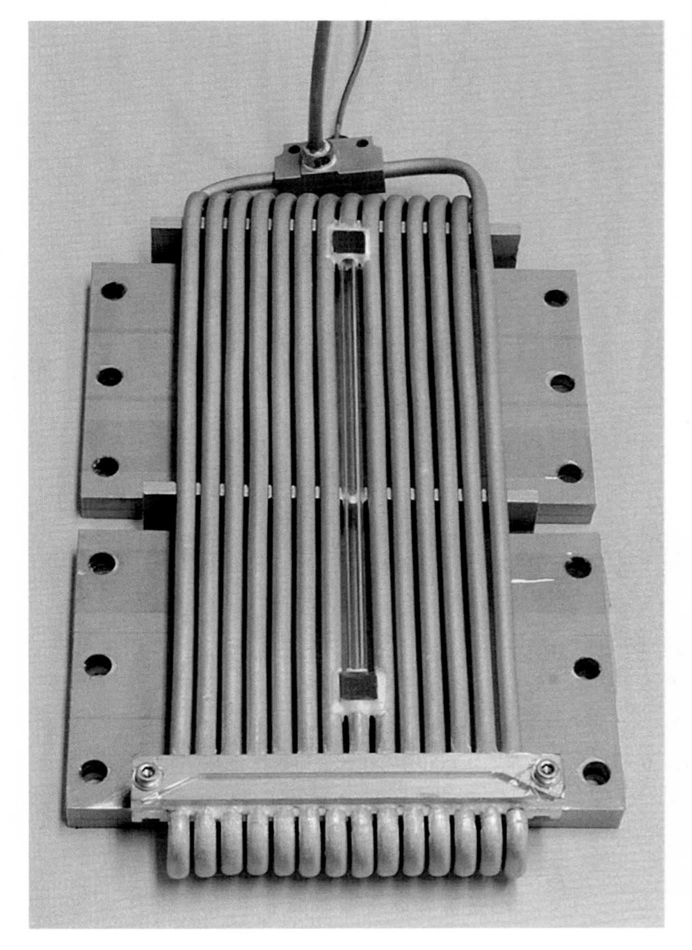

**FIGURE 7.18** Closed loop PHP test piece for space applications, 80 × 200 × 30 mm (W × L × T) [18]. *Courtesy: Prof. M. Marengo, University of Brighton.*

### 7.1.8 Thermal ground plane – coefficient of thermal expansion matched vapour chamber

Thermal ground planes (TGPs) are thin vapour chambers (1–3 mm), which are deployed to thermally spread high heat fluxes generated in wide band gap (WBG) electronic components such as IGBTs, power amplifiers and LEDs. Through a DARPA TGP development programme, Boyd has developed CTE (coefficient of thermal expansion) matched TGPs that allow direct attachment to the substrate and dissipation of high heat fluxes [19]. By deploying copper–molly–copper (CMC) laminate material and by varying the copper flash thickness, the CTE can be matched to the substrate material (silicon [Si], silicon carbide [SiC] and gallium nitride [GaN]). For example, the TGP illustrated in Fig. 7.19 has CTE ~7.1 PPM/K (CTE Copper = 16.7 PPM/K); therefore, it is matched to gallium arsenide (GaAs) (5.73 PPM/K) substrate material. The TGP also incorporates an enhanced hybrid sintered wick, which enables the dissipation of heat fluxes above 350 W/cm$^2$, with an effective thermal conductivity of 1200 W/m K (Fig. 7.20). Potential applications of this technology include disruptive photonic technologies such as lasers, waveguides, photodetectors, amplifiers, LEDs and optical fibres.

### 7.1.9 Ultra-thin vapour chambers

Transport and spreading of heat from discrete electronic components within confined spaces or narrow gaps can be achieved by flattening conventional circular heat pipes to thicknesses of 1.0–1.5 mm or by deploying two-phase vapour chamber heat pipes in the same thickness range, to spread the waste heat over a larger heat dissipation area. The rapid evolution of smaller, thinner, lightweight mobile electronic devices requires higher performance thermal solutions, to handle high heat fluxes generated by the miniaturised electronic components.

**FIGURE 7.19** High heat flux thermal ground plane (TGP), CTE matched to the heat source substrate material. *Courtesy: Boyd.*

**FIGURE 7.20** High heat flux hybrid wick within CTE matched TGP. *Courtesy: Boyd.*

Ultra-thin vapour chambers [13,20,21] enable a thickness reduction and high thermal performance with the ability to handle high heat flux bursts and high-performance modes of mobile devices (Fig. 7.21). Ultra-thin vapour chambers offer a thickness reduction to ~0.4 mm in copper and approaching 0.3 mm in stainless-steel variants, with application-matched, customised wick structures incorporated into the design. For a typical mobile phone application (100 mm × 50 mm; at 60°C), the 0.4 mm thickness copper and 0.3 mm stainless-steel ultra-thin vapour chambers have similar thermal conductances of 5700 W/m K and 6000 W/m K, respectively, versus 1300 W/m K for a 0.3 mm thick graphene thermal spreader.

Ultra-thin vapour chambers offer the potential benefits of increased reliability and lifetime, reduced mass and reduced touch temperatures, which are important considerations in wearable electronics. The high tensile strength of stainless-steel ultra-thin vapour chambers allows dual functionality as a structural component, potentially enabling further size and mass reductions of the electronic devices.

### 7.1.10 Flexible heat pipes

Fig. 7.22 is a photograph of a family of flexible heat pipes for aircraft applications [22]. The adiabatic section is made from flexible hose. This allows freedom when mounting the heat pipe to the device being cooled and the heat sink. In addition, the flexible section can accommodate relative motion between the heat source and the sink. The evaporator section mounts to electronics on an actuator and the condenser attaches to the aircraft structure. The actuator moves whilst the condenser remains stationary. A second type of flexible heat pipe is referred to as 'bellows heat pipes' due to the inclusion of a bellows component within the heat pipe adiabatic section (Fig. 7.23). This type of heat pipe can accommodate misalignment in the final assembly and/or vibration loading that occurs in the application.

Intel considered current and future laptop computers [23], the first heat pipe being used in such a computer in 1994. Conventional laptop heat pipe assemblies transport heat from a CPU to the edge of the laptop, normally the side or rear panel, where a finned heat sink on the condenser section can be forced air-cooled. Boyd proposed that a 'thermal hinge'

**FIGURE 7.21** Ultra-thin vapour chamber examples for miniaturised electronics applications [13]. *Courtesy: Boyd.*

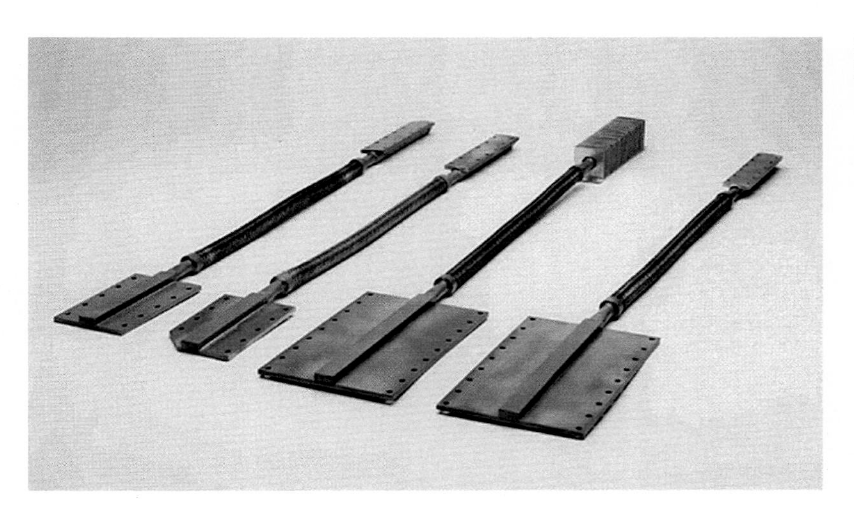

**FIGURE 7.22** A family of flexible heat pipes used in aircraft [22]. *Courtesy: Boyd.*

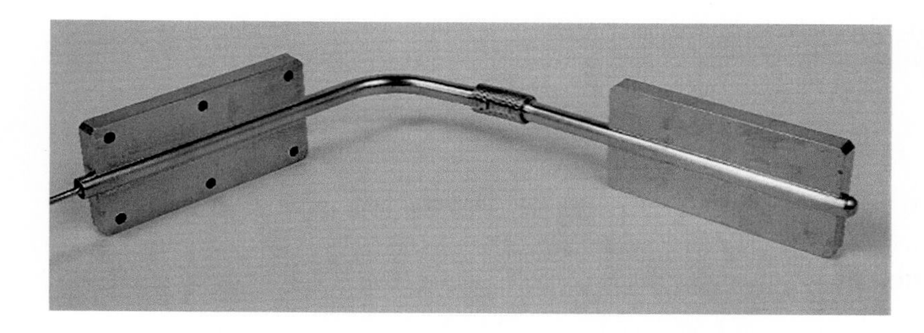

**FIGURE 7.23** Flexible bellows heat pipe assembly. *Courtesy: Boyd.*

**FIGURE 7.24** Furukawa electric bendable sheet heat pipe [24].

be used to allow the heat pipe to be extended into the lid of the laptop (which houses the screen), enabling natural convection from the lid surface, allowing the fan to be discarded. Tests to simulate opening and closing the lid showed that the effective performance was maintained over 50,000 bending cycles.

Flexible heat pipes, also be referenced as 'bendable' heat pipes, that may be bent only once, to fit it within a piece of equipment, can also be a specific requirement. An example of a bendable sheet heat pipe, permanently bent to shape is shown by Furukawa Electric in Japan (Fig. 7.24) [24]. The sheet heat pipe has dimensions of 20 × 150 mm, a thickness of 0.6 mm thickness and a transport capacity of 10 W, targeted at lightweight, miniaturised consumer electronics applications.

### 7.1.11 Miscellaneous systems

There are a number of other fluid flow mechanisms used in heat pipes, which can benefit their application in electronics thermal management. Two mechanisms that have been studied by Boyd include the combined pulsating and capillary transport (CPCT) system, and the use of graded wicks [21]. The CPCT-type system has the potential of achieving heat flux capabilities of over 300 W/cm$^2$.

The CPCT unit, a variant of the PHPs described in Chapter 6, uses both pulsating fluid motion and capillary forces to take the working fluid to high heat flux regions. The grooves serve as channels for the pulsating fluid, and supplementary sintered wicks are provided for extra nucleation sites. Another PHP design is described in Ref. [25]; this Canadian research leads to optimum fill ratio criteria.

The graded wick concept, a derivative of composite wicks described in Chapters 3 and 5, uses relatively open pores for liquid transport between the evaporator and the condenser but fine pores (sintered) close to the heat input and output walls of the heat pipe. Thus the pressure drop along most of the pipe is modest. Performance was slightly better than conventional units but pore size optimisation could lead to improvements. Reference [21] includes theoretical analyses of the systems for those wishing to follow up the concepts in depth.

An additional heat pipe type is the integrated heat pipe, where multifunctionality can be achieved by incorporating the thermal, electrical and mechanical requirements of the application into a single component. An example was presented by the University of Twente [26], where a small form factor vapour chamber heat pipe was integrated directly into the PCB structure. The study development focused on manufacturability and led to a highly cost-effective solution.

## 7.2 Applications

In this section, a small number of applications are used to illustrate the features of several heat pipe types.

### 7.2.1 Heat pipes to cool a concentrated heat source

Fig. 7.25 shows a heat pipe assembly design to remove heat from a concentrated high heat flux source and spread it to a large finned area where the heat can be more effectively removed. Heat pipe assembly applications are typical examples of the deployment of heat pipes in electronics thermal management applications.

### 7.2.2 Multikilowatt heat pipe assembly

Fig. 7.26 is a photograph of a heat pipe assembly designed to remove up to 10 kW of heat from power electronics such as IGBTs. These assemblies are used in motor drives and traction applications. In traction applications, an interesting design challenge may arise when the train is stationary, where airflow through the fin stack reverts from forced air convection created by the motion of the train, to natural convection.

### 7.2.3 Surgical heat pipe applications

Applications of heat pipes in the thermal management of medical electronics applications are broad including haematology/dialysis equipment, PCR test equipment, medical imaging scanners (CT, MRI, X-Ray, ultrasound) and LEDs, cameras and lasers utilised in robotic surgical and cosmetics applications. An interesting handheld electrosurgical instruments application is the deployment of heat pipes to thermally manage electrode tips in cautery bipolar forceps, enabling efficient and safe coagulation or cutting of organic tissue [27,28], as overheating of organic tissues to temperatures above 45°C may lead to soft tissue necrosis, fistulas or burns. Moreover, tissue temperatures above 41.8°C within the brain may lead to temperature regulation issues such as hyperthermia, the electrical input power in previous surgical equipment (without thermal control) was reduced, to minimise the risk of excessive tissue temperatures. This in turn extended procedure durations and reduced accuracy. The incorporation of heat pipes within each leg of the cautery bipolar forceps (Fig. 7.27), to thermally manage the electrode tip temperatures, allows an increase in electrical input

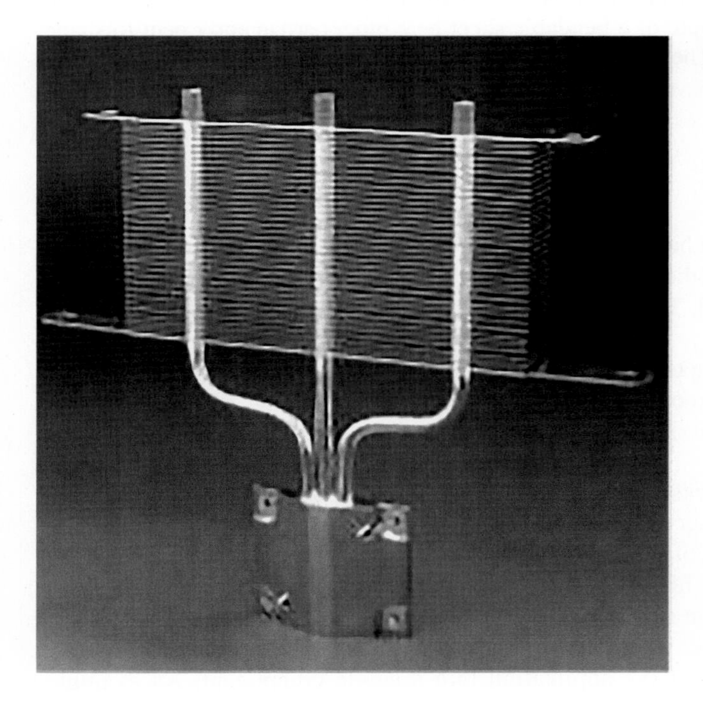

**FIGURE 7.25** Heat pipe assembly for removing heat from a concentrated heat source. *Courtesy: Boyd.*

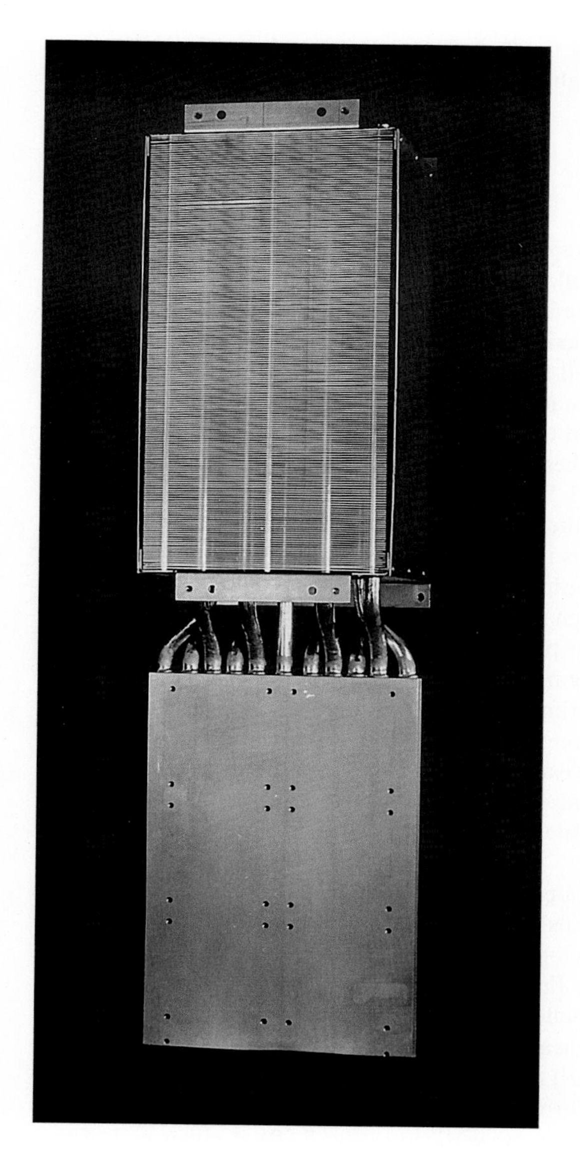

**FIGURE 7.26** Multikilowatt heat pipe assembly. *Courtesy: Boyd.*

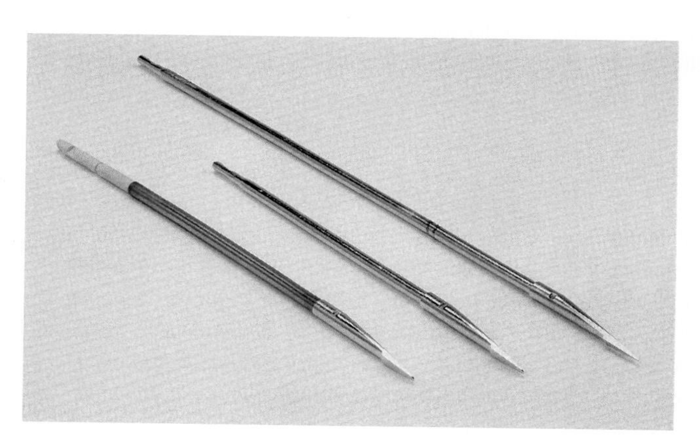

**FIGURE 7.27** Heat pipe integrated with surgical electrode for cautery bipolar forceps application. *Courtesy: Boyd.*

power to be applied, speeding up the surgical procedure and improving the accuracy of the procedure. The technology enables cauterisation immediately after the incision, for example in highly vascularised organs such as the liver, brain and spleen.

## 7.3 Electric vehicle cell cooling applications

With the drive towards carbon neutrality, the EV market is growing rapidly and is set to displace combustion engine vehicle technology within this decade. As presented by Worwood et al. [8], effective EV and hybrid electric vehicle (HEV) battery cell thermal management, to maintain uniform temperatures across cylindrical cells and to provide cooling to keep the cells in an optimal temperature range, is critical in maximising the electrical performance and overall lifetime of the cells. Heat pipes are proposed as a potential option in cell thermal management of cylindrical 18,650 and 32,113 cell formats with a graphite anode and lithium cobalt aluminium oxide cathode, targeted at high-performance electric vehicles and hybrid electric vehicles. The heat pipe is proposed to form the cell mandrel, minimising the internal cell thermal resistance and maximising transport to the cell tabs, where the heat is dissipated. It is predicted that the integration of heat pipes can reduce the cell thermal resistance by 51.5% and 49.1% for the 18,650 and 3211 cells, respectively, versus a reduction of 20.3% and 49.1% for copper mandrel equivalents. A range of heat pipe diameters and cell cooling types is considered in the study. Transient load analysis is also included.

A review of experimental investigations of the deployment of heat pipes as part of the EV and HEV's battery pack/module is presented by Abdelkareem et al. [29]. The review is thorough and considers multiple heat pipe types including copper–water sintered heat pipes, PHPs, LHPs and ultra-thin heat pipes and enhanced nanoparticle working fluid heat pipes that are integrated with secondary thermal management systems, such as forced air convection, liquid cooling and phase-change material thermal storage. Predominantly, the experimental investigations target module and pack-level thermal management, to transport heat away from the heat collectors that interface with the surfaces of lithium-ion battery cells. The objectives of the studies include (1) dissipating excessive heat during fast charging and discharging, (2) optimising the cell's thermal operating conditions to maximise power output & efficiency by reducing cell operating temperatures and achieving near-isothermal temperature profiles across the cell surfaces, and to (3) maximise the lifetime of the battery cells.

The deployment of flat heat pipe 'heat mats' to thermally manage a prototype prismatic battery module, incorporating 16 commercially manufactured lithium-titanate cells, was investigated by Jouhara et al. [30,31]. The 36 V prototype prismatic module, with energy storage capacity of 0.85 kWh, was developed by Vantage Power for an EV bus charging application, which aims to contribute towards decarbonisation of the fleet of buses in England and Wales (currently ~5700). In the first iteration of the system, a single heat mat was mounted horizontally (Fig. 7.28) that interfaced with the base of the prismatic module and transported 60% of the waste heat generated by the cells to a cooling channel (water or refrigerant) built into the edge of the heat mat. Test conducted on the prototype were representative of a real-world duty cycle (C 83%–22%) that simulated ~7.2 miles of urban driving, followed by a rapid charge at ~3.5 kW. The second iteration of the system deployed two heat mat heat pipe systems vertically (Fig. 7.29) at either end of the cells (tab end cooling), which both enhanced thermal performance and provided a level of redundancy to the system, that may prevent thermal runaway, should a cell or heat mat fail in the end application.

Testing of the system determined the maximum cell temperature to be maintained below 28°C in the horizontal orientation and below 24.5°C for the dual vertical heat mat arrangement. Therefore both configurations enable the battery module to operate within the optimum temperature range of 15°C–35°C for Li-ion batteries. The temperature uniformity across the heat mat was maintained at ± 1°C, with the temperature difference across the height of the cells at ~6°C and ~2°C for the horizontal and vertical orientations, respectively.

## 7.4 Telecommunications applications

Over the last decade, the capability of telecom transmitter technology and the associated thermal challenges have advanced significantly. As pointed out at the beginning of this chapter, heat dissipation loads have progressed from 120 W per PCB (*Heat Pipes 5th Edition*) to 120 W per chip (10 W/cm$^2$) (*Heat Pipes 6th Edition*). Telecom transmitters are now predicted to exceed 350 W per PCB, with each card deploying multiple miniaturised 80 W (40 W/cm$^2$) microprocessors in future 6G applications. The following section recaps the 2G heat pipe technology presented in *Heat Pipes, 6th Edition* and introduces emerging thermal technology that addresses the thermal requirements of 6G and 7G telecom applications.

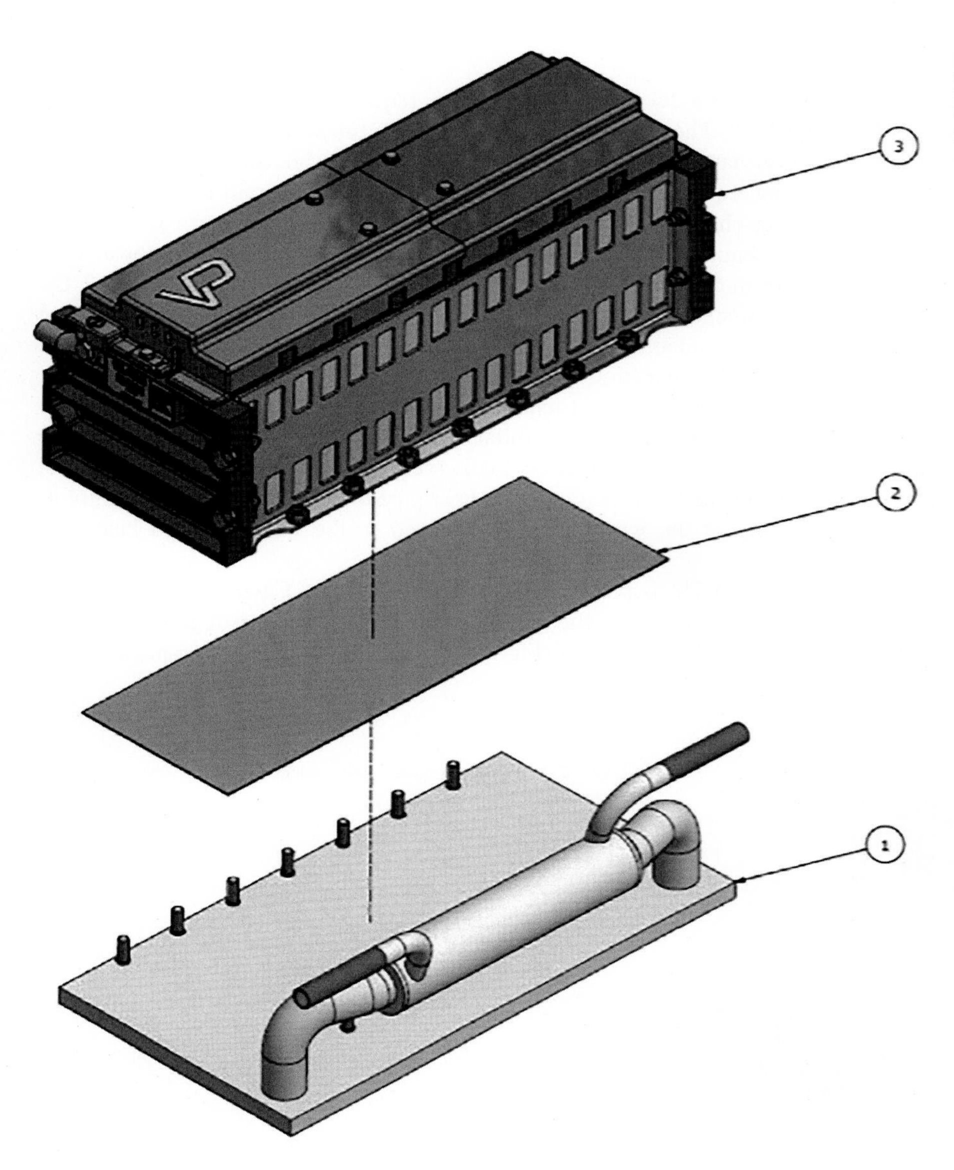


**FIGURE 7.28** Prismatic cell module, horizontal heat pipe heat mat assembly. (1) Heat pipe heat mat. (2) Thermal interface material. (3) Prismatic cell module with 16 × Li-ion cylindrical cells [30].

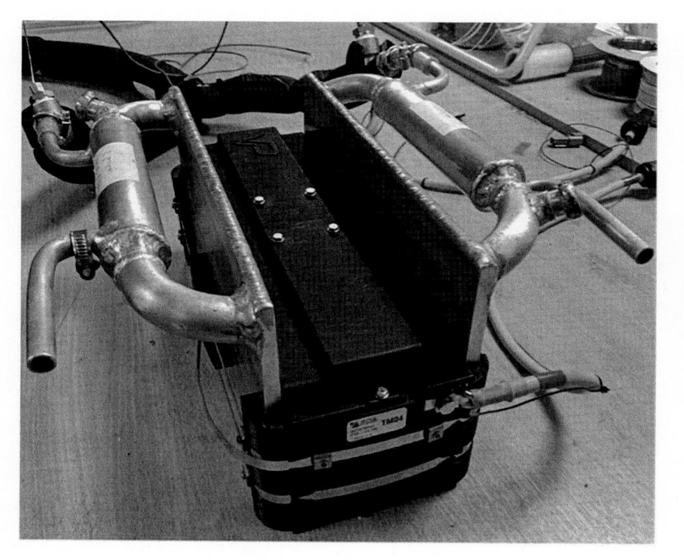

**FIGURE 7.29** Prismatic cell module configured with two vertical heat pipe heat mat systems [31].

### 7.4.1 Previous 2G telecom heat pipe applications

The deployment of pole-mounted telecom transmitters for a 2G telecom application, requiring a total heat dissipation of 240 W, was presented in *Heat Pipes, 6th Edition*, Ch0.8. The transmitter was required to function over an ambient temperature range from −40°C to +50°C as a passive, natural convection, heat pipe assembly. It was discussed that in addition to the maximum ambient temperature, the increase in heat dissipation had introduced a second worst-case operating condition, where in subzero conditions, water-charged heat pipes would freeze, leading to thermal runaway. Traditionally, the low dissipative heat loads had not caused any issues below 0°C; however, the increased power load of the 2G application led to exceeding the maximum microchip temperature by 15°C (135°C at an ambient of −17°C). This challenge was overcome by deploying a dual heat sink heat pipe assembly, where a small, local heat sink mounted directly to the chip took advantage of the much larger available junction to ambient temperature difference to provide cooling whilst the heat pipes were frozen. A heat pipe connection to transport the heat to a much larger, remotely located heat sink was used to dissipate the heat at the upper ambient temperature limit. These features were incorporated into the final version of the assembly (Fig. 7.30). An example of a similar heat pipe assembly for a 4G application is shown in Fig. 7.31 [32]. In this application, the increased dissipation loads were transported by seven methanol charged heat pipes to an external natural convection fin stack, formed onto the heat pipe condensers. Using methanol as the working fluid overcame the challenge of freezing of the working fluid observed in the water-charged heat pipe applications.

A second telecom system consisting of a two-phase LTS, is presented by the University of Caen Basse Normandie and France Telecom [33]. The LTS evaporator was positioned in the high-temperature convection region above the electronics enclosure and was used to transport heat outside the cabinet (Fig. 7.32). In the system tested, *n*-pentane was used as the LTS working fluid.

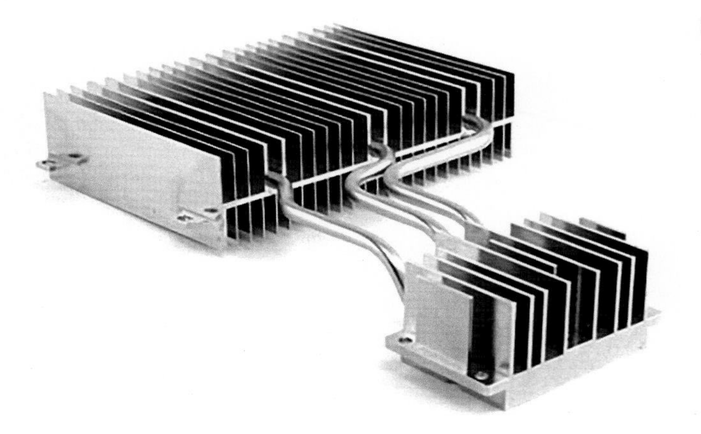

**FIGURE 7.30** Pole-mounted 2G telecom heat pipe assembly. *Courtesy: Boyd.*

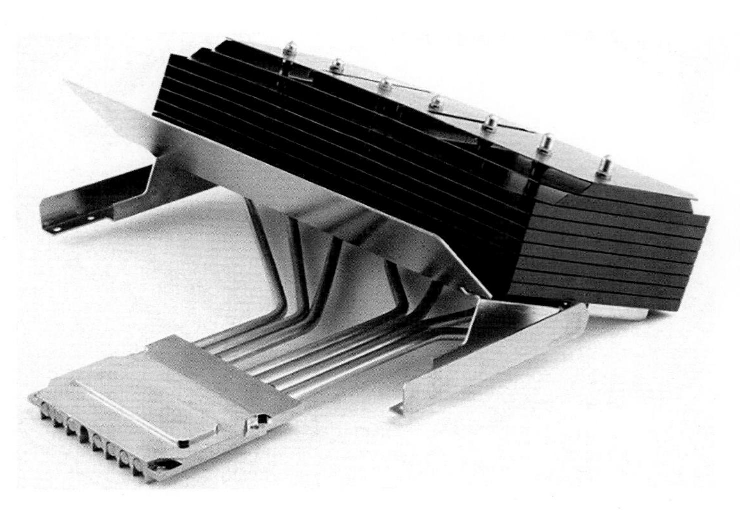

**FIGURE 7.31** 4G pole-mounted transmitter application [32]. *Courtesy: Boyd.*

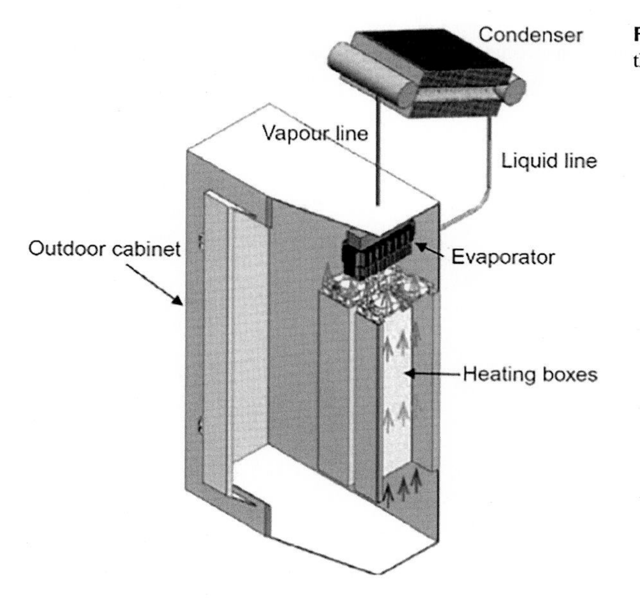


**FIGURE 7.32** LTS developed in France for telecommunications equipment thermal control [33].

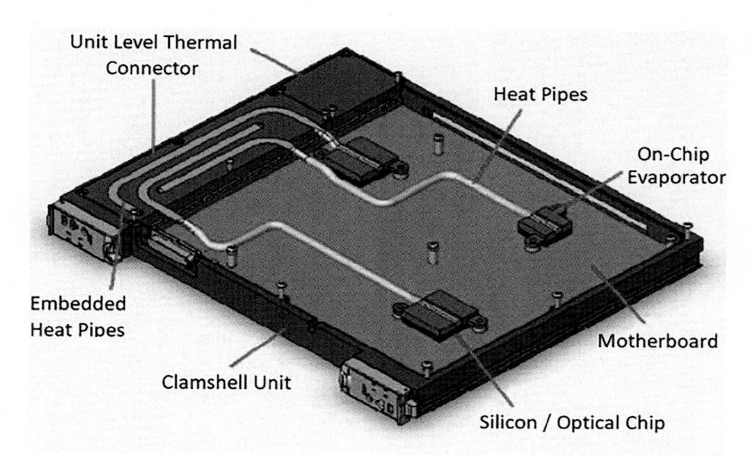


**FIGURE 7.33** Telecom chassis containing thermal assembly with three heat pipes [34].

A third PCB-level integrated high-power telecommunications thermal management system was also presented (*Heat Pipes, 6th Edition*), where both heat pipes and LTS were deployed within the electronic box, in direct contact with the high heat dissipation microprocessors. The chassis shown in Fig. 7.33 was required to maintain maximum component temperatures of $\leq 70°C$ in air at 55°C ambient temperature [34]. The chassis formed part of an overall thermal-bus (Therma-Bus) system, where an LTS evaporator in direct contact with the unit-level thermal connector transported heat out of the system.

## 7.4.2 Two-phase heat sink fins in telecom transmitter applications

In current 5G pole-mounted outdoor radio applications, it is common to deploy an aluminium heat sink that forms the structure of the electronics enclosure, enabling close proximity with the electronics mounted directly onto the heat sink. To reduce the thermal resistance of the heat sink base and reduce the overall mass of the system (base thickness reduction), heat pipes and vapour chambers can be deployed. In high-performance applications, the thermal efficiency of the heat sink can be improved further by integrating a two-phase thermosyphon within each fin (Fig. 7.34). Thermosyphons enable two-phase heat transfer from the fin base to the tip and 2D heat spreading along

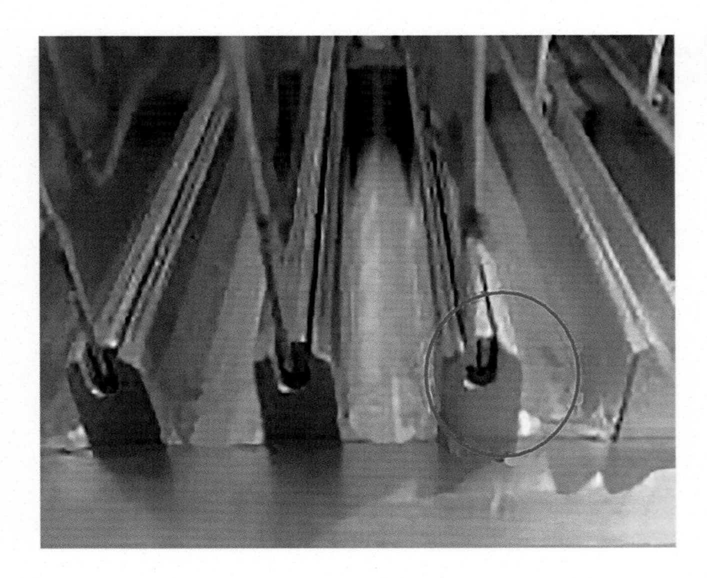

**FIGURE 7.34** Example of a two-phase thermosyphon fin. Fins can be bonded or mechanically swaged to the heat sink base. *Courtesy: Boyd.*

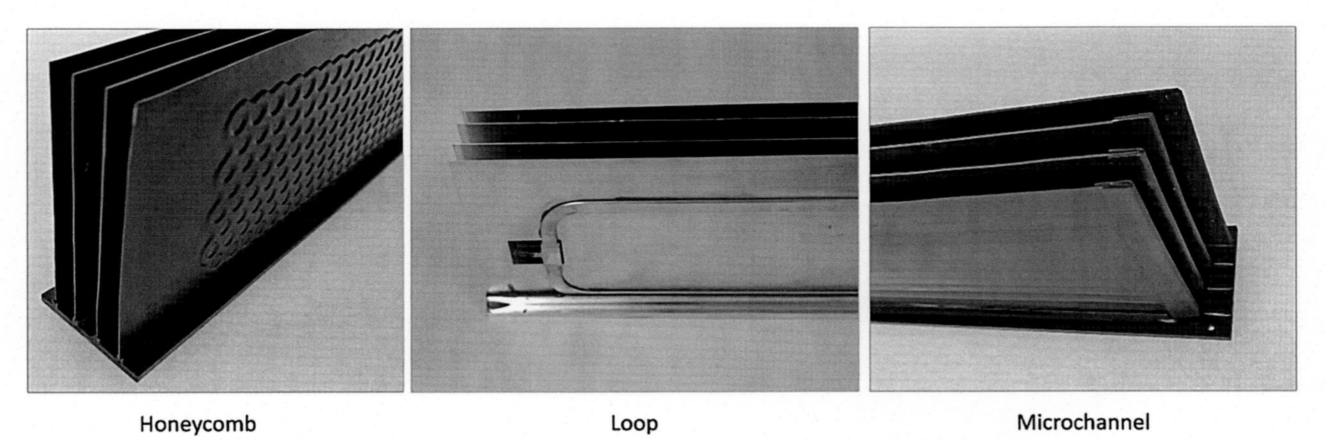


**FIGURE 7.35** Two-phase thermosyphon heat sink fin variations. *Courtesy: Boyd.*

the length of the heat sink base. Three main types of thermosyphon fins deploying honeycomb, loop and microchannel thermosyphons are shown in Fig. 7.35. Table 7.1 compares each fin type against conventional extruded and bonded fins.

### 7.4.3 3D vapour chambers

To meet the demands of future high-power, high heat flux applications, 3D vapour chamber heat exchangers (Fig. 7.36) were developed in the DARPA Research and Development programme entitled Microtechnologies for Air-Cooled Exchangers (MACE). Interconnected two-phase flow passages within a vapour chamber base, with parallel thermosyphon blades, create a 3D flow network and an isothermal heat sink [32,35]. The high surface area of augmented folded fins (louvred fins) between the heat sink blades enhances heat dissipation from the system by forced air convection. The overall 3D vapour chamber has exceptionally low thermal resistance and high coefficients of performance of ~0.02 K/W at 150 CFM (Fig. 7.37).

3D vapour chambers are targeted at high heat dissipation commercial and military applications where space is limited and a high level of integration is required, such as power electronics, power converters, telecom and medical equipment, military radars and microwave/RF applications. Typically copper and water are used as the vessel materials and working fluid; however, aluminium with alternative fluids can be used in extreme temperatures or thermal cycling environments, such as aerospace applications and UAVs (unmanned air vehicles). The technology has been demonstrated in a series of standard 'unit' sizes from 3U to 6U, with a power load range of 250–2000 W. An example of the application of a 4U vapour chamber and thermal performance testing of the unit at 1000 W is shown in Figs 7.38 and 7.39.

**TABLE 7.1** Comparison of thermosyphon fin technology against conventional fin types.

| Fin technology | Heat transport/ performance | Fin customisation, (holes/notches/ cut-outs) | Vapour space customisation | Weight optimisation | ΔMass compared to extruded/ painted (g (%)) |
|---|---|---|---|---|---|
| Extruded aluminium | Conduction | Difficult | N/A | Limited | Baseline |
| Bonded aluminium | Conduction − | Yes | N/A | Moderate | +225.3 (+11%) |
| Honeycomb | Two-phase + | Yes, impacts vapour space | Channel locations can be customised. Edge seal requirements prevent evaporator from being inserted into base groove | Moderate | −250.7 (−11%) |
| Loop | Two-phase ++ | Yes, adjust loop location as needed, does not need to cover entire fin | Loop location/angle can be customised and evaporator may be inserted into base groove | Highly flexible | −313.2 (−18%) |
| Microchannel | Two-phase ++ | No | Fixed vapour space, but evaporator is inserted into base groove by default | Limited | −141.1 (−8%) |

*Note:* This type of comparison is typically application specific and based on factors important to the customer.
*Source*: Adapted from Boyd.

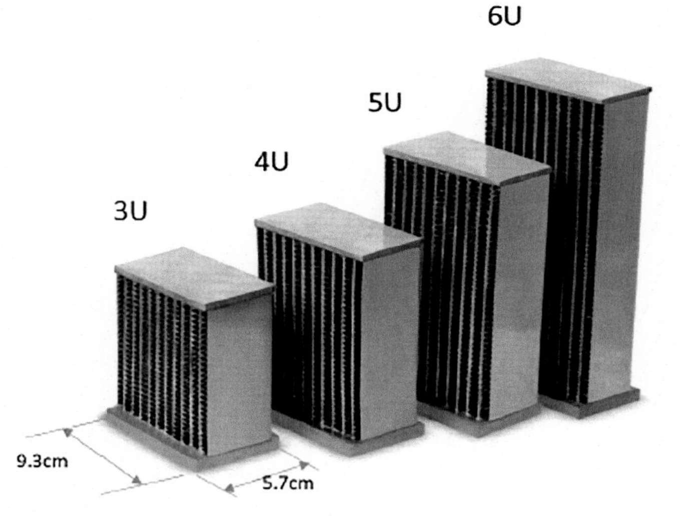


**FIGURE 7.36** 3D vapour chamber (HEX type) [32,35]. *Courtesy: Boyd.*

## 7.4.4 Direct contact 3D loop thermosyphons for future 6G and 7G telecom applications

As noted at the beginning of this chapter, future telecom transmitters are predicted to deploy PCBs with total heat dissipation in excess of 350 W, with multiple miniaturised 80 W/(40 W/cm$^2$) processors deployed onto the PCB. To address the increased heat dissipation requirements of future 6G and 7G radio transmitter applications, 3D LTS technology is under development by Boyd, which offers a reduction in thermal resistance of $\approx$30% and mass reduction

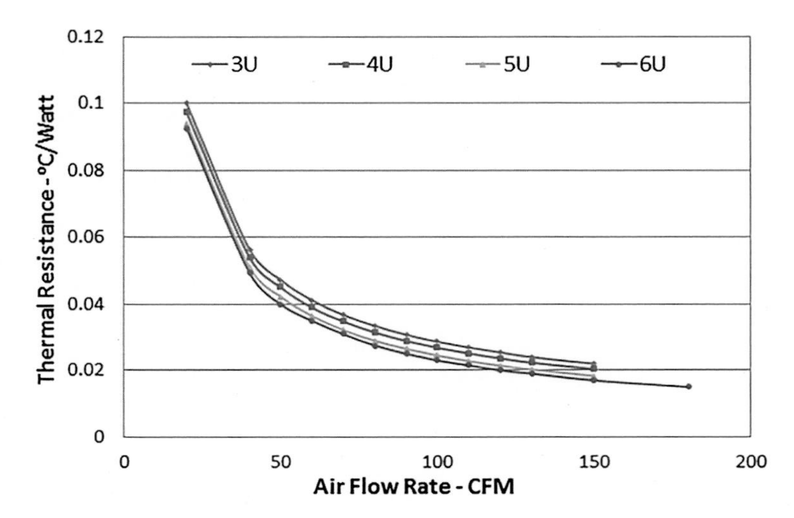


**FIGURE 7.37** Thermal resistance versus airflow rate for a range of 3D vapour chambers, with 95 × 55 mm heat input area [35]. *Courtesy: Boyd.*

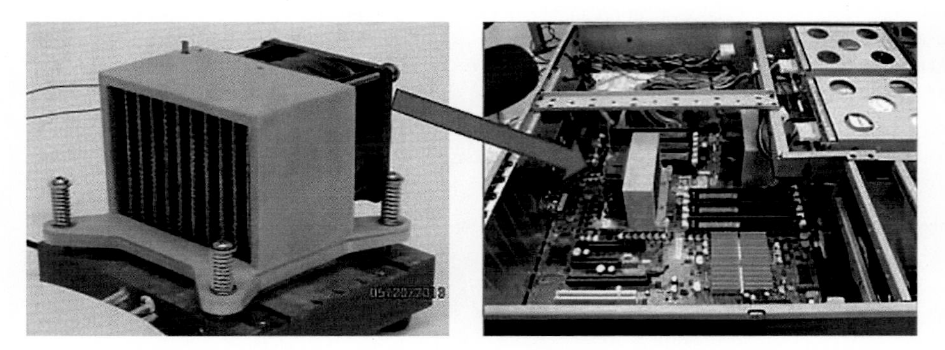

**FIGURE 7.38** 4U 3D Vapour Chamber for an 800 W electronics application. *Courtesy: Boyd.*

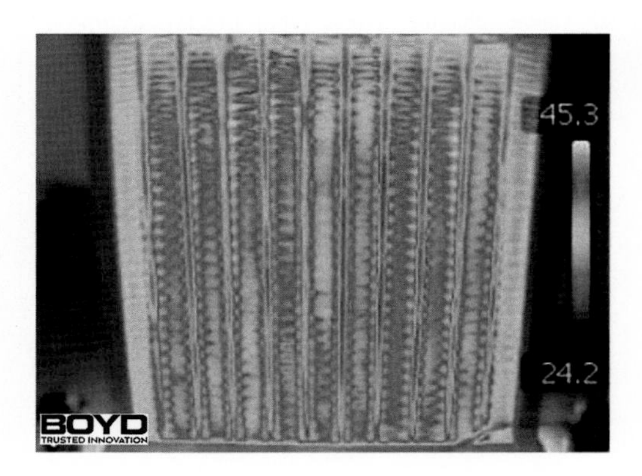


**FIGURE 7.39** Infrared image showing testing of a 4U 3D vapour chamber at 1000 W [35]. *Courtesy: Boyd.*

of ≈75% compared to an equivalent natural convection extruded heat sink [36]. For example, the thermal resistance of a 3D LTS tested in a 990 W lighting application was reported $R^{th}$ = 0.102 K/W, versus 0.156 K/W for a heat pipe heat sink.

The 3D LTS has similar functionality to 3D vapour chambers but offers increased performance by integrating two-phase loops into the system. The 3D LTS is formed by interconnecting multiple parallel two-phase fins (thin web MPEs) to upper and low manifolds (Fig. 7.40) [37,38]. Heat is input into the leading edge of the fins by conduction from aluminium bars that make a direct contact with the discrete components (Fig. 7.41). The heat flows vertically upwards within microchannel passages at the leading edge of each fin. The vapour is distributed by an upper manifold throughout additional microchannels at the middle and rear of each fin, where the heat is dissipated to the air, causing the vapour to be condensed back to a liquid. The condensate flows by gravity into the lower manifold, completing the two-phase loop. IR camera images efficient thermal spreading throughout a natural convection 3D LTS for a 500 W

FIGURE 7.40 3D thermosyphon for transmitter applications [38]. *Courtesy: Boyd.*

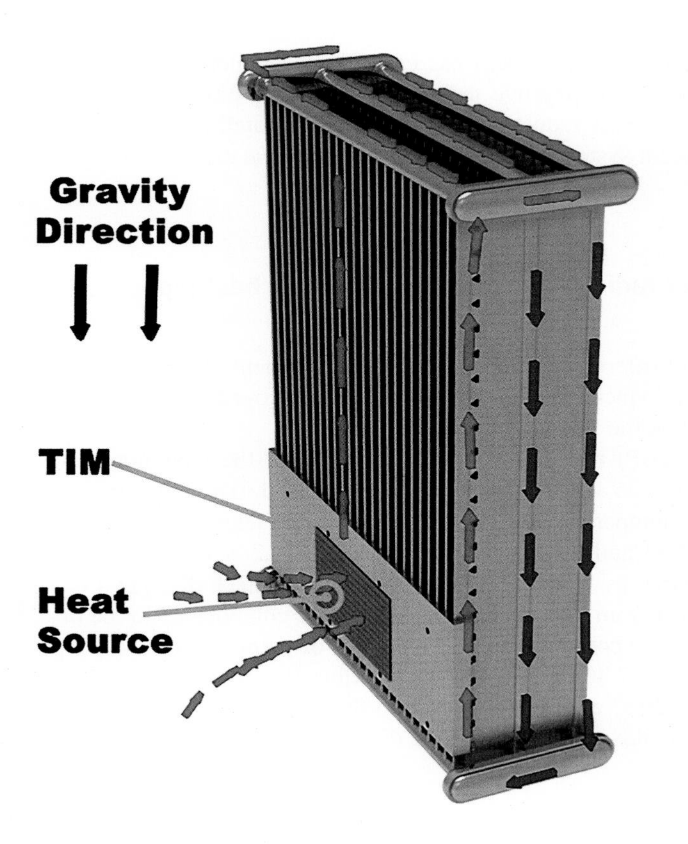


FIGURE 7.41 3D loop thermosyphon internal fluid flow directions [38]. *Courtesy: Boyd.*

RRU application, with a maximum temperature of 87.8°C (Fig. 7.42). As the vapour and liquid pressure drops act in the same direction around the loop, an increase in the maximum heat transport capacity is achieved by the 3D LTS versus the 2D counterflow heat sink blade thermosyphons reported earlier in the chapter.

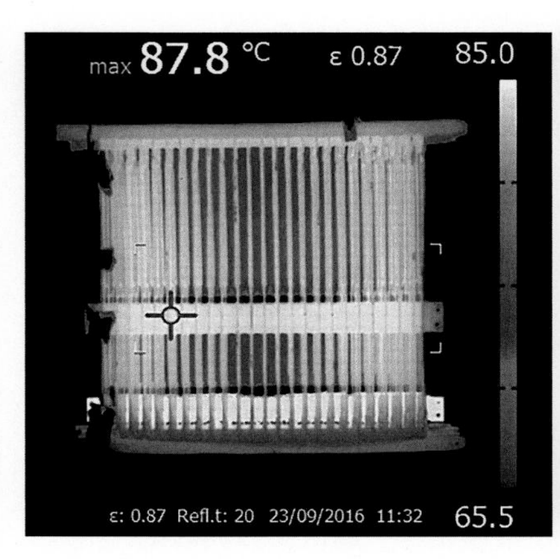


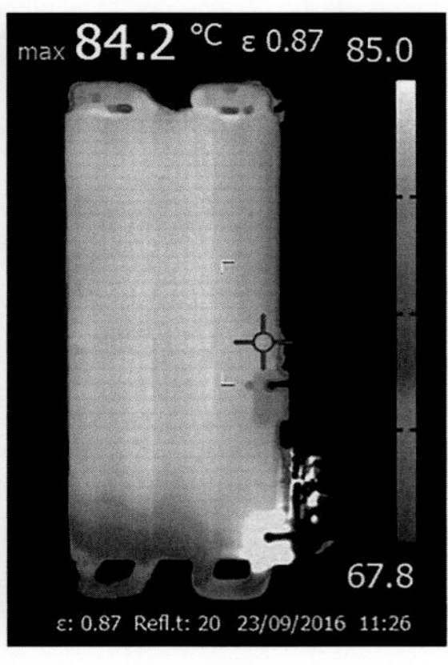


**FIGURE 7.42** 3D loop thermosyphon infrared thermal test imaging. *Courtesy: Boyd.*

## 7.5 Space applications

As the time of writing this text, NASA's Space Launch System (SLS), the most powerful rocket in the world, has lifted off from Launch Pad 30B at the Kennedy Space Centre, Florida, propelling the Orion Spacecraft on its 25.5-day ARTEMIS 1 mission, travelling ~40,000 miles beyond the moon, and has returned to the Earth. As described by NASA Administrator, Bill Nelson, 'the uncrewed flight will test Orion to the limits, in the rigours of deep space, in preparation for human exploration of the moon and ultimately Mars' [39]. This global achievement signifies the level of innovation and engineering within the space sector.

Within the commercial space sector, innovation is also buoyant. For example, the European Space Agency's (ESA) NEOSAT activity (next-generation geostationary telecommunication platform) has enabled engagement of the supply chain with major European satellite manufacturers, to develop and improve technologies that are now in orbit [40]. NEOSAT targeted at 30% performance enhancement and 30% mass reduction in the 3–6 tonne satellite range, with a major focus area being improvement of the thermal management system. The following sections give examples of space heat pipes in commercial and exploration applications.

### 7.5.1 Australian Square Kilometre Array Pathfinder radio-telescope copper–water heat pipe application

In addition to in-orbit applications, the terrestrial space sector offers opportunities for the deployment of advanced heat pipe systems, such as radio-telescope receiver applications. Typically, radio telescopes are deployed in remote low-population regions to minimise the impact of human activity on the functionality of the telescope.

The Australian Square Kilometre Array Pathfinder (ASKAP) radio telescope constructed by the Commonwealth Science and Industrial Research Organization (CSIRO) [41] has 36 12-m reflector antennas, located in a severe environment with low night-time temperatures and daytime ambient temperatures up to 55°C. Each radio telescope deploys an advanced phased array feed (PAF) receiver module (Figs 7.43 and 7.44) that incorporates a Ø1.2 × 6 mm thick heat pipe chassis disc (Fig. 7.45) [13,42], designed and manufactured by Boyd in the United Kingdom. Each disc [13,43,44] has 117 embedded copper–water heat pipes that transport heat from the 188 individual receiver components to the disc perimeter. Heat is dissipated from the system by eight secondary cooling systems consisting of perpendicular heat pipe thermal links and forced air convection heat pipe fin stacks (Fig. 7.46). To enable functionality in a 55°C ambient, thermoelectric coolers (TECs) that subcool the disc perimeter to 20°C are incorporated into the secondary cooling system, preventing the thermal failure of the electronics. The TECs add an additional heat dissipation load of 560 W to the overall system requirement. Fig. 7.43

### 7.5.2 Aluminium–ammonia constant conduction heat pipes for space applications

Aluminium–ammonia constant conduction heat pipes (CCHPs), with axially grooved wicks (AGHP), have been deployed for a number of decades in geostationary earth orbit (GEO) telecom satellite applications as surface heat pipes and linking heat pipes, to transport heat from an interface with the electronics chassis to the radiator panels, and as

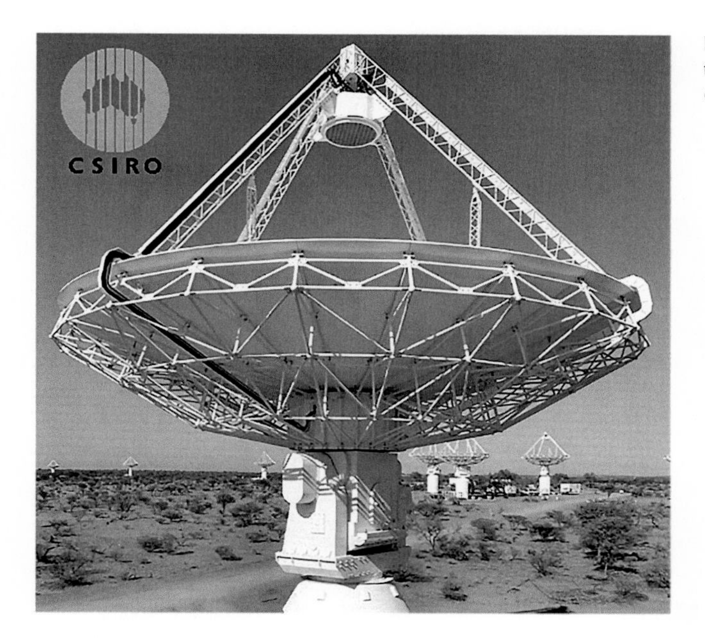


**FIGURE 7.43** Photograph of a CSIRO ASKAP radio telescope with the location of the PAF receiver module highlighted [13,42]. *Courtesy: CSIRO.*

**FIGURE 7.44** Close-up view of the CSIRO PAF receiver module with integrated Boyd UK heat pipe receiver disc [13]. *Courtesy: CSIRO.*

embedded heat pipes within the radiator panels. An example of CCHPs for cooling a high-power active phased array antenna is shown in Fig. 7.47 [45].

The antenna deployed over 100 waveguide-type antenna elements, each with an individual T-module (transmitter module), giving a heat dissipation requirement of 578 W, which including a safety factor limit of 1.25, increased the design power to 723 W. Four 832.5 mm long CCHPs were deployed to collect and transport heat from the T-module framework to a heat dissipation panel. The CCHPs were designed and tested to ESA space standards for two-phase heat transport equipment [46]. Ground test pieces were tested over an operating temperature range of −20°C to 80°C, at power loads of 153 W and 191.25 W. The tests were conducted with the CCHP at an adverse tilt angle of −0.24°C (against gravity) and at an angle of +5°C (with gravity-aided return of the condensate), as per the ground test orientation of the overall antenna assembly. Over the last decade, the increase in performance and packing density of

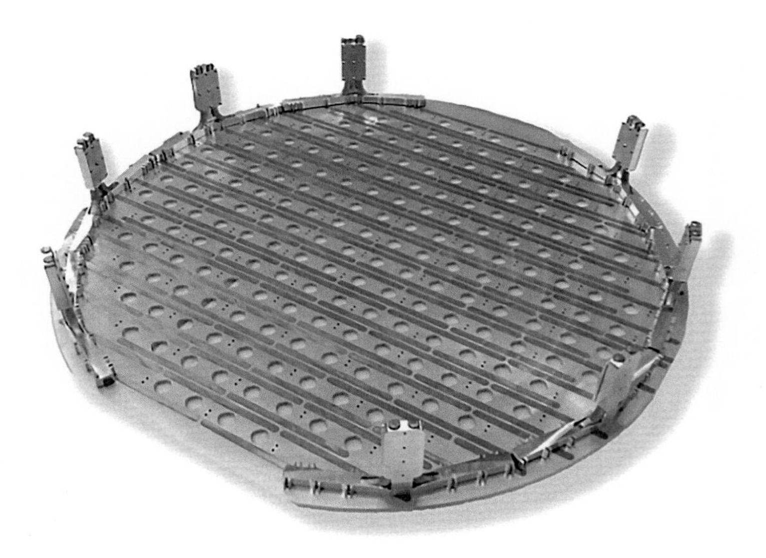

**FIGURE 7.45** Ø1.2 m heat pipe receiver disc for ASKAP radio telescope [13]. *Courtesy: Boyd.*

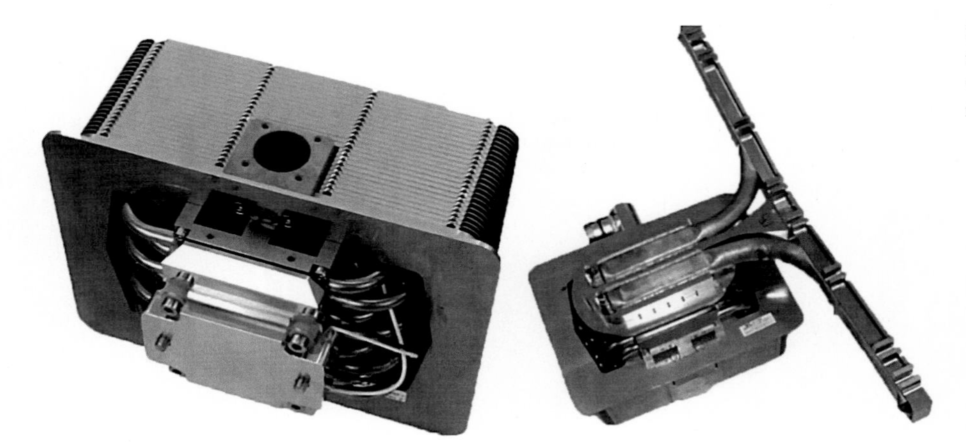

**FIGURE 7.46** Secondary cooling system: heat pipe assembly and forced convection heat pipe fin stack for ASKAP radio telescope [13]. *Courtesy: Boyd.*

**FIGURE 7.47** Aluminium CCHP extrusion profiles for a waveguide application [45].

microelectronics within satellite payloads has led to an increase in overall heat dissipation from ~5 kW to >20 kW per satellite platform. The quantity of CCHPs deployed in GEO satellites has increased from ~150 to ~300 CCHPs per platform. With typical CCHP lengths ranging from <1 m to >4 m, the overall CCHP mass is significant and impacts the overall cost of ownership of the satellite and return on investment. The overall performance of the thermal system from the microprocessor to the radiator is a critical design requirement. This requirement can be addressed by introducing direct thermal management of heat sources, as discussed in the next section.

However, the emerging low earth orbit (LEO) constellation market aims to address the challenges of large GEO satellites by deploying constellations of up to several hundred small or medium-sized satellites. Although the quantity of CCHPs deployed per satellite is low (e.g. 10–60 CCHPs), it is a step-increase in production capacity as the overall requirement per LEO constellation is in excess of the current global capacity (e.g. 10,000–40,000 CCHPs per constellation). The global supply chains are in the process of building capacity, with Boyd creating capacity both in North America and the United Kingdom.

### 7.5.3 Space copper–water heat pipes (TRL9)

As the increase in overall satellite heat dissipation is driven by increasing dissipation from discrete semiconductors mounted within the payload electronics chassis, direct contact copper–water heat pipes offer a significant reduction in the thermal resistance between the discrete microelectronics components and the chassis level thermal interface with the platform level thermal management system (TMS). For example, Boyd UK has developed and qualified copper–water heat pipe technology and has achieved TRL9 (technology readiness level), which is deployed both in direct cooling of the microelectronics and as embedded heat pipes at the chassis level [47]. The development was completed in collaboration with Thales Alenia Space's (TAS) thermal engineering team based in Toulouse, France, as part of an ESA European Component Initiative (ECI) activity AO-7623 aimed at developing advanced cooling technologies compatible with novel flip-chip and high pin count semiconductors. The qualification test vehicle consisting of three stacked slice chassis is shown in Fig. 7.48.

Each chassis incorporates advanced PCB technology, direct flip-chip thermal management utilising two copper–water heat pipe assemblies (Fig. 7.49) and encapsulated graphite solid conduction technology (600–1000 W/m K) both in the chassis base and to thermally manage components on the rear of the PCB. The qualification test vehicle was qualified in alignment with the ESA ECSS requirements for space two-phase heat transfer equipment to both the ESA AO-7623 and TAS test specifications. Test files included thermal characterisation tests, ageing and life tests and thermomechanical testing. The heat pipe qualification test pieces were required to transport 20 W from the microprocessor to the enhanced k-Core chassis base, bypassing the high thermal resistance conduction path through the PCB and solid aluminium chassis material. However, the heat transport capacity of current heat pipe assemblies has increased considerably. Direct two-phase thermal management gives the advantage of enabling the deployment of higher performance, higher heat dissipation semiconductors, whilst minimising the impact on the overall thermal budget. The reduction in thermal resistance within the chassis can be utilised to operate the space radiators at a higher temperature, enabling higher heat dissipation from the same radiation surface area, as the heat dissipation capacity of radiation heat transfer is driven by the $T^4$ term in the Stefan–Boltzmann equation ($Q = \varepsilon\sigma A T^4$). Both mini-heat pipe assemblies and embedded

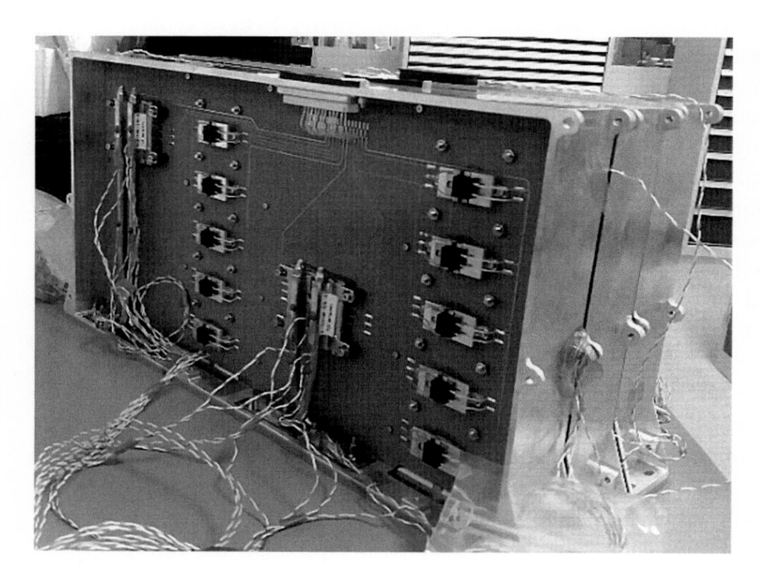

**FIGURE 7.48** Mini-heat pipe TMS space flight qualification test vehicle [47]. *Courtesy: Boyd and Thales Alenia Space.*

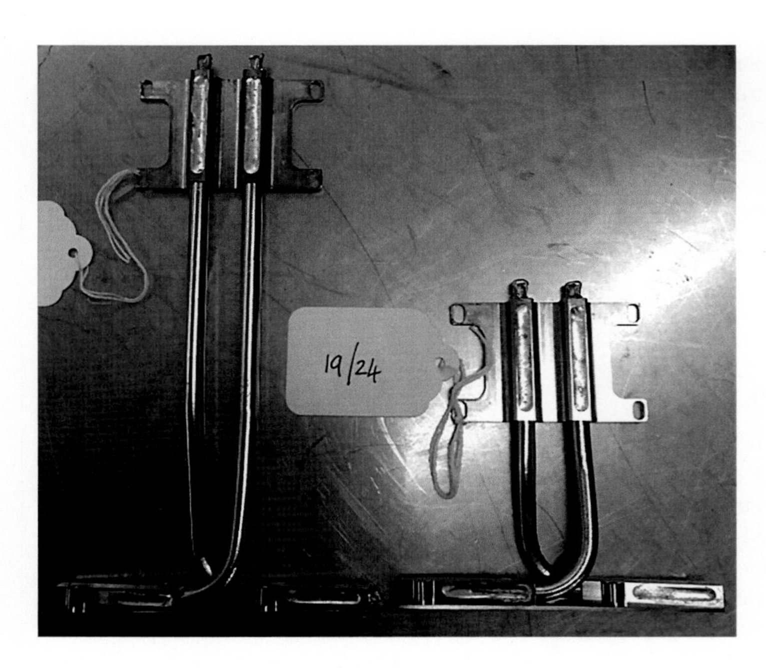


**FIGURE 7.49** Copper-Water Mini-Heat Pipe TMS's qualification test pieces, short and long versions (mass = 92 and 68 g; length = 176.5 and 96.5 mm, respectively) [47]. *Courtesy: Boyd and Thales Alenia Space.*

heat pipe chassis have in-orbit flight heritage and are at technology readiness level 9 (TRL9). Mini-heat pipe technology is now trending towards dissipation powers in excess of 100 W and multifunctionality is being designed into components.

### 7.5.4 Additive manufacture heat pipe and vapour chambers for space aplications

The advancement of additive layer manufacturing processes such as LPBF has enabled the manufacture of metal components with complex geometry and integrated multifunctionality. An author of this book, in collaboration with the University of Liverpool, has pioneered the development of heat pipe and vapour chamber technology with integrated additive-manufactured capillary wick structures and awarded patents, with priority date in 2012 [14]. Predominantly the developments have focused on LPBF of titanium powders, to produce heat pipes and vapour chambers for space applications. Following proof-of-concept feasibility studies [15], such as the LHP primary wick presented earlier in this chapter, the major development activity was conducted through the ESA ARTES project '370–400 K *g*-friendly heat pipes with low operating temperature'. The objectives of the ARTES activity were to provide an alternative to axially grooved wick, aluminium–ammonia heat pipes, that, by deploying an advanced capillary wick, are able to make a direct thermal connection into the electronics payload and improve functionality against gravity to ease ground test of satellites [13,48]. This was achieved by developing laser parameters to create miniaturised titanium alloy lattice structures, capable of capillary pumping liquid phase ammonia against gravity. The target transport capacity was 30 W. Various different vessel materials and wick combinations were tested; however, the main focus was on a conventionally constructed aluminium–ammonia heat pipe assembly with screen mesh wick and an additive-manufactured titanium–ammonia heat pipe with integrated miniaturised lattice wick structure. For consistency, both sets of heat pipes were 8 mm diameter × 200 mm long circular heat pipes, epoxied to machined aluminium saddles (Fig. 7.50). Both assembly types completed a series of qualification tests, including thermal performance testing, thermal cycle testing, pressure testing, lifetime at temperature testing and vibration testing to simulate launch. The transport capability was tested at various angles against gravity. It was found that the additive-manufactured heat pipe was functional to an adverse tilt angle of −15 degrees (−20 degrees for 1 × heat pipe), versus −2 degrees for the screen mesh wicked heat pipes.

Further development was completed in the Innovate UK 'CLASS' project, advanced developed laser parameter sets for a novel AM titanium powder, which were used to construct titanium–water vapour chambers that were capable of capillary pumping vertically against gravity to the maximum height of the vapour chamber (78 mm). Figs 7.51 and 7.52 show the internal vapour chamber structure and Fig. 7.53 shows the external features of a set of integrated chassis vapour chambers on an AM build plate. The vapour chamber was tested to a power of 50 W, mounted vertically against gravity in the test rig. The additive-manufactured heat pipe and vapour chamber shows promise for future space,

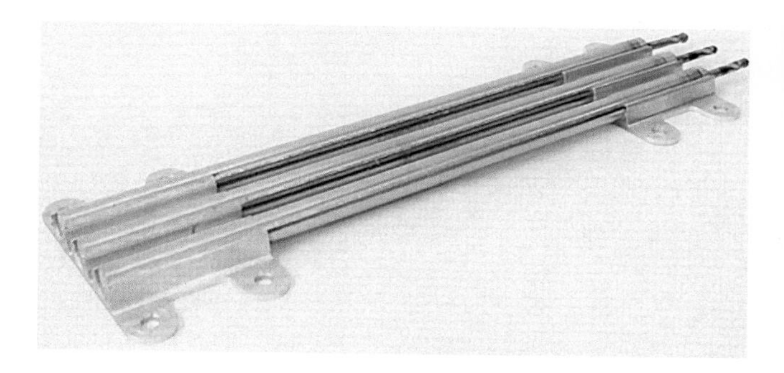

**FIGURE 7.50** Additive-manufactured titanium–ammonia heat pipe assembly breadboard test piece [48]. *Courtesy: Boyd.*

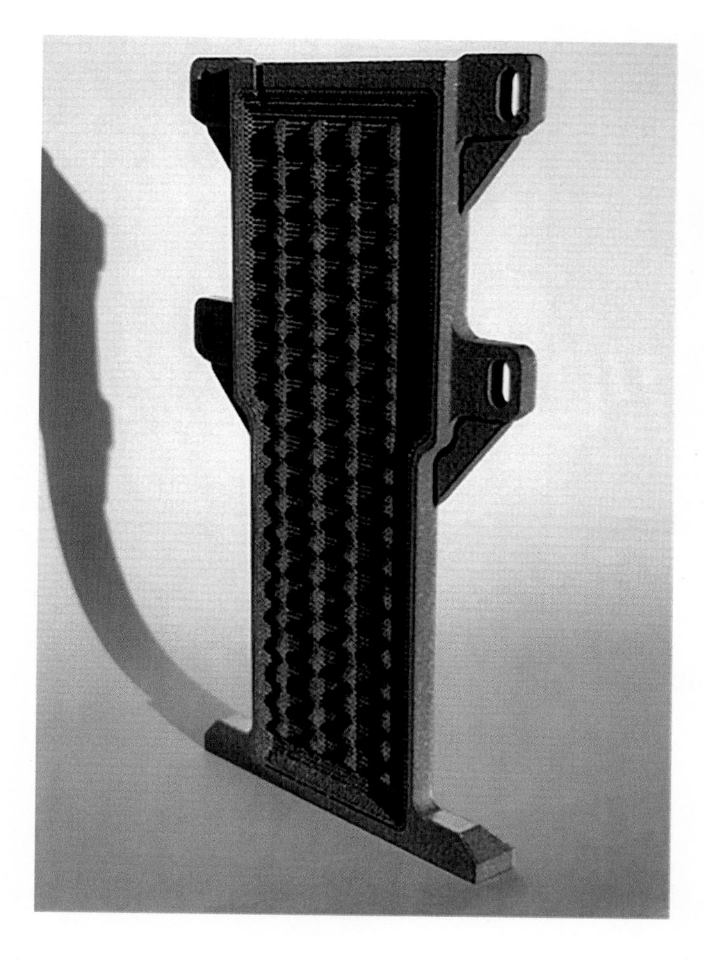

**FIGURE 7.51** Additive-manufactured vapour chamber cross-section showing internal lattice wick structure [13]. *Courtesy: Boyd.*

aerospace and terrestrial applications, where integration of the thermal component, mass reduction and multifunctionality are desirable. Therefore a higher operating temperature than for ammonia, with increased heat flux across the surface of the evaporator, is required. For ground testing, as these heat pipes are expected to be deployed perpendicular to the axially grooved heat pipes, they must also have the ability to function against gravity.

### 7.5.5 NASA Mars 2020 Perseverance Rover SuperCam copper–methanol heat pipes

Deployment on the Mars 2020 Perseverance Rover is a pinnacle of the achievement of heat pipe technology [49]. Having landed on 18 February 2021, the Perseverance Rover is carrying out its mission 'to search for signs of previously or currently habitable conditions on Mars and for signs of past microbial life'. The SuperCam module includes spectroscopy and imaging instrumentation, which enables the Rover to complete critical imaging and chemical and

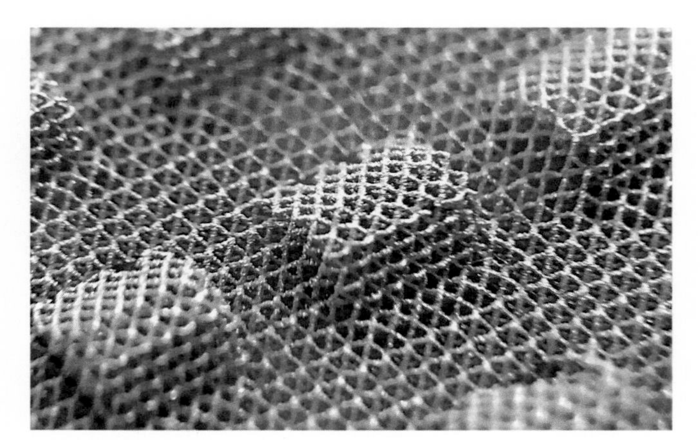

**FIGURE 7.52** Additive-manufactured titanium lattice wick structure. *Courtesy: Boyd.*

**FIGURE 7.53** Additive-manufactured vapour chambers on build plate. *Courtesy: Boyd.*

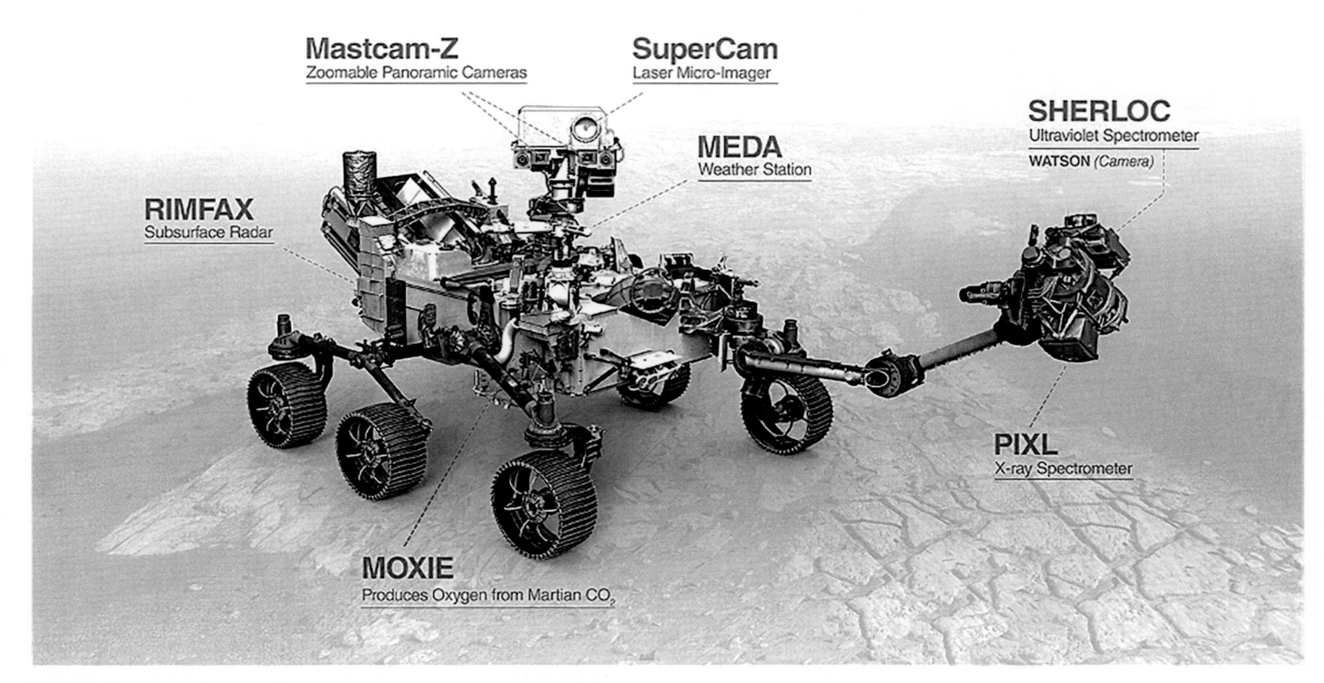


**FIGURE 7.54** NASA 2020 Perseverance Rover SuperCam with Boyd heat pipe assemblies onboard [49,50].

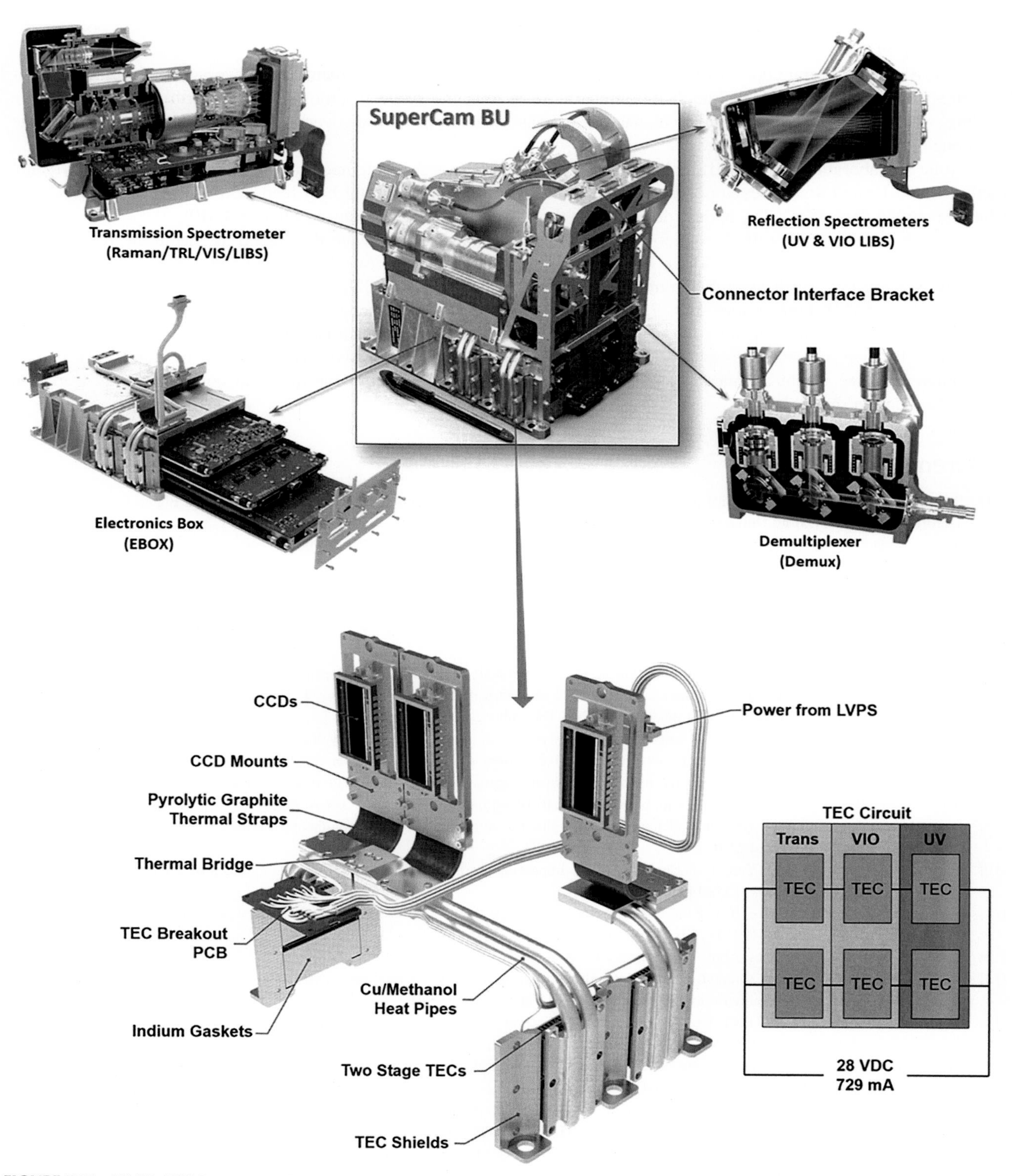


**FIGURE 7.55** NASA 2020 Perseverance Rover SuperCam module of two copper–ethanol heat pipe assemblies onboard [49,53].

mineral analysis, to determine the physical properties of the environment and samples within the Rover's robotic workspace (Fig. 7.54) [50].

The Rover must operate continuously without maintenance in harsh environments, with average environmental temperatures of ~ -62.2°C [51]. Although the temperature is low, the reduced atmosphere (vs Earth) limits convective cooling, therefore enhanced thermal management was needed to avoid overheating the SuperCam electronic components.

A joint collaboration between Los Alamos National Laboratory and engineers at Boyd's Lancaster Pennsylvania facility in the United States developed heat pipe technology to transport a total heat load of 6 W from the SuperCam's Charge-coupled devices (CCDs, optical detectors) to the Rover Accessory Mounting Plate (RAMP).

As water is frozen below 0°C, copper–methanol heat pipes assemblies were developed to enable the functionality of the Rover's temperature range. Three heat pipe assemblies, each with two 5 mm diameter copper–methanol heat pipes, were deployed. Each assembly transported 2 W of heat from the CCDs to TECs, enabling a stable ~20°C of cooling below RAMP temperature. Fig. 7.55 shows the position of a copper–methanol heat pipe assembly mounted onto the SuperCam module. At the time of writing, the Mars 2020 Perseverance Rover is 660 days into its mission, has travelled a distance of 8.66 miles across the surface of Mars and is in the proximity of the Jezero crater [52]. It is of great interest to see where advanced applications and the emergence of quotable quantum computing may lead heat pipe technology in the future.

## References


[1] R.J. McGlen, et al., Integrated thermal management techniques for high power electronic devices, in: Proceedings of Eighth UK National Heat Transfer Conference, Oxford University, 2003.

[2] Intel® Xeon® Platinum 8380 Processor. <https://medium.com/intel-tech/what-80-million-i-o-per-second-with-a-standard-2u-intel-xeon-system-f62422d0873d/> (accessed 22.05.22).

[3] Intel® Xeon® Platinum 8380 Processor. <https://medium.com/intel-tech/what-80-million-i-o-per-second-with-a-standard-2u-intel-xeon-system-f62422d0873d/> (accessed 22.05.22).

[4] DARPA wants electronics with radically novel liquid cooling technology. <http://www.networkworld.com/community/blog/darpa-wants-electronics-radically-novel-liquid-cooling-technology/> (accessed 16.03.13).

[5] T.D. Swanson, G.C. Birur, NASA thermal control technologies for robotic spacecraft, Appl. Therm. Eng. 23 (2003) 1055–1065.

[6] R. Wrobel, R.J. McGlen, Heat pipes in thermal management of electrical machines – a review, J. Therm. Sci. Eng. Prog. 26 (2021) 101053. Available from: https://doi.org/10.1016/j.tsep.2021.101053. 1 December 2021.

[7] P. Mathial, Global microchip shortage in automotive industry reinforces need for better supply chain planning, Diginomica, 8 November 2021. <https://diginomica.com/global-microchip-shortage-automotive-industry-reinforces-need-better-supply-chain-planning/> (accessed 24.05.22).

[8] D. Worwood, et al., A new approach to the internal thermal management of cylindrical battery cells for automotive applications, J. Power Sources 346 (2017) 151–166. Available from: https://doi.org/10.1016/j.jpowsour.2017.02.023. 1 April 2017.

[9] Boyd Corporation, Heat Pipe Brochure, October 2022. [Online]. <https://info.boydcorp.com/hubfs/Thermal/Two-Phase-Cooling/Boyd-Heat-Pipes-Brochure.pdf/> (accessed 23.10.22).

[10] S.H. Moon, G. Hwang, H.G. Yun, T.G. Choy, Y. Kang, Improving thermal performance of miniature heat pipe for notebook PC cooling, Microelectron. Reliab. 42 (2002) 135–140.

[11] S.H. Moon, G. Hwang, S.C. Ko, Y.T. Kim, Experimental study on the thermal performance of micro-heat pipe with cross-section of polygon, Microelectron. Reliab. 44 (2004) 315–321.

[12] Y.F. Maydanik, Loop heat pipes, Appl. Therm. Eng. 25 (2005) 635–657.

[13] R.J. McGlen, An introduction to additive manufactured heat pipe technology andadvanced thermal management products, Therm. Sci. Eng. Prog. 25 (2021) 100941. Available from: https://doi.org/10.1016/j.tsep.2021.100941.

[14] R.J. McGlen, J.G. Thayer, PATENT: Capillary device for use in heat pipe and method of manufacturing such capillary device, European Patent No. 2715265, US Patent No. US20220128312, International Publication No. WO 2012/160128 (29/11/2021 Gazette 2012/48).

[15] M. Ameli, B. Agnew, P.S. Leung, B. Ng, C.J. Sutcliffe, J. Singh, et al., A novel method for manufacturing sintered aluminium heat pipes (SAHP), Appl. Therm. Eng. 52 (2013) 498–504. January.

[16] D. Khrustalev, Loop thermosyphons for cooling of electronics, in: Eighteenth Annual IEEE Semiconductor Thermal Measurement and Management Symposium, Proceedings 2002 (Cat. No. 02CH37311), 2002, pp. 145–150, doi: 10.1109/STHERM.2002.991360.

[17] G. Burban, V. Ayel, A. Alexandre, P. Lagonotte, Y. Bertin, C. Romestant, Experimental investigation of a pulsating heat pipe for hybrid vehicle applications, Appl. Therm. Eng. 50 (1) (2013) 94–103.

[18] P. Luca, L. Cattani, F. Bozzoli, M. Mameli, S. Filippeschi, S. Rainieri, et al., Thermal characterization of a multi-turn pulsating heat pipe in microgravity conditions: Statistical approach to the local wall-to-fluid heat flux, Int. J. Heat. Mass. Transf. 169 (2021) 120930ISSN 0017-9310. Available from: https://doi.org/10.1016/j.ijheatmasstransfer.2021.120930.

[19] Boyd Corporation, Thermal ground plane for wide band gap semiconductor thermal management, October 2022 [Online]. <https://info.boydcorp.com/hubfs/Thermal/Two-Phase-Cooling/Boyd-Thermal-Ground-Plane-Technical-Datasheet.pdf\> (accessed 22.10.22).

[20] Boyd Corporation, Ultra-thin titanium vapor chamber, December 2022 [Online]. <https://www.boydcorp.com/thermal/two-phase-cooling/vapor-chambers/ultra-thin-vapor-chamber.html#menu1/> (accessed 17.12.22).

[21] Z.J. Zuo, M.T. North, Miniature high heat flux heat pipes for cooling of electronics. Available from: <http://www.boydcorp.com/>.

[22] Boyd, Flexible Heat Pipe and Flexible Heat Pipe Cold Plate [Online]. <https://info.boydcorp.com/hubfs/Thermal/Two-Phase-Cooling/Boyd-Flexible-Heat-Pipe-Technical-Datasheet.pdf/> (accessed 03.02.23).

[23] A. Ali, R. DeHoff, K. Grubb, Advanced thermal solutions for higher power notebook computers. Available from: <http://www.thermacore.com/>, 1999.

[24] Furukawa Electric, Ultra-thin sheet-shaped heat pipe "pera-flex®", Furukawa Rev. (25)(2004) 64–66.

[25] K. Nikkanen et al., Pulsating heat pipes for microelectronics cooling, in: Proceedings of 2005 AIChE Spring National Meeting, 2005, pp. 2421–2428.

[26] W.W. Wits, F.J.A.M. van Houten, Improving system performance through an integrated design approach, CIRP Ann. Manuf. Technol. 60 (1) (2011) 187–190.

[27] United States Patent No. US 6,800,077 B1, Thermal Corp., Heat Pipe for Cautery Surgical Instrument, 2004.

[28] Integra LifeSciences Services, Codman Speciality Surgical, Codman® Elecgtrosurgery Bipolar Forceps, October 2022 [Online]. <https://www.integralife.com/file/general/1528811867.pdf/> (accessed 30.10.22).

[29] M. Ali Abdelkareem, H. Jouhara, et al., Thermal management systems based on heat pipes for batteries in EVs/HEVs, J. Energy Storage 51 (2022) 104384. Available from: https://doi.org/10.1016/j.est.2022.104384. ISSN 2352-152X.

[30] H. Jouhara, et al., Investigation, development and experimental analyses of a heat pipe based battery thermal management system, Int. J. Thermofluids 1–2 (2020) 100004. Available from: https://doi.org/10.1016/j.ijft.2019.100004. ISSN 2666-2027.

[31] H. Jouhara, et al., Heat pipe based battery thermal management: evaluating the potential of two novel battery pack integrations, Int. J. Thermofluids 12 (2021) 100115. Available from: https://doi.org/10.1016/j.ijft.2021.100115. ISSN 2666-2027.

[32] Boyd Corporation, Two phase thermal solution guide, February 2022 [Online]. <https://info.boydcorp.com/hubfs/Thermal/Two-Phase-Cooling/Boyd-Two-Phase-Cooling-Guide-v2.pdf/> (accessed 05.11.22).

[33] A. Samba, H. Louahlia-Gualous, S. Le Masson, D. Norterhauser, Two-phase thermosyphon loop for cooling outdoor telecommunications equipment, Appl. Therm. Eng. 50 (2013) 1351–1360.

[34] R.J. McGlen, R. Jachuck, S. Lin, Integrated thermal management techniques for high power devices, Appl. Therm. Eng. 24 (2004) 1143–1156.

[35] Boyd Corporation, Compact, high performance air-cooled heat sinks, technical data sheet. 2020 [Online]. <https://info.boydcorp.com/hubfs/Thermal/Two-Phase-Cooling/Boyd-3D-Vapor-Chamber-MACE-Technical-Datasheet.pdf/> (accessed 05.11.22).

[36] Boyd Corporation, Thermosiphons vs heat pipes, 30 November 2021 [Online]. <https://www.boydcorp.com/resources/resource-center/blog/thermosiphons-vs-heat-pipes.html/> (accessed 12.11.22).

[37] Boyd Corporation, Thermosiphons. [Online]. <https://www.boydcorp.com/thermal/two-phase-cooling/thermosiphons/thermosiphon-heat-transport-assemblies.html/> (accessed 12.11.22).

[38] Boyd Corporation, 3D direct contact loop thermosiphon (Online Video). <https://info.boydcorp.com/hubfs/Videos/3D-Direct-Contact-Loop-Thermosiphon.mp4/> (accessed 12.11.22).

[39] K. Hambleton, T. Fairley, L. Cheshire, S. Potter, (NASA) Liftoff! NASA's Artemis I Mega Rocket Launches Orion to Moon, 16 November 2022 (Online). <https://www.nasa.gov/press-release/liftoff-nasa-s-artemis-i-mega-rocket-launches-orion-to-moon/> (accessed 18.11.22).

[40] European Space Agency, NEOSAT/ARTES 14 – The next generation of geostationary telecommunication platform. <https://artes.esa.int/next-generation-platform-neosat/> (accessed 13.11.22).

[41] CSIRO, Australia Telescope National Facility, CSIRO, 15 10 2018 [Online]. <https://www.atnf.csiro.au/the_atnf/index.html/> (accessed 11.03.20).

[42] L. Ball, L. Harvey-Smith, 2015, ASKAP Early Science Programme, CSIRO, 07 09 2015 [Online]. <https://www.atnf.csiro.au/projects/askap/ASKAP_EarlyScienceProgram.pdf/> (accessed 03.02.23).

[43] CSIRO Astronomy and Space Science, ASKAP Industry Engagement Case Study: Manufacturing, [Online]. <https://www.atnf.csiro.au/projects/askap/ASKAP_industry_1.pdf/> (Accessed 11.03.20).

[44] Photonics Media, Aavid Thermacore Develops Thermal Management System for Australian Telescope, Photonics Media. <https://www.photonics.com/Articles/Aavid_Thermacore_Develops_Thermal_Management/a62032/>, 2017.

[45] M. Parlak. R.J. McGlen, 2015, Cooling of high power active phased array antenna using axially grooved heat pipe for a space application, in: Conference: 2015 7th International Conference on Recent Advances in Space Technologies (RAST), doi:10.1109/RAST.2015.7208439.

[46] ESA-ESTEC, ECSS-E-ST-31-02C Rev. 1 Space Engineering, Two-phase heat transport equipment, 15 March 2017.

[47] R.J. McGlen, F. Michard, C. Conton, S. Cochrane, R. Waterston, Qualification for space flight of two-phase copper–water heat pipes with sintered wick structure and k-core encapsulated graphite technology for thermal management of next generation space electronics, in: Conference: 2020, International Conference on Environmental Systems (ICES), paper no. ICES-2020-301. <https://ttu-ir.tdl.org/handle/2346/86293/>, 2020.

[48] R.J. McGlen, C.J. Sutcliffe, Additive manufactured titanium–ammonia heat pipes for thermal management of space electronic devices, in: Conference: 2020, International Conference on Environmental Systems (ICES), paper no. ICES-2020-306. <https://ttu-ir.tdl.org/handle/2346/86294/>, 2020.

[49] Boyd, Boyd develops cooling solution for the Critical SuperCam Assembly of the Perseverance Mars Rover, 1 July 2020 [Online]. <https://www.boydcorp.com/resources/resource-center/blog/mars-rover-perseverance-supercam-heat-pipe-cooling.html/> (accessed 30.10.22).

[50] NASA, Mars 2020 Mission, Perseverance Rover Instruments [Online]. <https://mars.nasa.gov/mars2020/spacecraft/instruments/> (accessed 24.02.23).

[51] NASA, What Is Mars?, 10 August 2020 [Online]. <https://www.nasa.gov/audience/forstudents/5-8/features/nasa-knows/what-is-mars-58.html/> (accessed 28.12.22).

[52] NASA, Where is Perseverance? [Online]. <https://mars.nasa.gov/mars2020/mission/where-is-the-rover/> (accessed 28.12.22).

[53] R.C. Wiens, S. Maurice, S.H. Robinson, et al., The SuperCam instrument suite on the NASA Mars 2020 Rover: body unit and combined system tests, Space Sci. Rev. 217 (2021) 4. Available from: https://doi.org/10.1007/s11214-020-00777-5.

Appendix 1

# Working fluid properties

Fluids listed (in order of appearance):

| | |
|---|---|
| Helium | Water |
| Nitrogen | Flutec PP9 |
| Ammonia | High-temperature organics |
| Pentane | Mercury |
| Acetone | Caesium |
| Methanol | Potassium |
| Ethanol | Sodium |
| Flutec PP2 | Lithium |
| Heptane | |

Properties listed:

| | |
|---|---|
| Latent heat of evaporation | Vapour dynamic viscosity |
| Liquid density | Vapour pressure |
| Vapour density | Vapour-specific heat |
| Liquid thermal conductivity | Liquid surface tension |
| Liquid dynamic viscosity | |

**Helium**

| Temperature (°C) | Latent heat (kJ/kg) | Liquid density (kg/m$^3$) | Vapour density (kg/m$^3$) | Liquid thermal conductivity (W/m°C × $10^{-2}$) | Liquid viscosity (cP × $10^2$) | Vapour viscosity (cP × $10^3$) | Vapour pressure (bar) | Vapour-specific heat (kJ/kg°C) | Liquid surface tension (N/m × $10^3$) |
|---|---|---|---|---|---|---|---|---|---|
| −271 | 22.8 | 148.3 | 26.0 | 1.81 | 3.90 | 0.20 | 0.06 | 2.045 | 0.26 |
| −270 | 23.6 | 140.7 | 17.0 | 2.24 | 3.70 | 0.30 | 0.32 | 2.699 | 0.19 |
| −269 | 20.9 | 128.0 | 10.0 | 2.77 | 2.90 | 0.60 | 1.00 | 4.619 | 0.09 |
| −268 | 4.0 | 113.8 | 8.5 | 3.50 | 1.34 | 0.90 | 2.29 | 6.642 | 0.01 |

**Nitrogen**

| Temperature (°C) | Latent heat (kJ/kg) | Liquid density (kg/m$^3$) | Vapour density (kg/m$^3$) | Liquid thermal conductivity (W/m°C) | Liquid viscosity (cP) | Vapour viscosity (cP × $10^2$) | Vapour pressure (bar) | Vapour-specific heat (kJ/kg°C) | Liquid surface tension (N/m × $10^2$) |
|---|---|---|---|---|---|---|---|---|---|
| −203 | 210.0 | 830.0 | 1.84 | 0.150 | 2.48 | 0.48 | 0.48 | 1.083 | 1.054 |
| −200 | 205.5 | 818.0 | 3.81 | 0.146 | 1.94 | 0.51 | 0.74 | 1.082 | 0.985 |
| −195 | 198.0 | 798.0 | 7.10 | 0.139 | 1.51 | 0.56 | 1.62 | 1.079 | 0.870 |
| −190 | 190.5 | 778.0 | 10.39 | 0.132 | 1.26 | 0.60 | 3.31 | 1.077 | 0.766 |
| −185 | 183.0 | 758.0 | 13.68 | 0.125 | 1.08 | 0.65 | 4.99 | 1.074 | 0.662 |
| −180 | 173.7 | 732.0 | 22.05 | 0.117 | 0.95 | 0.71 | 6.69 | 1.072 | 0.561 |
| −175 | 163.2 | 702.0 | 33.80 | 0.110 | 0.86 | 0.77 | 8.37 | 1.070 | 0.464 |
| −170 | 152.7 | 672.0 | 45.55 | 0.103 | 0.80 | 0.83 | 1.07 | 1.068 | 0.367 |
| −160 | 124.2 | 603.0 | 80.90 | 0.089 | 0.72 | 1.00 | 19.37 | 1.063 | 0.185 |
| −150 | 66.8 | 474.0 | 194.00 | 0.075 | 0.65 | 1.50 | 28.80 | 1.059 | 0.110 |

**Ammonia**

| Temperature (°C) | Latent heat (kJ/kg) | Liquid density (kg/m$^3$) | Vapour density (kg/m$^3$) | Liquid thermal conductivity (W/m°C) | Liquid viscosity (cP) | Vapour viscosity (cP × $10^2$) | Vapour pressure (bar) | Vapour-specific heat (kJ/kg °C) | Liquid surface tension (N/m × $10^2$) |
|---|---|---|---|---|---|---|---|---|---|
| −60 | 1343 | 714.4 | 0.03 | 0.294 | 0.36 | 0.72 | 0.27 | 2.050 | 4.062 |
| −40 | 1384 | 690.4 | 0.05 | 0.303 | 0.29 | 0.79 | 0.76 | 2.075 | 3.574 |
| −20 | 1338 | 665.5 | 1.62 | 0.304 | 0.26 | 0.85 | 1.93 | 2.100 | 3.090 |
| 0 | 1263 | 638.6 | 3.48 | 0.298 | 0.25 | 0.92 | 4.24 | 2.125 | 2.480 |
| 20 | 1187 | 610.3 | 6.69 | 0.286 | 0.22 | 1.01 | 8.46 | 2.150 | 2.133 |
| 40 | 1101 | 579.5 | 12.00 | 0.272 | 0.2 | 1.16 | 15.34 | 2.160 | 1.833 |
| 60 | 1026 | 545.2 | 20.49 | 0.255 | 0.17 | 1.27 | 29.8 | 2.180 | 1.367 |
| 80 | 891 | 505.7 | 34.13 | 0.235 | 0.15 | 1.40 | 40.9 | 2.210 | 0.767 |
| 100 | 699 | 455.1 | 54.92 | 0.212 | 0.11 | 1.60 | 63.12 | 2.260 | 0.500 |
| 120 | 428 | 374.4 | 113.16 | 0.184 | 0.07 | 1.89 | 90.44 | 2.292 | 0.150 |

**Pentane**

| Temperature (°C) | Latent heat (kJ/kg) | Liquid density (kg/m$^3$) | Vapour density (kg/m$^3$) | Liquid thermal conductivity (W/m°C) | Liquid viscosity (cP) | Vapour viscosity (cP $\times 10^2$) | Vapour Pressure (bar) | Vapour-specific heat (kJ/kg °C) | Liquid surface tension (N/m $\times 10^2$) |
|---|---|---|---|---|---|---|---|---|---|
| −20 | 390.0 | 663.0 | 0.01 | 0.149 | 0.344 | 0.51 | 0.10 | 0.825 | 2.01 |
| 0 | 378.3 | 644.0 | 0.75 | 0.143 | 0.283 | 0.53 | 0.24 | 0.874 | 1.79 |
| 20 | 366.9 | 625.5 | 2.20 | 0.138 | 0.242 | 0.58 | 0.76 | 0.922 | 1.58 |
| 40 | 355.5 | 607.0 | 4.35 | 0.133 | 0.200 | 0.63 | 1.52 | 0.971 | 1.37 |
| 60 | 342.3 | 585.0 | 6.51 | 0.128 | 0.174 | 0.69 | 2.28 | 1.021 | 1.17 |
| 80 | 329.1 | 563.0 | 10.61 | 0.127 | 0.147 | 0.74 | 3.89 | 1.050 | 0.97 |
| 100 | 295.7 | 537.6 | 16.54 | 0.124 | 0.128 | 0.81 | 7.19 | 1.088 | 0.83 |
| 120 | 269.7 | 509.4 | 25.20 | 0.122 | 0.120 | 0.90 | 13.81 | 1.164 | 0.68 |

**Acetone**

| Temperature (°C) | Latent heat (kJ/kg) | Liquid density (kg/m$^3$) | Vapour density (kg/m$^3$) | Liquid thermal conductivity (W/m°C) | Liquid viscosity (cP) | Vapour viscosity (cP $\times 10^2$) | Vapour pressure (bar) | Vapour-specific heat (kJ/kg °C) | Liquid surface tension (N/m $\times 10^2$) |
|---|---|---|---|---|---|---|---|---|---|
| −40 | 660.0 | 860.0 | 0.03 | 0.200 | 0.800 | 0.68 | 0.01 | 2.00 | 3.10 |
| −20 | 615.6 | 845.0 | 0.10 | 0.189 | 0.500 | 0.73 | 0.03 | 2.06 | 2.76 |
| 0 | 564.0 | 812.0 | 0.26 | 0.183 | 0.395 | 0.78 | 0.10 | 2.11 | 2.62 |
| 20 | 552.0 | 790.0 | 0.64 | 0.181 | 0.323 | 0.82 | 0.27 | 2.16 | 2.37 |
| 40 | 536.0 | 768.0 | 1.05 | 0.175 | 0.269 | 0.86 | 0.60 | 2.22 | 2.12 |
| 60 | 517.0 | 744.0 | 2.37 | 0.168 | 0.226 | 0.9 | 1.15 | 2.28 | 1.86 |
| 80 | 495.0 | 719.0 | 4.30 | 0.160 | 0.192 | 0.95 | 2.15 | 2.34 | 1.62 |
| 100 | 472.0 | 689.6 | 6.94 | 0.148 | 0.170 | 0.98 | 4.43 | 2.39 | 1.34 |
| 120 | 426.1 | 660.3 | 11.02 | 0.135 | 0.148 | 0.99 | 6.70 | 2.45 | 1.07 |
| 140 | 394.4 | 631.8 | 18.61 | 0.126 | 0.132 | 1.03 | 10.49 | 2.50 | 0.81 |

**Methanol**

| Temperature (°C) | Latent heat (kJ/kg) | Liquid density (kg/m$^3$) | Vapour density (kg/m$^3$) | Liquid thermal conductivity (W/m°C) | Liquid viscosity (cP) | Vapour viscosity (cP $\times 10^2$) | Vapour pressure (bar) | Vapour-specific Heat (kJ/kg °C) | Liquid surface tension (N/m $\times 10^2$) |
|---|---|---|---|---|---|---|---|---|---|
| −50 | 1194 | 843.5 | 0.01 | 0.210 | 1.700 | 0.72 | 0.01 | 1.20 | 3.26 |
| −30 | 1187 | 833.5 | 0.01 | 0.208 | 1.300 | 0.78 | 0.02 | 1.27 | 2.95 |
| −10 | 1182 | 818.7 | 0.04 | 0.206 | 0.945 | 0.85 | 0.04 | 1.34 | 2.63 |
| 10 | 1175 | 800.5 | 0.12 | 0.204 | 0.701 | 0.91 | 0.10 | 1.40 | 2.36 |
| 30 | 1155 | 782.0 | 0.31 | 0.203 | 0.521 | 0.98 | 0.25 | 1.47 | 2.18 |
| 50 | 1125 | 764.1 | 0.77 | 0.202 | 0.399 | 1.04 | 0.55 | 1.54 | 2.01 |
| 70 | 1085 | 746.2 | 1.47 | 0.201 | 0.314 | 1.11 | 1.31 | 1.61 | 1.85 |
| 90 | 1035 | 724.4 | 3.01 | 0.199 | 0.259 | 1.19 | 2.69 | 1.79 | 1.66 |
| 110 | 980 | 703.6 | 5.64 | 0.197 | 0.211 | 1.26 | 4.98 | 1.92 | 1.46 |
| 130 | 920 | 685.2 | 9.81 | 0.195 | 0.166 | 1.31 | 7.86 | 1.92 | 1.25 |
| 150 | 850 | 653.2 | 15.9 | 0.193 | 0.138 | 1.38 | 8.94 | 1.92 | 1.04 |

**Ethanol**

| Temperature (°C) | Latent heat (kJ/kg) | Liquid density (kg/m$^3$) | Vapour density (kg/m$^3$) | Liquid thermal conductivity (W/m°C) | Liquid viscosity (cP) | Vapour viscosity (cP $\times 10^2$) | Vapour pressure (bar) | Vapour-specific heat (kJ/kg °C) | Liquid surface tension (N/m $\times 10^2$) |
|---|---|---|---|---|---|---|---|---|---|
| −30 | 939.4 | 825.0 | 0.02 | 0.177 | 3.40 | 0.75 | 0.01 | 1.25 | 2.76 |
| −10 | 928.7 | 813.0 | 0.03 | 0.173 | 2.20 | 0.80 | 0.02 | 1.31 | 2.66 |
| 10 | 904.8 | 798.0 | 0.05 | 0.170 | 1.50 | 0.85 | 0.03 | 1.37 | 2.57 |
| 30 | 888.6 | 781.0 | 0.38 | 0.168 | 1.02 | 0.91 | 0.10 | 1.44 | 2.44 |
| 50 | 872.3 | 762.2 | 0.72 | 0.166 | 0.72 | 0.97 | 0.29 | 1.51 | 2.31 |
| 70 | 858.3 | 743.1 | 1.32 | 0.165 | 0.51 | 1.02 | 0.76 | 1.58 | 2.17 |
| 90 | 832.1 | 725.3 | 2.59 | 0.163 | 0.37 | 1.07 | 1.43 | 1.65 | 2.04 |
| 110 | 786.6 | 704.1 | 5.17 | 0.160 | 0.28 | 1.13 | 2.66 | 1.72 | 1.89 |
| 130 | 734.4 | 678.7 | 9.25 | 0.159 | 0.21 | 1.18 | 4.30 | 1.78 | 1.75 |

**Flutec PP2**

| Temperature (°C) | Latent heat (kJ/kg) | Liquid density (kg/m$^3$) | Vapour density (kg/m$^3$) | Liquid thermal conductivity (W/m°C) | Liquid viscosity (cP) | Vapour viscosity (cP $\times 10^2$) | Vapour pressure (bar) | Vapour-specific heat (kJ/kg °C) | Liquid surface tension (N/m $\times 10^2$) |
|---|---|---|---|---|---|---|---|---|---|
| −30 | 106.2 | 1942 | 0.13 | 0.637 | 5.200 | 0.98 | 0.01 | 0.72 | 1.90 |
| −10 | 103.1 | 1886 | 0.44 | 0.626 | 3.500 | 1.03 | 0.02 | 0.81 | 1.71 |
| 10 | 99.8 | 1829 | 1.39 | 0.613 | 2.140 | 1.07 | 0.09 | 0.92 | 1.52 |
| 30 | 96.3 | 1773 | 2.96 | 0.601 | 1.435 | 1.12 | 0.22 | 1.01 | 1.32 |
| 50 | 91.8 | 1716 | 6.43 | 0.588 | 1.005 | 1.17 | 0.39 | 1.07 | 1.13 |
| 70 | 87.0 | 1660 | 11.79 | 0.575 | 0.720 | 1.22 | 0.62 | 1.11 | 0.93 |
| 90 | 82.1 | 1599 | 21.99 | 0.563 | 0.543 | 1.26 | 1.43 | 1.17 | 0.73 |
| 110 | 76.5 | 1558 | 34.92 | 0.550 | 0.429 | 1.31 | 2.82 | 1.25 | 0.52 |
| 130 | 70.3 | 1515 | 57.21 | 0.537 | 0.314 | 1.36 | 4.83 | 1.33 | 0.32 |
| 160 | 59.1 | 1440 | 103.63 | 0.518 | 0.167 | 1.43 | 8.76 | 1.45 | 0.01 |

**Heptane**

| Temperature (°C) | Latent heat (kJ/kg) | Liquid density (kg/m$^3$) | Vapour density (kg/m$^3$) | Liquid thermal conductivity (W/m°C) | Liquid viscosity (cP) | Vapour viscosity (cP $\times 10^2$) | Vapour pressure (bar) | Vapour-specific heat (kJ/kg°C) | Liquid surface tension (N/m $\times 10^2$) |
|---|---|---|---|---|---|---|---|---|---|
| −20 | 384.0 | 715.5 | 0.01 | 0.143 | 0.69 | 0.57 | 0.01 | 0.83 | 2.42 |
| 0 | 372.6 | 699.0 | 0.17 | 0.141 | 0.53 | 0.60 | 0.02 | 0.87 | 2.21 |
| 20 | 362.2 | 683.0 | 0.49 | 0.140 | 0.43 | 0.63 | 0.08 | 0.92 | 2.01 |
| 40 | 351.8 | 667.0 | 0.97 | 0.139 | 0.34 | 0.66 | 0.20 | 0.97 | 1.81 |
| 60 | 341.5 | 649.0 | 1.45 | 0.137 | 0.29 | 0.70 | 0.32 | 1.02 | 1.62 |
| 80 | 331.2 | 631.0 | 2.31 | 0.135 | 0.24 | 0.74 | 0.62 | 1.05 | 1.43 |
| 100 | 319.6 | 612.0 | 3.71 | 0.133 | 0.21 | 0.77 | 1.10 | 1.09 | 1.28 |
| 120 | 305.0 | 592.0 | 6.08 | 0.132 | 0.18 | 0.82 | 1.85 | 1.16 | 1.10 |

**Water**

| Temperature (°C) | Latent heat (kJ/kg) | Liquid density (kg/m$^3$) | Vapour density (kg/m$^3$) | Liquid thermal conductivity (W/m°C) | Liquid viscosity (cP) | Vapour viscosity (cP $\times 10^2$) | Vapour pressure (bar) | Vapour-specific heat (kJ/kg °C) | Liquid surface tension (N/m $\times 10^2$) |
|---|---|---|---|---|---|---|---|---|---|
| 20 | 2448 | 998.2 | 0.02 | 0.603 | 1.00 | 0.96 | 0.02 | 1.81 | 7.28 |
| 40 | 2402 | 992.3 | 0.05 | 0.630 | 0.65 | 1.04 | 0.07 | 1.89 | 6.96 |
| 60 | 2359 | 983.0 | 0.13 | 0.649 | 0.47 | 1.12 | 0.20 | 1.91 | 6.62 |
| 80 | 2309 | 972.0 | 0.29 | 0.668 | 0.36 | 1.19 | 0.47 | 1.95 | 6.26 |
| 100 | 2258 | 958.0 | 0.60 | 0.680 | 0.28 | 1.27 | 1.01 | 2.01 | 5.89 |
| 120 | 2200 | 945.0 | 1.12 | 0.682 | 0.23 | 1.34 | 2.02 | 2.09 | 5.50 |
| 140 | 2139 | 928.0 | 1.99 | 0.683 | 0.20 | 1.41 | 3.90 | 2.21 | 5.06 |
| 160 | 2074 | 909.0 | 3.27 | 0.679 | 0.17 | 1.49 | 6.44 | 2.38 | 4.66 |
| 180 | 2003 | 888.0 | 5.16 | 0.669 | 0.15 | 1.57 | 10.04 | 2.62 | 4.29 |
| 200 | 1967 | 865.0 | 7.87 | 0.659 | 0.14 | 1.65 | 16.19 | 2.91 | 3.89 |

**Flutec PP9**

| Temperature (°C) | Latent heat (kJ/kg) | Liquid density (kg/m$^3$) | Vapour density (kg/m$^3$) | Liquid thermal conductivity (W/m°C) | Liquid viscosity (cP) | Vapour viscosity (cP $\times 10^2$) | Vapour pressure (bar) | Vapour-specific heat (kJ/kg °C) | Liquid surface tension (N/m $\times 10^2$) |
|---|---|---|---|---|---|---|---|---|---|
| −30 | 103.0 | 2098 | 0.01 | 0.060 | 5.77 | 0.82 | 0.00 | 0.80 | 2.36 |
| 0 | 98.4 | 2029 | 0.01 | 0.059 | 3.31 | 0.90 | 0.00 | 0.87 | 2.08 |
| 30 | 94.5 | 1960 | 0.12 | 0.057 | 1.48 | 1.06 | 0.01 | 0.94 | 1.80 |
| 60 | 90.2 | 1891 | 0.61 | 0.056 | 0.94 | 1.18 | 0.03 | 1.02 | 1.52 |
| 90 | 86.1 | 1822 | 1.93 | 0.054 | 0.65 | 1.21 | 0.12 | 1.09 | 1.24 |
| 120 | 83.0 | 1753 | 4.52 | 0.053 | 0.49 | 1.23 | 0.28 | 1.15 | 0.95 |
| 150 | 77.4 | 1685 | 11.81 | 0.052 | 0.38 | 1.26 | 0.61 | 1.23 | 0.67 |
| 180 | 70.8 | 1604 | 25.13 | 0.051 | 0.30 | 1.33 | 1.58 | 1.30 | 0.40 |
| 225 | 59.4 | 1455 | 63.27 | 0.049 | 0.21 | 1.44 | 4.21 | 1.41 | 0.01 |

**High-temperature organic (diphenyl-diphenyl oxide eutectic)**

| Temperature (°C) | Latent heat (kJ/kg) | Liquid density (kg/m$^3$) | Vapour density (kg/m$^3$) | Liquid thermal conductivity (W/m°C) | Liquid viscosity (cP) | Vapour viscosity (cP $\times 10^2$) | Vapour pressure (bar) | Vapour-specific heat (kJ/kg °C) | Liquid surface tension (N/m $\times 10^2$) |
|---|---|---|---|---|---|---|---|---|---|
| 100 | 354.0 | 992.0 | 0.03 | 0.131 | 0.97 | 0.67 | 0.01 | 1.34 | 3.50 |
| 150 | 338.0 | 951.0 | 0.22 | 0.125 | 0.57 | 0.78 | 0.05 | 1.51 | 3.00 |
| 200 | 321.0 | 905.0 | 0.94 | 0.119 | 0.39 | 0.89 | 0.25 | 1.67 | 2.50 |
| 250 | 301.0 | 858.0 | 3.60 | 0.113 | 0.27 | 1.00 | 0.88 | 1.81 | 2.00 |
| 300 | 278.0 | 809.0 | 8.74 | 0.106 | 0.20 | 1.12 | 2.43 | 1.95 | 1.50 |
| 350 | 251.0 | 755.0 | 19.37 | 0.099 | 0.15 | 1.23 | 5.55 | 2.03 | 1.00 |
| 400 | 219.0 | 691.0 | 41.89 | 0.093 | 0.12 | 1.34 | 10.90 | 2.11 | 0.50 |
| 450 | 185.0 | 625.0 | 81.00 | 0.086 | 0.10 | 1.45 | 19.00 | 2.19 | 0.03 |

**Mercury**

| Temperature (°C) | Latent heat (kJ/kg) | Liquid density (kg/m$^3$) | Vapour density (kg/m$^3$) | Liquid thermal conductivity (W/m°C) | Liquid viscosity (cP) | Vapour viscosity (cP $\times 10^2$) | Vapour pressure (bar) | Vapour-specific heat (kJ/kg °C) | Liquid surface tension (N/m $\times 10^2$) |
|---|---|---|---|---|---|---|---|---|---|
| 150 | 308.8 | 13 230 | 0.01 | 9.99 | 1.09 | 0.39 | 0.01 | 1.04 | 4.45 |
| 250 | 303.8 | 12 995 | 0.60 | 11.23 | 0.96 | 0.48 | 0.18 | 1.04 | 4.15 |
| 300 | 301.8 | 12 880 | 1.73 | 11.73 | 0.93 | 0.53 | 0.44 | 1.04 | 4.00 |
| 350 | 298.9 | 12 763 | 4.45 | 12.18 | 0.89 | 0.61 | 1.16 | 1.04 | 3.82 |
| 400 | 296.3 | 12 656 | 8.75 | 12.58 | 0.86 | 0.66 | 2.42 | 1.04 | 3.74 |
| 450 | 293.8 | 12 508 | 16.80 | 12.96 | 0.83 | 0.70 | 4.92 | 1.04 | 3.61 |
| 500 | 291.3 | 12 308 | 28.60 | 13.31 | 0.80 | 0.75 | 8.86 | 1.04 | 3.41 |
| 550 | 288.8 | 12 154 | 44.92 | 13.62 | 0.79 | 0.81 | 15.03 | 1.04 | 3.25 |
| 600 | 286.3 | 12 054 | 65.75 | 13.87 | 0.78 | 0.87 | 23.77 | 1.04 | 3.15 |
| 650 | 283.5 | 11 962 | 94.39 | 14.15 | 0.78 | 0.95 | 34.95 | 1.04 | 3.03 |
| 750 | 277.0 | 11 800 | 170.00 | 14.80 | 0.77 | 1.10 | 63.00 | 1.04 | 2.75 |

**Caesium**

| Temperature (°C) | Latent heat (kJ/kg) | Liquid density (kg/m$^3$) | Vapour density (kg/m$^3$) | Liquid thermal conductivity (W/m°C) | Liquid viscosity (cP) | Vapour viscosity (cP $\times 10^2$) | Vapour pressure (bar) | Vapour-specific heat (kJ/kg °C) | Liquid surface tension (N/m $\times 10^2$) |
|---|---|---|---|---|---|---|---|---|---|
| 375 | 530.4 | 1740 | 0.01 | 20.76 | 0.25 | 2.20 | 0.02 | 1.56 | 5.81 |
| 425 | 520.4 | 1730 | 0.01 | 20.51 | 0.23 | 2.30 | 0.04 | 1.56 | 5.61 |
| 475 | 515.2 | 1720 | 0.02 | 20.02 | 0.22 | 2.40 | 0.09 | 1.56 | 5.36 |
| 525 | 510.2 | 1710 | 0.03 | 19.52 | 0.20 | 2.50 | 0.16 | 1.56 | 5.11 |
| 575 | 502.8 | 1700 | 0.07 | 18.83 | 0.19 | 2.55 | 0.36 | 1.56 | 4.81 |
| 625 | 495.3 | 1690 | 0.10 | 18.13 | 0.18 | 2.60 | 0.57 | 1.56 | 4.51 |
| 675 | 490.2 | 1680 | 0.18 | 17.48 | 0.17 | 2.67 | 1.04 | 1.56 | 4.21 |
| 725 | 485.2 | 1670 | 0.26 | 16.83 | 0.17 | 2.75 | 1.52 | 1.56 | 3.91 |
| 775 | 477.8 | 1655 | 0.40 | 16.18 | 0.16 | 2.28 | 2.46 | 1.56 | 3.66 |
| 825 | 470.3 | 1640 | 0.55 | 15.53 | 0.16 | 2.90 | 3.41 | 1.56 | 3.41 |

**Potassium**

| Temperature (°C) | Latent heat (kJ/kg) | Liquid density (kg/m$^3$) | Vapour density (kg/m$^3$) | Liquid thermal conductivity (W/m°C) | Liquid viscosity (cP) | Vapour viscosity (cP $\times 10^2$) | Vapour pressure (bar) | Vapour-specific heat (kJ/kg °C) | Liquid surface tension (N/m $\times 10^2$) |
|---|---|---|---|---|---|---|---|---|---|
| 350 | 2093 | 763.1 | 0.002 | 51.08 | 0.21 | 0.15 | 0.01 | 5.32 | 9.50 |
| 400 | 2078 | 748.1 | 0.006 | 49.08 | 0.19 | 0.16 | 0.01 | 5.32 | 9.04 |
| 450 | 2060 | 735.4 | 0.015 | 47.08 | 0.18 | 0.16 | 0.02 | 5.32 | 8.69 |
| 500 | 2040 | 725.4 | 0.031 | 45.08 | 0.17 | 0.17 | 0.05 | 5.32 | 8.44 |
| 550 | 2020 | 715.4 | 0.062 | 43.31 | 0.15 | 0.17 | 0.10 | 5.32 | 8.16 |

| | | | | | | | | | |
|---|---|---|---|---|---|---|---|---|---|
| 600 | 2000 | 705.4 | 0.111 | 41.81 | 0.14 | 0.18 | 0.19 | 5.32 | 7.86 |
| 650 | 1980 | 695.4 | 0.193 | 40.08 | 0.13 | 0.19 | 0.35 | 5.32 | 7.51 |
| 700 | 1969 | 685.4 | 0.314 | 38.08 | 0.12 | 0.19 | 0.61 | 5.32 | 7.12 |
| 750 | 1938 | 675.4 | 0.486 | 36.31 | 0.12 | 0.20 | 0.99 | 5.32 | 6.72 |
| 800 | 1913 | 665.4 | 0.716 | 34.81 | 0.11 | 0.20 | 1.55 | 5.32 | 6.32 |
| 850 | 1883 | 653.1 | 1.054 | 33.31 | 0.10 | 0.21 | 2.34 | 5.32 | 5.92 |

**Sodium**

| Temperature (°C) | Latent heat (kJ/kg) | Liquid density (kg/m$^3$) | Vapour density (kg/m$^3$) | Liquid thermal conductivity (W/m°C) | Liquid viscosity (cP) | Vapour viscoity. (cP $\times$ 10) | Vapour pressure (bar) | Vapour-specific heat (kJ/kg °C) | Liquid surface tension (N/m $\times 10^2$) |
|---|---|---|---|---|---|---|---|---|---|
| 500 | 4370 | 828.1 | 0.003 | 70.08 | 0.24 | 0.18 | 0.01 | 9.04 | 1.51 |
| 600 | 4243 | 805.4 | 0.013 | 64.62 | 0.21 | 0.19 | 0.04 | 9.04 | 1.42 |
| 700 | 4090 | 763.5 | 0.050 | 60.81 | 0.19 | 0.20 | 0.15 | 9.04 | 1.33 |
| 800 | 3977 | 757.3 | 0.134 | 57.81 | 0.18 | 0.22 | 0.47 | 9.04 | 1.23 |
| 900 | 3913 | 745.4 | 0.306 | 53.35 | 0.17 | 0.23 | 1.25 | 9.04 | 1.13 |
| 1000 | 3827 | 725.4 | 0.667 | 49.08 | 0.16 | 0.24 | 2.81 | 9.04 | 1.04 |
| 1100 | 3690 | 690.8 | 1.306 | 45.08 | 0.16 | 0.25 | 5.49 | 9.04 | 0.95 |
| 1200 | 3577 | 669.0 | 2.303 | 41.08 | 0.15 | 0.26 | 9.59 | 9.04 | 0.86 |
| 1300 | 3477 | 654.0 | 3.622 | 37.08 | 0.15 | 0.27 | 15.91 | 9.04 | 0.77 |

**Lithium**

| Temperature (°C) | Latent heat (kJ/kg) | Liquid density (kg/m$^3$) | Vapour density (kg/m$^3$) | Liquid thermal conductivity (W/m°C) | Liquid viscosity (cP) | Vapour viscosity (cP $\times 10^2$) | Vapour pressure (bar) | Vapour-specific heat (kJ/kg °C) | Liquid surface tension (N/m $\times 10^2$) |
|---|---|---|---|---|---|---|---|---|---|
| 1030 | 20 500 | 450 | 0.005 | 67 | 0.24 | 1.67 | 0.07 | 0.532 | 2.90 |
| 1130 | 20 100 | 440 | 0.013 | 69 | 0.24 | 1.74 | 0.17 | 0.532 | 2.85 |
| 1230 | 20 000 | 430 | 0.028 | 70 | 0.23 | 1.83 | 0.45 | 0.532 | 2.75 |
| 1330 | 19 700 | 420 | 0.057 | 69 | 0.23 | 1.91 | 0.96 | 0.532 | 2.60 |
| 1430 | 19 200 | 410 | 0.108 | 68 | 0.23 | 2.00 | 1.85 | 0.532 | 2.40 |
| 1530 | 18 900 | 405 | 0.193 | 65 | 0.23 | 2.10 | 3.30 | 0.532 | 2.25 |
| 1630 | 18 500 | 400 | 0.340 | 62 | 0.23 | 2.17 | 5.30 | 0.532 | 2.10 |
| 1730 | 18 200 | 398 | 0.490 | 59 | 0.23 | 2.26 | 8.90 | 0.532 | 2.05 |

An addition to the working fluid properties is a table on the properties of potential thermosyphon working fluids at 25°C.

**Properties of potential thermosyphon working fluids at 25°C[a]**

| Fluid | Source of data | Merit number ($Wm^2$) | Problem | Pressure at 80°C (bar) | Pressure at 80°C (MPa) | Pressure at 100°C (MPa) | $T_c$ (°C) | $P_c$ (MPa) |
|---|---|---|---|---|---|---|---|---|
| Water (718) | [1] | 4809 | | 0.4737 | 0.04737 | 0.10132 | | |
| Ammonia | [1] | 2619 | C | 41.418 | 4.1418 | 6.2553 | | |
| Flutec PP2 | [1] | 2512 | d | | | | | |
| R290 (propane) | [14] | 1827 | F, P | 31.317 | 3.1317 | | | 4.2477 |
| Propene | [14] | 1824 | F | | | | | |
| Methanol | [1] | 1784 | T, F | 1.7805 | 0.17805 | 0.35342 | | |
| Acetone | [1] | 1586 | F | 2.1513 | 0.21513 | 0.37403 | | |
| R32 | [15] | 1423 | F, P | | | | 78.41 | 5.8579 |
| RE170 (dimethylether) | [1,14,15] | 1421 | F | 22.667 | 2.2667 | 3.33305 | | |
| Ethanol | [1] | 1210 | F | 1.0132 | 0.10132 | 0.20265 | | |
| Pentane | [1] | 1117 | F | 3.676 | 0.3676 | 0.59047 | | |
| R152a | [15] | 1114 | F | 23.441 | 2.3441 | 3.5084 | | |
| Heptane | [1] | 1019 | F | 0.5703 | 0.05703 | 0.1061 | | |
| Co2(744) | [15] | 1002 | P | | | | 30.98 | 7.3748 |
| Toluene | [1,14,15] | 985 | F | 0.2666 | 0.02666 | 0.0666 | | |
| R600 | [15] | 966 | F | 10.13 | 1.013 | 1.5269 | | |
| R600a | [14] | 930 | F | 13.419 | 1.3419 | 1.984 | | |
| R134a | [15] | 922 | | 26.331 | 2.6331 | 3.9721 | | |
| R245fa | [16] | 826 | | 7.89 | 0.789 | 1.261 | | |
| Isopropyl alcohol | [1,15] | 826 | T, F | 1.0132 | 0.10132 | 0.20265 | | |
| R125 | [15] | 751 | | | | | 66 | 3.6 |
| R113 | [15] | 706 | Banned under Montreal Protocol | 2.6559 | 0.26559 | 0.43764 | | |
| Hexane | [17,26] | | F [26] | 1.013 | 0.1013 | 0.2026 | | |
| R227ea | [18] | | | | | | 102.8 | 2.987 |
| R1270 (propylene) | [15] | | F | 37.238 | 3.7238 | | | 4.6646 |
| R365mfc | [19] | | F | | | | | |

*Notes*: All temperatures in °C. Merit number at 25°C. Problem: Most of the fluids have some drawbacks. C, compatibility; £, cost; F, flammable; P, high vapour pressure; T, toxic.
Pressures: These are given for 80°C and 100°C. In some cases, this exceeds the critical point, and where this is the case, $T_c$ indicates the critical temperature in °C and $P_c$ the critical pressure in MPa.
References used in this table:

[1] Properties from tables in Appendix 1.
[14] http://www.boconline.co.uk/en/products-and-supply/refrigerant-gases/natural-refrigerants/care10-r600a/care10-r600a.html > (accessed 03.11.13).
[15] ASHRAE, ASHRAE Handbook - Fundamentals, American Society of Heating, Refrigerating and Air-Conditioning Engineers, Inc., Atlanta, U.S.A., 1997.
[17] Wikipedia, Hexane. <http://en.wikipedia.org/wiki/>, 2008 (accessed 03.03.12). Original source: G.F. Carruth, R. Kobayashi, Vapor pressure of normal paraffins ethane through n-decane from their triple points to about 10 mmHg, J. Chem. Eng. Data 18 (2) (1973) 115–126.
[18] R227ea data: <http://coolprop.sourceforge.net/Fluids/R227EA.html> (accessed 03.11.13).
[19] R365mfc data: <http://coolprop.sourceforge.net/Fluids/R365MFC.html> (accessed 03.11.13).
[26] <http://www.chemspider.com/Chemical-Structure.7767.html> (accessed 03.11.13).

[a]Taken from R. MacGregor, P.A. Kew, D.A. Reay, Investigation of low global warming potential working fluids for a closed two-phase thermosyphon, J. Appl. Therm. Eng. 51 (2013), 917–925.

*Note*: CFCs and HCFCs are omitted from the list now, although one will find experiments still being carried out using these fluids. The HFCs are used as working fluids in heat pipes, but are now subject to limitations on their use, as they are global warming gases. 'Natural' working fluids include hydrocarbons – used in refrigeration and heat pump cycles – and of course ammonia, a very useful heat pipe working fluid at appropriate vapour temperatures. For data on the 'refrigerants' now being used, the ASHRAE databases are ideal sources of properties, and designers will find properties in databanks of design software. It will be interesting to see what heat pipe/thermosyphon users do with carbon dioxide!.

Appendix 2

# Thermal conductivity of heat pipe container and wick materials

| Materials | Thermal conductivity (W/m°C) |
|---|---|
| Aluminium | 205 |
| Brass | 113 |
| Copper (0°C–100°C) | 394 |
| Glass | 0.75 |
| Nickel (0°C–100°C) | 88 |
| Magnesium | 156 |
| Mild steel | 45 |
| Stainless steel (type 304) | 17.3 |
| Teflon | 0.17 |

Appendix 3

# A selection of heat pipe–related websites

http://mscweb.gsfc.nasa.gov/545web/ (Interesting NASA website devoted to activities of the Thermal Engineering Branch – not just heat pipes).

http://www.globalspec.com/ (The Engineering Search Engine – useful for identifying heat pipe manufacturers – currently early 2005 – lists 21, all in the United States, so not fully comprehensive!).

http://tmrl.mcmaster.ca/ (Gives information on research and related publications). Thermal Management Research Laboratory, McMaster University, Canada.

http://www.aascworld.com/Thermal-Control-Structures/service-1121794310/program.html (Data on heat pipe panels for spacecraft, constructed by AASC). Applied Aerospace Structures Corp., Stockton, California, United States.

http://www.1-act.com Advanced Cooling Technologies, Inc., US manufacturer of heat pipes.

http://www.alibaba.com/showroom/heat-pipe.html (Lists many Chinese suppliers, principally solar-collector related units).

http://www.amecthermasol.co.uk/AmecThermasolContact.html (Amec Thermasol is a supplier of heat management and thermal control solutions for electronic applications).

http://www.apricus.com/ (Manufacturer of heat pipe–based solar collectors). Nanjing, China.

http://www.atherm.com (Atherm is a French thermal engineering company that includes in its portfolio heat pipes for cooling).

http://www.atk.com/ (The website of Alliant Techsystems, Inc. – heat pipe manufacturer, including loop heat pipes). United States.

http://www.cast.cn/CastEN/index.asp (Research institute working on cryogenic heat pipes and VCHPs for space applications in China). Chinese Academy of Space Technology, China.

www.commerce.com.tw/products/EN/H/Heat_Pipe.htm (Website listing heat pipe suppliers in Asia and areas. Lists many large suppliers and a number of minor players that may supply locally). Based in Taiwan.

www.crtech.com (Modelling using CFD and so on). Colorado, United States.

http://www.econotherm.eu/ (Econotherm is a leading United Kingdom–based manufacturer of heat pipe waste heat recuperators, economisers, pre-heaters and steam condensers).

http://www.enertron-inc.com/enertron-products/heat-pipe-selection.php (Enertron Inc. is a thermal management and engineering company with expertise in the utilisation and integration of heat pipes).

www.europeanthermodynamics.com (Heat pipes for electronics cooling). Leicestershire, United Kingdom.

www.fujikura.co.uk (Website of the UK branch of Fujikura. Heat pipe supplier). Chessington, Surrey, United Kingdom.

www.furukawa.co.jp/english/ (Manufacturer of a range of heat pipe systems – English website with access to technical articles). The Furukawa Electric Company, Japan.

http://www.globalsources.com/manufacturers/Solar-Heat-Pipe.html (List of solar heat pipe manufacturers and suppliers).

www.heatsink-guide.com/heatpipes.shtml (Includes information and descriptions of heat sinks using heat pipes).
www.heat-pipes.com (Site of CRS Engineering, manufacturers of a wide range of heat pipes, including units for injection moulding and electronic thermal control) CRS Engineering Ltd., Alnwick, Northumberland, United Kingdom

www.heatpipe.com (Heat pipe manufacturer). Gainesville, Florida, United States.

www.heatpipeindia.com (Moulding and waste heat recovery applications). Golden Star Technical Services, Pune, India.

www.hsmarston.co.uk/ (Heat pipe supplier and also makes other compact heat exchangers). H.S. Marston Ltd., Wolverhampton, United Kingdom.

www.kellysearch.com/ (A product search engine. Lists 116 heat pipe suppliers (Spring 2005) and is better at identifying international players than Globalspec, but over-emphasises Chinese SMEs).

http://www.lanl.gov/science-innovation/capabilities/engi-neering/index.php (Site of Los Alamos Laboratory – includes heat pipe data). Los Alamos, New Mexico, United States.

http://www.manufacturers.com.tw/electronics/heat-pipe.html (Lists many Taiwanese and Chinese heat pipe suppliers).

www.mjm-engineering.com (Heat pipes and consultancy on thermal design based upon use of heat pipes). MJM Engineering Co., Naperville, Illinois, United States.

http://www.norenproducts.com/heat-pipes (The heat pipe catalogue of Noren Products). Noren Products, Inc., Menlo Park, California, United States.

www.nottingham.ac.uk/sbe/ (The website of the School of the Built Environment at Nottingham University, where heat pipes related to renewable energy and other uses are researched). Nottingham University, United Kingdom.

www.pipcar.co.uk (Heat pipes for core cooling in injection moulding etc.). Tonbridge, Kent, United Kingdom.

http://www.quick-cool-shop.de/ (A German supplier of heat pipes).

www.silverstonetek.com (Heat pipes for electronics thermal control, including sintered units). SilverStone Technology Co., Ltd.

www.spcoils.co.uk (Manufacturer of heat pipe heat exchangers and dehumidifiers). S & P Coil Products Ltd., Leicester, United Kingdom.

https://www.boydcorp.com/aavid.html (The UK website of Boyd Corporation, a major manufacturer and supplier of heat pipes and related systems). Boyd Technologies Ashington UK Ltd, Ashington, Northumberland, United Kingdom.

www.transterm.ro/ (The website of the Romanian heat pipe manufacturer, Transterm). Transterm, 2200 Brasov, Romania.

www.itmo.by/ (The website of the Luikov Heat & Mass Transfer Institute, where much research on innovative heat pipes is carried out). P. Brovka, Minsk, Belarus.

Appendix 4

# Conversion factors

| Physical quantity | | |
|---|---|---|
| Mass | 1 lb | = 0.4536 kg |
| Length | 1 ft | = 0.3048 m |
| | 1 in | = 0.0254 m |
| Area | 1 $ft^2$ | = 0.0929 $m^2$ |
| Force | 1 lbf | = 4.448 N |
| Energy | 1 Btu | = 1.055 kJ |
| | 1 kWh | = 3.6 MJ |
| Power | 1 hp | = 745.7 W |
| Pressure | 1 lbf/$in^2$ | = 6894.76 N/$m^2$ |
| | 1 bar | = 105 N/$m^2$ |
| | 1 atm | = 101.325 kN/$m^2$ |
| | 1 torr | = 133.322 N/$m^2$ |
| Dynamic viscosity | 1 Poise | = 0.1 Ns/$m^2$ |
| Kinematic viscosity | 1 stoke | = $10^{-4}$ $m^2$/s |
| Heat flow | 1 Btu/h | = 0.2931 W |
| Heat flux | 1 Btu/$ft^2$/h | = 3.155 W/$m^2$ |
| Thermal conductivity | 1 Btu/$ft^2$h°F/ft | = 1.731 W/$m^2$°C/m |
| Heat transfer coefficient | 1 Btu/$ft^2$h°F | = 5.678 W/$m^2$°C |

Appendix 5

# Mass calculations

The source and vapour temperature of any heat pipe can be correlated by:

$$T_s = T_v + R_s Q \tag{1}$$

where $T_s$ is the source temperature, $T_v$ is the vapour temperature, $Q$ is the heat load and $R_s$ is the total heat transfer resistance between the heat source and vapour space. As the source temperature varies with heat load, we can deduce Eq. 1 to:

$$\frac{dT_s}{dQ} = \frac{dT_v}{dQ} + R_s \tag{2}$$

In heat pipes where self-control exists, the term $\frac{dT_v}{dQ}$ will equate to zero due to the large storage volume. The term itself will never be negative. In scenarios where the thermal resistance between the source and the pipe is small, the source temperature will remain relatively constant, although this is rarely the case even if a constant vapour temperature is achieved. As the differential forms, a variation in source temperature dependant on changes in load and sink conditions, no general terms exist to control the available heat rejection. We can assume:

1. The non-condensing gas obeys the ideal gas low.
2. The conditions are at steady state.
3. A sharp interface exists between both gas and vapour phases.

The simultaneous equation of energy and mass conservation of the non-condensable and auxiliary fluid, force balance and auxiliary functions in relation to temperature, and the total pressure and pressure temperature relationships can be denoted by:

$$dT_s = \frac{\left\{\frac{\partial T_v}{\partial Q} + R_s(1 + S + S_1)\right\}dQ + \left\{\frac{\partial T_v}{\partial T_o} + S\varphi_o\right\}dT_o + S\varphi_{ST}dT_{ST}}{1 + S + S_1 + S_2} \tag{3}$$

where $T_s$ defines the change in source temperature as a function of changes in the heat load ($Q$), $T_o$ is the heat sink temperature, $T_{ST}$ is the gas storage temperature. $S$, $S_1$ and $S_2$ are control parameters expressed as functional form. $S$ is a control parameter related to the use of a non-condensable gas and a fixed storage control volume (for active and passive control). $S_1$ is applicable to a variable storage unit (0 for active control). $S_2$ is applicable when passive controls are used to sense the source temperature (0 for active control). $\varphi$ (with subscripts O (sink) and ST (storage condition)) establishes the conservation of mass of the non-condensable within the inactive part of the condenser and within the storage volume and can be defined as:

$$\varphi = \frac{1}{(P\alpha)_v}\left[P_{vis}\gamma\left(\frac{\partial P_v}{\partial P_{vis}}\right) + \frac{\partial P_g}{\partial T}\right] \tag{4}$$

where $P_v$ is the vapour pressure, $P_{vis}$ is the vapour pressure in the inactive condenser region or the storage region.

And:

$$\alpha_v = \partial \ell n P_v / \partial T_v \tag{5}$$

$$\gamma = \partial \ell n P_{vis} / \partial T \tag{6}$$

If an active system is applied, Eq. 3 can be deduced to:

$$dT_s = \frac{\left\{\frac{\partial T_v}{\partial Q} + R_s(1 + S)\right\}dQ + \left\{\frac{\partial T_v}{\partial T_o} + S\varphi_o\right\}dT_o + S\varphi_{ST}dT_{ST}}{1 + S} \tag{7}$$

The storage volume requirements for an active control system can be determined from the mass conservation considerations such as:

$$M_g = \frac{(P_v - P_{vi})}{R_g T_o} V_{ic} + \frac{(P_v - P_{vST}) V_{ST}}{R_g T_{ST}} \quad (8)$$

where $M_g$ is the mass of the non-condensable gas. $P_v$ is the vapour pressure. $P_{vi}$ is the vapour pressure in the inactive part of the condenser. $V_{ic}$ is the volume of the inactive part of the condenser. $V_{ST}$ is the volume of the storage volume. $P_{vST}$ is the vapour pressure in the storage volume. $T_o$ is the heat sink temperature. $T_{ST}$ is the storage temperature. $R_g$ represents the gas constant.

To determine the storage requirement, the examination of heat pipe extremes must be evaluated. At higher power conditions, the non-condensable gas should be contained within the storage volume. This implies that the storage volume should be at the lowest temperature it can achieve (maximum sink temperature), and the working fluid of low vapour temperature at this temperature can be used. Therefore the following equation can be deduced:

$$M_g = (P_v - P_{vST})_H \frac{V_{ST}}{R_g T_{OH}} \quad (9)$$

## The subscript H refers to high power calculations

At lower power conditions, all non-condensable should be in the condenser section. This implies that the partial pressure of the vapour in the storage volume approaches the system vapour pressure. In a practical application, the storage temperature should be lower than the heat pipe vapour temperature to allow a concentration of non-condensable in the storage volume, which reduces the potential for mass diffusion. Therefore the following equation can be deduced for low power calculations:

$$M_g = \left(\frac{P_v - P_{vi}}{R_g T_o}\right)_L V_{ic} + \left(\frac{P_v - P_{vST}}{R_g T_{st}}\right)_L V_{ST} \quad (10)$$

## The subscript L refers to low power calculations

If the storage temperature is equal to the vapour temperature, the $V_{ST}$ in Eq. 10 can be set to zero. By solving both Eq. 9 and Eq. 10 as simultaneous equations, the result can determine the storage volume requirements.

When the storage volume is dependent on the heat load as defined in Eq. 1. The variation of the vapour temperature can be calculated if the source heat transfer resistance can be calculated. As shown in Eq. 11. One key observation from Eq. 11 is that the storage requirements will be larger for a working fluid with a low vapour pressure at a nominal operating condition. To minimise storage requirements, a fluid with a high vapour pressure at the operating conditions should be selected. By solving Eq. 11, the term VST can further be applied in Eq. 8 and Eq. 9 to solve the mass of non-condensable.

$$\frac{V_{ST}}{V_{ic}} = \frac{\left(1 - \frac{P_{vi}}{P_v}\right)_L \left(\frac{T_{ST}}{T_o}\right)_L}{\left(\frac{P_{vH}}{P_{vL}}\right)\left(1 - \frac{P_{vST}}{Pv}\right)_H \left(\frac{T_L}{T_H}\right)_{ST} - \left(1 - \frac{P_{vST}}{P_v}\right)_L} \quad (11)$$

# Index

*Note*: Page numbers followed by "*f*" and "*t*" refer to figures and tables, respectively.

Printed in the United States
by Baker & Taylor Publisher Services